TODAY'S TECHNICIAN ™

SHOP MANUAL

For Automatic Transmissions & Transaxles

SEVENTH EDITION

TODAY'S TECHNICIAN™

SHOP MANUAL

For Automatic Transmissions & Transaxles

SEVENTH EDITION

Ken Pickerill
Southern Illinois University at Carbondale

Jack Erjavec
Series Editor

 Cengage

Australia • Brazil • Canada • Mexico • Singapore • United Kingdom • United States

Today's Technician™: Automatic Transmissions & Transaxles, Seventh edition

Ken Pickerill

Jack Erjavec

SVP, GM Skills & Global Product Management: Jonathan Lau

Product Director: Matthew Seeley

Senior Product Manager: Katie McGuire

Product Assistant: Kimberly Klotz

Production Service: SPi-Global

Executive Director, Content Design: Marah Bellegarde

Director, Learning Design: Juliet Steiner

Learning Designer: Mary Clyne

Vice President, Strategic Marketing Services: Jennifer Ann Baker

Marketing Director: Sean Chamberland

Marketing Manager: Andrew Ouimet

Senior Director, Content Delivery: Wendy Troeger

Director, Content Delivery: Alexis Ferraro

Senior Content Manager: Cheri Plasse

Digital Delivery Lead: Amanda Ryan

Senior Designer: Angela Sheehan

Cover image(s): Umberto Shtanzman/ Shutterstock.com

For product information and technology assistance, contact us at **Cengage Customer & Sales Support, 1-800-354-9706 or support.cengage.com.**

For permission to use material from this text or product, submit all requests online at **www.copyright.com.**

Library of Congress Control Number: 2018964359

ISBN: 978-1-337-79217-2
Package ISBN: 978-1-337-79215-8

Cengage
200 Pier 4 Boulevard
Boston, MA 02210
USA

Cengage is a leading provider of customized learning solutions with employees residing in nearly 40 different countries and sales in more than 125 countries around the world. Find your local representative at: **www.cengage.com.**

To learn more about Cengage platforms and services, register or access your online learning solution, or purchase materials for your course, visit **www.cengage.com.**

Printed in the United States of America

Print Number: 06   Print Year: 2024

CONTENTS

PHOTO SEQUENCES

JOB SHEETS

The *Today's Technician*™ series features textbooks and digital learning solutions that cover all mechanical and electrical systems of automobiles and light trucks. The content corresponds to the 2017 ASE Education Foundation program accreditation requirements. They are specifically correlated to the Task Lists contained in each level of program accreditation; Maintenance and Light Repair (MLR), Automotive Service Technology (AST), and Master Service Technology (MAST).

Additional titles include remedial skills and theories common to all of the certification areas and advanced or specific subject areas that reflect the latest technological trends. *Today's Technician: Automatic Transmissions and Transaxles, 7e*, is designed to give students a chance to develop the same skills and gain the same knowledge that today's successful technician has. This edition also reflects the most recent changes in the guidelines established by the ASE Education Foundation.

The purpose of the ASE Education Foundation program accreditation is to evaluate technician training programs against standards developed by the automotive industry and recommend qualifying programs for accreditation. Programs can earn accreditation upon the recommendation of ASE Education Foundation. These national standards reflect the skills that students must master. ASE Education Foundation accreditation ensures that certified training programs meet or exceed industry-recognized, uniform standards of excellence.

HIGHLIGHTS OF THIS EDITION—CLASSROOM MANUAL

The text has been updated throughout to include the latest developments. Some of these new topics include the various transmission designs used in hybrid vehicles, the Lepelletier gear setup (which results in five- to nine-speed transmissions), power split units, and continuously variable transmission designs. There is also more information on current electronic control systems, including the various protocols used for multiplexing.

Chapter 1 introduces the purpose of automatic transmissions and how they link to the rest of the vehicle. This chapter also describes the purpose and location of the subsystems, as well as the major components of the system and subsystems. The goal of this chapter is to establish a basic understanding for students to base their learning on. All systems and subsystems that are discussed in detail later in the text are introduced, and their primary purpose described. Chapter 2 covers the underlying basic theories of operation. This is valuable to the students and the instructors because it covers the theories that other textbooks assume the reader knows. All related basic physical, chemical, and thermodynamic theories are covered in this chapter.

Chapter 3 applies those theories to the operation of an automatic transmission. The emphasis has been moved from hydraulics to electronic controls to reflect changes in the operation of transmissions from strictly hydraulic to electronic "clutch-to-clutch" operation. The chapter goes on to provide the foundation of torque converter and power flow operation in preparation for an expanded discussion later in the text. Chapter 4 goes deeply into the electronics involved in today's transmissions. This is a chapter that has been heavily updated, since manufacturers are constantly adding more sophisticated electronics to transmissions.

Chapter 4 reflects the fact that transmissions have been transformed into devices that are as much electrical as they are mechanical and hydraulic. Basic networking is described to explain how the TCM works with the vehicle network to provide very precise and smooth shifts.

The chapters that follow cover the major components of an automatic transmission and transaxle, such as torque converters, pumps, hydraulic circuits, gears and shafts, and reaction and friction units. The last chapter takes a look at the commonly used transmissions and transaxles. This includes their mechanical, hydraulic, and electronic systems. The models used reflect the most commonly found transmissions in most school settings.

Current transmission models are used as examples throughout the text. Many are discussed in detail. This includes five- to nine-speed and continuously variable transmissions. This new edition also has more information on nearly all automatic-transmission-related topics. Finally, the art has been updated throughout the text to enhance comprehension and improve visual interest.

HIGHLIGHTS OF THIS EDITION—SHOP MANUAL

Along with the Classroom Manual, the Shop Manual has been updated to match current trends. Service information related to the new topics covered in the Classroom Manual is included in this manual. In addition, new photo sequences have been added and others updated. The purpose of these detailed photos is to show students what to expect when they perform the same procedure. They can also provide students a familiarity with a system or type of equipment that their school may not have. Although the main purpose of the textbook is not to prepare someone to successfully pass an ASE exam, all of the information required to do so is included in the textbook.

Chapters 1 and 2 cover the need-to-know transmission-related information about tools, safety, and typical service procedures. Chapter 1 covers safety issues. To stress the importance of safe work habits, one full chapter is dedicated to safety. This chapter includes common shop hazards, safe shop practices, safety equipment, and the legislation concerning the safe handling of hazardous materials and wastes. Chapter 2 covers the basics of what a transmission technician does to earn a living, including basic diagnostics. This chapter also includes those tools and procedures that are commonly used to diagnose and service automatic transmissions and transaxles.

Chapters 3 covers basic diagnostic techniques, starting with interviewing the customer and proceeding to proper inspection and testing techniques.

Chapter 4 covers electrical diagnostic tools and techniques that are essential for a technician to be efficient in his or her diagnosis.

The rest of the chapters have been thoroughly updated. Much of the updating focuses on the diagnosis and service to new systems, as well as those systems instructors have said they need more help in.

New photo sequences on reprogramming a TCM and overhauling five- and six-speed transmissions have been added. Currently accepted service procedures are used as examples throughout the text. These procedures also serve as the basis for new job sheets that are included in the text. Finally, the art has been updated throughout the text to enhance comprehension and improve visual interest.

CLASSROOM MANUAL

Features of the Classroom Manual include the following:

Cognitive Objectives

These objectives define the contents of the chapter and define what the student should have learned upon completion of the chapter.

Each topic is divided into small units to promote easier understanding and learning.

Cross-References to the Shop Manual

Reference to the appropriate page in the Shop Manual is given whenever necessary. Although the chapters of the two manuals are synchronized, material covered in other chapters of the Shop Manual may be fundamental to the topic discussed in the Classroom Manual.

Marginal Notes

These notes add "nice-to-know" information to the discussion. They may include examples or exceptions, or may give the common trade jargon for a component.

Terms to Know Definitions

Many of the new terms are pulled out into the margin and defined.

Author's Notes

This feature includes simple explanations, stories, or examples of complex topics. These are included to help students understand difficult concepts.

A Bit of History

This feature gives the student a sense of the evolution of the automobile. This feature not only contains nice-to-know information, but also should spark some interest in the subject matter.

Summaries

Each chapter concludes with a summary of key points from the chapter. These are designed to help the reader review the chapter contents.

Terms To Know List

A list of new terms that appear within the chapter.

Review Questions

Short-Answer Essays, Fill in the Blanks, and Multiple-Choice questions are found at the end of each chapter. These questions are designed to accurately assess the students' competence in the objectives stated at the beginning of the chapter.

A BIT OF HISTORY

All the hydraulic principles that are applied in a brake system are based on the work of a seventeenth-century scientist named Blaise Pascal. Pascal's work is known as Pascal's law. Pascal's law says that pressure at any one point in a confined liquid is the same in every direction and applies equal force on equal areas. Additionally, liquids take the path of least resistance. One of the most important results of Pascal's work was the discovery that fluids can be used to increase force. Pascal was the first person to demonstrate the relationships of pressure, force, and motion and the inverse relationship of motion and force. On an automobile, Pascal's laws describe the work in the hydraulic system of an automatic transmission and other systems. Pascal's laws are even at work in the movement of liquid fuel from a tank to the fuel-injection system on the engine.

CHAPTER 2
SPECIAL TOOLS AND PROCEDURES

Upon completion and review of this chapter, you should be able to:

- Describe the use of common diagnostic tools for automatic transmissions and transaxles.
- Describe the use of common pneumatic, electrical, and hydraulic power tools found in an automotive service department.
- Describe some of the special tools used to service automatic transmissions and the driveline.
- Identify the major measuring instruments and devices used by technicians.
- Describe the proper procedure for measuring with a micrometer.
- Describe the measurements normally taken by a technician while working on a vehicle's drivetrain.
- Describe the different sources for service information that are available to technicians.
- Describe the different types of fasteners used in the automotive industry.
- Describe the requirements for ASE certification as an automotive technician and a master auto technician.

Terms To Know

ASE
Backlash
Bolt head
Bolt shank
Circuit tester
Diagnosis
Dial caliper
Dial indicator
Digital multimeter (DMM)
Digital volt/ohmmeter (DVOM)
Duty cycle
Endplay
Fillet
Flat rate
Frequency
Grade marks
Graphing multimeter (GMM)
Machinist's rule
Micrometer
Oscilloscope
Preload
Press-fit
Pulse width
Repair order
Safety stands
Scan tool
Technical Service Bulletin (TSB)
Thread pitch
Torque wrench
Vehicle identification number (VIN)
Vernier caliper
Waveform

INTRODUCTION

Many different special tools and testing and measuring devices are used to service automatic transmissions. The most commonly used tools are the topic of the following paragraphs. Although you need to be more than familiar with, and will be using, common hand tools, they are not part of this discussion. You should already know what they are and how to use and care for them.
...gears... planetary gearsets are... the Ravigneaux gearset, which has two... two sets of planet gears, and a common ring gear.

REVIEW QUESTIONS

Short-Answer Essays

1. How does direct drive result in a planetary gearset?
2. What do the input shaft and the output shaft connect to?
3. What are the three common designs of compound planetary gearsets? How do they differ?
4. How is reverse gear accomplished in a planetary gearset?
5. How can a gear reduction be obtained from a planetary gearset?
6. What is the most important function of a differential?
7. What is the purpose of a transfer case?
8. In basic terms, explain how a helical gear, constant-mesh automatic transmission works.
9. What happens in a planetary gearset when no member is held? Explain your answer.
10. When a transmission is described as having two planetary gearsets in tandem, what does that mean?

Fill in the Blanks

1. When an _____ gear is in mesh with an _____ gear, the two gears will rotate in the same direction.
2. Gear reduction can occur if the _____ gear is held and the _____ gear is the input gear.
3. In the Lepelletier gear set, the _____ gears and the _____ of the compound planetary gearset can be _____ at different speeds.

SUMMARY

- The major components of a transaxle are the same as those of a transmission, except the transaxle assembly includes the final drive and differential gears in addition to the transmission.
- The input shaft drives a planetary gearset, which provides the various forward gears, a neutral position, and one reverse gear.
- Power flow through the gears is controlled by multiple-friction disc packs, one-way clutches, and friction bands. These hold a member of the gearset when force is applied to them, thereby making different gear ratios possible.
- An automatic transmission or transaxle selects ratios according to engine speed, engine load, vehicle speed, and other operating conditions.
- Gear ratio is an expression of the number of revolutions made by a driving gear compared to the number of revolutions made by...
- The torque converter automatically engages and disengages power from the engine to the transmission in relation to engine speed.
- The main components of the torque converter are the cover, turbine, impeller... connected to the flywheel... from the engine to the... transmission's oil pump... the input shaft... fluid from the... flow of fluid...
- A single planet... planet carrier... gears...

TRANSMISSIONS

...ssion is the servo assembly. The... d. The band must tightly hold... city of the band is determined... to it. This pressure or holding... es the force through hydraulic... re of 70 psi applied to it, the... he servo is increased further... and. The total force applied... connected to a planetary... d gearset members. This... fluid pressure applied to... ton is 5 inches, the force... uses a Belleville spring...

connector between the engine and the transmission, removes heat from the transmission, and lubricates the transmission's moving parts. To remove the heat from the fluid the ATF is directed through a cooler, which dissipates its heat.
- Typical reaction devices are bands, multiple-disc clutches, and overrunning clutches.
- Central to the hydraulic control system is the valve body assembly, which is typically made of two or three main parts: a valve body, separator plate, and transfer plate.
- Valve bodies are normally fitted with three types of valves: spool valves, check-ball... poppet valves. Valve bodies can also... pressure control solenoids and... the transmission fluid... the transmission's oil pump... oil under pressure to... nsmission. ...automatic... ane-type

SHOP MANUAL

To stress the importance of safe work habits, the Shop Manual also dedicates one full chapter to safety. Other important features of this manual include the following:

Performance-Based Objectives

These objectives define the contents of the chapter and define what the student should have learned upon completion of the chapter. These objectives also correspond to the list of required tasks for ASE certification.

Although this textbook is not designed to simply prepare someone for the certification exams, it is organized around the ASE task list. These tasks are defined generically when the procedure is commonly followed and specifically when the procedure is unique for specific vehicle models. Imported and domestic model automobiles and light trucks are included in the procedures.

Customer Care

This feature highlights those little things a technician can do or say to enhance customer relations.

Cautions and Warnings

Throughout the text, warnings are given to alert the readers to potentially hazardous materials or unsafe conditions. Cautions are given to advise the students of things that can go wrong if instructions are not followed or if a nonacceptable part or tool is used.

CHAPTER 8
GEAR AND SHAFT SERVICE

Upon completion and review of this chapter, you should be able to:

- Inspect, measure, and replace thrust washers and bearings.
- Inspect and replace bushings.
- Inspect and measure a planetary gear assembly and replace parts as necessary.
- Inspect and replace shafts.
- Remove and install the various types of shaft seals.

- Inspect, repair, or replace transaxle drive chains, sprockets, gears, bearings, and bushings.
- Inspect and replace parking pawl, shaft, spring, and retainer.
- Inspect, measure, repair, adjust, or replace transaxle final drive components.
- Inspect and service the transmission housing and cases.

Basic Tools
Basic mechanic's tool set
Clean lint-free rags
Appropriate service information
Clean ATF

Terms To Know
Etching
Selective thrust washers
Thrust load

INTRODUCTION

The overall reliability of a transmission depends on the structural integrity of the shafts and gears. Therefore these parts, as well as all thrust washers, bearings, and bushings (**Figure 8-1**), should be carefully inspected and replaced if damaged or worn.

THRUST WASHERS, BUSHINGS, AND BEARINGS

A thrust washer is designed to support a thrust load and keep parts from rubbing together, preventing premature wear on parts such as planetary gearsets (**Figure 8-2**) and transfer and final drive assemblies. Selective thrust washers come in various thicknesses to take up clearances and adjust end play at many different locations. Some transmissions have a selective thrust washer between the front and the rear multiple-disc assemblies (**Figure 8-3**). Thrust washers and thrust bearings are used wherever rotating parts must have their end play maintained. To control the end play of nonrotating parts, selective shims or spacers are used.

Thrust load is a load placed in parallel with the center of an axis.

Thrust washers are often referred to as thrust plates.

The oil pan can ...
indicate a more serious problem, ...
Be sure to use the specified fluid and check for the latest ...
problem. Some rebuilders may recommend that an oil additive be added to the ...
additive is designed to improve or alter the friction capabilities of regular ATF.

Special Tools
Short piece of hose
Graduated container
OSHA-approved air nozzle

CUSTOMER CARE After replacing a converter because of clutch problems, be sure to explain to the customer that other engine or transmission problems may become more evident with the new converter clutch. Because the old converter didn't fully engage, the fluid in the converter may have dampened many of the vibrations from the engine and transmission. The new converter will now connect the engine to the transmission so engine and transmission vibrations can be mechanically transmitted through the drivetrain.

Torque Converter-Related Cooler Problems

Besides causing transmission overheating problems, the vehicle may stall when the transmission is shifted into reverse gear. The cause of this problem may be plugged transmission cooler lines, or the cooler itself may be plugged. Fluid normally flows from the torque converter through the transmission cooler. If the cooler passages are blocked, fluid is unable to exhaust from the torque converter and the converter clutch piston remains engaged. When the clutch is engaged, there is no vortex flow in the converter and therefore little torque multiplication takes place in the converter.

To verify that the transmission cooler is plugged, disconnect the cooler return line from the radiator or cooler (**Figure 6-5**). Connect a short piece of hose to the outlet of the cooler and allow the other end of the hose to rest inside an empty container. Start the engine and measure the amount of fluid that flows into the container after 20 seconds. Normally, 1 quart of fluid should flow into the container. If less than that filled the container, a plugged cooler is indicated.

To correct a plugged transmission cooler, disconnect the cooler lines at the transmission and the radiator. Blow air through the cooler, one end at a time, then through the cooler lines. The air will clear large pieces of debris from the transmission cooler. Always use low air pressure of no more than 50 psi. Higher pressures may damage the cooler. If there is little air flow through the cooler, the radiator or external cooler must be removed

Shop Manual
page 202

Classroom Manual
Chapter 6, page 195

SERVICE TIP Stall testing is not recommended on many late-model transmissions. This test places extreme stress on the transmission and should only be conducted if recommended by the manufacturer.

STALL TESTING THE TORQUE CONVERTER

Two methods are commonly used to check the operation of the stator's one-way clutch: the stall test and the bench testing. Bench testing a converter will be covered later in this chapter. For now, we are still trying to determine if the converter is the cause of the customer's complaint. To bench test a converter, it must be removed from the vehicle and there is no need to do that unless we know the converter is at fault.

Caution
If a stall test is not correctly conducted, the converter and/or transmission can be damaged.

WARNING Make sure no one is around the engine or in front of the vehicle while a stall test is being conducted. A lot of stress is put on the engine, transmission, and brakes during the test. If something lets go, somebody can be seriously hurt.

Basic Tools Lists

Each chapter begins with a list of the basic tools needed to perform the tasks included in the chapter.

Terms to Know Definitions

Many of the new terms are pulled out into the margin and defined.

Service Tips

Whenever a shortcut or special procedure is appropriate, it is described in the text. These tips are generally those things commonly done by experienced technicians.

Photo Sequences

Many procedures are illustrated in detailed Photo Sequences. These detailed photographs show the students what to expect when they perform particular procedures. They also can provide the students a familiarity with a system or type of equipment that their school may not have.

Cross-References to the Classroom Manual

Reference to the appropriate page in the Classroom Manual is given whenever necessary. Although the chapters of the two manuals are synchronized, material covered in other chapters of the Classroom Manual may be fundamental to the topic discussed in the Shop Manual.

Special Tools Lists

Whenever special tools are required to complete a task, they are listed in the margin next to the procedure.

Marginal Notes

These notes add "nice-to-know" information to the discussion. They may include examples or exceptions, or may give the common trade jargon for a component.

Job Sheets

Located at the end of each chapter, the Job Sheets provide a format for students to perform procedures covered in the chapter. A reference to the ASE Education Foundation task addressed by the procedure is included on the Job Sheet.

Case Studies

Case Studies concentrate on the ability to properly diagnose the systems. Beginning with Chapter 3, each chapter ends with a Case Study in which a vehicle has a problem, and the logic used by a technician to solve the problem is explained.

ASE-Style Review Questions

Each chapter contains ASE-style review questions that reflect the performance-based objectives listed at the beginning of the chapter. These questions can be used to review the chapter as well as to prepare for the ASE certification exam.

CASE STUDY

A transmission technician was investigating a complaint from a customer who indicated that his vehicle had a very hard downshift at times when going around corners. The technician drove the vehicle with the customer since the complaint was intermittent. The technician let the customer drive, and the situation occurred. The technician made a note that the transmission did make a hard downshift under a load going uphill and slowing down to take a corner. The technician was able to determine that the downshift in question was from third to second gears. The technician was happy that he took the time to go with the customer to know exactly when and where the complaint occurred. The technician looked for technical bulletins when he arrived back at the shop and found that a problem in the valve body had been identified with a harsh 3-2 downshift. The bulletin called for a cleaning of the shift valve and replacement of the return spring, while checking for freedom of movement in the valve. The technician did not see any apparent problems, but performed the recommended procedure anyway. The technician was a bit wary that the transmission would be fixed when he test drove the vehicle, but to his relief the transmission performed flawlessly. The customer took the vehicle the same day. The customer also called the service manager the next day and expressed his gratitude to the technician for resolving his problem. It seemed that the customer had been to two other shops where the problem could not be duplicated.

ASE-STYLE REVIEW QUESTIONS

1. *Technician A* says electrical components from the valve body must not be soaked in parts cleaner. *Technician B* says a lint-free rag is a must when wiping down valves.
   Who is correct?
   A. A only
   B. B only
   C. Both A and B
   D. Neither A nor B

2. *Technician A* says a scratched separator plate must be replaced. *Technician B* says make sure all drain back openings in the spacer plate are clear.
   Who is correct?
   A. A only
   B. B only
   C. Both A and B
   D. Neither A nor B

3. *Technician A* says over-torqueing the hold-down bolts of the valve body can cause the valves to stick in their bores. *Technician B* says flat-filing the surface of the valve body will allow the valve body to seal properly and prevent cross leaks.
   Who is correct?
   A. A only
   B. B only
   C. Both A and B
   D. Neither A nor B

4. While assembling a valve body after cleaning it: *Technician A* lubricates all parts with clean ATF. *Technician B* says it is important that the valve-retaining plugs or caps be placed in the correct bore and in the correct direction.
   Who is correct?
   A. A only
   B. B only
   C. Both A and B
   D. Neither A nor B

5. *Technician A* says never use a magnet to remove the check balls. *Technician B* says be sure to do something that will help you remember where the balls were, such as photographing or drawing their location.
   Who is correct?
   A. A only
   B. B only
   C. Both A and B
   D. Neither A nor B

6. When inspecting valves from the valve body: *Technician A* says the edges of the spools should be damage-free but should not be sharp. *Technician B* says to identify a bent valve, install it into its bore and rotate it. If it binds or drags while it is being rotated, it is bent.
   Who is correct?
   A. A only
   C. Both A and B

CHAPTER 2
SPECIAL TOOLS AND PROCEDURES

Upon completion and review of this chapter, you should be able to:

- Describe the use of common diagnostic tools for automatic transmissions and transaxles.
- Describe the use of common pneumatic, electrical, and hydraulic power tools found in an automotive service department.
- Describe some of the special tools used to service automatic transmissions and the driveline.
- Identify the major measuring instruments and devices used by technicians.
- Describe the proper procedure for measuring with a micrometer.
- Describe the measurements normally taken by a technician while working on a vehicle's drivetrain.
- Describe the different sources for service information that are available to technicians.
- Describe the different types of fasteners used in the automotive industry.
- Describe the requirements for ASE certification as an automotive technician and a master auto technician.

Terms To Know List

A list of new terms that appear within the chapter.

Terms To Know

ASE	Endplay	Pulse width
Backlash	Fillet	Repair order
Bolt head	Flat rate	Safety stands
Bolt shank	Frequency	Scan tool
Circuit tester	Grade marks	Technical Service Bulletin (TSB)
Diagnosis	Graphing multimeter (GMM)	Thread pitch
Dial caliper	Machinist's rule	Torque wrench
Dial indicator	Micrometer	Vehicle identification number (VIN)
Digital multimeter (DMM)	Oscilloscope	Vernier caliper
Digital volt/ohmmeter (DVOM)	Preload	Waveform
Duty cycle	Press-fit	

INTRODUCTION

Many different special tools and testing and measuring devices are used to service automatic transmissions. The most commonly used tools are the topic of the following paragraphs. Although you need to be more than familiar with, and will be using, common hand tools, they are not part of this discussion. You should already know what they are and how to use and care for them.

ASE Challenge Questions

Each technical chapter ends with five ASE challenge questions. These are not mere review questions; rather, they test the students' ability to apply general knowledge to the contents of the chapter.

ASE CHALLENGE QUESTIONS

1. While discussing the condition of engine mounts:
   Technician A says engine movement may change the effective length of the shift cable and therefore affect engagement of the gears.
   Technician B says shifting and vibration problems are seldom caused by worn, loose, or broken engine and transmission mounts.
   Who is correct?
   A. A only
   B. B only
   C. Both A and B
   D. Neither A nor B

2. *Technician A* says a leaking torque converter may be indicated by the presence of transmission fluid over the entire shell of the converter.
   Technician B says transmission oil on the converter's rear shell indicates a leaking torque converter.
   Who is correct?
   A. A only
   B. B only
   C. Both A and B
   D. Neither A nor B

3. Typically a transmission has two sensors mounted externally to the transmission case.
   Technician A says one of these sensors is the Input shaft speed sensor.
   Technician B says the other is the MAP sensor.
   Who is correct?
   A. A only
   B. B only
   C. Both A and B
   D. Neither A nor B

4. *Technician A* says a new terminal can be used to check for proper connection fit by comparing the drag on each pin of the connector.
   Technician B says some sensors can affect both engine and transmission operation.
   Who is correct?
   A. A only
   B. B only
   C. Both A and B
   D. Neither A nor B

5. Pressure testing reveals low pressure in all gears.
   Technician A says using a reverse stall test forces the pump output to maximum pressure.
   Technician B says a reverse stall test will determine whether the pump output is low.
   Who is correct?
   A. A only
   B. B only
   C. Both A and B
   D. Neither A nor B

ASE Practice Examination

A 50-question ASE practice exam, located in the appendix, is included to test students on the contents of the Shop Manual.

APPENDIX A
ASE PRACTICE EXAMINATION

Final Exam Automatic Transmission/Transaxle A2

1. Which of the following is the *least* likely cause for a buzzing noise from a transmission in Park?
   A. Improper fluid level or condition
   B. Defective oil pump
   C. Defective flexplate
   D. Damaged planetary gearset

2. A vehicle experiences engine flare in low gear only.
   Technician A says the torque converter lockup clutch is slipping.
   Technician B says the transmission oil pump is not providing the required pressure.
   Who is correct?
   A. A only
   B. B only
   C. Both A and B
   D. Neither A nor B

3. The results of a hydraulic pressure test are being discussed:
   Technician A says low idle pressure may be caused by a defective exhaust gas recirculation system.
   Technician B says low neutral and park pressures may indicate a fluid leakage past the clutch and servo seals.
   Who is correct?
   A. A only
   B. B only
   C. Both A and B
   D. Neither A nor B

4. The vehicle creeps in NEUTRAL.
   Technician A says a too high engine idle speed could be the cause.
   Technician B says a too tight clutch pack may be the problem.
   Who is correct?
   A. A only
   B. B only
   C. Both A and B
   D. Neither A nor B

5. *Technician A* says overtorqued valve body fasteners may cause erratic shifting.
   Technician B says debris in the valve body can cause erratic shifting.
   Who is correct?
   A. A only
   B. B only
   C. Both A and B
   D. Neither A nor B

6. The transmission's output shaft and its sealing components are being discussed:
   Technician A says the shaft and all of its sealing components must be replaced if nicks and scratches are found in the shaft's sealing area.
   Technician B says all of the shaft's seals and rings must be replaced during a rebuild.
   Who is correct?
   A. A only
   B. B only
   C. Both A and B
   D. Neither A nor B

7. *Technician A* says harsh shifts could be caused by a transmission that has not been properly "relearned" after TCM replacement.
   Technician B says a harsh shift could be caused by using the incorrect fluid type.
   Who is correct?
   A. A only
   B. B only
   C. Both A and B
   D. Neither A nor B

8. The vehicle does not move in any gear.
   Technician A says a misadjusted shift cable could be the cause.
   Technician B says leakage at the oil pump and/or valve body could cause this condition.
   Who is correct?
   A. A only
   B. B only
   C. Both A and B
   D. Neither A nor B

SUPPLEMENTS

Instructor Resources

The *Today's Technician* series offers a robust set of instructor resources, available online at Cengage's Instructor Resource Center and on DVD. The following tools have been provided to meet any instructor's classroom preparation needs:

- An Instructor' Guide provides lesson plans, teaching tips, and complete answers to end-of-chapter questions.
- Power Point presentations include images, videos, and animations that coincide with each chapter's content coverage.
- Cengage Testing Powered by Cognero® delivers hundreds of test questions in a flexible, online system. You can choose to author, edit, and manage test bank content from multiple Cengage Learning solutions and deliver tests from your LMS, or you can simply download editable Word documents from the DVD or Instructor Resource Center.
- An Image Gallery includes photos and illustrations from the text.
- The Job Sheets from the Shop Manual are provided in Word format.
- End-of-Chapter Review Questions are also provided in Word format, with a separate set of text rejoinders available for instructors' reference.
- To complete this powerful suite of planning tools, correlation guides are provided to the ASE Education Foundation tasks and to the previous edition.

MindTap Automotive for Today's Technician™: Automatic Transmissions and Transaxles, 7th edition

MindTap Automotive for Automatic Transmissions and Transaxles provides a customized learning experience with relevant assignments that will help students learn and apply concepts while it allows instructors to measure skills and outcomes with ease. MindTap Automotive meets the needs of today's automotive classroom, shop, and student with a robust suite of interactive learning tools, including:

- Cengage Automotive Theory simulations covering topics in Electricity/Electronics, Brakes, Engine Performance, Engine Repair, Steering and Suspension, Heating & Air Conditioning, Automatic Transmissions, and Manual Transmissions.
- S/P2® training modules covering Safety, Pollution Control, and soft skills for career preparation.
- On-line job sheets based on ASE Education Foundation tasks.

MindTap Automotive also offers students the opportunity to practice diagnostic techniques in a safe environment while strengthening their critical thinking and troubleshooting skills with the inclusion of diagnostic scenarios from Delmar Automotive Training Online (DATO). Additional engaging activities include videos, animations, matching exercises, and gradable assessments.

REVIEWERS

The author and publisher would like to extend a special thanks to the individuals who reviewed this text and offered their invaluable feedback:

Matthew Bockenfeld
Vatterott College Joplin, MO

Michael Grooms
Austin Community College
Austin, Texas

Shih-Wei Jack Chen
Utah State University Eastern
Price, UT

Sumner Huckaby
Greenville Technical College
Greenville, SC

Cody R. LeMasters
Vatterott College
Joplin, MO

Thomas R. Sypherd
The University of Northwestern Ohio
Lima, Ohio

Todd Richardson
Utah State University Eastern
Price, UT

SPECIAL THANKS

The author and publisher would like to extend a special thanks to Mike Whitcraft and Robert Callahan of Ranken Technical College for their help with pictures and the loaning of transmissions used in the Photo Sequences.

CHAPTER 1
SAFETY

Upon completion and review of this chapter, you should be able to:

- Explain how safety practices are part of professional behavior.
- Explain the basic principles of personal safety, including protective eyewear, clothing, gloves, shoes, and hearing protection.
- Inspect equipment and tools for unsafe conditions.
- Properly work around batteries.
- Explain precautions that need to be followed while working around rotating pulleys and belts.
- Recognize the hazards of working around supplemental restraint systems (air bags).
- Understand the importance of safety and accident prevention in an automotive shop.
- Explain the procedures and precautions for safely using tools and equipment.
- Explain the precautions that need to be followed to safely raise a vehicle on a lift.
- Properly lift heavy objects.
- Explain some of the precautions for working on high-voltage systems.
- Explain the procedures for responding to an accident.
- Recognize fire hazards.
- Extinguish common classes of fires.
- Identify substances that could be regarded as hazardous materials.
- Describe the purpose of the laws concerning hazardous wastes and materials, including the right-to-know laws.

Terms To Know

Bloodborne pathogens	Class D fire	Safety Data Sheet (SDS)
Carbon monoxide (CO)	Corrosivity	OSHA
Category III multimeters	EP toxicity	Reactivity
Caustic	Hazardous waste	Shank
Class A fire	Hydrocarbon	Spontaneous combustion
Class B fire	Ignitability	Volatility
Class C fire		

INTRODUCTION

Safety is everyone's job. You should work safely to protect yourself and the people around you. Perhaps the best single safety rule is "Think before you act." Too often people working in shops gamble by working in an unsafe way. Often they win and no one gets hurt, but all it takes is one accident and all of their past winnings are lost. By gambling and, perhaps, saving five minutes, you can lose an eye or a hand. By acting first and thinking last, you can ruin your back and lose your career. Accidents in a shop can be prevented by others and by *you*. Safe work habits also prevent damage to the vehicles and equipment in the shop.

It has been said that 50 percent of all shop accidents could have been prevented by a single individual: the technician.

Working on automobiles can be dangerous. It can also be fun and very rewarding. To keep the fun and rewards rolling in, you need to try to prevent accidents by working safely. In an automotive repair shop, there is great potential for serious accidents, simply because of the nature of the business and the equipment used. Through carelessness, the automotive repair industry can be one of the most dangerous occupations.

An *accident* is something that happens unintentionally.

However, the chances of your being injured while working on a car are close to nil if you learn to work safely and use common sense. Shop safety is the responsibility of everyone in the shop: you, your fellow students or employees, and your employer or instructor. Everyone must work together to protect the health and welfare of all who work in the shop.

PERSONAL SAFETY

Personal safety involves those precautions you take to protect yourself from injury. These include wearing protective gear (**Figure 1-1**), dressing for safety, working professionally, and correctly handling tools and equipment.

When you have neat work habits, you display a professional attitude. Actually, neat habits are also safe habits. Cleaning up spills and keeping equipment and tools out of the path of others prevent accidents. True professionals take time to clean their tools and work area. With a professional attitude, you do not clown around in the shop, do not throw items in the shop, and do not create an unsafe condition for the sake of saving time. Rather than ignoring basic safety rules to save time, a professional saves time by performing effective diagnostics and repair procedures.

Professionals also take pride in their work, treat customers and their vehicles with respect, and try to stay current with all technical, safety, and environmental concerns.

Figure 1-1 Proper dress prevents injuries.

Eye Protection

Your eyes can become infected or permanently damaged by many things in a shop. Some procedures, such as grinding, result in tiny particles of metal and dust being thrown off at very high speeds. These metal and dirt particles can easily get into your eyes, causing scratches or cuts on your eyeball. Pressurized gases and liquids escaping a hose or hose fitting can spray a great distance. If these chemicals get into your eyes, they can cause blindness. Dirt and sharp bits of corroded metal can easily fall into your eyes while you are working under a vehicle.

Eye protection should be worn whenever you are exposed to these risks. To be safe, you should wear it whenever you are working in the shop. There are many types of eye protection (**Figure 1-2**). Safety glasses have lenses made of safety glass and some sort of side protection. Regular prescription glasses should not be worn as a substitute for safety glasses. When prescription glasses are worn in a shop, they should be fitted with side shields.

Wearing safety glasses at all times is a good habit to get into. Choose glasses that fit well and feel comfortable. Some shops require that everyone wear eye protection whenever they are in the work area.

Some procedures may require that you wear additional eye protection. For example, when you are cleaning parts with a pressurized spray, you should wear a face shield. The face shield not only gives added protection to your eyes but it also protects the rest of your face.

If chemicals such as battery acid, fuel, or solvents get into your eyes, flush them continuously with clean water. Have someone call a doctor and get medical help immediately.

Many shops have eyewash stations (**Figure 1-3**) or safety showers that should be used whenever you or someone else has been sprayed or splashed with a chemical.

Figure 1-2 Different types of eye protection worn by automotive technicians: (A) safety glasses, (B) goggles, and (C) face shield.

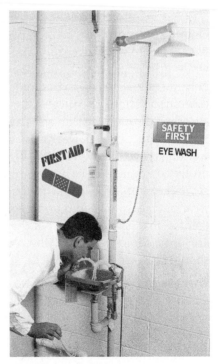

Figure 1-3 Using an eyewash station.

Clothing

Clothing that hangs freely, such as shirttails, can create a safety hazard and cause serious injury. Nothing you wear should be allowed to dangle in the engine compartment or around equipment. Shirts should be tucked in and buttoned and long sleeves buttoned or carefully rolled up. Your clothing should be well fitted and comfortable but made of strong material. Loose, baggy clothing can easily be caught in moving parts and machinery. Neckties should not be worn. Some technicians prefer to wear coveralls or shop coats to protect their personal clothing.

Long hair and loose, hanging jewelry can create the same type of hazard as loose-fitting clothing. They can get caught in moving engine parts and machinery. If you have long hair, tie it back or tuck it under a cap.

Rings, necklaces, bracelets, and watches should not be worn while working. A ring can rip your finger off, a watch or bracelet can cut your wrist, and a necklace can choke you. This is especially true when working with or around electrical wires. The metal used to make jewelry conducts electricity very well and can easily cause a short, through you, if it touches a bare wire.

> An electrical short is basically an alternative path for the flow of electricity.

Keep your clothing clean. If you spill gasoline or oil on yourself, change that item of clothing immediately. Oil that is against your skin for a prolonged period of time can produce rashes or other allergic reactions. Gasoline can irritate cuts and sores.

Foot Protection

You should also protect your feet. Tennis and jogging shoes provide little protection if something falls on your foot. Boots or shoes made of leather or a material that approaches the strength of leather offer much better protection from falling objects. There are many designs of safety shoes and boots that have steel plates built into the toe and **shank** to protect your feet. Many also have soles that are designed to resist slipping on wet surfaces. Foot injuries are not only quite painful but can also put you out of work for some time.

> The **shank** of a shoe is that portion of the shoe that protects the ball of your foot. It is the narrow part of a shoe between the heel and sole.

Hand Protection

Good hand protection is often overlooked. A scrape, cut, or burn can seriously impair your ability to work for many days. A well-fitted pair of heavy work gloves should be worn while grinding, welding, or handling chemicals or high-temperature components. Polyurethane or vinyl gloves should be worn when handling strong and dangerous **caustic** chemicals. These chemicals can easily burn your skin. Also, some chemicals can be absorbed through the skin and cause paralysis.

> A **caustic** material has the ability to destroy or eat through something. Caustic materials are considered extremely corrosive.

Many technicians wear thin, surgical-type latex or nitrile gloves whenever they are working on vehicles (**Figure 1-4**). These offer little protection against cuts but do offer protection against disease and grease buildup under and around your fingernails. These gloves are comfortable and are quite inexpensive. Latex gloves are more comfortable to wear but weaken when they are exposed to gas, oil, or solvents. Nitrile gloves are not as comfortable as latex gloves but they are not affected by gas, oil, or solvents. Nitrile gloves are also a choice for those who may be allergic to latex. Your choice of hand protection should be based on what you are doing.

Disease Prevention. When you are ill with something that may be contagious, see a doctor and do not go to work or school until the doctor says there is little chance of someone else contracting the illness from you.

You should also be concerned with and protect yourself and others from bloodborne pathogens. **Bloodborne pathogens** are pathogenic micro-organisms that are present in human blood and can cause disease. These pathogens include, but are not limited to, hepatitis B virus (HBV) and human immunodeficiency virus (HIV). For everyone's protection, any injury that causes bleeding should be dealt with as a threat to others. You should avoid contact with the blood of another. If you need to administer some form of first aid, make sure

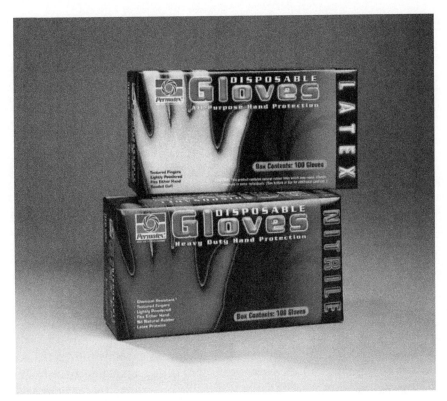

Figure 1-4 Disposable latex and nitrile gloves.

you put on hand protection before you do so. You should also wear gloves and other protection when handling the item that caused the cut. This item should be sterilized immediately. Most importantly, as with all injuries, report the accident to your instructor or supervisor.

You also need to protect your hands and other body parts when working with pressurized fluids. There is a risk of blood poisoning (sepsis) due to the inadvertent introduction of toxins (solvents, diesel fuel, and so on) into the bloodstream from a pressurized source. The best protection for this is working carefully and wearing heavy gloves.

Ear Protection

Exposure to very loud noise levels for extended periods of time can lead to a loss of hearing. Air wrenches, air hammers, engines running under a load, and vehicles running in enclosed areas can all generate annoying and harmful levels of noise. Simple earplugs or earphone-type protectors should be worn in environments that are constantly noisy.

Respiratory Protection

It is not uncommon for a technician to work with chemicals that have toxic fumes. Air or respiratory masks should be worn whenever you will be exposed to toxic fumes. Cleaning parts with solvents and painting are the most common activities for which respiratory masks should be worn. Masks should also be worn when handling hazardous materials.

FIRE HAZARDS AND PREVENTION

The presence of gasoline is so common that its dangers are often forgotten. A slight spark or an increase in heat can cause a fire or explosion. Gasoline fumes are heavier than air. Therefore, when an open container of gasoline is sitting about, the fumes spill out over the sides of the container. These fumes are more flammable than liquid gasoline and can easily explode.

 Caution

Never siphon gasoline or diesel fuel with your mouth. These liquids are poisonous and can make you sick or fatally ill.

Figure 1-5 Flammable liquids should be stored in safety-approved containers.

Never smoke around gasoline or in a shop filled with gasoline fumes. If a vehicle has a gasoline leak or you have caused a leak by disconnecting a fuel line, wipe it up immediately and stop the leak. Make sure that any grinding or welding that may be taking place in the area is stopped until the spill is totally cleaned up and the floor has been flushed with water. The rags used to wipe up the gasoline should be taken outside to dry, then stored in an approved dirty rag container. If vapors are present in the shop, have the doors open and turn on the ventilating system. It takes only a small amount of fuel mixed with air to cause combustion.

Gasoline should always be stored in approved containers and never in a glass bottle or jar (**Figure 1-5**). If the glass jar were knocked over or dropped, a terrible explosion could take place.

The **volatility** of a substance is how easily the substance vaporizes or explodes.

Diesel Fuel. Diesel fuel is not as **volatile** as gasoline, but should be stored and handled in the same way. It is also not as refined as gasoline and tends to be a very dirty fuel. It normally contains many impurities, including active microscopic organisms that can be highly infectious. If diesel fuel happens to get into an open cut or sore, thoroughly wash it immediately.

Solvents. Cleaning solvents are also not as volatile as gasoline, but they are still **flammable**. These should be stored and treated in the same way as gasoline.

The **flammability** of a substance is how well the substance supports combustion.

Handle all solvents (and any liquids) with care to avoid spillage. Keep all solvent containers closed, except when pouring. Proper ventilation is very important in areas where volatile solvents and chemicals are used. Solvent and other combustible materials must be stored in approved and designated storage cabinets or rooms with adequate ventilation. Never light matches or smoke near flammable solvents and chemicals, including battery acids.

A **hydrocarbon** is a substance composed of hydrogen and carbon molecules.

Rags. Oily or greasy rags can also be a source for fires. These rags should be stored in an approved container and never thrown out with normal trash (**Figure 1-6**). Like gasoline, oil is a **hydrocarbon** and can ignite with or without a spark or flame. This ignition without an external source of heat or fire is called **spontaneous combustion**. For this reason place the oily rags in a container as soon as you are finished using them.

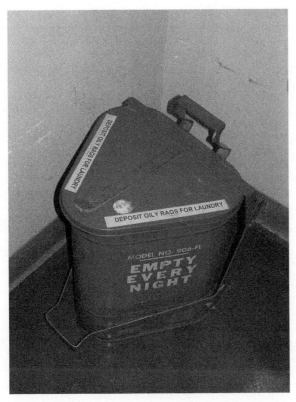

Figure 1-6 Dirty rags must be stored in a special container.

Fire Extinguishers

In case of a fire you should know the location of the fire extinguishers and fire alarms in the shop and should also know how to use them. You should also be aware of the different types of fires and the fire extinguishers used to put out these types of fires (**Figure 1-7**).

Figure 1-7 Different types of fire extinguishers.

TABLE 1-1   GUIDE TO EXTINGUISHER SELECTION

	Class of Fire	Typical Fuel Involved	Type of Extinguisher
Class △A Fires (green)	**For Ordinary Combustibles** Put out a Class A fire by lowering its temperature or by coating the burning combustibles.	Wood Paper Cloth Rubber Plastics Rubbish Upholstery	Water*[1] Foam* Multipurpose dry chemical[4]
Class ▪B Fires (red)	**For Flammable Liquids** Put out a Class B fire by smothering it. Use an extinguisher that gives a blanketing, flame-interrupting effect; cover whole flaming liquid surface.	Gasoline Oil Grease Paint Lighter fluid	Foam* Carbon dioxide[5] Halogenated agent[6] Standard dry chemical[2] Purple K dry chemical[3] Multipurpose dry chemical[4]
Class ●C Fires (blue)	**For Electrical Equipment** Put out a Class C fire by shutting off power as quickly as possible and by always using a nonconducting extinguishing agent to prevent electric shock.	Motors Appliances Wiring Fuse boxes Switchboards	Carbon dioxide[5] Halogenated agent[6] Standard dry chemical[2] Purple K dry chemical[3] Multipurpose dry chemical[4]
Class ★D Fires (yellow)	**For Combustible Metals** Put out a Class D fire of metal chips, turnings, or shavings by smothering or coating with a specially designed extinguishing agent.	Aluminum Magnesium Potassium Sodium Titanium Zirconium	Dry powder extinguishers and agents only

*Cartridge-operated water, foam, and soda-acid types of extinguishers are no longer manufactured. These extinguishers should be removed from service when they become due for their next hydrostatic pressure test.

Notes:
(1) Freezes in low temperatures unless treated with antifreeze solution, usually weighs over 20 pounds (9 kg), and is heavier than any other extinguisher mentioned.
(2) Also called ordinary or regular dry chemical (sodium bicarbonate).
(3) Has the greatest initial fire-stopping power of the extinguishers mentioned for class B fires. Be sure to clean residue immediately after using the extinguishers so sprayed surfaces will not be damaged (potassium bicarbonate).
(4) The only extinguishers that fight A, B, and C classes of fires. However, they should not be used on fires in liquefied fat or oil of appreciable depth. Be sure to clean residue immediately after using the extinguisher so sprayed surfaces will not be damaged (ammonium phosphates).
(5) Use with caution in unventilated, confined spaces.
(6) May cause injury to the operator if the extinguishing agent (a gas) or the gases produced when the agent is applied to a fire is inhaled.

Basically, there are four classes of fires: **Class A fires** are those in which wood, paper, and other ordinary materials are burning; **Class B fires** are those involving flammable liquids, such as gasoline, diesel fuel, paint, grease, oil, and other similar liquids; and **Class C fires** are electrical fires. **Class D fires** are a unique type of fire in which the burning material is a metal. An example of this is a burning "mag" wheel. The magnesium used in the construction of the wheel is a flammable metal and will burn brightly when subjected to high heat.

Using the wrong type of extinguisher may cause the fire to grow instead of putting it out. All extinguishers are marked with a symbol or letter to signify what class of fire they were intended for (**Table 1-1**). You should know the location and rating of each fire extinguisher in the shop before you need one.

Using a Fire Extinguisher

Remember, during a fire, never open doors or windows unless it is absolutely necessary; the extra draft will only make the fire worse. Make sure the fire department is contacted before or during your attempt to extinguish a fire. To extinguish a fire, stand 6–10 feet from the fire. Each type of extinguisher requires a slightly different use (**Figure 1-8**).

Figure 1-8 Each type of fire extinguisher requires a slightly different operation.

For most, you should follow the guidelines shown in **Photo Sequence 1**. Before releasing the agent from the extinguisher, hold the extinguisher firmly in an upright position. Aim the nozzle at the base and use a side-to-side motion, sweeping the entire width of the fire. Stay low to avoid inhaling the smoke. If it gets too hot or too smoky, get out. Remember, never go back into a burning building for anything.

To help remember how to use an extinguisher, remember the word PASS:

Pull the pin from the handle of the extinguisher.
Aim the extinguisher's nozzle at the base of the fire.
Squeeze the handle.
Sweep the entire width of the fire with the contents of the extinguisher.

If there is not a fire extinguisher handy, a blanket or fender cover may be used to smother the flames. Be careful when doing this as the heat of the fire may burn you and

P1-1 Multipurpose dry chemical fire extinguisher.

P1-2 Hold the fire extinguisher in an upright position.

P1-3 Pull the safety pin from the handle.

P1-4 Free the hose from its retainer and aim it at the base of the fire.

P1-5 Stand 8 feet from the fire. Do not go any closer to the fire.

P1-6 Squeeze the lever while sweeping the hose from side to side. Keep the hose aimed at the base of the fire.

the blanket. If the fire is too great to smother, move everyone away from the fire and call the local fire department. A simple under-the-hood fire can cause the total destruction of the car and the building and can take lives. You must be able to respond quickly and precisely to avoid a disaster.

Never put water on a gasoline fire. The water will just spread the fire—the proper fire extinguisher smothers the flames.

ROTATING BELTS AND PULLEYS

The thermostatic switch for an electric cooling fan may also be disconnected to prevent the fan from coming on.

Be very careful around belts, pulleys, wheels, chains, or any other rotating mechanism. When working around an engine's drive belts and pulleys, make sure hands, shop towels, or loose clothing do not come in contact with the moving parts. Hands and fingers can be quickly pulled into a revolving belt or pulley even at engine idle speeds.

⚠ WARNING **Be careful when working around electric engine cooling fans. These fans are controlled by a thermostat and can come on without warning, even when the engine is not running. Whenever you must work around these fans, disconnect the electrical connector to the fan motor before reaching into the area around the fan (Figure 1-9).**

Radiator support—upper

Front

Front cross member

Motor
coolant fan

Figure 1-9 When possible, disconnect the connector to the cooling fan
before reaching into or working near the area around the fan.

TOOL AND EQUIPMENT SAFETY

When you work with any equipment, make sure you use it properly and that is set up according to the manufacturer's instructions. All equipment should be properly maintained and periodically inspected for unsafe conditions. Frayed electrical cords or loose mountings can cause serious injuries. All electrical outlets should be equipped to allow for the use of three-pronged electrical cords. The third prong allows for a safety ground connection (**Figure 1-10**). All equipment with rotating parts should be equipped with safety guards that reduce the possibility of the parts coming loose and injuring someone. Do not depend on someone else to inspect and maintain equipment. Check it out before you use it! If you find the equipment unsafe, disconnect it from the power source, put a sign on it to warn others, and notify the person in charge.

Hand Tool Safety

Hand tools should always be kept clean and be used only for the purpose for which they were designed. Oily hand tools can slip out of your hand and cause broken fingers or at least cut or skinned knuckles. Your tools should also be inspected for cracks, broken parts, or other dangerous conditions before you use them.

Damaged
cord

Missing
ground
terminal

Defective   **Correct**

Figure 1-10 Make sure all three prongs on an electrical connector are in good condition before plugging in the cord.

Knives, chisels, and scrapers must be used in a motion that will keep the point or blade moving away from your body. Always hand a pointed or sharp tool to someone else with the handle toward the person you are handing the tool to.

Power Tool Safety

Power tools are operated by an outside source of power, such as electricity, compressed air, or hydraulic pressure. Safety around power tools is very important. Serious injury can result from carelessness. Always wear safety glasses when using power tools.

If a tool is electrically powered, make sure it is properly grounded. Also, when using electrical power tools, never stand on a wet or damp floor. Disconnect the power source before doing any work on the machine or tool. Before plugging in any electric tool, make sure its switch is in the off position. When you are done using the tool, turn it off and unplug it. Never leave a running power tool unattended.

Before using the tool, check the power cord for bare wires or cracks in the insulation. If the cord is damaged, do not plug it into the wall outlet. Repair the cord before using the tool or use another tool.

When using power equipment on a small part, never hold the part in your hand. Always mount the part in a bench vise or use vise-grip pliers. Never try to use a machine or tool beyond its stated capacity or for operations requiring more than the rated power of the tool.

When working with larger power tools, such as a bench or floor grinding wheel, check the machine and/or the grinding wheels for signs of damage before using them. If a wheel is damaged, it should be replaced and not used. Check the speed rating of the wheel and make sure it matches the speed of the machine. Never spin a grinding wheel at a speed higher than it is rated for.

A safety guard is a protective cover over a moving part. Be sure to place all safety guards in position (**Figure 1-11**). Although safety guards are designed to prevent injury, you should still wear safety glasses and/or a face shield while using the machine. Make sure there are no people or parts around the machine before starting it. Keep your hands and clothing away from the moving parts. Maintain a balanced stance while using the machine.

Compressed Air Equipment Safety

Compressed air is used to inflate tires, apply paint, and drive tools. Compressed air can be dangerous when it is not used properly. When using compressed air, safety glasses and/or a face shield should be worn. Particles of dirt and pieces of metal, blown by the high-pressure air, can penetrate your skin or get into your eyes.

Figure 1-11 Before using a bench grinder, make sure all safety guards are in place.

Before using a compressed air tool, check all hose connections. Pneumatic tools must always be operated at the pressure recommended by the manufacturer.

Always hold an air nozzle or air control device securely when starting or shutting off the compressed air. A loose nozzle can whip suddenly and cause serious injury. Never point an air nozzle at anyone. Never use compressed air to blow dirt from your clothes or hair. Never use compressed air to clean the floor or workbench. Also, never spin bearings with compressed air. If the bearing is damaged, one of the steel balls or rollers might fly out and cause serious injury.

Lift Safety

Always be careful when raising a vehicle on a lift or a hoist. Adapters and hoist plates must be positioned correctly on twin-post and rail-type lifts to prevent damage to the underbody of the vehicle. There are specific lift points that allow the weight of the vehicle to be evenly supported by the adapters or hoist plates. The correct lift points can be found in the vehicle's service information. **Figure 1-12** shows typical locations for unibody and frame cars. Always follow the manufacturer's instructions. Before operating any lift or hoist, carefully read the operating manual and follow the operating instructions.

Once the lift supports are properly positioned under the vehicle, raise the lift until the supports contact the vehicle. Then, check the supports to make sure they are in full contact with the vehicle. Shake the vehicle to make sure it is securely balanced, then raise it to the desired working height. Before working under a vehicle, make sure the lift's locking devices are fully engaged.

The Automotive Lift Institute (ALI) is an association concerned with the design, construction, installation, operation, maintenance, and repair of automotive lifts. Their primary concern is safety. Every lift approved by ALI has a safety label. It is a good idea to read through the safety tips included on this label before using a lift.

Figure 1-12 Typical lift points for (A) unibody and (B) frame-body vehicles.

Jack and Jack Stand Safety

A vehicle can be raised by a hydraulic jack. A handle on the jack is moved up and down to raise part of a vehicle and a valve is turned to release the hydraulic pressure in the jack to lower the part. At the end of the jack is a lifting pad. The pad must be positioned under an area of the vehicle's frame or at one of the manufacturer's recommended lift points. Never place the pad under the floor pan or under steering and suspension components. Always position the jack so the wheels of the vehicle can roll as the vehicle is being raised. Also make sure that the wheels of the jack can roll while the vehicle is being raised. If the wheels of the jack cannot roll the jack pad may be pulled out of position and the vehicle may fall.

⚠ **WARNING Never use a lift or jack to move something heavier than it is designed for. Always check the rating before using a lift or jack. If a jack is rated for two tons, do not attempt to use it for a job requiring five tons. It is dangerous for you and the vehicle.**

Safety (jack) stands are placed under a sturdy chassis member, such as the frame or axle housing, to support the vehicle. Once the safety stands (**Figure 1-13**) are in position, the hydraulic pressure in the jack should be slowly released until the weight of the vehicle is on the stands. Jack stands have a capacity rating. Always use the correct rating of jack stand.

Never move under a vehicle when it is only supported by a jack; rest the vehicle on safety stands before moving under the vehicle. The jack should be removed after the jack stands are set in place. This eliminates potential hazards, such as a jack handle sticking out into a walkway.

Chain Hoist and Crane Safety

Heavy parts of the automobile, such as engines, are removed with chain hoists or cranes. (Another term for a chain hoist is *chain fall*.) To prevent serious injury, chain hoists and cranes must be properly attached to the parts being lifted (**Figure 1-14**). Always use bolts with enough strength to support the object being lifted. Place the chain hoist or crane directly over the assembly. Then, attach the chain or cable to the hoist.

Figure 1-13 Jack or safety stands are placed under the vehicle after it has been raised by a hydraulic jack and before work is done under the vehicle.

Figure 1-14 When using a chain hoist to pull an engine, the attachments to the engine should be secure and strong.

Cleaning Equipment Safety

Cleaning automotive parts can be divided into four basic categories: chemical cleaning, thermal cleaning, abrasive cleaning, and steam cleaning. Regardless of what method you use, you should always follow the precautions given by the manufacturer and you should always wear the recommended protective gear.

BATTERIES

When possible, you should disconnect the vehicle's battery before disconnecting any electrical wire or component. This prevents the possibility of a fire or electrical shock. It also eliminates the possibility of an accidental short, which can ruin the car's electrical system. Disconnect the negative or ground cable first (**Figure 1-15**), then disconnect the positive cable. Since electrical circuits require a ground to be complete, by removing the ground cable you eliminate the possibility of a circuit accidentally becoming completed. When reconnecting the battery, connect the positive cable first, then the negative.

The hydrogen gases that form in the top of a battery while it is being charged are very explosive. Never smoke or introduce any form of heat around a charging battery. An explosion will not only destroy the battery but may also spray sulfuric acid all over you, the car, and the shop. When connecting a battery charger to a battery, leave the charger off until all of its leads are connected. This will prevent electrical sparks and prevent a possible explosion.

The most dangerous battery is one that has been overcharged. It is hot and has been, or still may be, producing large amounts of hydrogen. Allow the battery to cool before working with or around it. Also, never use or charge a battery that has frozen electrolyte. Extreme amounts of hydrogen may be released from the electrolyte as it thaws.

⚡ Caution

The active chemical in a battery, the *electrolyte*, is basically sulfuric acid. Sulfuric acid can cause severe skin burns and permanent eye damage, including blindness. If battery acid gets on your skin, wash it off immediately and flush your skin with water for at least five minutes. If the electrolyte gets into your eyes, immediately flush them out with water, then immediately see a doctor. *Never* rub your eyes, just flush them well and go to a doctor. Working with and around batteries is an obvious time to wear safety glasses or goggles.

Figure 1-15 Always disconnect a battery by removing the negative cable first.

Hybrid vehicles generally have a high-voltage disconnect to isolate the high-voltage system (**Figure 1-18**). The high-voltage disconnect should be utilized anytime the technician might encounter the high-voltage system.

⚠ Caution

Always double-check the polarity of the battery charger's connections before turning the charger on. Incorrect polarity can damage the battery or cause it to explode.

VEHICLE SERVICE PRECAUTIONS

Working Safely on High-Voltage Systems

Electric drive vehicles (battery-operated, hybrid, and fuel cell electric vehicles) have high-voltage electrical systems (from 42 to 650 volts). These high voltages can kill you! Most high-voltage circuits are identifiable by size and color. The cables have thicker insulation and are typically colored orange (**Figure 1-16**). The connectors are also colored orange. On some vehicles, the high-voltage cables are enclosed in an orange shielding or casing. In addition, the high-voltage battery pack and most high-voltage components have "High Voltage" caution labels (**Figure 1-17**). Be careful not to touch these wires and parts.

Wear insulating gloves, commonly called "lineman's gloves," when working on or around the high-voltage system (**Figure 1-19**). These gloves must be class "0" rubber insulating gloves, rated at 1000 volts. Also, to protect the integrity of the insulating gloves, wear leather gloves over the insulating gloves while doing a service.

Figure 1-16 Most high-voltage circuits are identifiable by size and are typically colored orange.

Figure 1-17 High-voltage components may have caution notices to alert technicians of the possible dangers.

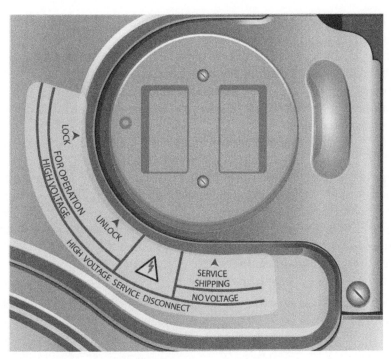

Figure 1-18 A high-voltage battery cutoff on a hybrid Ford Escape.

Figure 1-19 Special high-voltage gloves are worn inside leather outer gloves when working on the high-voltage electrical systems of hybrid vehicles.

Make sure they have no tears, holes, or cracks and are dry. Electrons can enter through the smallest of holes in your gloves. The integrity of the gloves should be checked before using them. To check the gloves, blow enough air into each one so they balloon out. Then fold the open end over to seal the air in. Continue to slowly fold that end of the glove toward the fingers. This will compress the air. If the glove continues to balloon as the air is compressed, it has no leaks. If any air leaks out, the glove should be discarded. All gloves, new and old, should be checked before they are used.

Other safety precautions that should always be adhered to when working on an electric drive vehicle are the following:

- Always adhere to the safety guidelines given by the vehicle's manufacturer.
- Obtain the necessary training before working on these vehicles.
- Perform each operation following the procedures defined by the manufacturer.
- Disable or disconnect the high-voltage system before servicing those systems.

- Anytime the engine is running, the generator is producing high voltage and care must be taken to prevent being shocked.
- Systems may have a large capacitor that must be discharged after the high-voltage system has been isolated. Wait for the prescribed amount of time (normally about 10 minutes) before working on or around the high-voltage system.
- After removing a high-voltage cable, cover the terminal with electrical tape.
- Always use insulated tools.
- Alert other technicians that you are working on the high-voltage system with a warning sign such as "high-voltage work: do not touch."
- Always install the correct circuit protection device into a high-voltage circuit.
- Many electric motors have a strong permanent magnet in them; individuals with a pacemaker should not handle these parts.
- When an electric drive vehicle needs to be towed into the shop for repairs, make sure it is not towed on its drive wheels. Doing this will drive the generator(s), which can overcharge the batteries and cause them to explode.
- Always use a **category III multimeter** when working on hybrid vehicles.

A special category III multimeter is required to work on hybrid vehicles. These multimeters are rated for the higher voltages encountered in a hybrid vehicle.

If the electric motor is sandwiched between the engine and the transmission, make sure you follow all procedures. The permanent magnet used in the motor is very strong and requires special tools to remove and install it.

Supplemental Inflatable Restraint (Air Bag) Precautions

Supplemental restraint systems (SRSs) have saved many people from serious injury or death, but the components can be dangerous to a technician if they are not handled correctly. First, make sure that the system is disabled according to the manufacturer's service information. Disabling usually involves removing system fuses or disconnecting the battery. Air bags are in several locations throughout the vehicle, including the steering wheel, instrument panel, seats, windshield, and roof trim. (**Figure 1-20**). In addition to the air bags, most vehicles also use seatbelt pretensioners (**Figure 1-21**). Both these components can be dangerous if handled incorrectly. It is important to read and follow the manufacturer's recommended procedures. In general, the following procedures should be followed:

- Always carry the air bag inflator module with the deployment door, trim cover, or tear seam pointed away from the body (**Figure 1-22**). This is so the inflator will not impact the technician if it should deploy accidently.

Figure 1-20 Typical airbag locations in the vehicle.

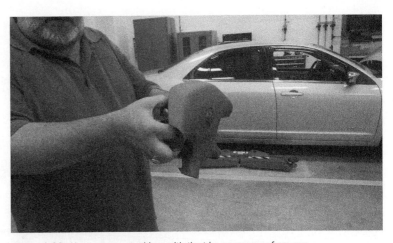

Figure 1-22 Always carry an airbag with the trim cover away from you.

Figure 1-21 A seat belt pretensioner can go off with considerable force and retract the seat belt if extended. If handling the pretensioner, make certain that the belt is not extended.

- Do not place an inflator module face down. If it should deploy, it could become airborne at a high speed (**Figure 1-23**).
- When a seat belt with pretensioner is handled, make sure that the seatbelt is not extended. If the pretensioner should happen to activate, the seatbelt could violently retract.
- If the electrical connectors of an inflator module are damaged, the inflator module must be replaced, not repaired (**Figure 1-24**).
- Whenever the air bags have deployed, make sure the system is serviced according to the manufacturer's instructions. There are many different procedures that vary greatly according to make and model of the vehicle involved.

Figure 1-23 Always place the airbag on a bench with the trim facing upwards.

Figure 1-24 NEVER attempt to repair an airbag electrical connector or wiring. If electrical connectors are damaged they must be replaced.

VEHICLE OPERATION

When a customer brings a vehicle in for service, certain driving rules should be followed to ensure your safety and the safety of those working around you. For example, before moving a car into the shop, buckle your safety belt. Make sure no one is near, the way is clear, and there are no tools or parts under the car before you start the engine. Check the brakes before putting the vehicle in gear. Then drive slowly and carefully in and around the shop.

When road testing the car, obey all traffic laws. Drive only as far as is necessary to check the automobile and verify the customer's complaint. Never make excessively quick starts, turn corners too quickly, or drive faster than conditions allow.

If the engine must be running while you are working on the car, block the wheels to prevent the car from moving. Place an automatic transmission into park and set the parking (emergency) brake. Never stand directly in front of or behind a running vehicle.

Run the engine only in a well-ventilated area to avoid the danger of poisonous **carbon monoxide (CO)** in the engine exhaust. CO is an odorless but deadly gas. Most shops have an exhaust ventilation system; always use it (**Figure 1-25**). Connect the hose from the vehicle's tailpipe to the intake for the vent system. Make sure the vent system is turned on before running the engine.

Exhaust contains an odorless, colorless, and deadly gas: **carbon monoxide (CO)**. This poisonous gas gives very little warning to the victim and can kill in just a few minutes.

Figure 1-25 When running an engine in a shop, always connect the exhaust to the ventilation system.

Figure 1-26 When lifting a heavy object, always use your legs and keep the object close to your body.

LIFTING AND CARRYING

When lifting a heavy object like a transmission, use a hoist or have someone else help you. If you must work alone, *always* lift heavy objects with your legs, not your back. Bend down with your legs, not your back, securely hold the object you are lifting, and then stand up, keeping the object close to you (**Figure 1-26**). Trying to "muscle" something with your arms or back can result in severe damage to your back and may end your career and limit what you do for the rest of your life.

ACCIDENTS

If there is an accident, you should first try to prevent any further injury. Your supervisor can try and determine the cause after any injured workers receive first aid.

Your entire work area should be kept clean and safe. Any oil, coolant, or grease on the floor can make it slippery. To clean up oil, use commercial oil absorbent. Most shops use absorbent mats in work areas to limit the effects of oil and fluid spills.

Keep all water off the floor. Water is slippery on smooth floors, and electricity flows well through water. Aisles and walkways should be kept clean and wide enough to move through easily. Make sure the work areas around machines are large enough to operate machines safely.

Make sure all drain covers are snugly in place. Open drains or covers that are not flush to the floor can cause toe, ankle, and leg injuries.

Handle all solvents (or any liquids) with care to avoid spillage. Keep all solvent containers closed, except when pouring. Proper ventilation is very important in areas where volatile solvents and chemicals are used. Solvents and other combustible materials must be stored in approved and ventilated storage cabinets or rooms.

Be extra careful when transferring flammable materials from bulk storage. Static electricity can build up enough to create a spark that could cause an explosion. Discard or clean all empty solvent containers. Never light matches or smoke near flammable solvents and chemicals, including battery acids.

Also, accidents can be prevented simply by the way you act. The following are some additional guidelines for working safely in a shop. This list does not include everything you should or shouldn't do; it merely gives some things to think about:

- Never smoke while working on a vehicle or working with any machine in the shop.
- Playing around is not fun when it sends someone to the hospital.

- To prevent serious burns, keep your skin away from hot metal parts such as the radiator, exhaust manifold, tailpipe, catalytic converter, and muffler.
- Always disconnect electric engine cooling fans when working around the radiator. These can turn on without warning and can easily chop off a finger or hand. Make sure you reconnect the fan after you have completed your repairs.
- When working with a hydraulic press, make sure the pressure is applied in a safe manner. It is generally wise to stand to the side when operating the press.
- Properly store all parts and tools by putting them away in a place, where people will not trip over them. This practice not only cuts down on injuries, but it also reduces time wasted looking for a misplaced part or tool.

First Aid

Your supervisor should be immediately informed of all accidents that occur in the shop. The work area should have a first-aid kit for treating minor injuries (**Figure 1-27**). Facilities for flushing eyes should also be near or in the shop area. Know where they are. Keep an up-to-date list of emergency telephone numbers clearly posted next to the telephone. These numbers should include a doctor, hospital, and fire and police departments.

It is also a good idea for you to know first aid. The knowledge of how to treat certain injuries can save someone's life. The American Red Cross offers many low-cost but thorough courses on first aid. You will realize the importance of these classes the first time you have to give first aid to someone or when someone must give it to you.

You should find out if there is a resident nurse in the shop or at the school, and know where the nurse's office is. If there are specific first-aid rules in your school or shop, make sure you are aware of them and follow them.

If someone is overcome by carbon monoxide, get him or her fresh air immediately. Burns should be cooled immediately by rinsing them with water. Whenever there is severe bleeding from a wound, try to stop the bleeding by applying pressure with clean gauze on or around the wound, and get medical help. Never move someone who may have broken bones unless the person's life is otherwise endangered. Moving that person may cause additional injury. Call for medical assistance.

Figure 1-27 A typical first-aid kit and its contents.

HAZARDOUS MATERIALS

A typical shop contains many potential health hazards for those working in it. These hazards can be classified as:

- Chemical hazards, caused by high concentrations of vapors, gases, or solids in the form of dust.
- Hazardous wastes, which are those substances left behind or result from a process or service.
- Physical hazards, including excessive noise, vibration, pressures, and temperatures.
- Ergonomic hazards, which are conditions that impede normal or proper body position and motion.

There are many federal agencies charged with ensuring safe work environments for all workers. These include the Occupational Safety and Health Administration (**OSHA**), the Mine Safety and Health Administration (MSHA), and the National Institute for Occupational Safety and Health (NIOSH). These entities, in addition to state and local governments, have instituted regulations that must be understood and followed. Everyone in a shop has the responsibility for adhering to these regulations.

OSHA

OSHA (Occupational Safety and Health Administration) is a branch of the U.S. Department of Labor. This branch was formed to write and enforce operational guidelines for all businesses to ensure that employees work in safe and healthy conditions. The guidelines include standards for cleanliness, air ventilation, fire prevention, emergency measures, equipment condition, and personal protective equipment.

It is the employer's responsibility to provide a place of employment that is free from all recognized safety and health hazards. OSHA regulates all safety and health issues concerning the automotive industry. Businesses may be periodically inspected by OSHA personnel. The owner of the business may be cited for any violations of the standards. The owner will be given a period of time to bring the work area into compliance. If the owner does not correct the situation or has been a repeat violator of the standards, OSHA will fine the owner.

Right-to-Know Laws

Every employee in a shop is protected by "right-to-know" laws. The general intent of these laws is that employers must provide a safe workplace with respect to hazardous materials. All employees must be trained about their rights under the legislation, the nature of the hazardous chemicals in their workplace, the labeling of chemicals, and the information about each chemical listed and described on **Safety Data Sheets (SDSs)** (**Figure 1-28**). These sheets are available in writing and/or online from the manufacturers and suppliers of the chemicals. They detail the chemical composition of and precautionary information for all products that can present health or safety hazards. The Canadian equivalents to the SDS are called Workplace Hazardous Materials Information Systems (WHMIS).

SDS sheets must be readily available to employees in their workplace area, either in binders or on computers. Back-up copies must be available in the case of a power outage. Employees must be familiar with the intended purposes of each substance, the recommended protective equipment, accident and spill procedures, and any other information regarding the safe handling of hazardous materials. This training must be given annually to employees and provided to new employees as part of their job orientation.

⚠ **Caution**

When handling any hazardous material, always wear the appropriate safety protection. Always follow the correct procedures while using the material and be familiar with the information given on the SDS for that material.

```
HEXANE
================================================================
MSDS Safety Information
================================================================
Ingredients
================================================================
Name: HEXANE (N_HEXANE)
% Wt: >97
OSHA PEL: 500 PPM
ACGIH TLV: 50 PPM
EPA Rpt Qty: 1 LB
DOT Rpt Qty: 1 LB
================================================================
Health Hazards Data
================================================================
LD50 LC50 Mixture: LD50:(ORAL,RAT) 28.7 KG/MG
Route Of Entry Inds _ Inhalation: YES
Skin: YES
Ingestion: YES
Carcinogenicity Inds _ NTP: NO
IARC: NO
OSHA: NO
Effects Of Exposure: ACUTE: INHALATION AND INGESTION ARE HARMFUL AND MAY BE FATAL.
INHALATION AND INGESTION MAY CAUSE HEADACHE, NAUSEA, VOMITING, DIZZINESS,
IRRITATION OF RESPIRATORY TRACT, GASTROINTESTINAL IRRITATION AND
UNCONSCIOUSNESS. CONTACT W/SKIN AND EYES  MAY CAUSE IRRITATION. PROLONGED SKIN
MAY RESULT IN DERMATITIS (EFTS OF OVEREXP)
Signs And Symptoms Of Overexposure: HLTH HAZ:CHRONIC:MAY INCLUDE CENTRAL
NERVOUS SYSTEM DEPRESSION.
Medical Cond Aggravated By Exposure: NONE IDENTIFIED.
First Aid: CALL A PHYSICIAN. INGEST: DO NOT INDUCE VOMITING. INHAL:REMOVE TO FRESH
AIR. IF NOT BREATHING, GIVE ARTIFICIAL RESPIRATION. IF BREATHING IS DIFFICULT, GIVE
OXYGEN. EYES:IMMED FLUSH W/PLENTY OF WATER FOR AT LEAST 15 MINS. SKIN:IMMED
FLUSH W/PLENTY OF WATER FOR AT LEAST 15 MINS WHILE REMOVING CONTAMD CLTHG &
SHOES. WASH CLOTHING BEFORE REUSE.
================================================================
Handling and Disposal
================================================================
Spill Release Procedures: WEAR NIOSH/MSHA SCBA & FULL PROT CLTHG. SHUT OFF
IGNIT SOURCES:NO FLAMES, SMKNG/FLAMES IN AREA. STOP LEAK IF YOU CAN DO SO W/OUT
HARM. USE WATER SPRAY TO REDUCE VAPS. TAKE UP W/SAND OR OTHER NON_COMBUST
MATL & PLACE INTO CNTNR FOR LATER (SU PDAT)
Neutralizing Agent: NONE SPECIFIED BY MANUFACTURER.
Waste Disposal Methods: DISPOSE IN ACCORDANCE WITH ALL APPLICABLE FEDERAL, STATE
AND LOCAL ENVIRONMENTAL REGULATIONS. EPA HAZARDOUS WASTE NUMBER:D001
(IGNITABLE WASTE)
Handling And Storage Precautions: BOND AND GROUND CONTAINERS WHEN TRANSFERRING
LIQUID. KEEP CONTAINER TIGHTLY CLOSED.
Other Precautions: USE GENERAL OR LOCAL EXHAUST VENTILATION TO MEET
TLVREQUIREMENTS. STORAGE COLOR CODE RED (FLAMMABLE).
================================================================
Fire and Explosion Hazard Information
================================================================
Flash Point Method: CC
Flash Point Text: _9F,_23C
Lower Limits: 1.2%
Upper Limits: 77.7%
Extinguishing Media: USE ALCOHOL FOAM, DRY CHEMICAL OR CARBON DIOXIDE. (WATER MAY
BE INEFFECTIVE).
Fire Fighting Procedures: USE NIOSH/MSHA APPROVED SCBA & FULL PROTECTIVE
  EQUIPMENT (FP N).
Unusual Fire/Explosion Hazard: VAP MAY FORM ALONG SURFS TO DIST IGNIT SOURCES & FLASH
BACK. CONT W/STRONG OXIDIZERS MAY CAUSE FIRE. TOX GASES PRDCED MAY INCL:CARBON
MONOXIDE, CARBON DIOXIDE.
================================================================
```

Figure 1-28 A sample Material Safety Data Sheet.

An SDS must be in a standard 16-section format that contains the following:

Section 1: Identification
This section provides the following information:
- The trade and chemical name of the product and the intended use.
- The manufacturer; responsible party or importer of the product; and contact information, including address and emergency phone contact information.
- All of the ingredients of the product.

Section 2: Hazard Communication
This section identifies the hazards of the chemical and the warning information associated with those hazards; the hazard classification of the chemical, such as flammable or corrosive; the hazard statement; hazard symbols, such as pictograms, used to identify the hazard; and the precautionary statement(s).

Section 3: Composition/Identification of Ingredients
This section identifies the ingredients in the product listed on the SDS, including additives and impurities.

Section 4: First-Aid Procedures
This section provides first-aid procedures described for untrained persons attempting to give first aid based on the route of exposure, such as inhalation, eye or skin contact, or swallowing; and information on symptoms and effects of exposure.

Section 5: Fire-Fighting Measures
This section provides information on appropriate equipment and measures to fight a fire that involves the material. It also includes hazards that may result from the combustion of the hazardous material.

Section 6: Accidental Release Measures
This section describes the necessary control measures if the substance is spilled or released into the atmosphere.

Section 7: Handling and Storage
This section provides safe handling and storage recommendations, including temperature and ventilation requirements for safe storage.

Section 8: Exposure Controls/Personal Protection
This section details the permissible exposure limits (PELs) recommended by the manufacturer or importer and the appropriate personal protective equipment such as respirators, eye protection, clothing, and glove type. This section also discusses such concerns as ventilation of the area where the material is in use.

Section 9: Physical and Chemical Properties
This section provides the following infloration:

Appearance in color and physical state	Boiling point and boiling range
Flammability characteristics	Flash point
Vapor pressure	Evaporation rate
Odor threshold	Flammability (solid, gas)
Vapor density	Solubility
pH	Auto-ignition temperature
Relative density	decomposition temperature
Melting point and freezing point	viscosity

Section 10: Stability and Reactivity
This section describes the reactivity hazards of the chemical and the chemical stability information. Reactivity data describes how reactive the material is to common substances

that might be encountered during the use of the chemical. Stability describes how stable the chemical is at room temperatures. If the chemical is instable, the necessary stabilizers are described. Also, conditions that should be avoided are described.

Section 11: Toxicological Information

This section describes how toxic the chemical is and the likely routes of exposure—such as inhalation, skin or eye contact, etc. The symptoms of exposure are described, along with the short- and long-term effects. In some cases, the toxicity of the material is unknown, which is also described in this section. The dose of the material to kill 50% of test animals (if known) is detailed. This section also explains whether or not this material is considered to be a carcinogen.

Section 12: Ecological Information (non-mandatory)

This section details the chemicals known effects to the environment, such as whether or not the material degrades naturally or persists for a long period of time.

Section 13: Disposal Considerations (non-mandatory)

This section details suggested disposal or recycle considerations, including any special considerations for landfills in dealing with the material (remember that the material may not be considered hazardous to have an SDS sheet).

Section 14: Transport Information (non-mandatory)

This section deals with transportation concerns such as the hazard transportation class.

Section 15: Regulatory Information (non-mandatory)

This section deals with any safety or health regulations not covered in the other sections of the SDS

Section 16: Other Information

This section lists the date of the SDS or any revisions that have been made.

Hazardous Materials Documentation

Shops must maintain documentation on the hazardous chemicals in the workplace, proof of training programs, records of accidents or spill incidents, satisfaction of employee requests for specific chemical information via the SDS, and a general right-to-know compliance procedure manual for use within the shop.

All hazardous material must be properly labeled, indicating what health, fire, or reactivity hazard it poses and what protective equipment is necessary when handling each chemical (**Figure 1-29**). The manufacturer of the hazardous material must provide all

es. DO NOT puncture or incinerate container. Do not store in
direct sunlight, other sources of heat or store at temperatures over
120 F. Wash hands after using product.

KEEP OUT OF THE REACH OF CHILDREN.

CONTAINS: Toluene (108-88-3), Menthanol (67-56-1), tillates (64742-89-8), Carbon Dioxide (124-38-9)

FIRST AID:
EYES: Flush with large amounts of water. Get medical attention if irritation persists.
SKIN: Wash exposed areas with soap and water.
INHALATION: If affected, remove to fresh air.
IF SWALLOWED: Call a physician or poison control center for instructions.

24-HOUR EMERGENCY NUMBER: 1-800-274-5263

DIRECTIONS: Spray directly on brake parts and allow

Figure 1-29 Carefully read the labels on all chemicals before using them.

relevant warnings and precautionary information, which must be read and understood by the user before application. Attention to all label precautions is essential for the proper use of the chemical and for prevention of hazardous conditions.

Hazardous Wastes

Many repair and service procedures generate what are known as **hazardous wastes**. Dirty solvents and cleaners are good examples of hazardous wastes. Something is classified as a hazardous waste by the Environmental Protection Agency (EPA) if it is on the EPA list of known harmful materials or has one or more of the following characteristics:

Ignitability. A liquid with a flashpoint below 140°F or a solid that can spontaneously ignite.

Corrosivity. A substance that dissolves metals and other materials or burns the skin.

Reactivity. Any material that reacts violently with water or other materials or releases cyanide gas, hydrogen sulfide gas, or similar gases when exposed to low pH acid solutions. This also includes material that generates toxic mists, fumes, vapors, and flammable gases.

EP (extreme pressure) toxicity. Materials that leach one or more of eight heavy metals in concentrations greater than 100 times primary drinking water standard concentrations.

> **Reactivity** is how easily a substance can cause or be part of a chemical reaction.

A complete EPA list of hazardous wastes can be found in the Code of Federal Regulations. It should be noted that no material is considered hazardous waste until the shop has finished using it and is ready to dispose of it.

In the United States, OSHA regulates the use of many of these materials. The EPA regulates the disposal of hazardous waste. A summary of these regulations follows:

- All businesses that generate hazardous waste must develop a hazardous waste policy.
- Each hazardous waste generator must have an EPA identification number.
- When waste is transported for disposal, a licensed waste hauler must be used to transport and dispose of the waste. A copy of a written manifest (an EPA form) must be kept by the shop.

Regulations on hazardous waste handling and generation have led to the development of equipment that is now commonly found in shops. Examples are thermal cleaning units, closed-loop steam cleaners, waste oil furnaces, oil filter crushers, refrigerant recycling machines, engine coolant recycling machines, and highly absorbent cloths.

⚠ **WARNING The shop is ultimately responsible for the safe disposal of hazardous wastes, even after the waste leaves the shop. Only licensed waste removal companies should be used to dispose of the waste. Make sure you know what the company is planning to do with the waste. Make sure you have a written contract stating what is supposed to happen with the waste. Leave nothing to chance. In the event of an emergency hazardous waste spill, contact the National Response Center (1-800-424-8802) immediately. Failure to do so can result in a $10,000 fine, a year in jail, or both.**

OSHA and the EPA have other strict rules and regulations that help to promote safety in the auto shop. These are described throughout this text whenever they are applicable.

Maintaining a vehicle involves handling and managing a wide variety of materials and wastes. Some of these wastes can be toxic to fish, wildlife, and humans when improperly managed. No matter the amount of waste produced, it is to the shop's legal and financial advantage to manage waste properly and, even more importantly, to prevent pollution.

Hazardous Waste Disposal

Hazardous wastes must be properly stored and disposed of. It is the shop's responsibility to hire a reputable, financially stable, and state-approved hauler, who will dispose of the shop wastes legally. Select a licensed hazardous waste hauler after seeking recommendations and reviewing the firm's permits and authorizations. If hazardous waste is dumped illegally, your shop may be held responsible.

Always keep hazardous waste separate, properly labeled, and sealed in the recommended containers. The storage area should be covered and may need to be fenced and locked if vandalism could be a problem.

Handling Shop Wastes

The following is a summary of the proper method of preparing and disposing of common hazardous wastes. When handling these classified wastes, make sure to wear the appropriate safety equipment (**Figure 1-30**). These are general guidelines; always follow the specific state and federal regulations for disposing of these items.

Oil. Recycle oil. Set up equipment, such as a drip table or screen table with a used oil collection bucket, to collect oil dripping off parts. Place drip pans under vehicles that are leaking fluids. Recycle the oil according to local regulations. Do not mix other wastes with used oil, except as allowed by your recycler.

ATF. Automatic transmission fluid should be collected in a separate container from other lubricants and collected by an approved recycler.

Oil Filters. Drain for at least 24 hours, crush, and recycle used oil filters.

Batteries. Recycle batteries by sending them to a reclaimer or back to the distributor. Store batteries in a water-tight, acid-resistant container. Inspect batteries for cracks and leaks when they come in. Treat a dropped battery as if it were cracked. Acid residue

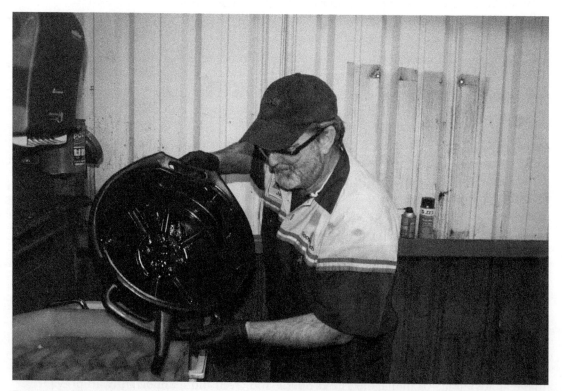

Figure 1-30 Wear the appropriate safety equipment when handling hazardous materials and waste.

is hazardous because it is corrosive and may contain lead and other toxics. Neutralize spilled acid by using baking soda or lime, and dispose of as hazardous material.

Containers. Cap, label, cover, and properly store above ground and outdoors any liquid containers and small tanks within a diked area and on a paved impermeable surface to prevent spills from running into surface or ground water.

Metal Residue from Machining. Collect metal filings when machining metal parts. Keep separate and recycle if possible. Prevent metal filings from falling into a storm sewer drain.

Other Solids. Store materials such as scrap metal, old machine parts, and worn tires under a roof or tarpaulin to protect them from the elements and to prevent the potential to create contaminated runoff. Consider recycling tires by retreading them.

Refrigerants. Recover or recycle refrigerants during the service and disposal of motor vehicle air conditioners and/or refrigeration equipment. You are not allowed to knowingly vent refrigerants into the atmosphere. Recovery or recycling during service must be performed by an EPA-certified technician using certified equipment and following specified procedures.

Solvents. Replace hazardous chemicals with less toxic alternatives that have equal performance. For example, substitute water-based cleaning solvents for petroleum-based solvent degreasers. To reduce the amount of solvent used when cleaning parts, use a two-stage process: dirty solvent followed by fresh solvent. Hire a hazardous waste management service to clean and recycle solvents. Store solvents in closed containers to prevent evaporation. Evaporation of solvents contributes to ozone depletion and smog formation. Properly label spent solvents and store on drip pans or in diked areas and only with compatible materials.

Liquid Recycling. Collect and recycle coolants from radiators. Store transmission fluids, brake fluids, and solvents containing chlorinated hydrocarbons separately, and recycle or dispose of them properly.

Shop Towels/Rags. Keep waste towels in a closed container marked "contaminated shop towels only." To reduce costs and liabilities associated with disposal of used towels, which can be classified as hazardous waste, investigate using a laundry service that is able to treat the wastewater generated from cleaning the towels.

Waste Storage. Always keep hazardous waste separate, properly labeled, and sealed in the recommended containers. The storage area should be covered and may need to be fenced and locked if vandalism could be a problem. Select a licensed hazardous waste hauler after seeking recommendations and reviewing the firm's permits and authorizations.

SUMMARY

- You have an important role in creating an accident-free work environment. Your appearance and work habits will go a long way toward preventing accidents. Common sense and concern for others will result in fewer accidents.
- It's everyone's job to periodically check for safety hazards in a shop.
- Dressing safely for work is very important. This includes snug-fitting clothing, eye and ear protection, protective gloves, strong shoes, and caps to cover long hair.
- Safety glasses should be worn whenever you are working under a vehicle or when you are using machining equipment, grinding wheels, chemicals, compressed air, or fuels.
- There are two areas of housekeeping you are responsible for: your work area and the rest of the shop.
- Dirty and oily rags should be stored in approved containers.
- All equipment should be inspected for safety hazards before being used.

- Gasoline and diesel fuel are highly flammable and should be kept in approved safety cans. Any gasoline spills or leaks should be cleaned up immediately.
- It is important to know when to use each of the various types of fire extinguishers. When fighting a fire, aim the nozzle at the base and use a side-to-side sweeping motion.
- Fire, heat, and sparks should be kept away from a battery. These could cause the battery to explode.
- Heavy objects should be lifted with your legs, not your back.
- Special care should be taken whenever using power tools, such as impact wrenches, gear pullers, and jacks.
- Always observe all relevant safety rules when operating a vehicle lift or hoist. Jacks, jack stands, chain hoists, and cranes can also cause injury if not operated safely.
- Use care whenever it is necessary to move a vehicle within the shop. Carelessness and playing around can lead to a damaged vehicle and serious injury.
- Carbon monoxide gas is a poisonous gas present in engine exhaust fumes. It must be properly vented from the shop using tailpipe hoses or other reliable methods.

- Accidents can be prevented by not having anything dangle near rotating equipment and parts.
- Accidents can happen, and when they do, you should respond immediately to prevent further injury or damage.
- All employees have the right to know what hazardous materials they are using to perform their job. Most of the needed information is contained on an SDS.
- Always strictly follow the service recommendations and precautions given by the manufacturer when working on a hybrid vehicle.
- Right-to-know laws began in 1983 and are designed to protect employees who must handle hazardous materials and waste on the job.
- Hazardous wastes must be properly disposed of. Typically, shops hire full-service waste management firms to remove and dispose of this waste.
- Proper air bag and seatbelt pretensioner handling are important to prevent injury for an unintended deployment and possible injury

ASE-STYLE REVIEW QUESTIONS

1. While discussing ways to prevent fires:
   Technician A says all dirty and oily rags should be stored in approved containers.
   Technician B says all dirty and oily rags should be kept in a pile until the end of the day, and then they should be moved to a suitable container.
   Who is correct?
   A. A only
   B. B only
   C. Both A and B
   D. Neither A nor B

2. *Technician A* says accidents can be prevented by not having anything dangle near rotating equipment and parts.
   Technician B says accidents can be prevented by having common sense.
   Who is correct?
   A. A only
   B. B only
   C. Both A and B
   D. Neither A nor B

3. Technician A says your supervisor should be immediately informed of all accidents that occur in the shop.
   Technician B says the work area should have a first-aid kit for treating minor injuries.
   Who is correct?
   A. A only
   B. B only
   C. Both A and B
   D. Neither A nor B

4. *Technician A* says safety glasses should be worn anytime there is a possibility of eye injury from working in the shop.
   Technician B says if regular prescription eyeglasses are worn in the shop, they must be equipped with side shields.
   Who is correct?
   A. A only
   B. B only
   C. Both A and B
   D. Neither A nor B

5. *Technician A* says unsafe equipment should have its power disconnected and be marked with a sign to warn others.

   Technician B says that all equipment should be inspected for safety hazards before being used.

   Who is correct?

   A. A only

   B. B only

   C. Both A and B

   D. Neither A nor B

6. *Technician A* says hazardous wastes must be properly disposed of.

   Technician B says shops hire full-service waste management firms to remove and dispose of hazardous wastes.

   Who is correct?

   A. A only

   B. B only

   C. Both A and B

   D. Neither A nor B

7. While discussing ways to safely use a vehicle lift.

   Technician A says each vehicle has specific lift points that can be found in service information.

   Technician B says before using a lift make sure to read the lift manufacturers instructions.

   Who is correct?

   A. A only

   B. B only

   C. Both A and B

   D. Neither A nor B

8. While discussing why exhaust fumes should be vented outdoors or drawn into a ventilation/filtration system:

   Technician A says exhaust gases contain amounts of carbon monoxide.

   Technician B says carbon monoxide is an odorless, colorless, and deadly gas.

   Who is correct?

   A. A only

   B. B only

   C. Both A and B

   D. Neither A nor B

9. While discussing the use of a fire extinguisher:

   Technician A says there are three classes of fires.

   Technician B says a Class A fire is a metal fire, such as a fire involving a magnesium wheel.

   Who is correct?

   A. A only

   B. B only

   C. Both A and B

   D. Neither A nor B

10. While discussing a car's electrical system:

    Technician A says you should always disconnect the negative or ground battery cable first, then disconnect the positive cable.

    Technician B says you should always connect the positive cable first, then the negative.

    Who is correct?

    A. A only

    B. B only

    C. Both A and B

    D. Neither A nor B

Name _____   **Date** _____

SHOP SAFETY EVALUATION

As a professional technician safety should be one of your first concerns. This job sheet will increase your awareness of shop safety rules and safety equipment. As you survey your shop area and answer the following questions, you will learn how to evaluate the safeness of your workplace.

ASE Education Foundation Correlation

This job sheet addresses the following **RST** tasks:

1.1   Identify general shop safety rules and procedures.

1.8   Identify the location and use of eye wash stations.

1.10   Comply with the required use of safety glasses, ear protection, gloves, and shoes during lab/shop activities.

1.11   Identify and wear appropriate clothing for lab/shop activities.

1.12   Secure hair and jewelry for lab/shop activities.

We Support
ASE | Education Foundation

Tools and Materials

• Copy of the shop rules from the instructor

Procedure

Your instructor will review your progress throughout this worksheet and should sign off the sheet when you complete it.

1. Have your instructor provide you with a copy of the shop rules and procedures.

   Have you read and understood the shop rules?   ☐ Yes   ☐ No

2. Before you begin to evaluate your work area, evaluate yourself. Are you dressed to work safely?   ☐ Yes   ☐ No

   If no, what is wrong?

3. Are your safety glasses OSHA approved?   ☐ Yes   ☐ No

   Do they have side protection shields?   ☐ Yes   ☐ No

4. Look around the shop and note any area that poses a potential safety hazard or is an area that you should be aware of.

   _____

   _____

   _____

   Any true hazards should be brought to the attention of the instructor immediately.

5. What is the air line pressure in the shop? _____ psi

   What should it be? _____ psi

6. Where is the first-aid kit(s) kept in the work area?

   _____

   _____

7. Ask the instructor to show the location of and demonstrate the use of the eyewash station. Where is it and when should it be used?

_____

_____

8. What is the shop's procedure for dealing with an accident?

_____

_____

9. Explain how to secure hair and jewelry while working in the shop.

_____

_____

10. List the phone numbers that should be called in the case of an emergency.

_____

_____

Problems Encountered

_____

_____

_____

Instructor's Response

_____

_____

_____

Name _____   **Date** _____

WORKING IN A SAFE ENVIRONMENT

This job sheet will help you work safely in the shop. The tools described are the hoist and floor jack, which are two of the most commonly used tools in the shop.

ASE Education Foundation Correlation ——————————————————————————————————

This job sheet addresses the following **RST** tasks:

1.2   Utilize safe procedures for handling of tools and equipment.

1.3   Identify and use proper placement of floor jacks and jack stands.

1.4   Identify and use proper procedures for safe lift operation.

1.5   Utilize proper ventilation procedures for working within the lab/shop area.

1.6   Identify marked safety areas.

1.9   Identify the location of the posted evacuation routes.

We Support

ASE | Education Foundation

Tools and Materials

• Vehicle for hoist and jack stand demonstration
• Service information

Describe the Vehicle Being Worked On

Year _____ Make _____ Model _____

VIN _____ Engine type and size _____

Vehicle Used

Make _____ Model _____ Year _____

Procedure

1. Are there safety areas marked around grinders and other machinery?   ☐ Yes   ☐ No

2. Are there shop emergency escape routes clearly marked?   ☐ Yes   ☐ No

3. Locate the exhaust ventilation system in the shop and describe how to operate the system. Explain the importance of the ventilation system.

   _____

   _____

4. What types of hoists are used in the shop? Are they hydraulic or electric?

   _____

   _____

5. Are the hoists asymmetrical or symmetrical? Explain the difference in the hoist types?

6. Find the location of the correct lifting points for the vehicle supplied by the instructor. On the rear of this sheet, draw a simple figure showing where these lift points are.

7. What is the difference in the positioning of a front-wheel-drive vehicle compared to a rear-wheel-drive vehicle?

8. Ask your instructor to demonstrate the proper use of the hoist.

Summarize the proper use of the hoist.

_____

_____

_____

_____

9. Demonstrate the proper use of jack stands with the help of your instructor.

Summarize the proper use of jack stands.

_____

_____

_____

10. Why should a floor jack never be used without jack stands?

Problems Encountered

_____

_____

_____

Instructor's Comments

_____

_____

_____

Name _____   **Date** _____

WORKING SAFELY AROUND AIR BAGS

Upon completion of this job sheet, you should be able to work safely around and with air bag systems.

ASE Education Foundation Correlation

This job sheet addresses the following **RST** tasks:

1.13   Demonstrate awareness of the safety aspects of supplemental restraint systems (SRS), electronic brake control systems, and hybrid vehicle high-voltage circuits.

1.14   Demonstrate awareness of the safety aspects of high-voltage circuits (such as high intensity discharge (HID) lamps, ignition systems, injection systems, etc.).

We Support

ASE | Education Foundation

Tools and Materials

- A vehicle(s) with air bag, HID lamp, and high-voltage ignition systems
- Safety glasses, goggles
- Service information appropriate to vehicle(s) used

Describe the Vehicle Being Worked On

Year _____ Make _____ Model _____

VIN _____ Engine type and size _____

Procedure

1. Locate the information about the air bag system in the service information. How are the critical parts of the system identified in the vehicle?

    _____

    _____

2. List the main components of the air bag system and describe their location.

    _____

    _____

3. There are some very important guidelines to follow when working with and around air bag systems. Look through the service information to find the answers to the questions and fill in the blanks with the correct words.

    a. Wear _____ when servicing an air bag system and when handling an air bag module.

    b. Wait at least _____ minutes after disconnecting the battery before beginning any service. The reserve _____ module is capable of storing enough energy to deploy the air bag for up to _____ minutes after battery voltage is lost.

    c. Never carry an air bag module by its _____ or _____, and, when carrying it, always face the trim and air bag _____ from your body. When placing a module on a bench, always face the trim and air bag _____.

d. Deployed air bags may have a powdery residue on them. _____ is produced by the deployment reaction and is converted to _____ when it comes in contact with the moisture in the atmosphere. Although it is unlikely that harmful chemicals will still be on the bag, it is wise to wear _____ and _____ when handling a deployed air bag. Immediately wash your hands after handling a deployed air bag.

e. A live air bag must be _____ before it is disposed. A deployed air bag should be disposed of in a manner consistent with the _____ and manufacturer's procedures.

f. Never use a battery- or AC-powered _____, _____, or any other type of test equipment in the system unless the manufacturer specifically says to. Never probe with a _____ for voltage.

4. Explain how an air bag sensor should be handled before it is installed on the vehicle.

Problems Encountered

_____

_____

_____

Instructor's Comments

_____

_____

_____

Name _____   Date _____

HYBRID VEHICLE HIGH-VOLTAGE CIRCUIT PRECAUTIONS

Upon completion of this job sheet, you will be able to locate and safely disconnect the high-voltage circuit on a hybrid vehicle.

ASE Education Foundation Correlation

This job sheet addresses the following **RST** tasks:

1.13   Demonstrate awareness of the safety aspects of supplemental restraint systems (SRS), electronic brake control systems, and hybrid vehicle high-voltage circuits.

We Support

ASE | Education Foundation

Tools and Materials

- Service information manual
- Vinyl tape for insulation
- Insulated gloves that are dry and are not cracked, ruptured, torn, or damaged in any way
- Hybrid vehicle
- Goggles or safety glasses with side shields

Describe the Vehicle Being Worked On

Year _____ Make _____ Model _____

VIN _____ Engine type and size _____

⚡ **WARNING** **Unprotected contact with any electrically charged ("hot" or "live") high-voltage component could cause serious injury or death. Manufacturers require their technicians to be trained in the hybrid vehicles they will be working on before they service them. This job sheet is not a substitute for factory training. It is the responsibility of the instructor to maintain a safe working environment in their particular situation.**

Procedure

1. Describe how this vehicle is identified as a hybrid.

   _____

   _____

   _____

After referring to the vehicle's service information, briefly describe the operation of the hybrid system.

_____

_____

_____

_____

At how many volts are the batteries rated?

_____

_____

_____

Where are the batteries located?

_____

_____

_____

How are the high-voltage cables labeled and identified?

_____

_____

_____

What is used to provide short-circuit protection in the high-voltage battery pack?

_____

_____

_____

What isolates the high-voltage system from the rest of the vehicle when the vehicle is shut off?

_____

_____

_____

Using the vehicle's service information as a guide, list at least five precautions that must be adhered to when servicing this hybrid vehicle.

_____

_____

_____

_____

_____

If the vehicle has been in an accident and some electrolyte from the batteries has leaked out, how should you clean up the spill?

_____

_____

_____

_____

Look under the hood and describe the components that are visible.

_____

_____

_____

_____

Without touching the high-voltage cables or components, describe the routing of these cables. Include what they appear to be connected from and to.

_____

_____

_____

_____

⚠ WARNING **Never assume that a hybrid vehicle is shut off simply because it is silent. Make sure the ignition key is in your pocket and not in the ignition switch.**

Describe the procedure for totally isolating the high-voltage system from the vehicle. This often involves the removal of a service plug. Be sure to include the location of this plug in your description of the procedure.

_____

_____

_____

_____

_____

_____

_____

_____

_____

_____

Make two signs saying, "Working on High-Voltage Parts. Do not touch!" Attach one to the steering wheel, and set the other one near the parts you are working on.

After the service plug has been removed, how long should you wait before working around or on the high-voltage system?

_____

1. Special gloves with liners are to be inspected before each use when working on a hybrid vehicle.
   a. How should the integrity of the gloves be checked?

   _____

   _____

   _____

   b. How often should the gloves be tested (at a lab) for recertification?

   _____

   _____

   _____

  c. What precautions must be taken when storing the gloves?

  _____

  _____

  _____

  d. When must the gloves be worn?

  _____

  _____

  _____

2. What color are the high-voltage cables on hybrid vehicles?

   _____

3. What must be done BEFORE disconnecting the main voltage supply cable?

   _____

   _____

   _____

4. Describe the safety basis of the "one hand rule."

   _____

   _____

   _____

5. Explain the procedure to disable high voltage on the vehicle you selected.

   _____

   _____

   _____

6. Explain the procedure to test for high voltage to ensure that the main voltage is disconnected.

   _____

   _____

Problems Encountered

_____

_____

_____

Instructor's Comments

_____

_____

_____

Name _____   Date _____

FIRE EXTINGUISHER INSPECTION AND USE

Upon completion of this job sheet, you will be able to demonstrate knowledge of the procedures for using fire extinguishers and other fire safety equipment, and identify the location of fire extinguishers in the shop.

ASE Education Foundation Correlation

This job sheet addresses the following **RST** tasks:

1.7   Identify the location and the types of fire extinguishers and other fire safety equipment; demonstrate knowledge of the procedures for using fire extinguishers and other fire safety equipment.

We Support

ASE | **Education Foundation**

NOTE: Never try to fight a large fire or one that is out of control. Get out and call 911.

Procedure

1. Identify the location of the fire extinguishers in the shop. Are they in a convenient location?

2. Are the fire extinguishers inspected annually? Do they have current tags?

3. What types of fires is your shop fire extinguisher rated to fight?

4. What types of fires should not be used with the shop fire extinguishers?

5. One way to remember the operation of a fire extinguisher is to remember the term PASS. Describe the meaning of PASS below.

   a. P _____

   b. A _____

   c. S _____

   d. S _____

Problems Encountered

_____

_____

_____

Instructor's Comments

_____

_____

_____

Name _____   Station _____   Date _____

SAFETY DATA SHEET USAGE

On completion of this job sheet, the student will be able to locate the SDS folder and describe the use of an SDS sheet on the job site.

ASE Education Foundation Correlation ——————————————————————

This job sheet addresses the following **RST** task:

1.15   Locate and demonstrate knowledge of material safety data sheets (MSDS).

We Support

ASE | Education Foundation

Tools and Materials

- Selection of chemicals from the shop
- SDS sheets

Procedure

1. Locate the SDS folder in the shop. It should be in a prominent location. Did you have any problems finding the folder?

   _____

   _____

2. Pick a common chemical from your tool room such as brake cleaner. Locate the chemical in the SDS folder. What chemical did you choose?

   _____

   _____

3. What is the flash point of the chemical? _____

4. Briefly describe why the flash point is important.

   _____

   _____

5. What is the first aid if the chemical is ingested?

   _____

   _____

6. Can this chemical be absorbed through the skin?

   _____

   _____

7. What are the signs of exposure to the chemical you selected?

   _____

   _____

8. What are the primary health hazards of the chemical?

   _____

   _____

9. What is the first-aid procedure for exposure to this chemical?

_____

_____

10. What are the recommendations for protective clothing?

_____

_____

Problems Encountered

_____

_____

_____

Instructor's Comments

_____

_____

_____

Name _____   Date _____

WORKING AROUND HIGH-VOLTAGE AND HIGH-PRESSURE AUTOMOTIVE SYSTEMS

ASE Education Foundation Correlation

This job sheet addresses the following **RST** tasks:

1.13   Demonstrate awareness of the safety aspects of supplemental restraint systems (SRS), electronic brake control systems, and hybrid vehicle high-voltage circuits.

1.14   Demonstrate awareness of the safety aspects of high voltage circuits (such as high intensity discharge (HID) lamps, ignition systems, injection systems, etc.).

We Support

ASE | Education Foundation

Tools and Materials

- Service information for HID Headlamps, Supplemental Restraints, electronic brake controller

Procedure:

Using the service information for the vehicle, answer the following questions.

1. List three precautions a technician should observe when working with HID headlamp systems.

   a. _____

   b. _____

   c. _____

2. A technician must never probe with a test lamp between the ballast and the bulb. Explain why.

   _____

   _____

   _____

HIGH-VOLTAGE IGNITION SYSTEM PRECAUTIONS

Describe the Vehicle Being Worked On (If Different From Above)

Year _____ Make _____ Model _____

Engine type and size _____

Procedure

Using the service information for the vehicle, answer the following questions concerning safety around high-voltage ignition systems.

1. High-voltage ignition systems can cause serious injury, especially for those who have heart problems. Name at least three precautions to take when working around high-voltage ignition systems.

   a. _____

   b. _____

   c. _____

HIGH-PRESSURE BRAKING SYSTEMS

Describe the Vehicle Being Worked On (If Different From Above)

Year _____ Make _____ Model _____

Engine type and size _____

NOTE: Many vehicles have a high-pressure accumulator or a high-pressure pump in their braking systems. Opening one of these systems can be hazardous due to the high pressures involved.

Procedure: Using service information for the vehicle, answer the following questions concerning high pressure safety.

1. Research the vehicle you have been assigned and describe the procedure that must be followed BEFORE opening the hydraulic braking system (de-pressurization procedure).

   _____

   _____

   _____

Problems Encountered

_____

_____

_____

Instructor's Comments

_____

_____

_____

CHAPTER 2
SPECIAL TOOLS AND PROCEDURES

Upon completion and review of this chapter, you should be able to:

- Describe the use of common diagnostic tools for automatic transmissions and transaxles.
- Describe the use of common pneumatic, electrical, and hydraulic power tools found in an automotive service department.
- Describe some of the special tools used to service automatic transmissions and the driveline.
- Identify the major measuring instruments and devices used by technicians.

- Describe the proper procedure for measuring with a micrometer.
- Describe the measurements normally taken by a technician while working on a vehicle's drivetrain.
- Describe the different sources for service information that are available to technicians.
- Describe the different types of fasteners used in the automotive industry.
- Describe the requirements for ASE certification as an automotive technician and a master auto technician.

Terms To Know

ASE
Backlash
Bolt head
Bolt shank
Circuit tester
Diagnosis
Dial caliper
Dial indicator
Digital multimeter (DMM)
Digital volt/ohmmeter (DVOM)
Duty cycle

Endplay
Fillet
Flat rate
Frequency
Grade marks
Graphing multimeter (GMM)
Machinist's rule
Micrometer
Oscilloscope
Preload
Press-fit

Pulse width
Repair order
Safety stands
Scan tool
Technical Service Bulletin (TSB)
Thread pitch
Torque wrench
Vehicle identification number (VIN)
Vernier caliper
Waveform

INTRODUCTION

Many different special tools and testing and measuring devices are used to service automatic transmissions. The most commonly used tools are the topic of the following paragraphs. Although you need to be more than familiar with, and will be using, common hand tools, they are not part of this discussion. You should already know what they are and how to use and care for them.

This chapter also discusses some of the everyday service procedures performed by an automatic transmission technician and the basic duties and responsibilities of all automotive technicians.

DIAGNOSTIC TOOLS

Automatic transmission repair requires much more than simply taking a unit apart and putting it back together. Perhaps your most important task when confronted with an automatic transmission problem is to properly identify its cause. Often the problem can be corrected without tearing down the unit. Many tools are available to help with diagnostics. These are discussed here. In addition, basic guidelines for diagnostics are presented.

Circuit Tester

Circuit testers are used to identify short and open circuits in any electrical circuit. Low-voltage testers are used to troubleshoot 6- to 12-volt circuits. A **circuit tester**, commonly called a test light, looks like a stubby ice pick. Its handle is transparent and contains a light bulb. A probe extends from one end of the handle and a ground clip and wire from the other end. When the ground clip is attached to a good ground and the probe touched to a live connector, the bulb in the handle will light up (**Figure 2-1**). If the bulb does not light, voltage is not available at the connector.

⚠ WARNING **Do not use a conventional 12-volt test light to diagnose components and wires in modern electronic systems. The current draw of these test lights may damage computers and system components. High-impedance test lights are available for diagnosing electronic systems.**

Self-Powered Test Light

A self-powered test light is called a continuity tester. It is used on non-powered circuits. It looks like a regular test light, except that it has a small internal battery. When the ground clip is attached to the negative side of a component and the probe touched to the positive side, the lamp will light if there is continuity in the circuit. If an open circuit exists, the lamp will not light. Do not use any type of test light or circuit tester to diagnose automotive air bag or computer systems.

Figure 2-1 The bulb of a test light will illuminate if there is power at a connector and the light is properly grounded.

Voltmeter

A voltmeter has two leads: a red positive lead and a black negative lead. The red lead should be connected to the positive side of the circuit or component. The black should be connected to ground or to the negative side of the component. Voltmeters should be connected across the circuit being tested (**Figure 2-2**).

The voltmeter measures the voltage available at any point in an electrical system. A voltmeter can also be used to test voltage drop across an electrical circuit, component, switch, or connector. A voltmeter can also be used to check for proper circuit grounding.

Ohmmeter

An ohmmeter measures the resistance to current flow in a circuit. In contrast to the voltmeter, which uses the voltage available in the circuit, the ohmmeter is battery powered. The circuit being tested must be open (**Figure 2-3**). If the power is on in the circuit, the ohmmeter will be damaged.

The meter sends current through the component and determines the amount of resistance based on the voltage dropped across the load. The scale of an ohmmeter reads from zero to infinity. A zero reading means there is no resistance in the circuit and may indicate a short in a component that should show a specific resistance. An infinite reading indicates a number higher than the meter can measure. This usually is an indication of an open circuit.

Ohmmeters are also used to trace and check wires or cables. Assume that one wire of a four-wire cable is to be found. Connect one probe of the ohmmeter to the known wire at one end of the cable and touch the other probe to each wire at the other end of the cable. Any evidence of resistance indicates the correct wire. Using this same method, you can check a suspected defective wire. If resistance is shown on the meter, the wire is sound. If no resistance is measured, the wire is defective (open). If the wire is okay, continue checking by connecting the probe to other leads. Any indication of resistance indicates that the wire is shorted to one of the other wires and that the harness is defective.

Ammeter

An ammeter measures current flow in a circuit. The ammeter must be placed into the circuit or in series with the circuit being tested (**Figure 2-4**). Normally, this requires disconnecting a wire or connector from a component and connecting the ammeter between

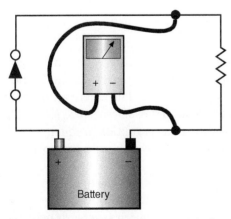

Figure 2-2 A voltmeter is used to measure available voltage and voltage drops.

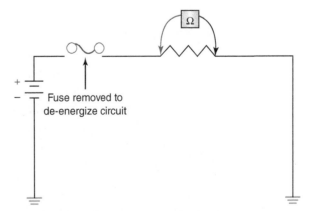

Figure 2-3 Always connect an ohmmeter to a component or circuit after the power has been disconnected.

Figure 2-4 An ammeter should always be connected in series with the circuit.

Figure 2-5 An inductive pickup for measuring amperes. This tool is often called a current probe.

the wire or connector and the component. The expected amperage should not exceed the amperage of the fuse in the multimeter. The red lead of the ammeter should always be connected to the side of the connector closest to the positive side of the battery and the black lead should be connected to the other side.

It is much easier to test current using an ammeter with an inductive pickup. The pickup clamps around the wire or cable being tested (**Figure 2-5**). These ammeters measure amperage based on the magnetic field created by the current flowing through the wire. This type of pickup eliminates the need to separate the circuit to insert the meter.

Because ammeters are built with very low internal resistance, connecting them in series does not add any appreciable resistance to the circuit. Therefore, an accurate measurement of the current flow can be taken.

Digital Multimeters

It's not necessary for a technician to own separate meters to measure volts, ohms, and amperes; a multimeter can be used instead. Top-of-the-line multimeters are multifunctional. Most test volts, ohms, and amperes in both DC and AC. Usually, there are several test ranges provided for each of these functions. In addition to these basic electrical tests, multimeters also test engine rpm, duty cycle, pulse width, diode condition, frequency, and even temperature. The technician selects the desired test range by turning a control knob on the front of the meter.

Multimeters are available with either analog or digital displays, but the only meter that should be used by technicians is the **digital volt/ohmmeter (DVOM)**, which is often referred to as a **digital multimeter (DMM)** (**Figure 2-6**). There are several drawbacks to using analog-type meters for testing electronic control systems. Many electronic components require very precise test results. Digital meters can measure volts, ohms, or amperes in tenths and hundredths. Another problem with analog meters is their low

Figure 2-6 A digital multimeter.

internal resistance (input impedance). The low input impedance allows too much current to flow through circuits and should not be used on delicate electronic devices.

Digital meters, on the other hand, have a high input impedance, usually at least 10 megohms (10 million ohms). Metered voltage for resistance tests is well below 5 volts, reducing the risk of damage to sensitive components and delicate computer circuits. A high-impedance digital multimeter must be used to test the voltage of some components and systems such as an oxygen (O_2) sensor circuit. If a low-impedance analog meter is used in this type of circuit, the current flow through the meter is high enough to damage the sensor.

DMMs have either an "auto range" feature, in which the appropriate scale is automatically selected by the meter, or they must be set to a particular range. In either case, you should be familiar with the ranges and the different settings available on the meter you are using. To designate particular ranges and readings, meters display a prefix before the reading or range (**Figure 2-7**). If the meter has a setting for milliamps, that means the

Prefix	Symbol	Relation to basic unit
Mega	M	1,000,000
Kilo	k	1000
Milli	m	.001 or $\frac{1}{1000}$
Micro	μ	.000001 or $\frac{1}{1,000,000}$
Nano	n	0.000000001
Pico	p	0.000000000001

Figure 2-7 The various symbols used to define values shown on a DMM.

readings will be given in milliamps or 1/1000th of an ampere. Ohmmeter scales are expressed as a multiple of tens or use the prefix k or M. k stands for Kilo or 1000. A reading of 10 kohms equals 10,000 ohms. An M stands for Mega or 1,000,000. A reading of 10 Mohms equals 10,000,000 ohms. When using a meter with an auto range, make sure you note the range being used by the meter. There is a big difference between 10 ohms and 10,000,000 ohms.

After the test range has been selected, the meter is connected to the circuit in the same way as if it were an individual meter.

When using the ohmmeter function, the DMM will show a zero or close to zero when there is good continuity. If the continuity is very poor, the meter will display an infinite reading. This reading is usually shown as a blinking "1.000," a blinking "1," or an "OL." Multimeters may also have the ability to measure duty cycle, pulse width, and frequency. All of these represent voltage pulses caused by the turning on and off of a circuit or the increase and decrease of voltage in a circuit. **Duty cycle** is a measurement of the amount of time something is on compared to the time of one cycle and is measured in percentage.

Pulse width is similar to duty cycle except that it is the exact time something is turned on and is measured in milliseconds. When measuring duty cycle, you are looking at the amount of time something is on during one cycle.

The number of cycles that occur in one second is called the **frequency**. The higher the frequency, the more cycles occur in a second. Frequencies are measured in Hertz. One Hertz is equal to one cycle per second.

Some DMMs have a pressure transducer as an attachment. These transducers can be used in place of a pressure gauge. The transducer converts pressure into an analog electrical signal. Pressure transducers are normally available with three types of electrical output: millivolt, amplified voltage, and low milliamps.

Lab Scopes

An **oscilloscope** is a visual voltmeter. An oscilloscope converts electrical signals to a visual image representing voltage changes over a specific period of time. This information is displayed in the form of a continuous voltage line called a **waveform** pattern or trace.

An upward movement of the voltage trace on an oscilloscope screen indicates an increase in voltage, and a downward movement of this trace represents a decrease in voltage. As the voltage trace moves across an oscilloscope screen, it represents a specific length of time (**Figure 2-8**).

The size and clarity of the displayed waveform is dependent on the voltage scale and the time reference selected. Most scopes are equipped with controls that allow voltage and time interval selection. It is important, when choosing the scales, to remember that a scope displays voltage over time.

Dual-trace oscilloscopes can display two different waveform patterns at the same time. This makes cause and effect analysis easier.

Graphing Multimeter

One of the latest trends in diagnostic tools is a **graphing multimeter (GMM)**. These meters display readings over time, similar to a lab scope. The graph displays the minimum and maximum readings on a graph, as well as displaying the current reading. By observing the graph, a technician can notice any undesirable changes during the transition from a low reading to a high reading, or vice versa. These glitches are some of the more difficult problems to identify without a graphing meter or a lab scope.

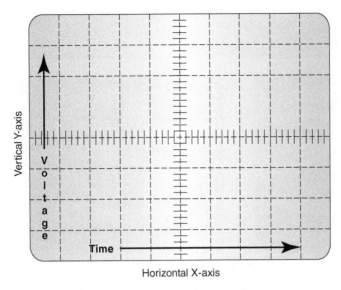

Figure 2-8 A scope displays changes in voltage over time.

Scan Tools

There are a variety of scan tools available to help diagnose the computers and their network on the vehicle. A **scan tool** is a microprocessor designed to communicate with the vehicle's computer (**Figure 2-9**). Connected to the computer through diagnostic connectors, a scan tool can access trouble codes, run tests to check system operations, and monitor the activity of the system. (To see how a scan tool is connected to a car, see **Photo Sequence 2**.)

Scan tools are capable of testing many onboard computer systems, such as climate controls, transmission controls, engine computers, antilock brake computers, air bag computers, and suspension computers, depending on the year and make of the vehicle and the type of scan tester. In many cases, the technician must select the computer system to be tested with the scanner after it has been connected to the vehicle.

Figure 2-9 A scan tool is used to diagnose electronic control systems.

PHOTO SEQUENCE 2
Connecting a Scan Tool

P2-1 With the ignition in the off position, connect the scan tool to the DLC.

P2-2 Turn the vehicles ignition and the scan tool on.

P2-3 The opening screen will appear. For this vehicle, we are going to choose the OBD II selection, also known as generic.

P2-4 The scan tool is communicating with the vehicle, press continue.

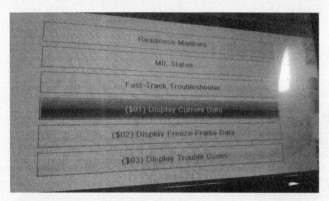

P2-5 The option menu comes up and the technician selects Current Data.

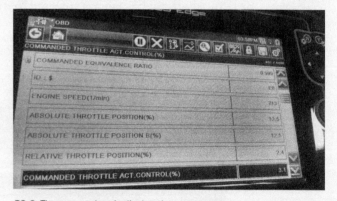

P2-6 The current data is displayed.

The scan tool is connected to the diagnostic connector on the driver's side of the vehicle under the dash. Most scan tools have the ability to store, or "freeze" data during a road test, and then play back this data when the vehicle is returned to the shop.

There are many different scan tools available. Some are a combination of other diagnostic tools, such as a lab scope and graphing multimeter. These may have the following capabilities:

- Retrieve diagnostic trouble codes (DTCs)
- Monitor system operational data

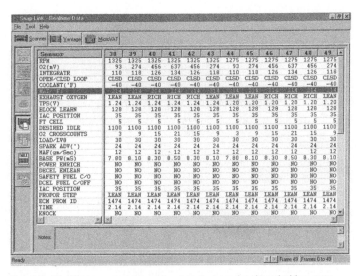

Figure 2-10 This is an example of the data that can be viewed with a scan tool. In this example, the information captured by the scan tool was sent to a personal computer.

- Reprogram the vehicle's electronic control modules
- Perform systems diagnostic tests
- Display appropriate service information, including electrical diagrams
- Display Technical Service Bulletins (TSBs)
- Troubleshooting instructions
- Easy tool updating through a personal computer (PC)

The vehicle's computer sets trouble codes when a voltage signal is entirely out of its normal range. The codes help technicians identify the cause of the problem when this is the case. If a signal is within its normal range but is still not correct, the vehicle's computer will not display a trouble code. However, a problem may still exist. Current and historic data is also available on most scan tools (**Figure 2-10**). This data allows a technician to look at what the system is experiencing when there is or is not a DTC.

Some scan tools work directly with a PC through un-cabled communication links, such as Bluetooth. Others use a laptop. There are even applications that will allow a cell phone to read data from the computer network. System tests help to determine what service the vehicle requires. Most of these scan tools also have the ability to do the following:

- Perform system and component tests
- Report test results of monitored systems
- View and print files on a PC
- Print DTC/Freeze frame
- Generate emissions reports
- View mode 6 information
- Display related TSBs
- Display full diagnostic code descriptions
- Observe live sensor data
- Graph live data
- Update the scan tool as a manufacturer's interfaces change

J2534 Interface Device

Many of the latest scan tools are PC based. The technician usually uses a laptop computer to communicate with the vehicle. Using a laptop frees the manufacturer from developing new hardware every few years. With the laptop, an interface device such as Honda's MVCI,

Figure 2-11 A J2534 interface tool to reprogram a module such as a TCM. This tool goes between a PC or scan tool and the DLC.

Ford's VCM II, or GM's MDI is used (**Figure 2-11**). These tools are sometimes called pass-through programming tools. The J2534 tool allows vehicle communications with the scan tool as well as with programming from the manufacturer via the Internet. If you work in an independent shop a subscription can be purchased from the manufacturer for a short or long term.

Breakout Box

On some vehicles and with some scan tools, it may be necessary to use a breakout box to monitor serial data. The breakout box is connected in series with Engine control module (ECM)/Powertrain control module (PCM) and the vehicle's wiring harness to the control unit. Often the breakout box comes with many different connectors so it can be installed in a variety of vehicles.

Any wire connected to the ECM/PCM can be monitored through the breakout box. A switch on the box allows a technician to select any pin at the computer's connectors. The scan tool will then display the activity at that pin. Most breakout boxes allow for a display of more than one pin at the same time.

Vacuum Gauge

A pressure lower than atmospheric pressure is considered a vacuum.

Measuring intake manifold vacuum is a way to determine the condition of an engine. This is important when diagnosing automatic transmissions. A transmission responds to engine load and engine load affects engine vacuum. As load increases, engine vacuum decreases. When an engine is worn, its ability to create a vacuum decreases. This means a poorly running engine will be seen by the transmission as having a load always on it. This results in poor shift quality and inefficient shift timing.

Manifold vacuum is tested with a vacuum gauge. Vacuum is formed on a piston's intake stroke. As the piston moves down, it lowers the pressure of the air in the cylinder if the cylinder is sealed. This lower cylinder pressure is called *engine vacuum*. Atmospheric pressure will force air into the cylinder when the intake valve of the cylinder with a vacuum opens. This is how the cylinder can be filled with air. The reason atmospheric pressure enters is that whenever there are both low and high pressure, the high pressure will always move toward the low pressure.

Vacuum is measured in inches of mercury (in./Hg) and in kilopascals (kPa) or millimeters of mercury (mm/Hg).

To measure vacuum, a flexible hose on the vacuum gauge is connected to a source of manifold vacuum, either on the manifold or at a point below the throttle plates. Sometimes this requires removing a plug from the manifold and installing a special fitting.

The test is made with the engine cranking (running). A good vacuum reading is typically at least 16 in./Hg. However, a reading of 15–20 in./Hg (50–65 kPa) is normally acceptable. Since the intake stroke of each cylinder occurs at a different time, the production of vacuum occurs in pulses. If the amount of vacuum produced by each cylinder is the same, the vacuum gauge will show a steady reading. If one or more cylinders are producing different amounts of vacuum, the gauge will show a fluctuating reading.

Hydraulic Pressure Gauge Set

An automatic transmission is a hydraulic machine. The amount of hydraulic pressure and the direction of fluid flow are the keys to proper operation. Therefore, observing the pressure and pressure changes is an excellent way to examine the operation of a transmission.

A common diagnostic tool for automatic transmissions is a hydraulic pressure gauge. A pressure gauge measures pressure in pounds per square inch (psi) or kilopascals (kPa). The gauge is normally part of a kit that contains various fittings and adapters (**Figure 2-12**). Most transmissions have at least one external fitting that allow for the connection of the pressure gauge to the interior of the transmission. Vehicle manufacturers also publish pressure charts so the results of the test can be compared to specifications.

Logical Diagnostics

The true measure of a good technician is their ability to find and correct the cause of a problem. Service information and other information sources will guide you through the diagnosis and repair of a problem. However, those guidelines will not always lead you to the exact cause of the problem. To do this you must use your knowledge and take a logical approach while

Figure 2-12 A hydraulic pressure gauge set with various adapters.

Diagnosis is a systematic study of something to determine the cause of improper operation or failure.

troubleshooting. **Diagnosis** is not guessing, and it is more than following a series of inter-related steps in order to find the solution to a specific problem. Diagnosis is a way of looking at systems that are not functioning the way they should and finding out why. It is knowing how the system should work and deciding if it is working correctly. Through an understanding of the purpose and operation of the system, you can accurately diagnose problems.

Most good technicians use the same basic diagnostic approach. Because this is such a logical approach, it can quickly lead to the cause of a problem. Logical diagnosis follows these steps:

1. Gather information about the problem.
2. Verify that the problem exists.
3. Thoroughly define what the problem is and when it occurs.
4. Research all available information and knowledge to determine the possible causes of the problem.
5. Isolate the problem by testing.
6. Continue testing to pinpoint the cause of the problem.
7. Locate and repair the problem, then verify the repair.

HYBRID TOOLS

A hybrid vehicle is an automobile and as such is subject to many of the same problems as a conventional vehicle. Most systems in a hybrid vehicle are diagnosed in the same way as well. However, a hybrid vehicle has unique systems that require special procedures and test equipment. It is imperative to have good information before attempting to diagnose these vehicles. Also, make sure you follow all test procedures precisely as they are given.

Category III DMM

An important diagnostic tool is a DMM. However, this is not the same DMM used on a conventional vehicle. The meter used on hybrids and other electric-powered vehicles should be classified as a category III (CAT III) meter (**Figure 2-13**). There are basically four categories for low-voltage electrical meters, each built for specific purposes and to meet certain standards. Low voltage, in this case, means voltages less than 1000 volts. The categories define how safe a meter is when measuring certain circuits. The standards for the various categories are defined by the American National Standards Institute (ANSI), the International Electrotechnical Commission (IEC), and the Canadian Standards Association (CSA). A CAT III meter is required for testing hybrid vehicles because of the high voltages, three-phase current, and the potential for high transient voltages. Transient voltages are voltage surges or spikes that occur in AC circuits. To be safe, you should have

Figure 2-13 Only meters with this symbol should be used on high-voltage vehicles.

a CAT III-1000 volts meter. A meter's voltage rating reflects its ability to withstand transient voltages. Therefore, a CAT III-1000 volts meter offers much more protection than a CAT III meter rated at 600 volts.

Insulation Resistance Tester

Another important tool is an insulation resistance tester. These can check for voltage leakage through the insulation of the high-voltage cables. Obviously, no leakage is desired, and any leakage can cause a safety hazard as well as damage to the vehicle. Minor leakage can also cause hybrid system-related drivability problems. This meter is not one commonly used by automotive technicians, but should be for anyone who might service a damaged hybrid vehicle, such as doing body repair. This should also be a CAT III meter and may be capable of checking resistance and voltage of circuits like a DMM.

To measure insulation resistance, system voltage is selected at the meter and the probes placed at their test position. The meter will display the voltage it detects. Normally, resistance readings are taken with the circuit de-energized unless you are checking the effectiveness of the cable or wire insulation. In this case, the meter is measuring the insulation's effectiveness and not its resistance.

⚠ WARNING **The probes for all meters used on high-voltage systems should have safety ridges or finger positioners. These help prevent physical contact between your fingertips and the meter's test leads. Additionally, you should be wearing the manufacturer's special hybrid gloves for even more protection.**

TORQUE WRENCHES

Torque is the twisting force used to turn a fastener against the friction between the threads and between the head of the fastener and the surface of the component. The fact that practically every vehicle and engine manufacturer publishes a list of torque recommendations is ample proof of the importance of using proper amounts of torque when tightening nuts or bolts. The amount of torque applied to a fastener is measured with a torque-indicating or torque wrench.

A **torque wrench** is basically a ratchet or breaker bar with some means of displaying the amount of torque exerted on a bolt when pressure is applied to the handle (**Figure 2-14**). Torque wrenches are available with the various drive sizes. Sockets are

Figure 2-14 The basic types of torque-indicating wrenches.

inserted onto the drive and then placed over the bolt. As pressure is exerted on the bolt, the torque wrench indicates the amount of torque.

The common types of torque wrenches are available with inch-pound, foot-pound, and newton-meter (metric) increments:

- A beam torque wrench is not highly accurate. It relies on a beam metal that points to the torque reading.
- A "click"-type torque wrench clicks when the desired torque is reached. The handle is twisted to set the desired torque reading.
- A dial torque wrench has a dial that indicates the torque exerted on the wrench. The wrench may have a light or buzzer that turns on when the desired torque is reached.
- A digital readout type displays the torque and is commonly used to measure turning effort, as well as for tightening bolts. Some designs of this type of torque wrench have a light or buzzer that turns on when the desired torque is reached.

POWER TOOLS

Power tools make a technician's job easier. They operate faster and with more torque than hand tools. However, power tools require greater safety measures. Power tools do not stop unless they are turned off. Power is furnished by air (on pneumatic tools), electricity, or hydraulic fluid.

Pneumatic tools are typically used by technicians because they have more torque, weigh less, and require less maintenance than electric power tools. However, electric power tools tend to cost less than pneumatic ones. Electric power tools can be plugged into most electric wall sockets, but to use a pneumatic tool, you must have an air compressor and an air storage tank. Cordless electrical tools are popular because they work without a cord or hose.

AIR WRENCHES

An impact wrench uses compressed air or electricity to hammer or impact a nut or bolt loose or tight (**Figure 2-15**). Light-duty impact wrenches are available in three drive sizes, ¼", ⅜", and ½", and two heavy-duty sizes, ¾" and 1".

⚠ WARNING **Impact wrenches should not be used to tighten critical parts or parts that may be damaged by the hammering force of the wrench.**

Figure 2-15 An air-impact wrench.

Air ratchets are often used during disassembly or reassembly work to save time. Because the ratchet turns the socket without an impact force, these wrenches can be used on most parts and with ordinary sockets. Air ratchets usually have a 1/4- or 3/8-inch drive. Air ratchets are not torque sensitive; therefore, a torque wrench should be used on all fasteners after snugging them up with an air ratchet.

Bench Grinder

This electric power tool is generally bolted to a workbench. A bench grinder should have safety shields and guards. Always wear face protection when using a grinder. A bench grinder is classified by wheel size. Six- to ten-inch wheels are the most common in auto repair shops. Three types of wheels are available with this bench tool.

- Grinding wheel. Used for a wide variety of grinding jobs, from sharpening cutting tools to de-burring.
- Wire wheel brush. Used for general cleaning and buffing, removing rust and scale, paint removal, de-burring, and so forth.
- Buffing wheel. For general-purpose buffing and polishing.

Shop Light

Adequate light is necessary when working under and around automobiles. A shop light can be battery powered (like a flashlight) or need to be plugged into a wall socket. Some shops have lights that pull down from a reel suspended from the ceiling. Because incandescent bulbs can pop and burn, it is highly recommended that you only use fluorescent or LED bulbs. Take extra care when using a shop light. Make sure the cord does not get caught in a rotating object. The bulb or tube is surrounded by a cage or enclosed in clear plastic to prevent accidental breaking and burning.

Blowgun

Blowguns are used for air-testing components and blowing off parts during cleaning (**Figure 2-16**). Never point a blowgun at yourself or someone else. A blowgun snaps into one end of an air hose and directs airflow when a button is pressed. Always use an OSHA-approved air blowgun. The OSHA-approved air blowgun has its output pressure limited to help prevent injury. Before using a blowgun, be sure it has not been modified to allow full pressure at the output.

> ⚠ **Caution**
>
> The sockets designed for impact wrenches are constructed of thicker but softer steel than other sockets, allowing them to withstand the force of the impact. Ordinary sockets must not be used with impact wrenches. They will crack or shatter because of the force and can cause injury.

Figure 2-16 Two examples of OSHA-approved blowguns.

LIFTING TOOLS

Lifting tools are necessary tools for many transmission service procedures. Typically, these tools are provided by the shop and are not the property of a technician. Correct operating and safety procedures should always be followed when using lifting tools.

Jacks

Jacks are used to raise a vehicle off the ground. They are available in two basic designs and in a variety of sizes. The most common jacks are hydraulic floor jacks. They are classified by the weights they can lift: 1, 2, and 3 tons, and so on. These jacks are controlled by moving the handle up and down. The other type of portable floor jack uses compressed air. Pneumatic jacks are operated by controlling air pressure at the jack.

Safety Stands

Safety stands are commonly called **jack stands.**

When a vehicle is raised by a jack, it should be supported by **safety stands**. Never work under a car with only a jack supporting it; always use safety stands. Hydraulic seals in the jack can let go and allow the vehicle to drop. Service information notes the proper locations for jacking and supporting a vehicle while it is raised from the ground. Always follow those guidelines.

Hydraulic Lift

The hydraulic floor lift is the safest lifting tool. It is able to raise the vehicle high enough to allow you to walk and work under it. Various safety features prevent a hydraulic lift from dropping if a seal does leak or if air pressure is lost. Before lifting a vehicle, make sure the lift is correctly positioned.

Engine hoists are often referred to as "cherry pickers."

Some technicians use pneumatic floor and transmission jacks. These jacks typically can raise the vehicle higher than hydraulic units. Before using this type of jack, make sure the unit is sealed and there are no air leaks.

Portable Crane

Often, to remove and install a transmission, the engine will be moved out of the vehicle with the transmission. To remove or install an engine, a portable crane is used. A crane uses hydraulic pressure that is converted to a mechanical advantage and lifts the engine from the vehicle. To lift an engine, attach a pulling sling or chain to the engine. Some engines have eye plates for use in lifting. If there are none, the sling must be bolted to the engine. The sling-attaching bolts must be large enough to support the engine and must thread into the block a minimum of 1 times the bolt diameter. Connect the crane to the chain. Raise the engine slightly and make sure the sling attachments are secure. Carefully lift the engine out of its compartment.

Lower the engine close to the floor so the transmission and torque converter can be removed from the engine, if necessary.

Transmission Jacks

Transmission jacks are designed to help you remove a transmission from under the vehicle (**Figure 2-17**). The weight of the transmission makes it difficult and unsafe to remove it without much assistance or a transmission jack. These jacks fit under the transmission and are typically equipped with hold-down chains. These chains are used to secure the transmission to the jack. The transmission's weight rests on the jack's saddle.

Transmission jacks are available in two basic styles. One is used when the vehicle is raised by a hydraulic jack and sitting on jack stands. The other style is used when the vehicle is raised on a lift.

Figure 2-17 A typical transmission jack.

Transaxle Removal and Installation Equipment

The removal and replacement (R&R) of transversely mounted engines may require tools not required for removing an RWD engine. The engines of some FWD vehicles are removed by lifting them from the top. Others must be removed from the bottom; this procedure requires different equipment. Make sure you follow the instructions given by the manufacturer and use the appropriate tools and equipment. The required equipment varies with manufacturer and vehicle model; however, most accomplish the same thing.

To remove the engine and transmission from under the vehicle, the vehicle must be raised. A crane or support fixture is used to hold the engine and transaxle assembly in place while the assembly is being readied for removal. When everything is set for removal of the assembly, the crane is used to lower the assembly onto a cradle. The cradle is similar to a hydraulic floor jack and is used to lower the assembly further so it can be rolled out from under the vehicle. The transaxle can be separated from the engine once it has been removed from the vehicle.

When the transaxle is removed as a single unit, the engine must be supported while it is in the vehicle before, during, and after transaxle removal. Special fixtures (**Figure 2-18**) mount to the vehicle's upper frame or suspension parts. These supports have a bracket that is attached to the engine. With the bracket in place, the engine's weight is now on the support fixture and the transmission can be removed.

Transmission/Transaxle Holding Fixtures

Special holding fixtures should be used to support the transmission or transaxle after it has been removed from the vehicle (**Figure 2-19**). These holding fixtures may be stand-alone units or may be bench mounted, allowing the transmission to be easily repositioned during repair work.

Figure 2-18 An engine-holding fixture for engine and transaxle removal in a FWD vehicle.

Figure 2-19 A typical holding fixture for a transaxle after it has been removed from the vehicle.

MEASURING TOOLS

Many of the procedures discussed in this manual require exact measurements of parts and clearances. Accurate measurements require the use of precision measuring devices that are designed to measure things in very small increments. Measuring tools are delicate instruments and should be handled with great care. Never strike, pry, drop, or force these tools. Also, make sure you clean them before and after every use.

There are many different measuring devices used by automotive technicians. This chapter will only cover those that are commonly used to service transmissions and other driveline components.

> ⚙ **SERVICE TIP** It is a good idea to check all measuring tools against known good equipment or tool standards. This will ensure that they are operating properly and capable of accurate measurement.

Machinist's Rule

The **machinist's rule** looks very much like an ordinary ruler. Each edge of this basic measuring tool is divided into increments based on a different scale. A typical machinist's rule based on the English system of measurement may have scales based on 1/8-, 1/16-, 1/1000-, and 1-inch intervals. Of course, metric machinist's rules are also available. Metric rules are usually divided into 0.5- and 1-mm increments (**Figure 2-20**).

Some machinist's rules may be based on decimal intervals. These are typically divided into ⅒-, ¹⁄₅₀-, and ¹⁄₁₀₀₀-inch (0.1, 0.03, and 0.01) increments. Decimal machinist's rules are very helpful when measuring dimensions that are specified in decimals; they make such measurements much easier.

A meter is 1/10,000,000 of the distance from the North Pole to the Equator, or 39.37 inches.

Vernier Caliper

A **vernier caliper** can make inside, outside, or depth measurements. It is marked in both British imperial and metric divisions called a vernier scale. A vernier scale consists of a stationary scale and a movable scale, in this case the vernier bar to the vernier plate. The length is read from the vernier scale.

A vernier caliper has a movable scale that is parallel to a fixed scale (**Figure 2-21**). These precision measuring instruments are capable of measuring outside and inside diameters and most will even measure depth. Vernier calipers are available in both

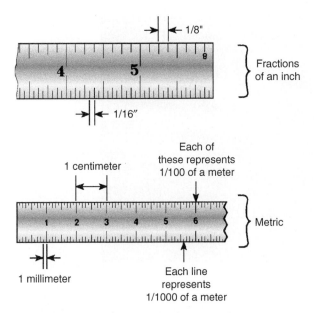

Figure 2-20 The graduations on a USCS and metric machinist's rule.

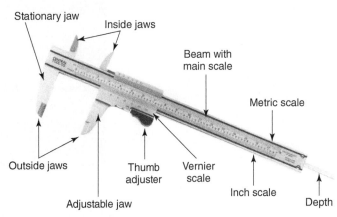

Figure 2-21 A vernier caliper.

Figure 2-22 The main scale of a caliper is divided into inches; each inch is divided into 10 parts, each equal to 0.100 inch. The area between the 0.100 marks is divided into four divisions, each equal to 0.025 inch.

imperial and metric scales. The main scale of the caliper is divided into inches; most measure up to 6 inches. Each inch is divided into 10 parts, each equal to 0.100 inch. The area between the 0.100 marks is divided into four. Each of these divisions is equal to 0.025 inch (**Figure 2-22**).

The vernier scale has 25 divisions, each one representing 0.001 inch. Measurement readings are taken by combining the main and the vernier scales. At all times, only one division line on the main scale will line up with a line on the vernier scale (Figure 2-22). This is the basis for accurate measurements.

To read the caliper, locate the line on the main scale that lines up with the zero (0) on the vernier scale. If the zero lined up with the 1 on the main scale, the reading would be 0.100 inch. If the zero on the vernier scale does not line up exactly with a line on the main scale, then look for a line on the vernier scale that does line up with a line on the main scale.

Dial Caliper

The **dial caliper** (**Figure 2-23**) is an easier-to-use version of the vernier caliper. Imperial calipers commonly measure dimensions from 0 to 6 inches. Metric dial calipers typically measure from 0 to 150 mm in increments of 0.02 mm. The dial caliper features a depth scale, bar scale, dial indicator, inside measurement jaws, and outside measurement jaws.

The main scale of an imperial dial caliper is divided into one-tenth (0.1) inch graduations. The dial indicator is divided into one-thousandth (0.001) inch graduations. Therefore, one revolution of the dial indicator needle equals one-tenth inch on the bar scale.

Figure 2-23 A dial caliper.

A metric dial caliper is similar in appearance but the bar scale is divided into 2-mm increments. Additionally, on a metric dial caliper, one revolution of the dial indicator needle equals 2 mm.

Both English and metric dial calipers use a thumb-operated roll knob for fine adjustment. When you use a dial caliper, always move the measuring jaws backward and forward to center the jaws on the object being measured. Make sure the caliper jaws lay flat on or around the object. If the jaws are tilted in any way, you will not obtain an accurate measurement.

Although dial calipers are precision measuring instruments, they are only accurate to plus or minus two-thousandths (60.002) of an inch. Micrometers are preferred when extremely precise measurements are desired.

Micrometers

The **micrometer** is used to measure linear outside and inside dimensions. Both outside and inside micrometers are calibrated and read in the same way. Measurements on both are taken with the measuring points of the tool in contact with the surfaces being measured.

The major components and markings of a micrometer include the frame, anvil, spindle, locknut, sleeve, sleeve numbers, sleeve long line, thimble marks, thimble, and ratchet (**Figure 2-24**). Micrometers are calibrated in either inch or metric graduations and are available in a range of sizes.

Photo Sequence 3 covers the basics for measuring with and reading a micrometer. Remember, always use the appropriately sized micrometer for the object being measured.

(A)   (B)

Figure 2-24 Major components of (A) an outside and (B) an inside micrometer.

PHOTO SEQUENCE 3
Typical Procedure for Using a Micrometer

P3-1 Micrometers can be used to measure the diameter and thickness of many objects. A common example for a transmission tech is a selective snap ring.

P3-2 Because the thickness of the snap ring is less than 1 inch, a 0–1-inch micrometer will be used to measure the snap ring.

P3-3 Each graduation on the sleeve represents 0.025 inch. To read a measurement on a micrometer, begin by counting the visible lines on the sleeve, then multiply that number by 0.025.

P3-4 The graduations on the thimble assembly define the area between the lines on the sleeve. The number indicated on the thimble is added to the measurement shown on the sleeve.

P3-5 Micrometer reading of 0.375 inch.

P3-6 Position the micrometer over the snap ring and slowly close the micrometer around the snap ring until a slight drag is felt while moving the micrometer back and forth over the snap ring.

P3-7 To prevent the reading from changing while you move the micrometer away from the stem, use your thumb to activate the lock lever.

P3-8 This reading (0.078) represents the thickness of the snap ring.

P3-9 Some micrometers are able to measure in 0.0001-inch increments. Use this type of micrometer when the specifications call for this degree of accuracy.

Typically, they measure an inch; therefore, the range covered by one micrometer would be 0–1 inch; another would measure 1–2 inches, and so on.

To measure small objects with an outside micrometer, open the jaws of the tool and slip the object between the spindle and the anvil. While holding the object against the anvil, turn the ratchet using your thumb and forefinger until the spindle contacts the object. The ratchet provides only enough pressure on the object to allow it to just fit between the tips of the anvil and spindle. The object should slip through with only a very slight resistance. When a satisfactory feel is reached, lock the micrometer.

Since each graduation on the sleeve represents 0.025 inch, begin reading the measurement by counting the visible lines on the sleeve and multiply that number by 0.025. The graduations on the thimble assembly define the area between the lines on the sleeve; therefore, the number indicated on the thimble should be added to the measurement shown on the sleeve. The sum is the outside diameter of the object.

Micrometers are available to measure in 0.0001 (ten-thousandths) of an inch. Use this type of micrometer if the specifications call for this much accuracy.

To measure larger objects, select a micrometer of the proper size, hold the frame of the micrometer, and slip it over the object. Turn the thimble while continuing to slip the micrometer over the object until you feel a very slight resistance. Rock the micrometer from side to side while doing this to make sure the spindle cannot be closed any farther. Then, lock the micrometer and take a measurement reading.

Some technicians use a digital micrometer, which is easier to read. These tools do not have the various scales described here; rather, the measurement is displayed and read directly off the micrometer (**Figure 2-25**).

A metric micrometer is read in the same way, except the graduations are expressed in the metric system of measurement (**Figure 2-26**). Each number on the sleeve represents 5 millimeters (mm) [0.005 meter (m)]. Each of the 10 equal spaces between each number, which have index lines alternating above and below the horizontal line, represents 0.5 mm (five-tenths of a millimeter). Therefore, one revolution of the thimble changes the reading one space on the sleeve scale (0.5 mm). The beveled edge of the thimble is divided into 50 equal divisions with every fifth line numbered: 0, 5, 10, … , 45. Since one complete revolution of the thimble advances the spindle 0.5 mm, each graduation on the thimble is equal to one-hundredth of a millimeter. As with the inch-graduated micrometer, the separate readings are added together to obtain the total reading.

Inside micrometers can be used to measure the inside diameter of a bore. To do this, place the tool inside the bore and extend the measuring surfaces until each end touches the bore's surface. If the bore is large, it might be necessary to use an extension rod to increase the micrometer's range. These extension rods come in various lengths. The inside micrometer is read in the same manner as an outside micrometer.

Figure 2-25 A digital micrometer.

Figure 2-26 Reading a metric micrometer: (A) 5 mm plus
(B) 0.5 mm plus (C) 0.01 mm, equals 5.51 mm.

A depth micrometer is used to measure the distance between two parallel surfaces. The sleeves, thimbles, and ratchet screws operate in the same way as other micrometers. Likewise, depth micrometers are read in the same way as other micrometers.

If a depth micrometer is used with a gauge bar, it is important to keep both the bar and the micrometer from rocking. Any movement of either part will result in an inaccurate measurement.

Telescoping Gauge. Telescoping gauges are used for measuring bore diameters and other clearances (**Figure 2-27**). They may also be called snap gauges. They are available in sizes ranging from fractions of an inch through 6 inches. Each gauge consists of two telescoping plungers, a handle, and a lock screw. Snap gauges are normally used with an outside micrometer.

To use the telescoping gauge, insert it into the bore and loosen the lock screw. This will allow the plungers to snap against the bore. Once the plungers have expanded, tighten the lock screw. Then, remove the gauge and measure the expanse with a micrometer.

Figure 2-27 A telescoping gauge is used to
measure the diameter of a bore.

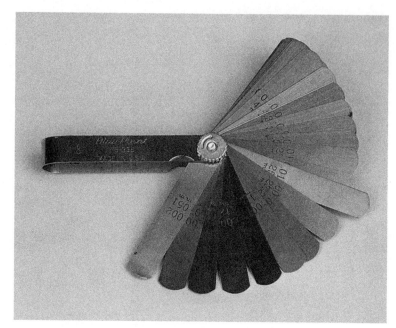

Figure 2-28 A typical feeler gauge pack.

Small-Hole Gauge. A small-hole or ball gauge works just like a telescoping gauge. However, it is designed for small bores. After it is placed into the bore and expanded, it is removed and measured with a micrometer. Like the telescoping gauge, the small-hole gauge consists of a lock, handle, and expanding end. The end expands or retracts by turning the gauge handle.

Feeler Gauge

A feeler gauge is a thin strip of metal or plastic of known and closely controlled thickness. Several of these metal strips are often assembled together as a feeler gauge set that looks like a pocketknife (**Figure 2-28**). The desired thickness gauge can be pivoted away from others for convenient use. A steel feeler gauge pack usually contains strips or leaves of 0.002–0.010-inch thickness (in steps of 0.001 inch) and leaves of 0.012–0.024-inch thickness (in steps of 0.002 inch).

A feeler gauge can be used by itself to measure piston clearance, fluid pump clearances, and other distances. It can also be used with a precision straightedge to check the flatness of a sealing surface.

Straightedge. A straightedge is no more than a flat bar machined to be totally flat and straight. Any surface that should be flat can be checked with a straightedge and feeler gauge set. The straightedge is placed across and at angles on the surface. At any low points on the surface, a feeler gauge can be placed between the straightedge and the surface. The size of the gauge that fills in the gap determines the amount of warpage or distortion.

Dial Indicator

A **dial indicator** is a measuring instrument that measures changes in dimension as the indicator is moved across an object or when the object is moved over the stem of the indicator.

The **dial indicator** is calibrated in 0.001-inch (one-thousandth–inch) increments (**Figure 2-29**). Metric dial indicators are also available. Both types are used to measure movement. Common uses of the dial indicator include measuring endplay (**Figure 2-30**), clutch pack clearances, backlash, and flexplate runout. Dial indicators are available with various face markings and measurement ranges to accommodate many measuring tasks.

To use a dial indicator, position the indicator rod against the object to be measured. Then, push the indicator toward the work until the indicator needle travels far enough around the gauge face to permit movement to be read in either direction. Move the object

Figure 2-29 A dial indicator.

Figure 2-30 A dial indicator is often used in automatic transmission service. This setup is used to measure the clearance of a multiple-friction disc pack.

to one extreme of its travel, then zero the indicator needle on the gauge. Always be sure the range of the dial indicator is sufficient to allow the amount of movement required by the measuring procedure. For example, never attempt to use a 1-inch indicator on a component that will move 2 inches.

TRANSMISSION TOOLS AND EQUIPMENT

Many different tools and equipment are used to service transmissions and drivelines. ASE Education Foundation has identified many of these and has said an automatic transmission technician must know what they are and how and when to use them. The tools and equipment listed by ASE Education Foundation and others are covered in the following discussion.

Presses

Many transmission and driveline repairs require the use of a powerful force to assemble or disassemble parts that are **press-fit** together. Removing and installing axle and final drive bearings, universal joint replacement, and transmission assembly work are just a few of the examples. Presses can be hydraulic, electric, air, or hand driven. Capacities range up to 150 tons of pressing force, depending on the size and design of the press. Smaller arbor and C-frame presses can be bench or pedestal mounted, while high-capacity units are freestanding or floor mounted.

⚠ Caution
Always wear safety glasses when using a press.

Bushing and Seal Pullers and Drivers

Another commonly used group of special tools comprise the various designs of bushing and seal drivers and pullers. Pullers are either threaded or slide hammer–type tools (**Figure 2-31**). Always make sure you use the correct tool for the job; bushings and seals are easily damaged if the wrong tool or procedure is used. Car manufacturers and specialty tool companies work closely together to design and manufacture special tools required to repair cars. Most of these special tools are listed in the appropriate service information.

Figure 2-31 Examples of slide handle pullers.

Figure 2-32 Examples of the jaws of a gear and bearing puller.

Gear and Bearing Pullers

Many tools are designed for a specific purpose. An example of a special tool is a gear and bearing puller (**Figure 2-32**). Many gears and bearings have a slight interference fit when they are installed on a shaft or in a housing. Something that has a press-fit has an interference fit. For example, the inside diameter of a bore is 0.001 inch smaller than the outside diameter of a shaft; when the shaft is fitted into the bore it must be pressed in to overcome the 0.001-inch interference fit. This press-fit prevents the parts from moving on each other. The removal of these gears and bearings must be done carefully to prevent damage to the gears, bearings, or shafts. Prying or hammering can break or bind the parts. A puller with the proper jaws and adapters should be used to remove gears and bearings. Using the proper puller, the force required to remove a gear or bearing can be applied with a slight and steady motion.

Retaining (Snap) Ring Pliers

Often, a transmission technician will run into many different styles and sizes of retaining or snap rings that hold subassemblies together or keep them in a fixed location. Using the correct tool to remove and install these rings is the only safe way to work with them (**Figure 2-33**). All transmission and driveline technicians should have an assortment of retaining ring pliers.

Clutch Tools

Multiple-disc clutch packs contain springs that must be depressed to disassemble and assemble a clutch pack. The spring compressors used to do this must be able to evenly compress the springs while allowing the technician to remove or install a retainer (**Figure 2-34**). Often these tools are specifically designed for a particular clutch assembly; however, some universal tools are available.

Special Tool Sets

Vehicle manufacturers and specialty tool companies work closely together to design and manufacture special tools required to repair transmissions. Most of these special tools are listed in the appropriate service information and are part of each manufacturer's essential tool kit.

Figure 2-33 Internal and external snap ring pliers.

Figure 2-34 A spring compressor used to remove the snap ring on a multiple-disc clutch pack.

SERVICE INFORMATION

Perhaps the most important tool you will use is service information. Service information is a necessary part of transmission and driveline service. There is no way a technician can remember all the procedures and specifications needed to repair an automobile correctly. Thus, a good technician relies on service information. Good information plus knowledge allows a technician to fix a problem with the least frustration and at the lowest cost to the customer. Service information also provides drawings and photographs that show where and how to perform certain procedures on the particular vehicle you are working on. Special tools or instruments required are listed and shown. Safety precautions are also provided, with regard to preventing injury and damaging parts.

Service information is now generally obtained from an electronic database, rather than a printed service information.

The primary source of repair and specification information for any car, van, or truck is the manufacturer. The manufacturer publishes service information on the Internet for every vehicle it builds. Information is typically divided into sections based on the major systems of the vehicle. In the case of transmissions, there is a section for each transmission that may be in the vehicle. Manufacturers' manuals cover all repairs, adjustments, specifications, detailed diagnostic procedures, and special tools required.

All a technician needs to do is enter vehicle information and then move to the appropriate part or system.

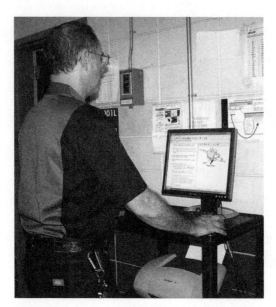

Figure 2-35 A technician using Internet-based service information.

Online data can be updated instantly and requires no physical storage space (**Figure 2-35**). These systems are easy to use and the information is quickly accessed and displayed. The computer's keyboard and mouse are used to make selections from the screen's menu. Once the information is retrieved, a technician can read it off the screen or print it out.

Technical Service Bulletins

Since many technical changes occur on specific vehicles each year, updates are published on the Internet as service bulletins (often referred to as **Technical Service Bulletins** or **TSBs**). These show changes in specifications and repair procedures during the model year. The car manufacturer provides these electronic bulletins to dealers and repair facilities on a regular basis.

General and Specialty Repair Manuals

Some service information is published on the Internet by independent companies rather than manufacturers. However, these publishers pay for and get most of their information from the carmakers. They may present component information, diagnostic steps, repair procedures, and specifications for several car makes on one website. Information is usually condensed and is more general in nature than that in manufacturers' manuals.

Many of the larger parts manufacturers have excellent guides to the various parts they manufacture or supply. They also provide updated service bulletins on their products. Other sources for up-to-date technical information are trade magazines and trade associations, which are also widely available in electronic form on the Internet.

Flat-Rate Manuals

Flat-rate manuals contain standards for the length of time a specific repair is supposed to require. Normally, they also contain a parts list with approximate or exact prices of parts. They are excellent for making cost estimates and are published by the manufacturers and independents. Most often, flat-rate manuals are contained in the shop's shop management software or are included in the service information available on the Internet.

Using Electronic Service Data

Today most technicians rely on the Internet to find specific repair information. This is due to the fact that service information can be obtained very quickly and easily. The information on the Internet can be easily updated. Another advantage of the Internet is that all of the information is easily accessible. A technician simply inputs what he or she needs, and a search engine finds the information. Also, to simplify things, the information is divided by specific systems.

The data available from most electronic service data providers includes the following information for each system of a specific vehicle:

- Adjustments
- Component locators
- Description and operation
- Diagnostic flowcharts
- Diagnostic trouble codes
- Diagrams
- Fluid types and volumes
- Maintenance schedules
- OEM wiring diagrams
- Parts and labor information
- Recall notices
- Required tools and equipment
- Service and repair procedures
- Service precautions
- Short cuts
- Specifications
- Technical Service Bulletins
- Testing and inspection procedures
- User tips

Vehicle Identification

Before performing any service, it is important for you to know exactly what type of vehicle you are working on. The best way to do this is to refer to the **vehicle's identification number (VIN)**. The VIN (**Figure 2-36**) is found on a plate behind the lower corner of the driver's side of the windshield, as well as at other locations on the vehicle. The VIN is made up of 17 characters and contains all pertinent information about the vehicle. The use of the 17-character alphanumeric (number and letter) code became mandatory beginning with 1981 vehicles and is used by all domestic and foreign manufacturers of vehicles.

Each character of a VIN has a particular purpose. The first character identifies the country where the vehicle was manufactured, for example:

- 1 or 4—U.S.A.
- 2—Canada
- 3—Mexico
- J—Japan
- K—Korea
- S—England
- W—Germany

The second character identifies the manufacturer, for example:

- A—Audi
- B—BMW
- C—Chrysler
- D—Mercedes Benz

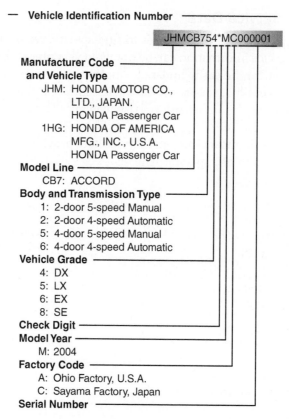

Figure 2-36 Basic makeup of a VIN.

- F—Ford
- G—General Motors
- H—Honda
- N—Nissan
- T—Toyota

The third character identifies the vehicle type or manufacturing division (passenger car, truck, bus, and so on). The fourth through eighth characters identify the features of the vehicle, such as the body style, vehicle model, and engine type.

The ninth character is used to identify the accuracy of the VIN and is a check digit. The 10th character identifies the model year, for example:

- 1—2001
- 3—2003
- 5—2005
- 7—2007
- 9—2009
- A—2010
- B—2011
- C—2012
- D—2013
- E—2014
- F—2015
- G—2016
- H—2017
- J—2018
- K—2019
- L—2020

Engine number
(A)

Transmission number
(automatic)   (B)

Transmission number
(manual)

Figure 2-37 Every manufacturer locates the ID of individual engines and transmission in a specific place. Always refer to the service information to determine what transmission you are working with. In this case, the engine serial number (A) is stamped on the cylinder block of the engine, and the transaxle serial number (B) is stamped on the housing.

The 11th character identifies the plant where the vehicle was assembled and the 12th to 17th characters identify the production sequence of the vehicle as it rolled off the manufacturer's assembly line.

The specifics needed to further decode the characters of the VIN can be found in a service information for the vehicle.

Transmission and Driveline Identification. The VIN will give you a general description of the transmission and driveline; however, to obtain the correct transmission specifications and other information, you must first identify the transmission you are working on. The transmission code (**Figure 2-37**) can be interpreted through information given in the service information. In addition, the manual may also help you identify the transmission through appearance, casting numbers, or markings on the housing.

Hotline Services

Hotline services provide answers to service concerns by telephone. Manufacturers provide help by telephone for technicians in their dealerships. Subscription hotline services enable independents to get repair information by phone. Some manufacturers also have a phone modem system that can transmit computer information from the car to another location. The vehicle's diagnostic link is connected to the modem. The technician in the service bay runs a test sequence on the vehicle. The system downloads the latest updated repair information on that particular model of car. If that does not repair the problem, a technical specialist at the manufacturer's location will review the data and propose a repair.

The International Automotive Technician's Network (iATN) is comprised of a group of thousands of professional automotive technicians from around the world. The technicians in this group exchange technical knowledge and information with other members. The web address for this group is http://www.iatn.net.

FASTENERS

The United States Customary System (USCS) is a system of weights and measures. Units of the imperial and USCS systems are equal except for some increments of volume and capacity.

Fasteners are things used to secure or hold parts of something together. Many types and sizes of fasteners are used by the automotive industry. Each fastener is designed for a specific purpose and condition. One of the most commonly used types of fastener is the threaded fastener. Threaded fasteners include bolts, nuts, screws, and similar items that allow a technician to install or remove parts easily.

Threaded fasteners are available in many sizes, designs, and threads. The threads can be either cut or rolled into the fastener. Rolled threads are 30 percent stronger than cut threads. They also offer better fatigue resistance because there are no sharp notches to create stress points. Fasteners are made to imperial or metric measurements. There are four classifications for the threads of imperial fasteners: Unified National Coarse (UNC), Unified National Fine (UNF), Unified National Extra Fine (UNEF), and Unified National Pipe Thread (UNPT or NPT). Metric fasteners are also available in fine and coarse threads.

Coarse threads are used for general-purpose work, especially where rapid assembly and disassembly is required. Finely threaded fasteners are used where greater holding force is necessary. They are also used where greater resistance to vibration is desired.

Bolts have a head on one end and threads on the other. Bolts are identified by defining the head size, shank diameter, thread pitch, length, and grade (**Figure 2-38**).

The abbreviation *SI* stands for the International System of weights and measures. The SI system is normally called the *metric system*.

Studs are rods with threads on both ends. Most often, the threads on one end are coarse while the other end has fine threads. One end of the stud is screwed into a threaded bore. A hole in the part to be secured is fitted over the stud and held in place with a nut screwed over the stud. Studs are used when the clamping pressures of a fine thread are needed and a bolt will not work. If the material the stud is being screwed into is soft (such as aluminum) or granular (such as cast iron), fine threads will not withstand a great amount of pulling force on the stud. Therefore, a coarse thread is used to secure the stud in the work piece and fine threaded nut is used to secure the other part to it. Doing this combines the clamping force of fine threads and the holding power of coarse threads.

Nuts are used with other threaded fasteners when a fastener is not threaded into a piece of work. Many different designs of nuts are found on today's cars. The most common one is the hex nut, which is used with studs and bolts and is tightened with a wrench.

Setscrews are used to prevent rotary motion between two parts, such as a pulley and shaft. Setscrews are either headless, requiring an Allen wrench or screwdriver to loosen and tighten them, or have a square head.

Figure 2-38 The basic dimensions of a bolt.

Bolt Identification

The **bolt head** is used to loosen and tighten the bolt; a socket or wrench fits over the head and is used to screw the bolt in or out. The size of the bolt head varies with the diameter of the bolt and is available in imperial and metric wrench sizes. Many confuse the size of the head with the size of the bolt. The size of a bolt is determined by the diameter of its shank. The size of the bolt head determines what size wrench is required.

Bolt diameter is the measurement across the major diameter of the threaded area or across the **bolt shank**. The length of a bolt is measured from the bottom surface of the head to the end of the threads.

The **thread pitch** of a bolt in the imperial system is determined by the number of threads that are in 1 inch of the threaded bolt length and is expressed in number of threads per inch. A UNF bolt with a-inch diameter would be a × 24 bolt. It would have 24 threads per inch. Likewise, a-inch UNC bolt would be called a × 16 bolt.

The distance, in millimeters, between two adjacent threads determines the thread pitch in the metric system. This distance will vary between 0.8 and 2.0 and depends on the diameter of the bolt. The lower the number, the closer the threads are placed and the finer the threads are.

The bolt's tensile strength, or *grade*, is the amount of stress or stretch it is able to withstand before it breaks. The type of material the bolt is made of and the diameter of the bolt determine its grade. In the imperial system, the tensile strength of a bolt is identified by the number of radial lines (**grade marks**) on the bolt's head. More lines mean higher tensile strength (**Table 2-1**). Count the number of lines and add two to determine the grade of a bolt.

A property class number on the bolt head identifies the grade of a metric bolt. This numerical identification comprises two numbers. The first number represents the tensile strength of the bolt. The higher the number, the greater the tensile strength. The second number represents the yield strength of the bolt. This number represents how much stress the bolt can take before it is not able to return to its original shape without damage. The second number represents a percentage rating. For example, a 10.9 bolt has a tensile strength of 1000 MPa (145,000 psi) and a yield strength of 900 MPa (90 percent of 1000). A 10.9 metric bolt is similar in strength to a SAE grade 8 bolt.

Nuts are graded to match their respective bolts, for example, a grade 8 nut must be used with a grade 8 bolt. If a grade 5 nut were used, a grade 5 connection would result. Grade 8 and critical applications require the use of fully hardened flat washers. These will not dish out when torqued as soft washers will.

> Always replace a bolt with the same size, thread pitch, and strength as the original bolt.

TABLE 2-1   STANDARD BOLT STRENGTH MARKINGS

Grade 2   Grade 5   Grade 8

Customary (inch) bolts—Identification marks correspond to bolt strength. Increasing numbers represent increasing strength.

4.6   4.8   5.8   8.8   9.8   10.9

Metric bolts—Identification class numbers correspond to bolt strength. Increasing numbers represent increasing strength.

Bolt heads can pop off because of **fillet** damage. The fillet is the smooth curve where the shank flows into the bolt head. Scratches in this area introduce stress to the bolt head, causing failure. Removing any burrs around the edges of holes can protect the bolt head. So can placing flat washers with their rounded, punched side against the bolt head and their sharp side against the work surface.

Fatigue breaks are the most common type of bolt failure. A bolt becomes fatigued from working back and forth when it is too loose. Under-tightening the bolt causes this problem. Bolts can also be broken or damaged by overtightening, being forced into a nonmatching thread, or bottoming out, which happens when the bolt is too long.

Tightening Bolts

Any fastener is nearly worthless if it is not as tight as it should be. When a bolt is properly tightened, it will be "spring loaded" against the part it is holding. This spring effect is caused by the stretch of the bolt when it is tightened. Normally a properly tightened bolt is stretched to 70 percent of its elastic limit. The elastic limit of a bolt is that point of stretch where the bolt will not return to its original shape when it is loosened. Not only will an overtightened or stretched bolt not have sufficient clamping force, but it will also have distorted threads. The stretched threads will make it more difficult to screw and unscrew the bolt or a nut on the bolt. Always check the service information to see if there is a torque specification for a bolt before tightening it. If there is, use a torque wrench and tighten the bolt properly.

BASIC GEAR ADJUSTMENTS

While a drivetrain is in operation, the gears, shafts, and bearings are subjected to loads and vibrations. Because of this, the drivetrain must normally be adjusted for the proper fit between parts. These adjustments require the use of precision measuring tools. Three basic adjustments are made when reassembling a unit or when a problem suggests that readjustment is necessary. Adjusting the clearance or play between two gears in mesh is referred to as *adjusting the backlash*. Endplay adjustments limit the amount of end-to-end movement of a gear shaft. Preload is an adjustment made to put a load on an assembly to offset the loads the assembly will face during operation.

Backlash in Gears

Backlash is the clearance between two gears in mesh (**Figure 2-39**). Excessive backlash can be caused by worn gear teeth, improper meshing of teeth, or bearings that do not support the gears properly. Excessive backlash can result in severe impact on the gear teeth from sudden stops or direction changes of the gears, which can cause broken gear teeth and gears. Insufficient backlash causes excessive overload wear on the gear teeth and can cause premature gear failure.

Backlash is measured with a dial indicator mounted so that its stem is in line with the rotation of the gear and perpendicular to the angle of the teeth. The gear is moved in both directions while the other gear it meshes with is held. The amount of movement on the dial indicator equals the amount of backlash present. The proper placement of shims on a gear shaft is the normal procedure for making backlash adjustments.

Endplay in Gears and Shafts

End clearance is often referred to as **endplay** and is the measurable axial (end-to-end) looseness of a bearing, assembly, or shaft.

Endplay refers to the measurable axial (end-to-end) looseness of a bearing. Endplay is always measured in an unloaded condition. To check endplay, a dial indicator is mounted against the side of a gear or the end of a shaft (**Figure 2-40**). The gear shaft is then pried in both directions and the readings noted. The difference between the two readings is the amount of endplay. Shims or adjusting nuts are widely used to adjust endplay.

Backlash

Figure 2-39 Backlash is the clearance between two gears in mesh.

Special tool

Dial indicator

Differential assembly

Move side gear up and down

Figure 2-40 This setup measures the endplay of a final drive gear assembly.

Choosing Selective Washers

A common procedure is the checking and correcting of subassembly end clearances. When these dimensions are correct, the transmission will have good shift characteristics and will wear normally. Look at the forward clutch assembly in **Figure 2-41**. The clearances in this clutch pack are critical and are controlled by the retaining snap ring. The specifications for this clearance are 0.046–0.068 inch (1.17–1.72 mm) and the snap rings are available in four different thicknesses: 0.060–0.064 inch (1.52–1.62 mm), 0.074–0.078 inch (1.87–1.98 mm), 0.086–0.106 inch (2.23–2.69 mm), and 0.102–0.106 inch (2.59–2.69 mm). To determine what size snap ring should be used, the end clearance of the pack needs to be measured.

Selective thrust washer

Sun gear driving shell

Output shell

Figure 2-41 A selective thrust washer is used between this front and rear multiple-friction disc assembly.

If the part number on a selective snap ring or washer can be read, its thickness can be determined by checking the number against specifications in the service information.

To do this, assemble the pack with the existing snap ring. Position a dial indicator on the clutch pack's pressure plate. Zero the dial indicator. Now push down on the clutch pack, release the pressure, and zero the indicator again.

Using the required special tool, lift up on the clutch pack until the pressure plate is fully seated against the retaining snap ring. Now read the dial indicator. The reading is the end clearance of the pack with the existing snap ring. If the measurement is not within specifications, a different thickness snap ring must be used. For example, if the end clearance measurement is 0.074 inch, a thicker snap ring must be installed. To determine how thick, remove the existing snap ring and measure its thickness. With a micrometer, the thickness was found to be 0.060 inch. By adding the measured clearance to the snap ring thickness, you know the clearance of the pack. In this case, that clearance is 0.134 inch. To determine what thickness snap ring is best to use, subtract the specification from this total clearance figure. Since the specification is a range, subtract the extremes of the range from the clearance ($0.134 - 0.046 = 0.08$ and $0.134 - 0.068 = 0.066$). What you find is that the correct snap ring is also within a range. Since the snap rings are available in four different sizes, the thickness that best meets the need is the one that should be used. In this case, a snap ring with a thickness of 0.074–0.078 inch is the best choice.

The endplay of or clearance between the shafts, clutches, and gears in some transmissions is set by the use of selective thrust washers or shims. The procedure for determining the correct size is the same as followed when selecting a selective snap ring.

Preloading of Geartrains

Preload is a fixed amount of pressure constantly applied to a component.

When normal operating loads are great, gear trains are often preloaded to reduce the deflection of parts. The amount of **preload** is specified in shop information and must be correct for the design of the bearings and the strength of the parts. If bearings are excessively preloaded, they will heat up and fail. When bearings are set too loose, the gears or shaft will wear rapidly due to the great amounts of deflection it will experience. Gear trains are preloaded by shims, thrust washers, adjusting nuts, or double-race bearings. Preload adjustments are normally checked by measuring turning effort or torque with a torque wrench (**Figure 2-42**).

Inch-pound
torque wrench

Extension

Special
tool

Figure 2-42 Turning torque is an indication of bearing preload and is checked with a torque wrench, normally a pound-inch wrench.

WORKING AS AN AUTO TECH

To be a successful automotive technician you need to have good training and a desire to succeed, and be committed to becoming a good technician and a good employee. A good employee works well with others and strives to make the business successful. The required training is not just in the automotive field. Good technicians need to have good reading, writing, and math skills. These skills will allow you to better understand and use the material found in service information and textbooks, and provide you with the basics for good communications with customers and others.

Repair Orders

A repair order (RO) is written for every vehicle brought into the shop for service. Repair orders may also be called service orders or work orders. They contain information about the customer, the vehicle, and the customer's concern or request; an estimate of the cost of the services; and when the services should be completed. Repair orders are legal documents that are used for many other purposes, such as payroll and general recordkeeping. Legally, an RO protects the shop and the customer. An RO is signed by the customer, who in doing so authorizes the service and accepts the terms noted on the RO. The customer, however, is protected against being charged more than the estimate given on the RO, unless he or she later authorizes a higher amount. Some states allow shops to be within 10 percent of the estimate, while others hold the shop to the amount that was estimated.

Today, most service facilities have shop management software in their computers. The information needed to complete an RO is input using the computer keyboard. The software package helps in the estimation of repair costs. The software also takes information from the RO and saves it in various files, each defined by its purpose.

Guidelines for Estimating Repair Costs. For legal reasons and to establish good customer relations, the projected repair costs must be calculated with as much accuracy as possible. To do this, do the following:

1. Make sure you have the correct information about the vehicle.
2. Always use the correct labor and parts guide or database for the specific vehicle.
3. Locate the exact service for that vehicle in the guide or database.
4. Using the guidelines provided in the guide or database, choose the proper time allocation listed for the service.
5. Multiply the allocated time by the shop's hourly flat rate.
6. Using the information given in the guide or database, identify the parts that will be replaced for that service.
7. Locate the cost of the parts in the guide or database or in the catalogs used by the shop.
8. Repeat the process for all other services required or requested by the customer.
9. Multiply the time allocations by the shop's hourly flat rate.
10. Add all of the labor costs together; this sum is the labor estimate for those services.
11. Add the cost of all the parts together; this sum is the estimate for the parts required for the services.
12. Add the total labor and parts costs together. If the shop charges a standard fee for shop supplies, add it to the labor and parts total. This sum is the cost estimate to present to the customer.

Compensation

Technicians are typically paid according to their abilities. Most often, new or apprentice technicians are paid by the hour. While being paid they are learning the trade and the business. Time is usually spent working with a master technician or doing low-skilled jobs.

As an apprentice learns more, he or she can earn more and take on more complex jobs. Once technicians have demonstrated a satisfactory level of skills, they can go on flat rate.

Flat rate is a pay system in which a technician is paid for the amount of work they do. Each job has a flat-rate time. Pay is based on that time, regardless of how long it took to complete the job. To explain how this system works, let us look at a technician who is paid $20.00 per flat rate. If a job has a flat-rate time of 3 hours, the technician will be paid $60.00 for the job, regardless of how long it took to complete it. Experienced technicians beat the flat-rate time nearly all of the time. Their weekly pay is based on the time "turned in," not on the time spent. If the technician turns in 60 hours of work in a 40-hour workweek, he or she actually earns $30.00 for each hour worked. However, if he or she turns in only 30 hours in the 40-hour week, the hourly pay is $15.00.

The flat-rate system favors good technicians who work in a shop that has a large volume of work. The use of flat-rate times allows for more accurate repair estimates to the customers. It also rewards skilled and productive technicians.

Employer–Employee Relationships

Being a good employee requires more than learning the skills for the job. When you begin a job, you enter into a business agreement with your employer. When you become an employee, you sell your time, skills, and efforts. In return, your employer pays you for these resources.

As part of the employment agreement, your employer also has certain responsibilities:

- *Instruction and supervision*—You should be told what is expected of you. A supervisor should observe your work, tell you if it is satisfactory, and offer ways to improve your performance.
- *A clean, safe place to work*—An employer should provide a clean and safe work area as well as a place for personal cleanup.
- *Wages*—You should know how much you are to be paid, what your pay will be based on, and when you will be paid before accepting a job.
- *Fringe benefits*—When you are hired, you should be told what benefits, such as paid vacations and employer contributions to health insurance and retirement plans, you can expect.
- *Opportunity*—This means you should be given a chance to succeed and possibly advance within the company.
- *Fair treatment*—All employees should be treated equally, without prejudice or favoritism.

On the other side of this business transaction, employees have responsibilities to their employers. Your obligations as an employee include the following:

- *Regular attendance*—A good employee is reliable. Businesses cannot operate successfully unless their workers are on the job.
- *Following directions*—As an employee, you are part of a team. Doing things your way may not serve the best interests of the company.
- *Responsibility*—You must be willing to answer for your behavior and work. You need to also realize that you are legally responsible for the work you do.
- *Productivity*—Remember that you are paid for your time as well as your skills and effort.
- *Loyalty*—Loyalty is expected. By being loyal, you will act in the best interests of your employer, both on and off the job.

Another responsibility you have as an employee is good customer relations. Learn to listen and communicate clearly. Be polite and organized, particularly when dealing with customers. Always be as honest as you possibly can.

Flat-rate manuals contain standards for the length of time a specific repair is supposed to require.

Look and present yourself as a professional, which is what automotive technicians are. Professionals are proud of what they do and they show it. Always dress and act appropriately and watch your language, even when you think no one is near.

Respect the vehicles on which you work. They are important to the lives of your customers. Always return a vehicle to the owner in a clean, undamaged condition. Remember, a car is the second largest expense a customer has. Treat it that way. It does not matter if you like the car. It belongs to the customer; treat it respectfully.

Explain the repair process to the customer in understandable terms. Whenever you are explaining something to a customer, make sure you do this in a simple way but without making the customer feel stupid. Always show the customers respect and be courteous to them. Not only is this the right thing to do, but it also leads to loyal customers.

ASE CERTIFICATION

The National Institute for Automotive Service Excellence (**ASE**) has established a voluntary certification program for automotive, heavy-duty truck, auto body repair, and engine machine shop technicians. In addition to these programs, ASE also offers individual testing in the areas of parts, alternate fuels, and advanced engine performance. This certification system combines voluntary testing with on-the-job experience to confirm that technicians have the skills needed to work on today's more complex vehicles. ASE recognizes two distinct levels of service capability—the automotive technician and the master automotive technician. The master automotive technician is certified by ASE in all eight major automotive systems. An automotive technician may have certification in only certain areas.

> Many **ASE** questions are in the *Technician A* and *Technician B* format, like the ones used in this manual.

To become ASE certified, a technician must pass one or more tests that stress diagnostic and repair problems. One of these areas is Automatic Transmissions and Transaxles. This test covers the diagnosis and repair of transmissions, transaxles, torque converters, and electronic control systems. You will find content on all of these topics in this manual, as well as the accompanying Classroom Manual.

After passing at least one exam and providing proof of two years of hands-on work experience, the technician becomes ASE certified. Retesting is necessary every five years to remain certified. A technician who passes one examination receives an automotive technician shoulder patch. The master automotive technician patch is awarded to technicians who pass all eight of the basic automotive certification exams (**Figure 2-43**).

> Passing the test and having the required work experience will enable you to be ASE certified in a particular area or areas.

You may receive credit for one of the two years by substituting relevant formal training in one, or a combination, of the following:

- High school training. Three years of training may be substituted for one year of experience.
- Post-high school training. Two years of post-high school training in a public or private trade school, technical institute, community or four-year college, or apprenticeship program may be counted as one year of work experience.
- Short courses. For shorter periods of post-high school training, you may substitute two months of training for one month of work experience.
- You may receive full credit for the experience requirement by satisfactorily completing a three- or four-year apprenticeship program.

Each certification test consists of 40–80 multiple-choice questions. The questions are written by a panel of technical service experts, including domestic and foreign vehicle manufacturers, repair and test equipment and parts manufacturers, working automotive technicians, and automotive instructors. All questions are pretested and quality-checked on a national sample of technicians before they are included in the actual test. Many test

Figure 2-43 My badge that says I am a certified Master Auto Technician.

questions force the student to choose between two distinct repair or diagnostic methods. The knowledge and skills needed to pass the tests includes:

- Basic technical knowledge—What is it? How does it work? This requires knowing what is in a system and how the system works. It also calls for knowing the procedures and precautions to be followed in making repairs and adjustments.
- Repair knowledge and skill—What is a likely source of a problem? How do you fix it? This requires you to understand and to apply generally accepted procedures and precautions for inspecting, disassembling, rebuilding, replacing, or adjusting components within a particular system.
- Testing and diagnostic knowledge and skill—How do you find what is wrong? How do you know you have corrected a problem? This requires that you be able to recognize that a problem does exist and to know what steps should be taken to identify the cause of the problem.

For further information on the ASE certification program, write to go to http://www.ase.com/Tests/ASE-Certification-Tests.aspx.

SUMMARY

- *Diagnosis* means finding the cause or causes of a problem. It requires a thorough understanding of the purpose and operation of the various automotive systems. Diagnostic charts found in service information can aid in diagnostics.
- Scan tools and vacuum and pressure gauges are common diagnostic tools for automatic transmission technicians.
- Torque wrenches are used to tighten fasteners to a specified torque.
- Always use the correct tool, in the correct way, for the job.

- Many special tools are available for special purposes. These tools are not normally part of a basic tool set but are purchased on an as-needed basis.
- Dial indicators are used to measure movement and are commonly used to measure the backlash and endplay of a set of gears.
- A micrometer can be used to measure the outside diameter of shafts and the inside diameter of holes. It may be calibrated in either imperial or metric graduations.

- The primary source of repair and specification information for any vehicle is the manufacturer's service information. Updates are published as service bulletins. These include changes made during the model year, which will not appear in the manual until the following year.
- Flat-rate manuals and software are ideal for making cost estimates. Published by manufacturers and independent companies, they contain figures showing how long specific repairs should take to complete, as well as a list of the necessary parts and their prices.
- *Gear backlash* reflects how tightly the teeth of two gears mesh.
- Whenever bolts are replaced, they should be replaced with exactly the same size, grade, and type as was installed by the manufacturer.

- Besides learning technical and mechanical skills, service technicians must learn to work as part of a team. As an employee, you will have certain responsibilities to your employer and your customers.
- Customer relations are an extremely important part of doing business. Professional, courteous treatment of customers and their vehicles is a must.
- The National Institute for Automotive Service Excellence (ASE) actively promotes professionalism within the industry. Its voluntary certification program for automotive technicians and master auto technicians helps guarantee a high level of quality service.

ASE-STYLE REVIEW QUESTIONS

1. While discussing automotive fasteners:

   Technician A says bolt sizes are listed by their appropriate wrench size.

   Technician B says whenever bolts are replaced, they should be replaced exactly the same size and type that was installed by the manufacturer.

   Who is correct?

   A. A only
   B. B only
   C. Both A and B
   D. Neither A nor B

2. While adjusting the end clearance of a clutch with a specified clearance of 0.044–0.051 inch and a measured clearance of 0.075 inch with the existing thrust washer thickness of 0.055 inch:

   Technician A uses a washer with a thickness between 0.079–0.086 inch to correct the end clearance.

   Technician B says the proper thickness for the washer is determined by adding the measured clearance to the thickness of the existing washer, then subtracting the specifications from that total.

   Who is correct?

   A. A only
   B. B only
   C. Both A and B
   D. Neither A nor B

3. *Technician A* says dial indicators are commonly used to measure the backlash of a set of gears.

   Technician B says dial indicators are commonly used to measure the endplay of a set of gears.

   Who is correct?

   A. A only
   B. B only
   C. Both A and B
   D. Neither A nor B

4. *Technician A* says pneumatic power tools require less maintenance than electric power tools.

   Technician B says cordless electrical tools are popular because they work without a cord or hose.

   Who is correct?

   A. A only
   B. B only
   C. Both A and B
   D. Neither A nor B

5. *Technician A* says that after an individual passes a particular ASE certification exam, he or she is certified in that test area.

   Technician B says all of the questions on an ASE certification exam are written as "Technician A and Technician B" questions.

   Who is correct?

   A. A only
   B. B only
   C. Both A and B
   D. Neither A nor B

6. While discussing the purpose of a torque wrench:

   Technician A says they are used to tighten fasteners to a specified torque.

   Technician B says they can be used to provide added leverage while loosening or tightening a bolt.

   Who is correct?

   A. A only   C. Both A and B
   B. B only   D. Neither A nor B

7. While discussing the purpose of micrometers:

   Technician A says micrometers are used to measure the diameter of an object.

   Technician B says outside micrometers are used to measure the outside diameter of an object, while inside micrometers are used to measure the inside diameter.

   Who is correct?

   A. A only   C. Both A and B
   B. B only   D. Neither A nor B

8. *Technician A* says an ohmmeter measures the resistance to current flow in a circuit.

   Technician B says the power to the circuit being measured must be off or the ohmmeter may be damaged.

   Who is correct?

   A. A only   C. Both A and B
   B. B only   D. Neither A nor B

9. *Technician A* looks at the ninth digit of the VIN to determine the exact year of the vehicle.

   Technician B says the information about the vehicle's engine can be found in the fourth through seventh characters of the VIN.

   Who is correct?

   A. A only   C. Both A and B
   B. B only   D. Neither A nor B

10. *Technician A* says finely threaded fasteners are used where greater resistance to vibration is desired.

    Technician B says coarse threads are used for general-purpose work, especially where rapid assembly and disassembly are required.

    Who is correct?

    A. A only   C. Both A and B
    B. B only   D. Neither A nor B

Name _____   Date _____

MEASURING TOOLS

Upon completion of this job sheet, you will be able to make measurements using micrometers, dial indicators, pressure gauges, and other measuring tools. You will be able to demonstrate the safe handling and use of appropriate tools, and demonstrate their proper use.

ASE Education Foundation Correlation ────────────────────────

This job sheet addresses the following **RST** tasks:

2.1   Identify tools and their usage in automotive applications.

2.2   Identify standard and metric designation.

2.3   Demonstrate safe handling and use of appropriate tools.

2.4   Demonstrate proper cleaning, storage, and maintenance of tools and equipment.

2.5   Demonstrate proper use of precision measuring tools
(i.e. micrometer, dial-indicator, dial-caliper).

We Support
ASE | Education Foundation

Procedure

The instructor should provide you with the following items to measure.

1. Item number one

   a. Obtain a measurement from an item with a micrometer.

   b. Describe the item measured _____

   c. Briefly describe the use of the tool _____

   d. Record the measurement _____ Standard or Metric

2. Item number two

   a. Obtain a measurement from an item with a dial indicator

   b. Describe the item measured _____

   c. Briefly describe the use of the tool _____

   d. Record the measurement _____ Standard or Metric

3. Item number three

   a. Obtain a measurement from an item with a pressure gauge

   b. Describe the item measured _____

   c. Briefly describe the use of the tool _____

   d. Record the measurement _____ Standard or Metric

4. Item number four

   a. Obtain a measurement from an item with a micrometer.

   b. Describe the item measured _____

   c. Briefly describe the use of the tool _____

   d. Record the measurement _____ Standard or Metric

5. Describe any special handling or safety precautions you need to follow while using the tools.

6. Describe any special cleaning that these tools might require.

Problems Encountered

_____

_____

_____

Instructor's Comments

_____

_____

_____

Name _____ Station _____ Date _____

PREPARING THE VEHICLE FOR SERVICE AND CUSTOMER

Upon completion of this job sheet, you will be able to prepare a service work order based on customer input, vehicle information, and service history. The student will also be able to describe the appropriate steps to take to protect the vehicle and delivering the vehicle to the customer after the repair.

ASE Education Foundation Correlation

This job sheet addresses the following **RST** tasks:

3.1   Identify information needed and the service requested on a repair order.

3.2   Identify purpose and demonstrate proper use of fender covers, mats.

3.3   Demonstrate use of the three C's (concern, cause, and correction).

3.4   Review vehicle service history.

3.5   Complete work order to include customer information, vehicle identifying information, customer concern, related service history, cause, and correction.

We Support

ASE | Education Foundation

Tools and Materials

- An assigned vehicle or the vehicle of your choice
- Service work order or computer-based shop management package
- Parts and labor guide

Work Order Source

Describe the system used to complete the work order. If a paper repair order is being used, describe the source.

_____

_____

Procedure   Task Completed

1. Describe the shop management software you will be using

   _____

   _____

2. Enter customer information, including name, address, and phone numbers onto the work order.   ☐

3. Locate and record the vehicle's VIN. Where did you find the VIN?

   _____

4. Enter the necessary vehicle information, including year, make, model, engine type and size, transmission type, license number, and odometer reading   ☐

5. Does the VIN verify that the information about the vehicle is correct?

   _____

6. Interview the customer to identify their concerns. Remember that the technician needs a good explanation to be able to duplicate the problem. (Role-play with another student if necessary.)

   Concern: _____

   _____

   _____

   _____

7. Prepare the vehicle for entering the service department. Add floor mats, seat covers, and steering wheel covers to the vehicle.   ☐

8. The history of service to the vehicle can often help diagnose problems as well as indicate possible premature part failure. Gathering this information from the customer can provide some of this information. For this job sheet assume the vehicle has not had a similar problem and was not recently involved in a collision. Service history is further obtained by searching files for previous service. Often this search is done by customer name, VIN, and license number. Check the files for any related service work.   ☐

9. Search for Technical Service Bulletins on this vehicle that may relate to the customer's concern. Did you find any? _____ If so, record the reference numbers here.

   _____

   _____

10. Based on the technician's findings, what is the cause of the concern?

    _____

    _____

11. Enter this information onto the work order.   ☐

12. Prepare to make a repair cost estimate for the customer. Identify all parts that will need to be replaced to correct the concern. List these here.

    _____

    _____

13. Describe the task(s) that will be necessary to replace the part.

    _____

    _____

14. Using the parts and labor guide, locate the cost of the parts that will be replaced and enter the cost of each item onto the work order at the appropriate place for creating an estimate. If the valve or cam cover is leaking, what part will need to be replaced?

    _____

15. Now, locate the flat-rate time for work required to correct the concern. List each task and its flat-rate time.

    _____

    _____

16. Multiply the time for each task by the shop's hourly rate and enter the cost of each item onto the work order at the appropriate place for creating an estimate. Ask your instructor what shop labor rate to use and record it here.

_____

17. Many shops have a standard amount they charge each customer for shop supplies and waste disposal. For this job sheet, use an amount of $10 for shop supplies.   ☐

18. Add the total costs and insert the sum as the subtotal of the estimate   ☐

19. Taxes must be included in the estimate. What is the sales tax rate and does it apply to both parts and labor, or just one of these?

_____

_____

20. Enter the appropriate amount of taxes to the estimate, and then add this to the subtotal. The end result is the estimate to give the customer.   ☐

21. By law, how accurate must your estimate be?

_____

_____

22. Generally speaking, the work order is complete and is ready for the customer's signature. However, some businesses require additional information; make sure you enter that information to the work order. On the work order there is a legal statement that defines what the customer is agreeing to. Briefly describe the contents of that statement.

_____

23. Now that the vehicle is completed and the vehicle is ready to be returned to the customer, what are the appropriate steps to take to deliver the vehicle to the customer? What would you do to make the delivery special? What should not happen when the vehicle is delivered to the customer?

_____

_____

Problems Encountered

_____

_____

_____

Instructor's Comments

_____

_____

_____

Name _____   **Date** _____

IDENTIFICATION OF SPECIAL TOOLS

Upon completion of this job sheet, you should be able to use the service information to determine what special tools are required to properly service a particular transmission.

ASE Education Foundation Correlation

This job sheet addresses the following **RST** task:

1.2   Utilize safe procedures for handling of tools and equipment.

This job sheet addresses the following **MLR** task:

A.1   Research vehicle service information including fluid type, vehicle service history, service precautions, and technical service bulletins. **(P-1)**

This job sheet addresses the following **AST/MAST** task:

A.2   Research vehicle service information including fluid type, vehicle service history, service precautions, and technical service bulletins. **(P-1)**

We Support
ASE | Education Foundation

Tools and Materials

• Appropriate service information

_____

Procedure

Your instructor will assign you a transmission type. You will use service information to identify and describe the special tools recommended for servicing this transmission. You will also identify any tools your shop has that can be used in place of the factory-specified tools.

1. The transmission assigned to you is:

    Transmission model _____ year _____

    Make _____ model _____

    VIN _____

2. The database you will use to find the information:

    _____

3. List the special tools referenced in the overhaul section of the service information for this transmission (include in your list what part or service each of the tools is used for).

    _____

    _____

    _____

    _____

    _____

    _____

    _____

    _____

4. List the special tools that are available for your use in the shop.

    _____

    _____

_____

_____

_____

_____

_____

5. List the available tools that can be used in place of the specified tools.

_____

_____

_____

_____

_____

_____

Instructor's Response

_____

_____

_____

Name _____   Date _____

GATHERING INFORMATION

Upon completion of this job sheet, you will be able to gather service information about a vehicle and its automatic transmission or transaxle.

ASE Education Foundation Correlation ⎯⎯⎯⎯⎯⎯⎯⎯⎯⎯⎯⎯⎯⎯⎯⎯⎯⎯⎯⎯

This job sheet addresses the following **MLR** tasks:

A.1   Research vehicle service information including fluid type, vehicle service history, service precautions, and technical service bulletins. (**P-1**)

This job sheet addresses the following **AST/MAST** tasks:

A.2   Research vehicle service information including fluid type, vehicle service history, service precautions, and technical service bulletins. (**P-1**)

We Support

ASE | Education Foundation

Tools and Materials

• Appropriate service information
• Computer

Describe the Vehicle Being Worked On

Year _____ Make _____ Model _____

VIN _____

Procedure

1. Using service information, briefly describe the automatic transmission in your assigned vehicle. How many forward gears are used? How many multiple-clutch disc units are used?

    _____

    _____

2. Does this transmission use bands and servos, or is it a clutch to clutch shifting transmission?

    _____

    _____

    _____

    _____

3. While looking in the engine compartment or under the vehicle, locate the identification tag on the transmission or transaxle. Describe where you found it.

    _____

    _____

    Summarize the information contained on this label.

    _____

    _____

    _____

4. Locate the information about the vehicle's automatic transmission. List the major components of the control system and describe the basic operation parameters.

_____

_____

_____

_____

_____

5. Using service information, locate and record all service precautions regarding the automatic transmission noted by the manufacturer.

_____

_____

_____

_____

6. Using the information that is available, locate and record the vehicle's service history. As an alternative, ask the owner about the vehicle's work history.

_____

_____

_____

_____

7. Using the information sources that are available, summarize all Technical Service Bulletins for this vehicle that relate to the automatic transmission and its control system.

_____

_____

_____

_____

Problems Encountered

_____

_____

_____

Instructor's Comments

_____

_____

_____

CHAPTER 3

DIAGNOSIS, MAINTENANCE, AND BASIC ADJUSTMENTS

Upon completion and review of this chapter, you should be able to:

- Listen to the driver's complaint, road test the vehicle, then determine the needed repairs.
- Diagnose unusual fluid usage, level, and condition problems.
- Replace automatic transmission fluid and filters.
- Diagnose noise and vibration problems.
- Inspect, replace, and align powertrain mounts.

- Perform oil pressure tests and determine needed repairs.
- Inspect and replace external seals and gaskets while the transmission is in the vehicle.
- Inspect, repair, and replace extension housing.
- Inspect, adjust, and replace external manual valve shift linkage, transmission range sensor/switch, and park/neutral position switch.

Terms To Know

Aeration

Electronic pressure control

Vehicle speed sensor

INTRODUCTION

Automatic transmissions are used in many rear-wheel-drive and four-wheel-drive vehicles. Automatic transaxles are most used on front-wheel-drive vehicles. The major components of a transaxle are the same as those in a transmission, except the transaxle assembly includes the final drive and differential gears. An automatic transmission/transaxle shifts automatically without the driver moving a gearshift or pressing a clutch pedal.

Because of the many similarities between a transmission and a transaxle, most of the diagnostic and service procedures are similar. Therefore, all references to a transmission apply equally to a transaxle unless otherwise noted. This rule also holds true for the questions on the certification test for automatic transmissions and transaxles.

Transmissions are strong and typically trouble-free units that require little maintenance. Maintenance normally includes fluid checks and scheduled linkage adjustments and oil changes.

All current automatic transmissions have electronic controls. These controls work with the hydraulic and mechanical systems of the transmission to provide reliable and efficient operation. Because of the variety of different electronic systems used by the manufacturers and the complexity of some of these systems, electrical and electronic transmission systems are covered in their own chapter (Chapter 4). The content of this chapter applies to transmissions with electronic controls.

Some of the procedures noted in this chapter are for both in-vehicle and out-of-vehicle service.

> ⚙️ **SERVICE TIP** Whenever you are diagnosing or repairing a transaxle or transmission, make sure you refer to the appropriate service information before you begin.

IDENTIFICATION

Prior to beginning diagnostics or any service on a transmission, its type must be properly identified. Proper identification ensures that the correct procedures will be followed, the correct replacement parts will be installed, and the correct information from Technical Service Bulletins can be retrieved. Correct identification of a transmission may not be possible just by looking at it or assuming that a certain transmission is always used in a particular model vehicle. Some transmission technicians rely on the shape of the oil pan to identify the transmission. This often is a worthwhile technique, but the shape may only indicate the basic model and not the exact design. Many transmissions have more than one model that looks identical from the outside or has only subtle differences which might cause problems if not properly identified.

The only positive way to identify the exact design and model of transmission is by locating its identification numbers. The transmission identification number (TIN) is typically listed on a sticker affixed to a late-model transmission (**Figure 3-1**). These stickers include a bar code and give information about the manufacturer of the transmission, the actual build date, the transmission part number, and the exact model of transmission.

Some examples of the location of and the information contained on the identification tags of common transmissions follow.

On Chrysler FWD vehicles, the transaxle can be identified by a barcode on top of the transaxle bellhousing. If the barcode is missing, there is a PK number stamp on the side of the transaxle (**Figure 3-2**). On RWD vehicles, the TIN is stamped on a pad on the left side of the transmission case's oil pan flange. Transmission operation requirements are different for each vehicle and engine combination. Some internal parts will differ between models. Always refer to the seven-digit part number for positive transmission identification when replacing parts.

Figure 3-1 Typical transmission identification label.

Figure 3-2 Chrysler (Transmission Identification Numbers) TIN locations.

Figure 3-3 A typical identification tag for a Ford Motor Company automatic transmission.

Late-model Ford RWD vehicles can be identified by identification tag numbers (**Figure 3-3**) located on the right side of the transmission case.

The transaxle in General Motors FWD vehicles can be identified by an identification number stamped on a plate attached to the rear face of the transaxle (**Figure 3-4**). The transmission/transaxle model is printed on the Service Parts Identification label located inside the vehicle or in the trunk.

The transmission in late-model RWD GM products can be identified by a letter code contained in the identification number. The ID number is stamped on the transmission case above the oil pan rail on the right rear side. The identification number contains information that must be used when ordering replacement parts.

The transaxle in many Hondas and Acuras can be identified by an identification number stamped on a white tag on the side of the transmission. The last four characters indicate the transmission model.

Aisin transmissions, which are used by Dodge, Hyundai, Ford, GM, and Lexus, can be identified by a plate attached to a side of the transmission case. The plate shows the transmission model number and serial number.

The transmission used in some models from Audi, Porsche, and Volkswagen can be identified by numbers cast into the top rear of the case. The transaxle model code is

Figure 3-4 Location of and information contained on a General Motors transaxle identification plate.

identified by figures stamped into the torque converter housing. These figures consist of the model code and build-date code. Some models have code letters and the date of manufacture stamped into the machined flat on the bellhousing rim. The valve body may also be stamped on its machined boss. The valve body identification tag is secured with valve body mounting screws. A torque converter code letter is stamped on the side of the attaching lug. Some of these model vehicles use a Mercedes transmission. These units have their model stamped on the right pan rail of the case.

Models that use Borg Warner transmissions have an identification number stamped on a plate attached to the torque converter housing near the throttle cable and distributor.

Jatco transmissions, which are manufactured by the Japan Automatic Transmission Company, are used by Chrysler, Infinity, Mitsubishi, Mazda, and Nissan. The model may be identified by a stamped metal plate attached to the right side of the transmission case. The model code appears on the second line and the serial number is on the bottom line.

Some Mazda, Mercedes-Benz, Toyota, and Subaru (Gunma) transmissions are best identified by the 11th character in the VIN, which is located at the top left of the instrument panel and on the transaxle flange on the exhaust side of the engine or the driver's door post. Most Mercedes-Benz transmissions have the identification number stamped into the pan rail on the right side of the case.

BMW, Peugeot, and some models of Volvo use ZF transmissions, which have an identification plate fixed to the left side of the transmission case. The lower left series of numbers on the plate indicate the number of gears, type of controls, type of gears, and torque capacity.

DIAGNOSTICS

The complexities involved in the operation of an automatic transmission (**Figure 3-5**) can make diagnosis quite complicated. This is especially true if a technician does not have a thorough understanding of the operation of a normally working transmission.

Figure 3-5 Basic components of a 6T70 transaxle.

Automatic transmission problems are usually caused by one or more of the following conditions:

- Poor engine performance
- Poor maintenance
- Problems in the hydraulic system
- Abuse resulting in overheating
- Mechanical malfunctions
- Electronic failures
- Improper adjustments

Diagnosis should begin by checking the condition of the fluid and its level, conducting a thorough visual inspection, and by checking the various control systems. On late-model vehicles, the initial inspection should include a check of the electronic control system. These control systems vary with the type of transmission and the manufacturer and require specific diagnostic and service procedures. Therefore, a separate chapter for these procedures is included in this book.

Poor engine performance can have a drastic effect on transmission operation. Low engine vacuum will cause a MAP sensor to sense a load condition when it actually is not present. This can cause delayed and harsh shifts. Delayed shifts can be caused by the engine running so badly that the throttle pedal must be pushed to the floor to keep it running or to get it going. This would tell the TCM that the driver was requesting full throttle and delay the shifting of the transmission.

Engine performance can also affect torque converter clutch operation. If the engine is running too poorly to maintain a constant speed, the converter clutch will engage and disengage at higher speeds. The customer complaint may be that the converter chatters, however the problem may be the result of engine misfires.

If the vehicle has an engine performance problem, the cause should be found and corrected before any conclusions on the transmission are made. Are there any DTCs set? Has there been any repair work done lately? Diagnosing a problem should follow a systematic procedure to eliminate every possible cause that can be corrected without removing the transmission. In order to properly diagnose a problem, you must totally understand the customer's concern or complaint. Make sure you know all you can about the conditions that exist when the problem occurs, such as the following:

- Cold or hot vehicle operating temperatures
- Cold or hot outside temperatures
- Loaded or unloaded vehicle
- City or highway driving
- Type of terrain
- Upshifting
- Downshifting
- Particular gear ranges
- Coasting
- Braking

CUSTOMER INTERVIEW

In order to properly diagnose a problem, you must totally understand the customer's concern or complaint. It is essential that you gather as much information from the customer as possible. The more details you have about the problem, the more accurate your diagnosis will be. Keep in mind that the problem may not be a transmission problem; it may be an engine or drivetrain problem. Ask the customer about any past repairs that may have been made, and check the service history to determine if there are any related problems. Gathering as much information as you can to describe the problem will help you decide if the problem is a transmission problem.

Get a good description of the problem from the customer. Find out when the problem was first noticed and if the problem is evident now. Sometimes taking a test drive with the customer and having them point out the noise or rough shift can help narrow down the problem area. A customer may have a driving style that contributes to a problem.

INSPECTION

Conducting a thorough inspection of the transmission and the rest of the vehicle is extremely important. Inspect the battery and all transmission-related wiring and connectors. Check the wheels and tires of the vehicle. Make sure they are the correct size for the vehicle; a change in tire diameter can affect transmission operation. Check the overall condition of the tires because excessive wear or damage can cause symptoms that the customer will blame on the transmission. Also, identify any and all non-factory-installed electrical accessories. Make sure these have been properly installed. If these are incorrectly connected to the system or connected into the control module's harness, the system will not operate properly.

Your inspection should also include a careful look at the transmission's fluid, the physical appearance of the transmission, the transmission's cooling system, and the engine and transmission mounts. All warning lamps in the instrument panel should be checked. If any of these remain on after the engine is started, the reason should be identified and corrected before you continue with your diagnosis. Lastly, use a scan tool to retrieve any fault codes that are held in the computer's memory.

FLUID CHECK

Transmissions without a Dipstick

Most late-model transmissions do not have a dipstick. The dipstick and filler tube were removed from these transmissions to prevent overfilling and contamination. Research has found that many transmission failures were caused by overfilling and/or using the wrong fluid. Without a dipstick, it is difficult to check the fluid level and condition. These transmissions have a vent/fill cap, typically located on the side of the transmission. Some also have a drain plug in the bottom of the pan; others have a plug in the pan that opens to an upright tube (a standpipe welded to the bottom of the pan and open at the top), and that is where the fluid level is checked. In addition, some transmissions are fitted with a fluid level sensor that will inform the driver when the fluid level is dangerously low.

To check the fluid level, the transmission must be warm and the vehicle must be on a level surface or raised on a lift. It is important that all four wheels are raised, because just lifting the front will put the vehicle on a slant and you will be unable to get an accurate reading. Find the fill plug and remove it. A small amount of fluid should leak out of the vent/fill opening if the fluid is at the correct level. If fluid does not come out, add fluid until it does. Always consult the repair manual for the location of the plug and position an oil drain pan under the area as a small amount of fluid may drip out. Also, this will usually be performed with the engine running so care must be exercised concerning hot exhaust parts. The procedure for checking a fluid level on a Chevrolet is covered in this chapter by **Photo Sequence 4**.

PHOTO SEQUENCE 4
Typical Procedure for Checking Transmission Fluid Level on Vehicles without a Dipstick

P4-1 Many modern vehicles are not equipped with a dipstick. These vehicles require a special check procedure to ensure proper fluid fill.

P4-2 To ensure the transmission is at the proper temperature, a scan tool is connected and set up for the vehicle.

P4-3 Ensure that the temperature is at 86°F–122°F before proceeding.

P4-4 With the engine idling and the brake pedal depressed, move the transmission lever through each gear, pausing for about 3 seconds at each selection. When this is finished, place the transmission back into PARK.

P4-5 Allow the engine to idle for 1 minute in PARK. Make sure temperature is still within the correct range.

P4-6 Raise the vehicle on the hoist.

P4-7 Remove the oil check plug carefully; fluid may run out rapidly if the unit is overfilled.

P4-8 Transmission fluid should drip from the hole as long as the temperature is correct. If fluid runs out in a stream the fluid level was too high and the level should be allowed to fall. If no fluid comes out, fluid should be added till the fluid begins to drip from the check hole.

Checking the Fluid When Equipped with a Dipstick

On most automobiles, the ATF level can be checked accurately only when the transmission is at operating temperature. The temperature of the transmission or transaxle is a vital factor when checking ATF levels. To check the ATF level, make sure the vehicle is on a level surface. Before removing the dipstick, wipe all dirt off the filler or dipstick tube and the dipstick handle. Fluid-level checking can be misleading because some transmissions have bimetallic elements that block fluid flow to the transmission cooler until a predetermined temperature is reached. The bimetallic element keeps the fluid inside the transmission hot. Also, if the vehicle has just been driven for a long trip or has been pulling a trailer, the fluid needs to cool before you can get an accurate reading.

Most manufacturers recommend that the engine should be running while you check the fluid level. Others recommend that the engine should not be running. Always refer to the vehicle's service information to identify the correct procedure. Also, make sure that the parking brake is engaged and take all necessary safety precautions while working under the hood.

When checking the fluid level, make sure the vehicle is on a level surface. If the transmission has a dipstick, wipe all dirt off the protective disc and the dipstick handle. Remove the dipstick and wipe it clean with a lint-free white cloth or paper towel. Reinsert the dipstick, remove it again, and note the reading. Markings on a dipstick indicate ADD levels, and on some models, FULL levels for cool, warm, or hot fluid (**Figures 3-6** and **3-7**). Always check the fluid level at least twice. If there are inconsistent readings, inspect the transmission's vent assembly to make sure it is clean and not clogged.

If the fluid level is low or off the crosshatch section of the dipstick, the problem could be external fluid leaks. Check the transmission case, oil pan, and cooler lines for evidence of leaks.

Low fluid levels may also be caused by poor maintenance. The level needs to be checked on a regular basis.

Low fluid levels can cause a variety of problems. Air can be drawn into the oil pump's inlet circuit and mixed with the fluid. This will result in aerated fluid, which causes slow pressure buildup, and low pressures, which cause slippage between shifts. Air in the pressure-regulator valve will cause a buzzing noise when the valve tries to regulate pump pressure. A low fluid level can also cause delayed shifting and slipping, which leads to overheating and accelerated clutch and band wear.

Classroom Manual
Chapter 3, page 90

If the dipstick has a "hot" level marked on it, this mark should be used only when the dipstick is hot to the touch.

Note: Do not overfill. It takes only one pint to raise level from "Add" to "Full" with a hot transmission.

Figure 3-6 Typical dipstick markings for an automatic transmission.

Note: Cold reading is above Full mark.

Figure 3-7 Typical dipstick markings for an automatic transaxle.

Excessively high fluid levels can also cause **aeration**. As the planetary gears rotate in high fluid levels, air can be forced into the fluid. Aerated fluid can foam, overheat, and oxidize. All of these problems can interfere with normal valve, clutch, and servo operation. Foaming may be evident by fluid leakage from the transmission's vent.

AUTHOR'S NOTE Air in any hydraulic system is bad news. You have probably experienced a brake hydraulic system with excessive air, a very low pedal, and very poor stopping ability. Imagine how the clutches apply with aerated fluid. Be very careful about adding fluid to a cold transmission to prevent overfilling. It can be surprising how much the fluid expands as it heats up.

CUSTOMER CARE Customers should be made aware that fluid level and condition should be checked at least every six months. Temperature fluctuations from summer to winter can cause a thermal breakdown of ATF. Even high-quality fluids can experience breakdown as a result of these frequent and extreme temperature changes. It is said that nearly 90 percent of all transmission failures are caused by fluid breakdown or oxidation.

 SERVICE TIP If the vehicle has a dipstick, some manufacturers recommend that a cold check of the fluid be done to make sure the transmission has enough fluid to operate safely until a hot check can be made. The cold check should be followed by a hot check as soon as possible. Use the cold check to check fluid level when the fluid temperature is between 80°F and 90°F (27°C and 32°C). The hot check is the most accurate way to check the fluid level.

The condition of the fluid should be checked while checking the fluid level. Examine the fluid carefully. The normal color of ATF is pink or red. If the fluid has a dark brownish or blackish color or a burnt odor, the fluid has been overheated. A milky color indicates that water has mixed with the fluid, possibly from engine coolant leaking into the transmission's cooler in the radiator. If there is any question about the condition of the fluid, drain out a sample for closer inspection.

If there is evidence that there is water or moisture in the fluid, the transmission must be completely disassembled and the following parts should be replaced:

- The torque converter
- All internal and external seals
- All transmission fluid filters
- All clutches and bands that have friction material
- All solenoids

Before the transmission is reinstalled, the transmission's fluid cooler(s) and its hoses and tubes must be flushed and cleaned.

After checking the ATF level and color, wipe the dipstick on absorbent white paper and look at the stain left by the fluid. Dark particles are normally band or clutch material,

⚠ Caution
If the fluid level is so low that it doesn't appear on the dipstick or is at the very bottom of the dipstick, the vehicle should not be driven until the level is brought up to normal. Driving with a very low level can cause internal transmission damage.

Figure 3-8 This oil pan has two magnets to collect metal shavings that may be suspended in the fluid.

Many transmission manufacturers and remanufacturers will not honor a warranty claim if the cooler has not been flushed or replaced. The cooler and lines can hold a substantial amount of contaminated fluid from a failed transmission.

Aeration is the process of mixing air with a liquid.

Oxidation occurs when something is mixed with oxygen to produce an oxygen-containing compound. Oxidation is also the term given to the chemical breakdown of a substance or compound caused by its combination with oxygen.

while silvery metal particles are normally caused by the wearing of the transmission's metal parts. If the dipstick cannot be wiped clean, it is probably covered with varnish, which results from fluid oxidation. Varnish will cause the spool valves to stick, leading to improper shifting speeds. Varnish or other heavy deposits indicate the need to change the transmission's fluid and filter.

Sometimes it is difficult to accurately define the condition of the fluid. At these times, it is wise to examine the fluid and the residue in the oil pan. In most cases, this should be done after a road test. Typically, the oil pan will be the collecting point for anything suspended in the fluid. It is quite common to find some residue in the pan, even with a fresh transmission. However, an examination of the residue will help decide whether the transmission has an internal problem or not. Also, many transmission pans have magnets (**Figure 3-8**) in them to collect metal shavings that may be floating in the fluid. Check the magnets because excessive metal particles indicate internal damage, but it should be noted that even these magnets will normally have at least some accumulated material from normal wear. The permanent magnet-type input and output speed sensors can also be removed and inspected. If metal particles have gathered on the sensors, an internal problem may be indicated. If the fluid is heavily contaminated, the transmission must be disassembled and all parts, including the transmission cooler, must be thoroughly cleaned or replaced. If the transmission fluid is contaminated with debris, the torque converter should be replaced. If the fluid is clean and there is only a small amount of residue in the fluid, the transmission problem is most likely caused by something external to the transmission.

If the fluid level is low and shows no signs of contamination, the transmission fluid level can be brought up to the desired level. However, if the transmission has a problem that is yet to be diagnosed, it is best to top off the fluid after road testing the vehicle. In general, it is best not to make any repairs or adjustments before the road test. Doing so may alter the symptoms reported by the customer.

If fluid needs to be added, add it in ½-pint (0.25-L) increments. Do not overfill the transmission fluid. Pour the fluid in slowly and carefully to avoid spilling it on the transmission and/or engine. Clean up any spills immediately; ATF can damage components in the engine compartment.

Final Drives. Some transaxles have a separate oil reservoir for the final drive assemblies. The fluid level in these units is done with a separate check. Normally, the fluid level is checked by removing the filler plug and observing the level of the fluid. The fluid should be level with the hole for the filler plug. Often, there is an additional plug at the bottom of the unit to allow for draining of the fluid.

> **⚙ SERVICE TIP** Synthetic ATF is normally a darker red than petroleum-based fluid. Synthetic fluids tend to look and smell burnt after normal use; therefore, the appearance and smell of these fluids is not a good indicator of the fluid's condition.

Transmission Fluids

Most ATF is dyed red so that it is not easily confused with engine oil. The exact fluid that should be used in an automatic transmission depends on the transmission design and the year the transmission was built. It is very important that the correct type of ATF be used. Always refer to the service or owner's manual for the correct type of fluid to use. Some transmission dipsticks are also marked with the type of ATF required.

Although there are many types of ATF available, the following are the most common:

- **ATF+4 (Chrysler Specification MS-7176E)**—This is a fluid formulated for Chrysler automatic transmissions where ATF+, ATF+2, or ATF+3 is recommended.
- **Dexron® VI/Mercon®**—This fluid is sometimes referred to as multi-purpose ATF because it is recommended for all GM and some Ford automatic transmissions (since 1983) requiring Dexron or Mercon transmission fluids. It is also suitable for most Mercedes-Benz passenger car automatic transmissions.
- **Matic S, K, D** JATCO transmissions: Subaru and Nissan.

> **⚙ SERVICE TIP** Contaminated fluid can sometimes be felt better than seen. Place a few drops of fluid between two fingers and rub them together. If the fluid feels dirty or gritty, it is contaminated with burned frictional material.

Some transmissions require the use of fluids not mentioned here. CVTs require a fluid that is much different than that used in automatic transmissions. Always use the fluid recommended by the manufacturer. The use of the wrong fluid may cause the transmission to operate improperly and/or damage the transmission.

Fluid Changes

> **CUSTOMER CARE** Abusive driving can overheat a transmission and cause fluid oxidation and breakdown. Inform your customers that they should always stay within the recommended towing load for their vehicle. Also tell them to avoid excessive rocking back and forth when they are stuck in snow or mud.

The transmission's fluid and filter should be changed whenever there is an indication of oxidation or contamination. **Photo Sequence 5** shows the correct procedure for performing an automatic transmission fluid and filter change. Periodic fluid and filter changes are also part of the preventive program for most vehicles. The frequency of this

 Caution

Make sure you always install the correct fluid in a transmission. The use of the wrong fluid can cause poor shifting or torque converter shudder and may cause damage to the transmission. It is important to note that often manufacturers recommend a special fluid for their transmissions. This is especially true of CVTs.

PHOTO SEQUENCE 5
Typical Procedure for Performing an Automatic Transmission Fluid and Filter Change

P5-1 Raise the vehicle to a good working height, safely positioned on the lift.

P5-2 Place the large-diameter oil drain pan under the transmission pan.

P5-3 Loosen all of the pan bolts except three at one end. This will allow some fluid to drain out.

P5-4 Support the pan with one hand and remove remaining bolts to remove the pan. Pour the fluid in the pan into the drain pan.

P5-5 Inspect the residue in the pan for indications of transmission problems. Then remove the old pan gasket and wipe the pan clean with a lint-free rag.

P5-6 Remove the filter from the transmission.

P5-7 Compare the replacement gasket and filter with the old ones to make sure the replacements are the right ones for this application.

P5-8 Install the new filter. Then lay the new gasket over the sealing surface of the pan.

P5-9 Install the pan onto the transmission. Install and tighten the bolts to specifications. Then lower the vehicle and pour new fluid into the transmission. Run the engine to circulate the new fluid, then turn it off and raise the vehicle. Check for fluid leaks.

service depends on the conditions that the transmission normally operates under. Severe usage requires that the fluid and filter be changed every 15,000 miles. Severe usage is defined as:

1. More than 50 percent operation in heavy city traffic during hot weather (above 90°F).
2. Police, taxi, commercial-type operation, or trailer towing.

The mileage interval that a manufacturer will recommend for a fluid and filter change will also depend on the type of transmission. For example, some General Motors transmissions use aluminum valves in their valve body. Since aluminum is softer than steel, aluminum valves are less tolerant of abrasives and dirt in the fluid. Therefore, to prolong the life of the valves, more frequent fluid changes are recommended.

CUSTOMER CARE Transmissions that operate at 175°F can go about 100,000 miles before the fluid starts oxidizing. However, fluid life is halved with every 20°F its temperature rises above 175°F. For example, fluid life expectancy drops to 50,000 miles at 195°F and just 25,000 miles at 215°F. If the vehicle is used for towing, ambient temperatures above 90°F or in heavy traffic the transmission fluid will not last nearly as long as the factory recommended intervals.

WARNING Be careful when draining the transmission fluid. It can be very hot and will tend to adhere to your skin, causing severe burns.

Change the fluid only when the engine and transmission are at normal operating temperatures. On most transmissions, you must remove the oil pan to drain the fluid. Some transmission pans on recent vehicles include a drain plug. A filter or screen is normally attached to the bottom of the valve body. Filters are made of paper or fabric and are held in place by screws, clips, or bolts (**Figure 3-9**). Filters should be replaced, not cleaned.

Beware of one-type-fits-all fluids. Transmission fluid is manufactured to very exacting tolerances. Transmission fluid has different frictional characteristics. The wrong fluid may damage the transmission clutch material. Make sure that you are using the recommended fluid for the vehicle.

Special Tools
Large drain pan
Clean white rags
Inch-pound torque wrench

Figure 3-9 Transmission fluid filters are attached to the transmission case by screws, bolts, retaining clips, and/or by the pickup tube.

Classroom Manual
Chapter 3, page 90

To drain and refill a typical transmission, the vehicle must be raised on a hoist. After the vehicle is safely in position, place a drain pan with a large opening under the transmission's oil pan. Then loosen the pan bolts and tap the pan at one corner to break it loose. Fluid will begin to drain from around the pan. When all fluid is drained, remove the oil pan.

After draining the fluid, carefully remove the pan. There will be some fluid left in the pan. Be prepared to dump it into the drain pan. Check the bottom of the pan for deposits. A small amount of blackish deposits from clutches and bands is normal. However, an excess amount along with a burnt smell indicates burned frictional linings. If the pan has goldish brown coating, the old fluid is oxidized. If there are many metal particles in the fluid or coating the pan, this is evidence of worn gears, washers, or bushings. Normally, the metal is gold looking, it is brass or bronze. These particles come from worn bushings or thrust washers. If the particles are steel or iron, they are from the gearsets, springs, or needle bearings. Plastic debris can be from a spring retainer or thrust washer.

Any abnormal contamination or debris means the transmission is worn and should be overhauled. If the residue appears normal, clean the oil pan and its magnet.

> ⚙ **SERVICE TIP** There are two filters in Chrysler's 68RFE. The primary filter is attached to the front pump and submerged in fluid at the bottom of the pan. The second filter is a spin-on type and is attached to the case next to the valve body. Both filters can be accessed by removing the pan.

Filters. Normally, the filter should be changed every 60,000–100,000 miles. The filters are coarse, meaning they can pass small debris through them. These particles can damage the transmission. It is best to replace the filter every time the fluid is changed or after the system has been flushed. If the old filter is kept in place during flushing, it may be able to trap some of the contaminants that were loosened during flushing. It is important to note that some transmissions do not have a filter, rather a screen is used to stop debris from circulating through the transmission.

Remove the filter and cut it open, and inspect it. Use a magnet to determine if metal particles are steel or aluminum. Steel particles indicate severe internal transmission wear or damage. If the metal particles are aluminum, they may be part of the torque-converter stator. Some torque converters use phenolic plastic stators, therefore metal particles found in these transmissions must be from the transmission itself. Filters are always replaced, whereas screens are cleaned. Screens are removed in the same way as filters. Clean the screen with fresh solvent and a stiff brush.

Some transmissions have a screw-on filter (**Figure 3-10**). These are similar to a conventional engine oil filter. Make sure the replacement filter is tightened properly and that the sealing gasket from the old filter is not stuck to the transmission case. Also apply some clean ATF to the seal before tightening the filter.

Other transmissions have a filter plate installed above the internal control module/solenoid assembly (**Figure 3-11**). Some transmissions also have an inline filter connected in the fluid line going to the fluid cooler (**Figure 3-12**).

Refilling the Assembly. Remove any traces of the old pan gasket on the case housing. Then, install a new filter and gasket on the bottom of the valve body and tighten the retaining bolts to the specified torque. If the filter is sealed with an O-ring, make sure it is properly installed.

Some transmissions have a reusable pan gasket (provided it is not damaged). These gaskets are usually rubber with a steel interior reinforcement. Check service information to be certain.

Figure 3-10 An external oil filter mounted to the bellhousing.

Bell housing

Fitting

Magnet

External filter

Figure 3-11 This control module/solenoid assembly is mounted to the side of the transmission and has a separate filter.

Filter

Control module/solenoid assembly

Figure 3-12 An inline filter for a fluid cooler.

Remove any traces of the old pan gasket from the oil pan and transmission housing. (Note that some pan gaskets are reusable. Check the service information to make sure.) Make sure the mounting flange of the oil pan is not distorted and bent. Oil pans are typically made of stamped steel. The thin steel tends to become distorted around the attaching bolt holes. These distortions can prevent the pan from fitting tightly against the transmission case. To flatten the pan, place the mounting flange of the pan on a block of wood and flatten one area at a time with a ball-peen hammer.

Verify that the new gasket is correct by comparing the old with the new. Then, re-install the pan using the gasket or sealant recommended by the manufacturer. Tighten the pan retaining bolts to the specified torque (**Figure 3-13**); often this specification is given in inch-pounds rather than foot-pounds.

Pan bolts—tighten to 20 lb.-ft.
(27 Nm)

Oil pan

Gasket

Figure 3-13 Always tighten the pan bolts to the correct torque specification.

Transmission gaskets should not be installed with any type of liquid adhesive or sealant, unless specifically noted by the manufacturer. If any sealer gets into the valve body, severe damage can result. Also, sealant can clog the oil filter. If a gasket is difficult to install, a thin coating of transmission assembly lube or petroleum jelly can be used to hold the gasket in place while assembling the parts.

One type of gasket that presents unique installation problems is the cork-type gasket. These gaskets tend to change shape and size with changes in humidity. If a cork gasket is slightly larger than it should be, soak it in water and lay it flat on a warm surface. Allow it to dry before installing it. If the gasket is slightly smaller than required, soak it in warm water prior to installation.

Whenever you are using a cork gasket to create a seal between two parts, make sure you properly tighten the two surfaces together. Tighten the attaching bolts or nuts, in a staggered pattern, to the specified torque so that the gasket material is evenly squeezed between the two surfaces. If too much torque is applied, the gasket may split. Never apply silicone RTV or another sealer to a cork gasket because the gasket will be squeezed out around all the bolt holes, quickly causing splitting and leakage.

With the pan securely in place, lower the vehicle. Now, add a little less than the required amount of fluid into the transmission. Always use the recommended type of ATF. The wrong fluid will alter the shifting characteristics of the transmission.

Start the engine and allow it to idle for at least one minute. Then, with the parking and service brakes applied, move the gear selector lever momentarily to each position, ending in park. Recheck the fluid level and add a sufficient amount of fluid to bring the level to about one-eighth of an inch below the "add" mark.

Run the engine until it reaches normal operating temperature. Then recheck the fluid level. It should be in the "hot" region on the dipstick. Make sure the dipstick is fully seated into the dipstick tube opening. This will prevent dirt from entering the transmission.

Using a Fluid Evacuator. The use of a fluid evacuator (**Figure 3-14**) makes it unnecessary to loosen and remove the oil pan. This makes the job simpler and certainly less messy. The evacuator draws the old fluid out of the transmission through the filler tube. Fluid evacuators create a vacuum in their reservoir. This low pressure is either created with an

⚠ **Caution**

Make sure you carefully check the specifications. The required torque is often given in pounds-inch. You can very easily break the bolts or damage the transmission case if you tighten the bolts to pounds-feet.

Some transaxles vent through the dipstick handle or tube.

Figure 3-14 An automatic transmission fluid exchanger.

air compressor or with a hand pump. The fluid in the transmission is at atmospheric pressure, therefore when it is exposed to the low pressure, it will move toward it. To use a typical evacuator tool, perform the following steps:

1. Bring the ATF's temperature up to its normal operating temperature.
2. Turn on the evacuator to create a vacuum in its reservoir.
3. After a suitable vacuum is present, insert the tool's tube into the filler tube as far as it can go.
4. Open the valve on the evacuator and collect the fluid in a container.
5. Properly dispose of the old fluid.
6. Remove the tool's tube from the filler tube.
7. Refill the transmission with new ATF.

Case porosity is caused by tiny holes that are formed by trapped air bubbles during the casting process.

Flushing. Draining the transmission does not remove all of the old fluid for the unit. Much fluid remains inside the torque converter, the transmission cooler, and the cooler lines. It is said that a drain and refill service replaces only 40 percent of the old contaminated ATF. In fact, when you drain the fluid through a drain plug, you end up draining only 20 percent of the fluid. In both cases, the new fluid mixes with the old and reduces the advantages of having new fluid.

⚙️ SERVICE TIP Many technicians recommend that the transmission not be flushed if the old fluid smells or looks burnt or if it has flakes in it. These indicate a need to overhaul the transmission.

To remove most of the old fluid and fill the system with fresh lubricant, the transmission should be flushed or the fluid totally exchanged. Normally, flushing also involves the use of a chemical that will dissolve varnish and other contaminants.

Flushing is done with a flushing tool. Some flushing tools have a pressure gauge that can be used to detect a restriction in the hydraulic circuits, filter, cooler, or lines and hoses. If the pressure reading is less than 8 psi, there is a restriction in the fluid path.

There are two basic types of flushing tools: a cooler line flush machine and a pump inlet flush machine. The cooler line flush machine connects into the line from the transmission to the transmission cooler. This type of machine may also be called a fluid exchange unit. The outlet from the machine is connected to the cooler. This means the fluid moves from the transmission into the machine and then to the cooler. The machine has a chamber separated by a diaphragm. The area above the diaphragm is filled with fresh fluid. With the engine running, fluid flows from the transmission into the lower part of the chamber. As the old fluid fills its portion within the chamber, it pushes up on the diaphragm and forces new fluid into the cooler. This continues until most of the old fluid is replaced by new fluid.

The pump inlet flush machine connects to the intake of the oil pump. The oil pan and filter are removed and a drain pan is set under the transmission. The machine provides fresh fluid to the pump inlet and as the fluid passes through the transmission it leaks from the bottom of the transmission into the drain pan. The new fluid has little chance of mixing with the old and the system is filled with fresh oil. This machine wastes a bit of fluid, however. It will cycle through about 20 quarts of fluid to flush out 15 quarts of old fluid. After the system is flushed, the filter and pan are reinstalled and the unit is filled with fluid.

FLUID LEAKS

Continue your diagnostics by conducting a quick and careful visual inspection. Check all drivetrain parts for looseness and leaks. If the transmission fluid was low or if there was no fluid, raise the vehicle and carefully inspect the transmission for signs of leakage. Leaks are often caused by defective gaskets or seals. Common sources of leaks are the oil pan seal, rear cover, final drive cover (on transaxles), extension housing, and electrical switches mounted into the housing (**Figure 3-15**). The housing itself may have a porosity problem, allowing fluid to seep through the metal. Case porosity may be repaired using an epoxy-type sealer.

Figure 3-15 Possible sources of fluid leaks on this transaxle.

During your inspection, check the area around the filler tube (if equipped) for signs of leakage. This type of leak is often corrected by replacing the seal or grommet between the tube and the housing. Also, check the fluid lines and hoses at the transmission and the oil cooler. At times, these leaks can be corrected by tightening the fittings. However, do not overtighten them; this can cause damage to the fittings or housing. If the fittings are not loose, the problem is usually corrected by installing new seals at the connections.

Sometimes the source of the leak is difficult to find. A careful cleaning of the transmission and the area surrounding it may help. After the transmission is cleaned, run the engine and with the brake pedal depressed. Move the gear selector through its gear ranges; this circulates the fluid throughout the transmission. Turn off the engine and inspect the transmission. The source of the leak may be now apparent.

If the leak is still not found, a fluorescent dye can be added to the fluid. With these products, the leak will be evident when an ultraviolet light passes by the source of the leak. The dye will also help to determine if ATF is leaking into the engine's cooling system. The following procedure is the typical way to conduct this type of leak check. Always follow the instructions of the manufacturer of the dye.

1. Add the recommended amount of dye to the transmission fluid.
2. Raise the vehicle and clean off all traces of oil from the top and bottom of the torque converter housing, the transmission case, the rear of the engine, and the engine's oil pan. Use compressed air to dry all the surfaces after they have been cleaned.
3. Lower the vehicle. Then start the engine and allow it to run until it reaches normal operating temperature.
4. Raise the vehicle with the engine still running and carefully use the ultraviolet light to find evidence of a leak.
5. Continue to run the engine until the probable source of leakage can be determined. Sometimes this process is sped up by running the engine at fast idle and occasionally shifting between drive and reverse.
6. When the leak is discovered, lower the vehicle and turn off the engine.

Oil Pan

A common cause of fluid leakage is the seal of the oil pan to the transmission housing. If there are signs of leakage around the rim of the pan, re-torquing the pan bolts may correct the problem. If tightening the pan does not correct the problem, the pan must be removed and a new gasket installed (refer back to Figure 3-13). Make sure the sealing surface of the pan's rim is flat and capable of providing a seal before reinstalling it.

Torque Converter

Torque converter problems can be caused by a leaking **converter** (**Figure 3-16**). This type of problem may be the cause of customer complaints of slippage and a lack of power. To check the converter for leaks, remove the converter access cover and examine the area around the torque converter shell. An engine oil leak may be falsely diagnosed as a converter leak. The color of engine oil is different from transmission fluid and may help identify the true source of the leak. However, if the oil or fluid has absorbed much dirt, both will look the same. An engine leak typically leaves an oil film on the front of the converter shell, whereas a converter leak will cause the entire shell to be wet. If the transmission's oil pump seal is leaking, only the back side of the shell will be wet. If the converter is leaking or damaged, it should be replaced.

If the torque converter hub seal is leaking, it should be replaced. To do this, the transmission must be removed. Normally, a slide hammer and a seal removal and installation tool designed for the transmission are required to replace the seal.

Classroom Manual
Chapter 3, page 91

Front pump
seal leak

Converter or front
pump gasket leak

Crankshaft
seal leak

Front pump-to-case
bolt leak

Converter drain plug
or converter-to-flywheel
stud leak

Figure 3-16 By determining the direction of fluid travel, the cause of a fluid leak around the torque converter can be identified.

Once the transmission has been removed, the torque converter is pulled out of the unit. Then the seal is pulled from the oil pump. Be careful not to damage the area where the seal is seated. When the converter is removed, make sure that the converter hub is not grooved or damaged, as that would damage the new seal. Additionally, when the seal is removed inspect the condition of the bushing in the front pump. If the bushing is damaged, the new seal will leak.

Position the new hub seal and drive it into position with the seal installation tool. Then install the torque converter. Make sure the converter is fully seated into the transmission by turning the converter and pushing the converter into place. The transmission must be fully engaged onto the stator support and front pump before the transmission is installed or damage to the unit will result. Finally, install the transmission into the vehicle.

Pump Seal and Gasket. If the fluid leak appears to be at the center of the torque converter, the oil pump seal or gasket may be leaking. If this is the case, the transmission must be removed to install new seals and gaskets.

Differential/Axle Seals. If oil is leaking from around where the axle shaft enters a transaxle, the seal is probably worn or damaged. Before proceeding to replace the seal, check the service information to see if the final drive unit has a separate oil reservoir from the transaxle. If it does, identify the type of fluid that must be used in the final drive.

To replace the seal, remove the axle shaft according to the procedure outlined in the service information. Once the shaft has been removed, use the proper tool to remove the seal, often a slide hammer–type puller is used (**Figure 3-17**).

To install a new seal, position the new seal on the installation tool designed for the transaxle. Then install the seal. Make sure it is squarely and securely positioned in its bore. Apply a light coat of lubricant on the end of the axle shaft and insert it into the transaxle housing.

⚠ **Caution**

Be careful not to damage the housing while removing the seal.

Figure 3-17 Using a slide hammer–type puller to remove an axle seal from the transaxle case.

Then, check the fluid level and add fluid as necessary. Never overfill the unit as that can cause damage to the transaxle.

Extension Housing

An oil leak stemming from the mating surfaces of the extension housing and the transmission case may be caused by loose bolts. To correct this problem, tighten the bolts to the specified torque. Also check for signs of leakage at the rear of the extension housing. Fluid leaks from the seal of the extension housing can be corrected with the transmission in the car. Often, the cause for the leakage is a worn extension housing bushing, which supports the sliding yoke of the driveshaft. Normally, bushing wear or damage is caused by faulty universal joints. When the driveshaft is installed, the clearance between the sliding yoke and the bushing should be minimal. If the clearance is satisfactory, a new oil seal will correct the leak. If the clearance is excessive, the repair requires that a new seal and a new bushing be installed. If the seal is faulty, the transmission vent should be checked for blockage.

Rear Oil Seal and Bushing Replacement

Procedures for the replacement of the rear oil seal and bushing vary little with each car model. General procedures for the replacement of the oil seals and bushings follow.

To replace the rear seal:

1. Remove the driveshaft.
2. Remove the old seal from the extension housing (**Figure 3-18**).
3. Lubricate the lip of the seal, then install the new seal in the extension housing.
4. Install the driveshaft.

To replace the rear bushing and seal:

1. Remove the driveshaft from the car. Inspect the yoke for wear and replace it if necessary.
2. Insert the appropriate puller tool into the extension housing until it grips the front side of the bushing.
3. Pull the seal and bushing from the housing.
4. Using the correct tools, drive a new bushing into the extension housing.
5. Install a new seal in the housing, and lubricate the inner rubber sealing surface with transmission assembly lube or the equivalent. Never install a seal dry (**Figure 3-19**).
6. Install the driveshaft.

Classroom Manual
Chapter 3, page 91

Special Tools

Slide hammer
Seal remover tool
Bushing removing tool
Seal driver
Bushing driver

Figure 3-18 Extension housing rear seals are removed with a slide hammer and a special removal tool.

Figure 3-19 Rear seals are installed into the extension housing with the proper driver.

Speed Sensors

Speed sensors can be the source of an oil leak. Typically, a transmission has two speed sensors fitted into the housing (**Figure 3-20**): the input shaft speed sensor (ISS; sometimes referred to as the TSS for turbine speed sensor) and the output shaft speed sensor (OSS), which on many transmissions is the same as the VSS.

Both of these sensors can be serviced with the transmission in the vehicle. Most are retained by a bolt that must be torqued to specifications. To remove them, put the gear selector in neutral and raise the vehicle. Then disconnect the connector to the sensor and remove the retaining bolt. The sensor can then be pulled from the housing. To install the sensor, apply a light coat of clean ATF on the seal and push it into place, then install and tighten the retaining bolt.

Figure 3-20 The location of the ISS and OSS in a typical transmission.

NOTE: Some late-model transmissions have the speed sensors and their harnesses incorporated into the valve body along with the TCM. This arrangement eliminates the external connections and wiring.

Electrical Connections

A visual inspection of the transmission should include a careful check of all electrical wires, connectors, and components. This inspection is especially important for today's electronically controlled transmissions. All faulty connectors, wires, and components should be repaired or replaced before continuing your diagnosis of the transmission.

Check all electrical connections to the transmission. Faulty connectors or wires can cause harsh or delayed and missed shifts. On transaxles, the connectors can normally be inspected through the engine compartment, whereas they can only be seen from under the vehicle on longitudinally mounted transmissions. To check the connectors, release the locking tabs and disconnect them, one at a time, from the transmission. Carefully examine them for signs of corrosion, distortion, moisture, and transmission fluid. A new terminal can be used to check for proper connection fit by comparing the drag on each pin of the connector. A connector or wiring harness may deteriorate if ATF reaches it. Also check the connector at the transmission. Using a small mirror and flashlight may help you get a good look at the inside of the connectors. Inspect the entire transmission wiring harness for tears and other damage. Road debris can damage the wiring and connectors mounted underneath the vehicle.

Classroom Manual
Chapter 3, page 101

The vehicle speed sensor generates a speed signal via a magnetic reluctance sensor and a toothed ring in the transmission.

Powertrain Mounts

The engine and transmission mounts on FWD cars are (**Figure 3-21**) important to the operation of the transaxle. Any engine movement may change the effective length of the shift cable and therefore affect engagement of the gears. Problems with transmission mounts may also affect the operation of a transmission on a RWD vehicle, but this type of problem will be less detrimental than the same type of problem on a FWD vehicle. Many shifting and vibration problems can be caused by worn, loose, or broken engine and transmission mounts. Visually inspect the mounts for looseness and cracks. To get a better look at the condition of the mounts, pull up and push down on the transaxle case while watching the mount. If the mount's rubber separates from the metal plate or if the case moves up but not down, replace the mount. If there is movement between the metal plate and its attaching point on the frame, tighten the attaching bolts to an appropriate torque.

Classroom Manual
Chapter 3, page 102

Then, from the driver's seat, apply the foot brake, set the parking brake, and start the engine. Put the transmission into a gear and gradually increase the engine speed to about 1500–2000 rpm. Watch the torque reaction of the engine on its mounts. If the engine's reaction to the torque appears to be excessive, broken or worn drivetrain mounts may be the cause.

If it is necessary to replace the transaxle mount, make sure you follow the manufacturer's recommendations for maintaining the alignment of the driveline. Failure to do this may result in poor gear shifting, vibrations, or broken cables. Some manufacturers recommend that a holding fixture or special bolt be used to keep the unit in its proper location.

When removing the transaxle mount, begin by disconnecting the battery's negative cable. Disconnect any electrical connectors that may be located around the mount. It may be necessary to move some accessories in order to service the mount without damaging some other assembly. Be sure to label any wires you remove to facilitate reassembly.

Install the engine support fixture and attach it to an engine hoist. Lift the engine just enough to take the pressure off of the mounts. Remove the bolts attaching the transaxle mount to the frame and the mounting bracket, then remove the mount.

Figure 3-21 The various mounts used to position the engine and transaxle in a typical FWD vehicle.

Even the slightest amount of corrosion can affect the output of a sensor. Increased resistance will always change the voltage signal of a circuit.

A **throttle position sensor** is most often referred to as a TP sensor.

A **manifold absolute pressure (MAP) sensor** senses engine vacuum. Some manufacturers require that the mount bolts be discarded and new ones installed during reassembly.

Classroom Manual Chapter 3, page 90

To install the new mount, position the transaxle mount in its correct location on the frame and tighten its attaching bolts to the proper torque. Install the bolts that attach the mount to the transaxle bracket. Prior to tightening these bolts, check the alignment of the mount. Once you have confirmed that the alignment is correct, tighten all loosened bolts to their specified torque. Remove the engine hoist fixture from the engine and reinstall all accessories and wires that may have been removed earlier.

Transmission Coolers

Transmission coolers are a possible source of fluid leaks. The efficiency of the coolers is also critical to the operation and longevity of the transmission. The vehicle may be equipped with the standard-type cooler in the radiator or may also have an auxiliary cooler. Both should be visually inspected and tested in the same way.

Follow these steps when inspecting the transmission cooler and associated lines and fittings:

1. Check the engine cooling system. The transmission cooler cannot be efficient if the engine's cooling system is defective. Repair all engine cooling system problems before continuing to check the transmission cooler.
2. Inspect the fluid lines and fitting between the cooler and the transmission (**Figure 3-22**). Check these for looseness, damage, signs of leakage, and wear. Replace any damaged lines. Replace or tighten any leaking fittings.
3. Some transmissions may have an in-line filter in the cooler lines. This filter should be checked for blockage and replaced as necessary.

Figure 3-22 Typical location of cooler line connections on a transmission.

⚙ **SERVICE TIP** In the procedure for removing the mount, some manufacturers recommend the use of a special alignment bolt, which is installed in an engine mount. This bolt serves as an indicator of powertrain alignment. If excessive effort is required to remove the alignment bolt, the powertrain must be shifted to allow for proper alignment.

⚡ **WARNING** **Do not remove the radiator cap on a warm engine. Wait until it cools down before proceeding with a coolant check. The hot coolant can burn you.**

4. Check the transmission's fluid for signs of engine coolant. Water or coolant will cause the fluid to appear milky with a pink tint. This is also an indication that the transmission cooler leaks and is allowing engine coolant to enter into the transmission fluid.

The cooler can be checked for leaks by disconnecting and plugging the transmission-to-cooler lines, at the radiator. Then remove the radiator cap to relieve any pressure in the system. Tightly plug one of the ATF line fittings at the radiator. Using the shop air supply with a pressure regulator, apply 50–70 psi of air pressure into the cooler at the other cooler line fitting. Look into the radiator. If you see bubbles, the cooler leaks.

⚙ **SERVICE TIP** The cooler lines for some transmissions have a seal inside the attaching bore in the transmission housing. A special tool is required to remove the old seal and install a new one (**Figure 3-23**). If the transmission is so equipped, make sure you replace the seal if the line is leaking or if the line has been disconnected from the housing.

AUTHOR'S NOTE The adhesive used to hold the clutch material in place is water based. If water or coolant enters the transmission the clutch material will fail. I once had a student who had a transmission stored in his garage for several years. Condensation had occurred over time and, even though the transmission was in great shape otherwise, the clutch material literally slid off the backing in every clutch pack.

Figure 3-23 The special tool required to remove a cooler line seal on some transmissions.

Checking Flow. The cooler system should also be checked for proper flow. To do this, make sure there are no fluid leaks. Start the engine and allow the transmission to reach normal operating temperature. Then turn the engine off and check the fluid level. Remove the transmission dipstick and install a funnel into the dipstick tube. Raise the vehicle up on a hoist. Then disconnect the cooler return line at the transmission. Securely fasten a rubber hose to the end of the cooler line and place the other end of the hose into the funnel. Now lower the vehicle and start the engine. Allow the engine to run at a fast idle and observe the flow of fluid into the funnel.

Another way to do this test is with the help of an assistant. Have the assistant pour the same amount of fluid into the filler tube as you are getting to flow into a bucket (**Figure 3-24**). If you get one pint, have the assistant pour in one pint. By replenishing the

Figure 3-24 Checking the flow of fluid through the cooler circuit.

fluid, your assistant is ensuring that you will not run out of fluid. Make sure the assistant keeps track of what is being poured in. Continue the test for 30 seconds.

The preferred way to check flow is with a flow meter. A flow meter is preferred because it has restrictions on the return side, which is closer to reality than an open hose. To use a flow meter, connect it into the cooler lines, start the engine, and observe.

Most manufacturers list specifications for transmission cooler flow rates. This is normally expressed in pints per second (such as 1 pint in 15–20 seconds). You should observe a steady flow of fluid. If the flow is erratic or weak, the cooler may be restricted, the transmission's pump may be weak, or there is an internal leak in the transmission.

> ⚠ **Caution**
>
> **Make sure the air pressure is regulated down to 50 to 70 psi before conducting this test. Most shops have air line pressures of nearly 150 psi. Pressures this high will damage the radiator or cooler.**

ROAD TESTING THE VEHICLE

Checking for DTCs

Prior to taking the vehicle on a road test, you should retrieve DTCs. At this stage of the diagnostic procedure, it is important to recognize what the control system is seeing. Although a detailed look at electronic control systems is given in Chapter 4, basic retrieval and interpretation information is necessary now. Connect the scan tool to the appropriate data line connector. Enter the information requested for the vehicle being tested. Then set it to retrieve the DTCs. Record all recovered DTCs. Clear the codes prior to your test drive. Use the scan tool after the test drive to see if the codes were reset during the test.

It is important that you do not correct the problems indicated by the DTCs before you take the vehicle on a road test. Correcting them may change your results and leave you with altered road test results. However, if the road test verifies the customer's initial complaint and anything else you found wrong, correct these problems before moving on with other repairs. All engine-related and body computer codes should be dealt with before you address the transmission codes. If the system displays only "pass" codes, continue your diagnosis according to the symptoms, using all available information to do this. Remember, the control system will only set a DTC when it senses that a sensor or output is not functioning the way it thinks it should. There are many things that can influence the function of the computer. All transmission complaints should be verified by road testing the vehicle and attempting to duplicate the customer's complaint. A knowledge of the exact conditions that cause the symptom and a thorough understanding of transmissions will allow you to accurately diagnose problems.

Classroom Manual
Chapter 3, page 95

Many problems that appear to be transmission problems may in fact be problems with the engine, the driveshaft, the U-joint or CV joint, the wheel bearings, wheel/tire imbalance, or other conditions. Make sure these are not the cause of the problem before you begin to diagnose and repair a transmission. Diagnosis becomes easy if you think about what is happening in the transmission when the problem occurs. If there is a shifting problem, think about the parts that are being engaged and what these parts are attempting to do.

> ⚙ **SERVICE TIP** Always refer to the service information to identify the particulars of the transmission you are diagnosing. It is also helpful to check for any Technical Service Bulletins that may be related to the customer's complaint.

Test Preparation

Road testing a vehicle to exactly duplicate the symptoms is essential to proper diagnosis. During the road test, the transmission should be operated in all possible modes and its operation noted. The customer's complaint should be verified and any other transmission

Range	Park	Reverse	Neutral	Drive 1st	Drive 2nd	Drive 3rd	Drive 4th	Drive 5th	Drive 6th
1–2–3–4 clutch				ON	ON	ON	ON		
3–5 clutch		ON				ON		ON	
4–5–6 clutch							ON	ON	ON
2–6 clutch					ON				ON
L/R clutch	ON	ON	ON						
Low sprag				ON					

Figure 3-25 Clutch and Sprag application chart for the GM 6L80 Transmission.

malfunctions should be noted. Your observations during the road test, your understanding of the operation of a transmission, and the information given in service informations, will help you identify the cause of any transmission problem. If conducted properly, road testing is diagnosis through a process of elimination.

Before beginning your road test, find and duplicate, from the service information, the chart which shows the shift solenoid, band, clutch, and sprag application for different gear selector positions (**Figure 3-25**). Using these charts will greatly simplify your diagnosis of automatic transmission problems. It is also wise to have a notebook or piece of paper to jot down notes about the operation of the transmission. If the transmission is electronically controlled, take a scan tool on the road test with you. It will be helpful to look at values related to the diagnosis. Make sure to take a partner with you to drive the vehicle while you watch the scan tool. Additionally, most scan tools have the capability of recording data that can be reviewed back at the shop. Do not try to drive and read the scan tool at the same time! Some manufacturers recommend the use of a symptoms chart or checklist that is provided in their service information (**Figure 3-26**).

Prior to road testing a vehicle, always check the transmission fluid level and condition of oil and correct any engine performance problems. Inspect the transmission for signs of fluid leakage. If leakage is evident, wipe off the leaking oil and dust. Then, note the location of the leaks in your notebook or on the symptoms chart.

Conducting the Test

Begin the road test with a drive at normal speeds to warm the engine and transmission. If a problem appears only when starting and/or when the engine and transmission is cold, record this symptom on the chart or in your notebook.

After the engine and transmission are warmed up, place the shift selector into the DRIVE position and allow the transmission to shift through all of its normal shifts. Any questionable transmission behavior should be noted.

During a road test, check for proper gear engagement as you move the selector lever to each gear position, including PARK. There should be no hesitation or roughness as the gears are engaging. Check for proper operation in all forward ranges, especially the upshifts and converter clutch engagement during light throttle operation. These shifts should be smooth and positive and occur at the correct speeds. These same shifts should feel firmer under medium to heavy throttle pressures. The torque converter clutch should be brought to the specified apply speed and their engagement noted. Again, record the operation of the transmission in these different modes in your notebook or on the diagnostic chart.

Force the transmission to downshift and record the quality of this shift and the speed at which it downshifts. These will be later compared to the specifications. Manual

Symptom — Engagement Concerns	Electrical	Hydraulic/Mechanical
No Forward	No electrical concerns	**Fluid** ■ Incorrect level ■ Condition Shift **Linkage** ■ Damaged or incorrectly adjusted **Incorrect Pressures** ■ Low forward clutch pressure, low line pressure **Fluid Filter and Seal Assembly** ■ Plugged, damaged ■ Filter seal damaged **Main Controls** ■ 3-4 shift valve, main regulator valve, manual valve—stuck, damaged ■ Bolts not tightened to specifications ■ 2-3 accumulator seals or gaskets damaged ■ Pressure-regulator valve **Pump Assembly** ■ Bolts not tightened to specifications ■ Porosity/cross leaks/ball missing or leaking, plugged hole ■ No. 3 and No. 4 seal rings damaged ■ Gaskets damaged **Forward Clutch Assembly** ■ Seals, piston damaged ■ Check balls damaged, missing, mislocated, not seating correctly ■ Friction elements damaged or worn **One-Way Clutch Assembly (Planetary)** ■ Worn, damaged or assembled incorrectly **Output Shaft** ■ Damaged

Figure 3-26 A portion of a typical manufacturer's diagnosis by symptom chart.

downshifts should also be made at a variety of speeds. The reaction of the transmission should be noted, as should all abnormal noises, and the gears and speeds at which they occur.

Most modern transmissions can be commanded to shift via a scan tool. If a problem was noted during a particular shift, if may be advantageous to check the shift concern of the transmission both under normal conditions and as commanded. Torque converter clutch can also be commanded on and off in most cases.

After the road test, check the transmission for signs of leakage. Any new leaks and their probable cause should be noted. Then compare your written notes from the road test to the information given in the service information to identify the cause of the malfunction. The service information usually has a diagnostic chart to aid you in this process.

The following problems and their causes are given as examples. The actual causes of these types of problems will vary with the different models of transmissions. Always refer to the appropriate band and clutch application chart while diagnosing shifting problems.

Using these will allow you to identify the cause of the shifting problems through the process of elimination.

These are typical causes of common problems:

- If the shifting for gears 1, 2, 3, and 4 is slipping or delayed, a possible problem with the 1–2–3–4 clutch or oil circuit is indicated.
- If 3, 5, and reverse are affected, then a problem with the 3–5–R clutch is indicated.
- If reverse slips, but 3 and 5 are okay, the Low/Reverse clutch may not be holding properly.

It is important to remember that delayed shifts or slippage may also be caused by leaking hydraulic circuits or sticking spool valves in the valve body. Since the application of bands and clutches are controlled by the hydraulic system and the network of modules on current automobiles, either an electrical or hydraulic problem can cause shifting problems. Examples include the TP sensor, which relays throttle position information, and the MAP sensor that relays engine load information. Network communications are a part of the transmission control system as well. If the ECM, TCM, and traction control modules do not communicate, the transmission may go into a default or limp mode.

Abnormal transmission noises and vibrations can be caused by faulty bearings, damaged gears, worn or damaged clutches and bands, or a bad oil pump as well as by contaminated fluid or improper fluid levels. Vibrations can also be caused by torque converter problems. All noises and vibrations that occur during the road test should be noted, as well as when they occur.

Transaxles have unique noises associated with them because of their construction. These noises can result from problems in the differential or drive axles. Use the following guide to determine if a noise is caused by the transaxle or by other parts in the drivetrain.

A knock at low speeds may indicate:

1. Worn drive axle CV joints or U-joints
2. Worn side gear hub counterbore

Noise that is most pronounced on turns may indicate:

1. Differential gear noise
2. Dry, worn CV joints

Clunk on acceleration or deceleration may indicate:

1. Loose engine mounts
2. Worn differential pinion shaft or side gear hub counterbore in the case
3. Worn or damaged drive axle inboard CV joints or U-joints

Clicking noise in turns may indicate:

1. Worn or damaged outboard CV joint (**Figure 3-27**)

Noises and Vibrations

Most vibration problems are caused by an unbalanced torque converter assembly, a poorly mounted torque converter, or a faulty output shaft. The key to determining the cause of the vibration is to pay particular attention to the vibration in relationship to engine and vehicle speed. If the vibration changes with a change in engine speed, the cause of the problem is most probably the torque converter. If the vibration changes with vehicle speed,

Noise
will be
noticeable

Noise
will be
loud

Noise will
be less
noticeable

Bad joint   Bad joint   Bad joint

Figure 3-27 A bad outside CV joint will have a different sound as the vehicle changes direction.

the cause is probably the output shaft or the driveline connected to it. The latter type of problem can be a bad extension housing bushing or universal joint, which would become worse at higher speeds.

Begin your diagnosis by determining if the cause of the problem is the driveline or the transmission. To do this, put the transmission in gear and apply the foot brakes. If the noise is no longer evident, the problem must be in the driveline or the output of the transmission. A common source of noise is worn wheel bearings. If the noise is still present, the problem must be in the transmission or torque converter.

Noise problems are also best diagnosed by paying a great deal of attention to the speed and the conditions at which the noise occurs. The conditions to pay most attention to are the operating gear and the load on the driveline. If the noise is engine speed related and is present in all gears, including PARK and NEUTRAL, the most probable source of the noise is the oil pump because it rotates whenever the engine is running. However, if the noise is engine related and is present in all gears except PARK and NEUTRAL, the most probable sources of the noise are those parts that rotate in all gears, such as the drive chain, the input shaft, and the torque converter.

Noises that only occur when a particular gear is operating must be related to those components responsible for providing that gear, such as a band or clutch. On some RWD vehicles, noise only in first gear can be caused by a defective rear transmission mount. If the noise is vehicle related, the most probable causes are the output shaft and final drive assembly. Often, the exact cause of noise and vibration problems can only be identified through a careful inspection of a disassembled transmission.

Whatever the exact fault in the transmission is the cause of a vibration or noise, the transmission will need to come out and the entire unit checked and carefully inspected. Proper diagnosis prior to disassembling the transmission will identify the specific areas that should be carefully looked at and can prevent unnecessary transmission removal and teardown.

Classroom Manual
Chapter 3, page 99

A transmission's oil circuit chart is often referred to as the transmission's flow chart.

DIAGNOSE HYDRAULIC CONTROL SYSTEMS

The best way to identify the exact cause of the problem is to use the results of the road test, logic, and oil circuit charts for the transmission being worked on. Before doing this, however, you should always check all sources for information about the symptom. Also, make sure you check the basics: trouble codes in the computer, fluid level and condition, leaks, and mechanical and electrical connections. Using the oil circuits, you can trace problems to specific shift solenoids, valves, servos, clutches, and bands.

The basic oil flow is the same for all transmissions. The oil pump supplies the fluid pressure that is used throughout the transmission. Fluid from the pump always goes to the pressure-regulating valve. From there, the fluid is directed to the manual shift valve. When the gear selector is moved, the manual valve directs the fluid to other valves and to the apply devices. By following the flow of the fluid on the oil circuit chart, you can identify which valves and apply devices should be operating in each particular gear selector position. Through a process of elimination, you can identify the most probable cause of the problem.

Most often, the valve body can be removed for service without removing the entire transmission. However, on some models, the transmission must be removed.

Mechanical controls can also contribute to shifting problems. The condition and adjustment of the various linkages and cables should be checked whenever there is a shifting problem. If all checks indicate that the problem is either an apply device or in the valving, an air pressure test can help identify the exact problem. Air pressure tests are also performed during disassembly to locate leaking seals and during reassembly to check the operation of the clutches and servos.

An air pressure test is conducted by applying clean, moisture-free air, at approximately 40 psi, through a rubber-tipped air nozzle. With the valve body removed, direct air pressure to the case holes that lead to the servo, accumulator, and clutch apply passages if possible (**Figure 3-28**). Cover the vent hole of the circuit being tested with a clean, lint-free shop towel; this will catch any fluid that may spray out. You should clearly hear or feel a dull thud that indicates the action of a holding device. If a hissing noise is heard, a seal is probably leaking in that circuit (**Figure 3-29**). If you cannot hear the action of the servo or clutch, apply air pressure again and watch the assembly to see if it is reacting to the air. A servo or accumulator should react immediately to the release of the apply air. If it does not, something is making it stick. Repair or replace the apply devices if they do not operate normally.

Figure 3-28 Using air pressure through a test plate to test hydraulic circuits.

Figure 3-29 Air testing points on a typical transmission.

Air pressure can normally be directed to the following circuits through the appropriate hole for each clutch and servo. Some manufacturers recommend the use of a specially drilled plate and gasket, which is bolted to the transmission case (**Figure 3-30**). This plate not only clearly identifies which passages to test but also seals off the other passages. Air is applied directly through the holes in the plate.

LEGEND

1. Converter bypass
2. Direct clutch
3. Forward clutch
4. 2D3 accumulator top
5. 2D3 accumulator bottom
6. Reverse servo
7. Overdrive servo apply
8. Overdrive servo release
9. Intermediate clutch
10. Reverse clutch
11. 1D2 accumulator apply

Figure 3-30 An example of a transmission air test plate.

> **SERVICE TIP** When a problem is not readily diagnosed, check the service information and Technical Service Bulletins before making assumptions about the problem or getting frustrated. Sometimes the cause of a problem is something you would least expect. For example, a fluid leak in a 2001–2013 Chevrolet Silverado 2500 or 3500 series equipped with an Allison transmission may occur when very cold (−15°F or colder). This leak can actually be caused by a low fluid level. The transmission pump can ingest air into the transmission fluid causing it to aerate. Once the fluid aerates, some fluid might be ejected from the transmission's vent tube. The cure is to make sure the fluid level is okay. In some cases a vented dipstick tube is also required. Imagine trying to troubleshoot this on your own!

PRESSURE TESTS

If you cannot identify the cause of a transmission problem from your inspection or road test, a pressure test should be conducted. This test measures the fluid pressure of the different transmission circuits during the various operating gears and gear selector positions. The number of hydraulic circuits that can be tested varies with the different makes and models of transmissions. However, most transmissions are equipped with pressure taps that allow the pressure test equipment to be connected to the transmission's hydraulic circuits (**Figures 3-31** and **3-32**). Pressure testing checks the operation of the oil pump and pressure-regulator valve. Pressure tests can be conducted on all transmissions, no matter how their pressures are regulated, or if they have hydraulic or electronic controls. Finally, late-model transmissions have the line pressure readings available as a parameter on the scan tool.

Classroom Manual
Chapter 3, page 99

A pressure test has its greatest value when the transmission does not shift smoothly or when the shift timing is wrong. Both of these problems may be caused by excessive line pressure, which can be verified by a pressure test.

Conducting a Pressure Test

Before conducting a pressure test on an electronic automatic transmission, correct and clear all trouble codes retrieved from the system. Also make sure the transmission fluid level and condition is okay and that the shift linkage is in good order and properly adjusted. To conduct a pressure test, use a tachometer, two pressure gauges (electronic gauges, if available), the correct manufacturer specifications (**Figure 3-33**).

First clutch pressure inspection hole

First hold clutch pressure inspection hole

Fourth clutch pressure inspection hole

Second clutch pressure inspection hole

Third clutch pressure inspection hole

Figure 3-31 Pressure taps on the outside of a typical Honda transaxle case.

Figure 3-32 Pressure taps on the outside of a typical Chrysler transaxle.

Gear Selector Position	Actual Gear	PRESSURE TAPS					
		Underdrive Clutch	Overdrive Clutch	Reverse Clutch	Torque Converter Clutch Off	2/4 Clutch	Low/Reverse Clutch
PARK[*] 0 mph	PARK	0–2	0–5	0–2	60–110	0–2	115–145
REVERSE[*] 0 mph	REVERSE	0–2	0–7	165–235	50–100	0–2	165–235
NEUTRAL[*] 0 mph	NEUTRAL	0–2	0–5	0–2	60–110	0–2	115–145
L[#] 20 mph	FIRST	110–145	0–5	0–2	60–110	0–2	115–145
3[#] 30 mph	SECOND	110–145	0–5	0–2	60–110	115–145	0–2
3[#] 45 mph	DIRECT	75–95	75–95	0–2	60–90	0–2	0–2
OD[#] 30 mph	OVERDRIVE	0–2	75–95	0–2	60–90	75–95	0–2
OD[#] 50 mph	OVERDRIVE WITH TCC	0–2	75–95	0–2	0–5	75–95	0–2

[*] Engine speed at 1500 rpm.
[#] CAUTION: Both front wheels must be turning at the same speed.

Figure 3-33 An oil pressure chart. These pressures are for an engine speed of 1500 rpm.

The pressure test is best conducted with three pressure gauges, but two will work. Two of the gauges should read up to 400 psi and the other to 100 psi. The two 400-psi gauges are usually used to check mainline and an individual circuit, such as mainline and direct or forward circuits. If a circuit is 10 psi lower than mainline pressure when they are both tested at exactly the same time, a leak is indicated. This is why it is best to use two 400-psi gauges. Next to the scan tool, a pressure gauge is the most valuable tool for automatic transmission diagnostics.

The pressure gauges are connected to the pressure taps in the transmission housing and routed so that the driver can see the gauges. The vehicle is then road tested and the gauge readings observed during the following operational modes: slow idle, fast idle, and wide-open throttle (WOT).

> **⚙ SERVICE TIP** When using two or more pressure gauges, be certain to put them on a "Tee" fitting at the same time to calibrate them. The gauges should be calibrated to read the same at 50, 75, 100, 125, and 150 psi.

WOT is a common acronym for wide-open throttle.

> **⚡ WARNING** **Always stop the engine when connecting and disconnecting the pressure gauges at the transmission. The fluid is under pressure and can easily spray at you and get in your eyes if a hose, fitting, or gauge leaks, so always wear safety goggles when using a pressure gauge.**

During the road test, observe the starting pressures and the steadiness of the increases that should occur with slight increases in load. The amount the pressure drops as the transmission shifts from one gear to another should also be noted. The pressure should not drop more than 15 psi between shifts.

Any pressure reading not within the specifications indicates a problem (**Figure 3-34**). Typically, when the fluid pressures are low, there is an internal leak, clogged filter, low oil pump output, or faulty pressure-regulator valve. If the fluid pressure increased at the wrong time or the pressure was not high enough, sticking valves or leaking seals are indicated. If the pressure drop between shifts is greater than approximately 15 psi, an internal leak at a seal is indicated. Always check the manufacturer's specifications for maximum drop-off before jumping to any conclusions.

To maximize the usefulness of a pressure test and to be better able to identify specific problems, begin the test by measuring line pressure. Mainline pressure should be checked in all gear ranges and at the three basic engine speeds. It is very helpful for you to make a quick chart such as the one shown in **Figure 3-35** to record the pressures during the test.

If the pressure in all operating gears is within specifications at slow idle, the pump and pressure regulator are working fine. If all pressures are low at slow idle, it is likely that there is a problem in the pump, a stuck-open pressure regulator, clogged filter, improper fluid level, or there is an internal pressure leak. To further identify the cause of the problem, check the pressure in the various gears while the engine is at a fast idle.

If the pressures at fast idle are within specifications, the cause of the problem is normally a worn oil pump; however, the problem may be an internal leak or a pressure regulator valve that is stuck open. Internal leaks typically are more evident in a particular gear range because that is when ATF is being sent to a particular device through a particular set of valves and passages. If any of these components leaks, the pressure will drop when that gear is selected or when the transmission is operating in that gear.

Further diagnostics can be made by observing the pressure change when the engine is operated at WOT in each gear range. A clogged oil filter will normally cause a gradual pressure drop at higher engine speeds because the fluid cannot pass through the filter fast enough to meet the needs of the transmission and the faster-turning oil pump. If the fluid

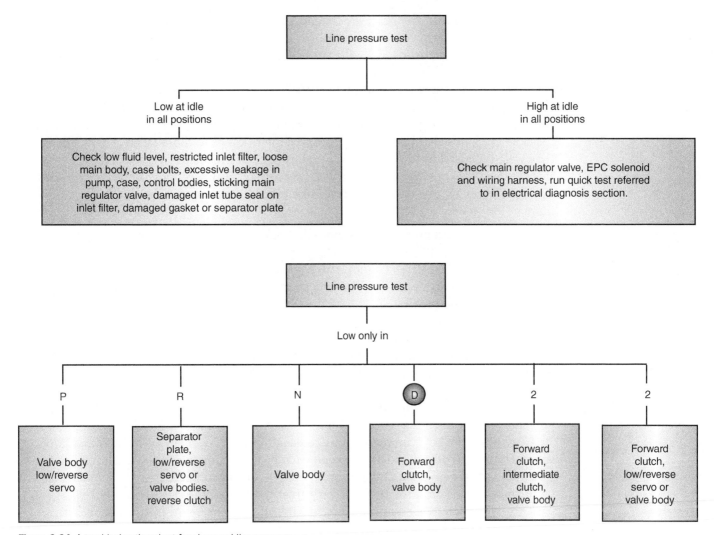

Figure 3-34 A troubleshooting chart for abnormal line pressure.

	Slow Idle	Fast Idle	WOT
P			/////
R			
N			/////
D			
3			
2			
1			

Figure 3-35 Pressure chart to aid in diagnostics.

pressure changed with the increase in engine speed, a stuck pressure regulator is the most probable cause of the problem. A stuck-open pressure regulator may still allow the pressure to build with increases in engine speed, but it will not provide the necessary boost pressures.

If the pressures are high at slow idle, a stuck-closed pressure regulator, a misassembled or stuck-open boost valve is indicated. If all pressures are low at WOT and are below specifications, the most likely cause of the problem is the pump or the control system.

If all pressures are high at WOT, compare the readings to those taken at slow idle. If they were high at slow idle and WOT, a faulty pressure regulator is indicated. To verify that the low pressures are caused by a weak or worn oil pump, conduct a *reverse stall test*. If the pressures are low during this test but are normal during all other tests, a weak oil pump is indicated.

Transmissions with an EPC Solenoid. On transmissions equipped with an **electronic pressure control** (EPC) solenoid, only one pressure gauge is needed along with a scan tool. To check the pressure, connect the scan tool and start the engine. Check the fluid level and correct it as necessary. Then check the scan tool for current and stored DTCs. Record any that are present. Then turn the engine off.

On Electronic pressure control transmissions, the PCM/TCM controls transmission line pressure by Pulse width modulation of the pressure control solenoid.

Carefully remove the plug for the line pressure test port. Then install the pressure gauge into that bore (**Figure 3-36**). Access the scan tool transmission output controls for the line PC solenoid. Start the engine. To have accurate line pressure readings, this procedure must be performed at least three times. Also, keep in mind that normally a scan tool is only able to control the line PC solenoid in PARK and NEUTRAL with the engine running below 1500 rpm. Allowing pressure control only in PARK and NEUTRAL prevents damage to clutches and seals by high pressure.

With the scan tool, increase and decrease the line PC solenoid in increments of approximately 15 psi (100 kPa). Between each of the increments allow the pressure to stabilize. Note the pressure readings on the scan tool and compare them to the readings on the pressure gauge. If the readings vary greatly, diagnose the cause.

Once the readings are noted, turn the engine off and remove the pressure gauge. Make sure to tighten the test plug properly.

Line pressure tap

Figure 3-36 The pressure gauge is installed in the housing at the EPC pressure tap.

	Transmission pressure with TP at 1.5 volts and vehicle speed above 8 km/h (5 mph)				
Gear	EPC Tap	Line Pressure Tap	Forward Clutch Tap	Intermediate Clutch Tap	Direct Clutch Tap
1	276–345 kPa (40–50 psi)	689–814 kPa (100–118 psi)	620–745 kPa (90–108 psi)	641–779 kPa (93–113 psi)	0–34 kPa (0–5 psi)
2	310–345 kPa (45–50 psi)	731–869 kPa (106–126 psi)	662–800 kPa (96–116 psi)	689–827 kPa (100–120 psi)	655–800 kPa (95–116 psi)
3	341–310 kPa (35–45 psi)	620–758 kPa (90–110 psi)	0–34 kPa (0–5 psi)	586–724 kPa (85–105 psi)	551–689 kPa (80–100 psi)

Figure 3-37 A pressure chart for a transmission equipped with an EPC.

EPC readings can also be taken during the road test. Make sure you meet the conditions recommended by the manufacturer before coming to any conclusion about the condition of the solenoid. If the EPC pressure is not within specifications (**Figure 3-37**), follow the instructions for checking the solenoid.

LINKAGES

Many transmission problems are caused by improper adjustment of the linkages. All transmissions have either a cable or a rod-type gear selector linkage.

Normal operation of a *neutral safety switch* provides a quick check for the adjustment of the gear selector linkage. To do this, move the selector lever slowly until it clicks into the park position. Turn the ignition key to the start position. If the starter operates, the park position is correct. After checking the park position, move the lever slowly toward the neutral position until the lever drops at the end of the N stop in the selector gate. If the starter also operates at this point, the gearshift linkage is properly adjusted. This quick test also tests the adjustment of the neutral safety switch. If the engine does not start in either or both of these positions, the neutral safety switch or the gear selector linkage needs adjustment or repair. Also, the starter should not operate in any other gear selector position. If it does, adjust the linkage.

The gear selector linkage is often referred to as the manual linkage because it controls the manual shift valve.

⚠ WARNING **Since you must work under the vehicle to adjust most shift linkages, make sure you properly raise and support the vehicle before working under it. Also, wear safety glasses or goggles while working under the vehicle.**

Neutral Safety Switch

Most neutral safety switches are part of a transmission range switch (**Figure 3-38**). The switch assembly can also be part of an internal mode switch (in GM cars) located inside the transmission (**Figure 3-39**). The neutral safety switch in most modern vehicles is used to inform the ECM or TCM that the transmission is in park or neutral so that the ECM knows to enable the starter. The internal mode switch (transmission range switch)

Figure 3-38 A digital transmission range (TR) sensor that serves as a neutral start switch, backup lamp switch, and as an input to the computer. It is located on the side of the transmission and is moved by the shift linkage.

Figure 3-39 The internal mode switch tells the PCM which range the driver has selected.

also informs the ECM or TCM which transmission range the driver has selected (Park, Reverse, Drive, etc.). The body control module also uses the signal to activate the backup lamps.

A scan tool can be used to check the switch for proper operation. With the scan tool installed, the technician can check for the proper operation of the transmission range switch.

To adjust a typical transmission range switch (externally mounted):

1. Place the shift lever in neutral.
2. Loosen the attaching bolts for the switch.
3. Using an aligning pin, move the switch until the pin falls into the hole in its rotor.
4. Tighten the attaching bolts.
5. Recheck the switch for continuity. If no voltage is present, replace the switch.

Gear Selector Linkage

A worn or misadjusted gear selector linkage will affect transmission operation. The transmission's manual shift valve must completely engage the selected gear (**Figure 3-40**). Partial manual shift valve engagement will not allow the proper amount of fluid pressure to reach the rest of the valve body. If the linkage is misadjusted, poor gear engagement, slipping, and excessive wear can result. If the linkage is out of adjustment, the manual shift valve will not be in its proper position and can cause internal fluid leaks. This can result in delayed shifting or slippage. The gear selector linkage should be adjusted so the manual shift valve detent position in the transmission matches the selector level detent and position indicator.

Classroom Manual
Chapter 3, page 81

To check the adjustment of the linkage, move the shift lever from the park position to the lowest drive gear. Detents should be felt at each of these positions. If the detent cannot be felt in either of these positions, the linkage needs to be adjusted. While moving the shift lever, pay attention to the gear position indicator. Although the indicator will move with an adjustment of the linkage, the pointer may need to be adjusted so that it shows the exact gear after the linkage has been adjusted.

Figure 3-40 Incorrect linkage adjustments may cause the manual shift valve to be positioned improperly in its bore and cause slipping during gear changes.

⚙️ SERVICE TIP If the proper prying tool is not used, the shift lever in some transmissions can easily break. If this tool is not available, apply some heat to the grommet. This should allow for easy separation of the cable or rod from the lever.

Linkage Adjustment

⚠️ WARNING **Always set the parking brake and block the wheels before moving the gear selector through its positions. This will prevent the vehicle from rolling over you or someone else.**

To adjust a typical floor-mounted gear selector linkage:

1. Place the shift lever into the drive position.
2. Loosen the locknuts and move the shift lever until drive is properly aligned and the vehicle is in the D range.
3. Tighten the locknut (**Figure 3-41**).

To adjust a typical cable-type linkage:

1. Place the shift lever into the park position.
2. Loosen the clamp bolt on the shift cable bracket.
3. Make sure the preload adjustment spring engages the fork on the transaxle bracket.
4. By hand, pull the shift lever to the front detent position (park), then tighten the clamp bolt. The shift linkage should now be properly adjusted.
5. If the cable assembly is equipped with an adjuster (**Figure 3-42**), move the shift lever into the park position and adjust the cable by rotating the adjuster into the lock position. Typically the adjuster will click when the lock is fully adjusted.

Classroom Manual
Chapter 3, page 82

Figure 3-41 Gearshift linkages normally attach directly to the shift lever at the transmission and are retained by a locknut.

Figure 3-42 A shift cable adjuster lock assembly.

To adjust a typical rod-type linkage:

1. Loosen or disconnect the shift rod at the shift lever bracket.
2. Place the gear selector into park and the manual shift valve lever into the park detent position.
3. With both levers in position, tighten the clamp on the sliding adjustment to maintain their relationship. On the threaded type of linkage adjustment, lengthen or shorten the connection as needed. On some vehicles, you may need to adjust the neutral safety switch after resetting the linkage.

After adjusting any type of shift linkage, recheck it for detents throughout its range. Make sure a positive detent is felt when the shift lever is placed into the park position, as a safety measure. If you are unable to make an adjustment, the levers' grommets may be badly worn or damaged and should be replaced (**Figure 3-43**). When it is necessary to disassemble the linkage from the levers, the plastic grommets used to retain the cable or rod should be replaced. Use a prying tool to force the cable or rod from the grommet, then

Figure 3-43 If the shift cable cannot be properly adjusted, the cable is distorted or some of the various grommets or brackets are worn and should be replaced.

cut out the old grommet. Pliers can be used to snap the new grommets into the levers and the cable or rod into the levers.

To check the adjustment of the linkage, make sure the transmission is in PARK when the selector is placed in that position. Also make sure the engine starts in PARK and NEUTRAL, and only in those positions. Check that the reverse lamps come on when the selector is placed in REVERSE. Check the gear selector indicator to make sure it displays the proper range.

Parking Pawl

Any time you have the oil pan off, you should inspect the transmission parts that are exposed. This is especially true of the parking pawl assembly (**Figure 3-44**). This component is typically not hydraulically activated; rather, the gearshift linkage moves the pawl into position to lock the output shaft of the transmission. Unless the customer's complaint indicates a problem with the parking mechanism, no test will detect a problem here.

Check the pawl assembly for excessive wear and other damage. Also check to see how firmly the pawl is in place when the gear selector is shifted into the park mode. If the pawl can be easily moved out, it should be repaired or replaced.

Figure 3-44 The entire parking pawl and gear assembly should be carefully inspected for wear or defects.

CASE STUDY

A technician had just finished rebuilding a 6L80 transmission from a Camaro. After the overhaul, the transmission would not move in reverse. After a complete search of the service information for possible causes, it was discovered that the wrong center support fluid passage seal was installed. The rebuild kit the technician had used included seals for all the variations of the 6L80 transmission. The technician did not compare the seals carefully before installing them into the transmission. Hopefully, the technician will pay more attention the next time!

ASE-STYLE REVIEW QUESTIONS

1. While diagnosing noises apparently coming from a transaxle assembly:

   Technician A says clunking on acceleration or deceleration can be caused by worn inboard CV joints.

   Technician B says a clicking noise heard when the vehicle is turning is probably caused by a worn or damaged outboard CV joint.

   Who is correct?

   A. A only

   B. B only

   C. Both A and B

   D. Neither A nor B

2. *Technician A* says that a clutch and band application chart shows which transmission components are applied in a particular gear.

   Technician B says that these charts can be very valuable in diagnosing a transmission problem.

   Who is correct?

   A. A only

   B. B only

   C. Both A and B

   D. Neither A nor B

3. While discussing the results of an oil pressure test:

   Technician A says when the fluid pressures are high, internal leaks, a clogged filter, low oil pump output, or a faulty pressure-regulator valve are indicated.

   Technician B says if the fluid pressure increased at the wrong time, sticking valves or leaking seals are indicated.

   Who is correct?

   A. A only   C. Both A and B
   B. B only   D. Neither A nor B

4. *Technician A* says normal operation of a neutral safety switch provides a quick check for the adjustment of the gear selector linkage.

   Technician B says the engine should start in PARK and NEUTRAL if the linkage and neutral safety switch are properly adjusted.

   Who is correct?

   A. A only   C. Both A and B
   B. B only   D. Neither A nor B

5. Technician A says low fluid levels in the transmission can cause aerated fluid.

   Technician B says slippage between shifts can be caused by aerated fluids.

   Who is correct?

   A. A only   C. Both A and B
   B. B only   D. Neither A nor B

6. *Technician A* says the transmission cooler cannot be efficient if the engine's cooling system is defective.

   Technician B says some transmissions also have an inline filter placed in the fluid line going to the fluid cooler.

   Who is correct?

   A. A only   C. Both A and B
   B. B only   D. Neither A nor B

7. While inspecting the condition of a vehicle's ATF:

   Technician A says if the fluid has a dark brownish or blackish color or a burnt odor much like burnt popcorn, the fluid has been overheated.

   Technician B says if the fluid has a milky color, engine coolant has been leaking into the transmission's cooler.

   Who is correct?

   A. A only   C. Both A and B
   B. B only   D. Neither A nor B

8. *Technician A* says that many shifting and vibration problems can be caused by worn, loose, or broken engine and transmission mounts.

   Technician B says if the mount's rubber stays attached to the metal plate when the transmission or engine moves, the mount is bad.

   Who is correct?

   A. A only   C. Both A and B
   B. B only   D. Neither A nor B

9. While discussing a pressure test:

   Technician A says to begin this test by checking line pressure.

   Technician B says the test can be run with a scan tool and a pressure gauge on transmissions that use an EPC solenoid.

   Who is correct?

   A. A only   C. Both A and B
   B. B only   D. Neither A nor B

10. While checking transmission seals;

   Technician A says the rear transmission oil seal can be replaced without removing the driveshaft.

   Technician B says that front seal replacement requires the transmission to be removed from the vehicle.

   Who is correct?

   A. A only   C. Both A and B
   B. B only   D. Neither A nor B

ASE CHALLENGE QUESTIONS

1. While discussing the condition of engine mounts:

   Technician A says engine movement may change the effective length of the shift cable and therefore affect engagement of the gears.

   Technician B says shifting and vibration problems are seldom caused by worn, loose, or broken engine and transmission mounts.

   Who is correct?

   A. A only   C. Both A and B

   B. B only   D. Neither A nor B

2. *Technician A* says a leaking torque converter may be indicated by the presence of transmission fluid over the entire shell of the converter.

   Technician B says transmission oil on the converter's rear shell indicates a leaking torque converter.

   Who is correct?

   A. A only   C. Both A and B

   B. B only   D. Neither A nor B

3. Typically a transmission has two sensors mounted externally to the transmission case.

   Technician A says one of these sensors is the Input shaft speed sensor.

   Technician B says the other is the MAP sensor.

   Who is correct?

   A. A only   C. Both A and B

   B. B only   D. Neither A nor B

4. *Technician A* says a new terminal can be used to check for proper connection fit by comparing the drag on each pin of the connector.

   Technician B says some sensors can affect both engine and transmission operation.

   Who is correct?

   A. A only   C. Both A and B

   B. B only   D. Neither A nor B

5. Pressure testing reveals low pressure in all gears.

   Technician A says using a reverse stall test forces the pump output to maximum pressure.

   Technician B says a reverse stall test will determine whether the pump output is low.

   Who is correct?

   A. A only   C. Both A and B

   B. B only   D. Neither A nor B

Name _____   Date _____

ROAD TESTING AN AUTOMATIC TRANSMISSION

Upon completion of this job sheet, you will be able to identify and interpret transmission concerns, ensure proper engine operation, and diagnose noise and vibration problems.

ASE Education Foundation Correlation ────────────────────────

This job sheet addresses the following **AST/MAST** task:

A.1   Identify and interpret transmission/transaxle concern, differentiate between engine performance and transmission/transaxle concerns; determine needed action. (**P-1**)

This job sheet addresses the following **MAST** task:

A.7   Diagnose noise and vibration concerns; determine needed action. (**P-2**)

We Support
ASE | Education Foundation

Tools and Materials

- Service information
- Clean rag
- Paper and pencil

Describe the Vehicle Being Worked On

Year _____ Make _____ Model _____

VIN _____ Engine type and size _____

Model and type of transmission _____

NOTE: Do not drive the vehicle alone, always take an assistant with you on the test drive.

Procedure

Task Completed

1. Determine the method of checking the fluid level in the transmission. Summarize the procedure here.   ☐

   _____

   _____

2. If the engine is not at normal operating temperature, start the engine and allow it to reach the operating temperature.   ☐

3. Check the level and condition of the fluid. Record your findings.   ☐

   _____

   _____

4. Describe the condition of the fluid and tell what the fluid indicates.

   _____

   _____

5. Check for any DTCs that might be present and record them here. Are there any that might cause or contribute to the customer's concern?

   _____

   _____

6. Inspect the transmission for leaks and other signs of wear or damage. Comments:

   _____

   _____

7. Find and duplicate the chart from the service information that shows the clutch application for the different gear selector positions. Using these charts will greatly simplify diagnosis of a problem. It is also wise to have paper and a pencil to jot down notes about the operation of the transmission.

☐

_____

_____

8. Drive the vehicle at normal speeds. Describe the overall behavior of the transmission and torque converter.

_____

_____

9. Place the shift selector into the Overdrive position. Allow the transmission to shift through all of its gears. Describe the operation of the transmission and torque converter. If something sounded or felt abnormal during any of the shifts or while operating in a particular gear, refer to the clutch application chart to identify possible causes of the problem.

_____

_____

10. Force the transmission to downshift and record the quality of that downshift and the speed at which it downshifted.

_____

_____

11. Manually cause the transmission to downshift. Record the quality of that shift.

_____

_____

12. Stop the vehicle and put the shift selector in the lowest gear. Begin driving and manually shift the transmission through the forward gears. Record the quality of those shifts and the conclusions you draw from this drive. Again, if a problem was evident, refer to the band and clutch application chart to identify possible causes.

_____

_____

13. Stop the vehicle and place the gear selector into reverse. Describe the feel of that shift.

_____

_____

14. With the vehicle stopped, place the gear selector into PARK. Describe the feel of that shift.

_____

_____

15. What are your recommendations for this transmission?

_____

_____

Problems Encountered

_____

_____

_____

Instructor's Comments

_____

_____

_____

Name _____ Station _____ Date _____

DIAGNOSING TRANSMISSION FLUID LOSS

Upon completion of this job sheet, you will be able to diagnose unusual fluid usage, level, and condition concerns.

ASE Education Foundation Correlation

This job sheet addresses the following **MLR** task:

A.4   Check transmission fluid condition; check for leaks. **(P-2)**

This job sheet addresses the following **AST/MAST** task:

A.3   Diagnose fluid loss and condition concerns; determine needed action. **(P-1)**

We Support
ASE | Education Foundation

Tools and Materials

- Service information
- Flashlight
- Protective clothing
- Goggles or safety glasses with side shields

Describe the Vehicle Being Worked On

Year _____ Make _____ Model _____

VIN _____ Engine type and size _____

Model and type of transmission _____

Procedure

1. Check the transmission case for any signs of damage or obvious leaks. Record your findings and recommendations.

   _____

   _____

2. Check the transmission driveshaft or extension housing seals, or, if you are inspecting a transaxle, the transaxle seal areas for signs of a leaking seal. Record your findings and recommendations.

   _____

   _____

3. Check for leaks at all connections where something is attached to the transmission case. Record your findings and recommendations.

   _____

   _____

4. Check the shift linkage for looseness, wear, and/or damage. Record your findings and recommendations.

   _____

   _____

5. Check any cables for binding, wear, and/or damage. Record your findings and recommendations.

   _____

   _____

6. Check the condition of all cooler lines and hoses. Record your findings and recommendations.

_____

_____

7. Locate and describe the location of the transmission's dipstick.

_____

_____

8. If the transmission does not have a dipstick, find the proper procedure for checking the transmission's fluid in the service information. Summarize that procedure.

_____

_____

9. Check the condition of the fluid; pay attention to the level, color, smell, and feel of the fluid. Record your findings and recommendations.

_____

_____

Problems Encountered

_____

_____

_____

Instructor's Comments

_____

_____

_____

Name _____   **Date** _____

CHECKING THE FLUID LEVEL OF A TRANSMISSION EQUIPPED WITH A DIPSTICK

Upon completion of this job sheet, you will be able to accurately determine the fluid level of an automatic transmission equipped with a dipstick.

ASE Education Foundation Correlation ————————————————————

This job sheet addresses the following **MLR** task:

A.2   Check fluid level in a transmission or a transaxle equipped with a dip-stick. (**P-1**)

This job sheet addresses the following **AST/MAST** task:

A.4   Check fluid level in a transmission or a transaxle equipped with a dip-stick. (**P-1**)

We Support

ASE | Education Foundation

Tools and Materials

- Service information
- Shop towel
- Flashlight
- Protective clothing
- Goggles or safety glasses with side shields

Describe the Vehicle Being Worked On

Year _____ Make _____ Model _____

VIN _____ Engine type and size _____

Model and type of transmission _____

Procedure

1. Using service information, determine the proper transmission fluid for this transmission. Report your results here.

   _____

   _____

2. According to the service information, what is the recommended fluid temperature to check the fluid level?

   _____

   _____

3. Is there a method of checking the transmission fluid cold? If so, detail the procedure here.

   _____

   _____

   Describe the "hot" fluid level check below.

   _____

   _____

_____

_____

_____

_____

_____

Problems Encountered

_____

_____

_____

Instructor's Comments

_____

_____

_____

Name _____   Date _____

CHECKING TRANSMISSION FLUID LEVEL IN A VEHICLE WITHOUT A DIPSTICK

Upon completion of this job sheet, you will be able to accurately determine the fluid level of an automatic transmission equipped with a dipstick.

ASE Education Foundation Correlation ————————————————————————

This job sheet addresses the following **MLR** task:

A.3   Check fluid level in a transmission or a transaxle not equipped with a dip-stick. (**P-1**)

This job sheet addresses the following **AST/MAST** task:

A.5   Check fluid level in a transmission or a transaxle not equipped with a dip-stick. (**P-1**)

We Support

ASE | Education Foundation

Tools and Materials
- Service information
- Scan tool
- Shop towel
- Flashlight
- Protective clothing
- Goggles or safety glasses with side shields

Describe the Vehicle Being Worked On

Year _____ Make _____ Model _____

VIN _____ Engine type and size _____

Model and type of transmission _____

Procedure

1. Using service information, determine the proper transmission fluid for this transmission. List your results here.

   _____

   _____

2. Is there a vent cap that has to be removed before checking the fluid level? Will the cap be replaced, or can it be reinstalled after use?

   _____

   _____

3. Describe the procedure to check the transmission fluid, make sure to include the temperature that the transmission fluid should be checked.

   _____

   _____

   _____

   _____

   _____

_____

_____

Problems Encountered

_____

_____

_____

Instructor's Comments

_____

_____

_____

Name _____   Date _____

DIAGNOSTIC PRESSURE TESTING OF AN AUTOMATIC TRANSMISSION

Upon completion of this job sheet, you will have demonstrated the ability to conduct a pressure test on a transmission.

ASE Education Foundation Correlation

This job sheet addresses the following **MAST** task:

A.6   Perform pressure tests (including transmissions/transaxles equipped with electronic pressure control); determine needed action. **(P-1)**

We Support
ASE | Education Foundation

Tools and Materials

- Hoist
- Tachometer/as required
- Pressure gauges with necessary adapters
- Scan tool

- Protective clothing
- Goggles or safety glasses with side shields
- Lint-free shop towels
- Service information

Describe the Vehicle Being Worked On

Year _____ Make _____ Model _____

VIN _____ Engine type and size _____

Model and type of transmission _____

Procedure   Task Completed

1. Check the fluid level and condition before starting the test. Record your findings here.

   _____

   _____

2. List the pressure specifications with the required operating conditions.

   _____

   _____

3. Is the transmission fitted with electronic pressure control (EPC)?

   _____

   _____

4. Are there any DTCs that relate to transmission pressures stored? If so, record them here.

5. If the transmission has an EPC, what special procedures must be followed while checking the operating pressures of the transmission?

6. If the transmission is equipped with an EPC, explain how the transmission pressure can be checked using a scan tool.

   _____

   _____

7. If the transmission has an EPC, what is the procedure to follow if the pressure on the gauge does not match the reading on the scan tool?

_____

_____

8. Have a fellow student sit in the vehicle to operate the throttle, brakes, and transmission during the test. Each vehicle tested may differ by make and model. Describe the procedure for your assigned vehicle. ☐

_____

_____

_____

9. Raise the vehicle on the hoist to a comfortable working height. ☐

10. Describe the location of the pressure tap(s) on the transmission.

_____

_____

11. Connect the pressure gauges to the appropriate service ports. ☐

12. Start the engine. Have your helper press firmly on the brake pedal and apply the parking brake. Then the helper should move the gear selector into the first test position. ☐

13. Run the engine at the specified speed. Move the gear selector as required by the service information. ☐

14. Observe and record the pressure readings at the various test conditions.

_____

_____

15. Turn off the engine and move the pressure gauges to the appropriate test ports for the next transmission range to be tested. Once again, some transmissions have only one pressure tap for line pressure, some have many depending on make and model. Describe the location and purpose of each pressure tap available.

_____

_____

16. If your transmission has more than one pressure tap, move the gauge to the next tap position. Restart the engine and have your helper move the gear selector into the range to be tested and increase the engine's speed to the required test speed. ☐

17. Observe and record the pressure readings in the various test conditions. (As applicable)

_____

_____

_____

18. Allow the engine to return to idle and turn it off. ☐

19. Repeat this sequence until all transmission ranges have been tested. ☐

20. Summarize the results of this test and compare them to specifications.

_____

_____

_____

_____

_____

21. Describe the importance of your findings as related to the diagnosis of this transmission.

_____

_____

Problems Encountered

_____

_____

_____

Instructor's Comments

_____

_____

_____

Name _____   Date _____

NOISE AND VIBRATION DIAGNOSTICS

Upon completion of this job sheet, you will be able to diagnose transmission noise and vibration concerns.

ASE Education Foundation Correlation ————————————————————————

This job sheet addresses the following **MAST** task:

A.7   Diagnose noise and vibration concerns; determine needed action. **(P-2)**

We Support
ASE | **Education Foundation**

Tools and Materials

- A vehicle with an automatic transmission
- Hoist
- Stethoscope
- Droplight or good flashlight
- Service information
- Protective clothing
- Goggles or safety glasses with side shields

Describe the Vehicle Being Worked On

Year _____ Make _____ Model _____

VIN _____ Engine type and size _____

Procedure

1. Check the fluid level of the transmission before starting work. Record the results here.

    _____

2. Start the vehicle and leave in PARK. Is the vibration or noise present at idle?

    _____

3. With the vehicle still in PARK, does the vibration or noise change with engine rpm? (If so go to number 4.)

    _____

4. To eliminate belt-driven components, remove the drive belt(s) and run the engine at the same rpm that the noise is present. (Limit the time the engine is run without the belt to a few seconds to prevent overheating.) Did the noise or vibration go away?

    _____

5. If the vibration is still present with the belt removed, what could be the source of the problem?

    _____

    _____

    _____

6. Describe your findings here. Were you able to determine the source of the vibration or noise?

    _____

    _____

    _____

    _____

Noise and/or Vibration while Driving

1. If the noise or vibration is present while driving, inspect the tires for any signs of wear that might produce a sound while driving, such as cupping, feathering, or broken belts. Look for obvious signs of a damaged rim. Describe your results here.

   _____

   _____

   _____

2. Place the vehicle on the rack and report on the condition of the following:

   a. Engine and transmission mounts _____

   b. Broken transmission bellhousing or missing bolts _____

   c. Cracked or broken flywheel _____

   d. Loose torque converter bolts _____

   e. Missing torque converter weights _____

   f. Damaged driveshaft or loose universal joints _____

   Describe your findings here._____

   _____

   _____

   _____

3. If no obvious cause of a noise/vibration is found while inspecting the vehicle on the rack, carefully test drive the vehicle and listen for a change in the sound while driving over different types of pavement, such as asphalt and concrete. Is there a change in the sound? Record your results here.

   _____

4. If the sound stays the same over different types of pavement, does the sound get louder with an increase in speed?

   _____

5. To help determine if the sound is from a wheel bearing, see if the noise changes on turns. What did you find?

   _____

6. If the noise appears to be from the driveline at a specific speed, try letting the vehicle coast down and then accelerate. Did the noise change with engine load?

   _____

   _____

7. If the noise/vibration does not change with engine load, observe the rpm that the noise occurs. See if the noise occurs at the same rpm in different gears. For instance if the noise occurs in fourth gear at 1800 rpm, does it occur at 1800 rpm in third gear?

8. Did you locate the source of the noise or vibration? If so, give an overview of how you found the problem here.

   _____

   _____

Problems Encountered

_____

_____

_____

Instructor's Comments

_____

_____

_____

Name _____   Date _____

AIR PRESSURE TESTING

Upon completion of this job sheet, you should be able to conduct an air pressure test on a transmission.

ASE Education Foundation Correlation

This job sheet addresses the following **MAST** task:

C.21   Air test operation of clutch and servo assemblies. (**P-1**)

We Support
ASE | Education Foundation

Tools and Materials

- An automatic transmission or transaxle on a bench, or a vehicle with an automatic transmission
- Compressed air and rubber-tipped air nozzle
- Pound-inch torque wrench
- Eye protection
- Air pressure testing plate
- Service information
- Lint-free shop towels

Describe the Transmission That Was Assigned to You:

Year _____ Make _____ Model _____

Model _____ Model and type of transmission _____

Procedure   **Task Completed**

1. If this job sheet will be completed on a transmission installed in a vehicle, raise the vehicle on a hoist. Then drain the transmission fluid and remove the oil pan, oil filter, and valve body.   ☐

2. If the job sheet will be completed with the transmission on a bench, remove the valve body.   ☐

3. Refer to the service information and identify the fluid passages required for testing.   ☐

4. If you have an air pressure testing plate, install it with the appropriate gaskets and tighten the plates to the manufacturer's specifications. The specified torque is _____.

5. Apply air pressure to the servo and clutch passages and listen for 5–10 seconds. Describe the sounds you hear and record what they indicate.

   _____

   _____

   _____

6. Release the air pressure and reapply pressure. Listen and record the results.

   _____

   _____

7. Remove the testing plate.   ☐

8. Summarize the results of this test.

_____

_____

_____

_____

Instructor's Response

_____

_____

_____

Name _____   Date _____

TRANSMISSION COOLER INSPECTION AND FLUSHING

Upon completion of this job sheet, you should be able to inspect, test, and flush the cooler system for an automatic transmission.

ASE Education Foundation Correlation

This job sheet addresses the following **AST/MAST** task:

C.2   Inspect, leak test, and flush or replace transmission/transaxle oil cooler, lines, and fittings. **(P-1)**

We Support

ASE | Education Foundation

Tools and Materials

- A vehicle with an automatic transmission or transaxle
- Line wrenches
- Tubing cutter
- Tubing flaring kit
- ATF pressure tester

- Various lengths of rubber hose
- Drain pan
- Compressed air and air nozzle
- Supply of clean solvent or mineral spirits
- Lint-free shop towels
- Service information

Describe the Vehicle That Was Assigned to You:

Year _____ Make _____ Model _____

Model _____ Model and type of transmission _____

Procedure

Task Completed

1. Remove the transmission dipstick and describe the condition, smell, and color of the fluid.

   _____

   _____

2. Open the radiator cap (after the engine is cooled down) and check the coolant for traces of ATF. Also check the cap and gasket for signs of ATF. Record your findings.

   _____

   _____

3. Replace the radiator cap.

4. Inspect the metal lines and fittings to and from the transmission cooler. Look for damage and signs of leakage. Describe your findings.

   _____

   _____

5. Summarize the results of your inspection.

   _____

   _____

6. Place a drain pan under the fittings that connect the cooler lines to the radiator.

7. Disconnect both cooler lines from the radiator. Plug or cap the lines from the transmission.

8. Plug or cap one fitting at the radiator.

9. Remove the radiator cap.

10. Hold the compressed air nozzle tightly against the open fitting.

11. Apply air pressure through the fitting (no more than 75 psi).

12. Check the coolant in the radiator for signs of air bubbles or air movement. Describe your findings.

_____

_____

13. Unplug the other fitting and cooler lines.

14. Reconnect the cooler lines to the radiator.

15. Summarize the results of the leak test.

_____

_____

_____

_____

16. Set the parking brake and start the engine. Allow the engine and transmission to reach normal operating temperature before proceeding. Turn off the engine.

17. Remove the transmission dipstick. Place a funnel in the dipstick tube.

18. Raise the vehicle on a hoist.

19. Disconnect the cooler return line at the point where it enters the transmission case.

20. Attach a rubber hose to the disconnected cooler return line.

21. Lower the vehicle and place the end of the hose into the funnel.

22. Start the engine and run it at about 1000 rpm.

23. Observe the flow of fluid into the funnel. Describe the flow and summarize what this indicates.

_____

_____

_____

_____

24. Turn off the engine.

25. Raise the vehicle and remove the rubber hose from the return line.

26. Reconnect the return line.

27. Lower the vehicle and reinstall the dipstick.

The remainder of this job sheet covers a procedure for flushing the transmission cooler. Check with your instructor before proceeding.

1. Raise the vehicle on a hoist.

2. Place a drain pan under the engine's radiator (or place it under the external transmission cooler if the vehicle is so equipped).

3. Disconnect both cooler lines at the ends of the lines and at the radiator or cooler.

4. Clean the fittings at the ends of the lines and at the radiator or cooler.

5. Connect a rubber hose to the inlet fitting at the radiator.

6. Place the other end of the hose into the drain pan.

7. Attach a hose funnel to the cooler return fitting. Install a funnel in the other end of this hose.

8. Pour small amounts of mineral spirits into the funnel. Observe the flow and the condition of the fluid moving into the drain pan. What does it look like?

   _____

   _____

9. If there is little flow, apply some air pressure to the hose connected to the return line.

10. Continue pouring mineral spirits into the funnel until the fluid entering the drain pan is clear.

11. Disconnect the cooler lines at the transmission and place the drain pan at one end of the lines.

12. Install the rubber hose and funnel to the other end of the lines.

13. Pour mineral spirits through the lines until the flow is clear. Describe your findings.

    _____

    _____

14. Pour ATF through the cooler lines to remove the mineral spirits.

15. Remove all hoses and reconnect the cooler lines to the transmission and radiator.

16. Check the transmission's fluid level and correct it if necessary.   ☐

    _____

Problems Encountered

_____

_____

_____

Instructor's Comments

_____

_____

_____

Name _____   Date _____

SERVICING LINKAGES

Upon completion of this job sheet, you will be able to inspect, adjust, or replace throttle valve and gear selector linkages and cables.

ASE Education Foundation Correlation

This job sheet addresses the following **MLR/AST/MAST** task:

B.1   Inspect, adjust, and replace external manual valve shift linkage, transmission range sensor/switch, and park/neutral position switch. **(P-2)**

We Support

ASE | Education Foundation

Tools and Materials

- Basic hand tools
- Service information
- Protective clothing
- Goggles or safety glasses with side shields

Describe the Vehicle Being Worked On

Year _____ Make _____ Model _____

VIN _____ Engine type and size _____

Procedure

1. If the gear selector linkage is misadjusted, poor gear engagement, slipping, and excessive wear can result. The gear selector linkage should be adjusted so the manual shift valve detent position in the transmission matches the selector level detent and position indicator. To check the adjustment of the linkage, move the shift lever from the PARK position to the lowest DRIVE gear. Detents should be felt at each of these positions. If the detent cannot be felt in either of these positions, the linkage needs to be adjusted. Describe your findings.

    _____

    _____

2. Move the gear selector lever slowly until it clicks into the PARK position. Turn the ignition key to the Start position. If the starter operates, the PARK position is correct. After checking the PARK position, move the lever slowly toward the NEUTRAL position until the lever drops at the end of the N stop in the selector gate. If the starter also operates at this point, neutral is okay. By completing this routine, what did you check?

    _____

    _____

3. If the engine does not start in P or N, the transmission range switch might be incorrectly adjusted or defective. The signal from the transmission range switch can be checked with the scan tool for validity, as well as trouble codes. Describe the scan tool data for the range switch and the functions it performs below.

    _____

    _____

4. Outline the procedure to adjust the transmission range switch below.

    _____

    _____

5. If your vehicle is equipped with a floor-mounted gear selector, outline the steps to adjust the gear selector linkage.

   Did you have any problems in adjusting a typical floor-mounted gear selector linkage?

   _____

   _____

6. If your vehicle is equipped with a cable-mounted linkage, outline the steps to adjust the gear selector linkage.

   _____

   _____

   Did you have any problems in adjusting the gear selector linkage?

   _____

   _____

   _____

   _____

7. After adjusting the shift linkage, recheck it for detents throughout its range. Make sure a positive detent is felt when the shift lever is placed into the PARK position, as a safety measure. Describe your findings.

   _____

   _____

8. If you are unable to make an adjustment, the levers' grommets may be badly worn or damaged and should be replaced. When it is necessary to disassemble the linkage from the levers, the plastic grommets used to retain the cable or rod should be replaced. Use a prying tool to force the cable or rod from the grommet, then cut out the old grommet. Pliers can be used to snap the new grommets into the levers and the cable or rod into the levers. What did you find?

   _____

   _____

Problems Encountered

_____

_____

_____

Instructor's Comments

_____

_____

_____

CHAPTER 4

ELECTRICAL AND ELECTRONIC SYSTEM DIAGNOSIS AND SERVICE

Upon completion and review of this chapter, you should be able to:

- Diagnose electronic control systems and determine needed repairs.
- Conduct a road test to determine if a fault is electrical or hydraulic.
- Use common electrical test instruments to locate problems.
- Conduct preliminary checks on transmission systems and determine needed repairs or service.
- Retrieve trouble codes from common electronically controlled automatic transmissions and determine needed repairs or service.
- Follow the prescribed diagnostic procedures according to DTC or symptom and determine needed repairs or service.

- Perform converter clutch system tests and determine needed repairs or service.
- Inspect, test, and replace electrical/electronic switches.
- Inspect, test, and replace electrical/electronic sensors.
- Inspect, test, bypass, and replace electrical/electronic solenoids.
- Inspect, test, adjust, or replace transmission-related electrical/electronic components.

🔧 Basic Tools

Basic mechanic's tool set

Scan tool

DMM

Jumper wires

Appropriate service information

Terms To Know

Amplitude

Cycle

Data link connector (DLC)

Default mode

Diagnostic trouble code (DTC)

Digital multimeter

Digital storage oscilloscope (DSO)

Flash programming

Freeze frame

Glitch

Impedance

Lab scope

Multimeter

On-board diagnostics

Open circuit

Parameter identification (PID)

Radio frequency interference (RFI)

Serial data

Short circuit

Sine wave

Sinusoidal

INTRODUCTION

One of the first tasks during diagnosis of an electronically controlled transmission is to determine if the problem is caused by the transmission or by electronics. To determine this, the transmission must be observed to see if it responds to commands given by the computer. Identifying whether the problem is related to the transmission or is electrical will determine what steps need to be followed to diagnose the cause of the problem.

Figure 4-1 The key electronic components of an AA80E transmission.

An electronically controlled transmission (**Figure 4-1**) will operate efficiently only if the commands it receives from the computer are valid. This holds true even if the hydraulic and mechanical parts of the transmission are normal. All diagnostics should begin with a scan tool to check for trouble codes in the system's computer. After the received codes are addressed, you can begin a more detailed diagnosis of the system and transmission. Your next step may be to manually activate the shift solenoids by connecting a jumper wire to them or by using a scan tool or transmission tester that allows you to manually activate the solenoids. Prior to doing this, the wiring to the solenoids should be studied to determine if the computer activates them by supplying voltage to them or by completing the ground circuit. Also you need to know what gear certain solenoids are activated in. This information can be found in the service information.

BASIC TRANSMISSION TESTING

Classroom Manual
Chapter 4, page 136

The best way to diagnose an electronically controlled transmission is to approach solving the problem in a logical way.

It is important that you totally understand what the complaint or problem is before you venture in and try to find the cause. This may include an interview or road test with the customer to thoroughly define the complaint and to identify when and where the problem occurs.

Since many transmission problems are caused by the basics, it is wise to conduct all of the preliminary checks required for a nonelectronically controlled transmission. In addition, you should conduct a thorough inspection of the electronic system. Whenever diagnosing a transmission, remember that an engine problem can and will cause the transmission to act abnormally.

Often, accurately defining the problem and locating related information in TSBs and other materials can indicate the cause of the problem. No manufacturer makes a perfect vehicle or transmission, so when the manufacturer recognizes common occurrences of a problem, it will issue a statement regarding the way to fix the problem. Also, for many DTCs and symptoms the service information will give a simple diagnostic chart or path for identifying the cause of the problem. These are designed to be followed step by step and will lead to a conclusion if you follow the path that corresponds exactly to the symptom. Check all available information before moving on in your diagnostics.

Sometimes the symptom will not match any of those described in the service information. This doesn't mean it is time to guess. It means it is time to clearly identify what is working correctly. By eliminating those circuits and components that are working correctly from the list of possible causes of the problem, you can identify what may be causing the problem and what should be tested further.

Although the first steps in troubleshooting include retrieving diagnostic codes, there are problems that will not be made evident by a code. These problems may be solved with the diagnostic charts or pure logic. However, this logic must be based on a thorough understanding of the transmission and its controls.

> A **TSB** is a Technical Service Bulletin in which the latest fixes and updates for a component or system are found.

> **Classroom Manual**
> Chapter 4, page 112

GUIDELINES FOR DIAGNOSING TRANSMISSION ELECTRICAL PROBLEMS

1. Make sure the battery has at least 12.6 volts before troubleshooting the transmission.
2. Check all fuses and identify the cause of any blown fuses.
3. Compare the wiring to all suspected components against the colors given in a wiring diagram.
4. When testing electronic circuits, always use a high-impedance test light or DMM (digital multimeter).
5. If an output device is not working properly, check its power circuit.
6. If an input device is not sending the correct signal back to the computer, check the reference voltage it is receiving and the voltage it is sending back to the computer.
7. Compare the voltages in and out of a sensor with the voltages the computer is sending out and receiving.
8. Make sure computer wiring harnesses do not run parallel with any high-current wires or harnesses. The magnetic field created by the high current may induce a voltage in the computer harness.
9. Take necessary precautions to prevent the possibility of static discharge while working with electronic systems.
10. While checking individual components, always check the voltage drop of the ground circuits. This is becoming more and more important because newer cars contain less material that conducts electricity well.
11. Make sure the ignition is off whenever you disconnect or connect an electronic device in a circuit.
12. All sensors should be checked in cold and hot conditions.
13. All wire terminals and connections should be checked for tightness and cleanliness. A new terminal can be used to perform a "drag test" to find loose terminals inside a connector.
14. Use electrical cleaning spray to clean all connectors and terminals.
15. Use dielectric grease at all connections to prevent future corrosion.
16. If you must break through the insulation of a wire to take an electrical measurement, make sure you tightly tape over the area after you are finished testing.

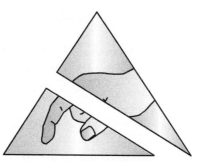

Figure 4-2 A manufacturer's warning symbol, used to show a circuit or component is sensitive to electrostatic discharge (ESD).

Electrostatic Discharge

Some manufacturers mark certain components and circuits with a code or symbol to warn technicians that they are sensitive to electrostatic discharge (**Figure 4-2**). Static electricity can destroy or render a component useless.

When handling any electronic part, especially those that are static sensitive, follow the guidelines below to reduce the possibility of electrostatic build up on your body and the inadvertent discharge to the electronic part. If you are not sure whether a part is sensitive to static, treat it as if it were.

1. Always touch a known good ground before handling the part. This should be repeated while handling the part and again after sliding across a seat, sitting down from a standing position, or walking a distance.
2. Avoid touching the electrical terminals of the part unless you are instructed to do so in the written service procedures. It is good practice to keep your fingers off all electrical terminals as the oil from your skin can cause corrosion. Use an anti-static strap, if one is available.
3. When you are using a voltmeter, always connect the negative meter lead first.
4. Do not remove a part from its protective package until it is time to install the part.
5. Before removing the part from its package, ground yourself and the package to a known good ground on the vehicle.

The static electricity that can be generated by a vinyl seat is 20,000 volts. It only takes 35 volts to destroy any electronic automotive device.

Although diagnostic trouble codes (DTCs) are very helpful during diagnosis, they also can become a stumbling block. This is especially true if more than one DTC is present. If two or more codes are present, you should look at the relationship of the codes to identify if they could have a common cause. Reacting to the individual codes may move your troubleshooting efforts beyond the common cause.

It is also important to remember that codes can be set by out-of-range signals. This doesn't mean the sensor or sensor circuit is bad. It could mean the sensor is working properly but there is a mechanical or hydraulic problem causing the abnormal signals. Not only can internal transmission problems cause codes to be set; so can basic electrical problems. Problems such as loose connections, broken wires, corrosion, and poor grounds will affect the signals in that circuit.

BASIC ELECTRICAL TESTING

Although the emphasis of this chapter is electronically controlled transmissions, basic electrical checks must be covered first. Many of the preliminary checks involve the use of common electrical test equipment. The proper use of test equipment is also necessary to test and identify the exact cause of problems after diagnostic codes are retrieved.

Ammeters

Measuring current flow can provide a summary of the activity in a circuit. If the current flow is lower than expected, there is some extra resistance in the circuit. This resistance can be the result of corrosion, frayed wires, or a loose connection. When current is higher than expected, a short or low resistance is indicated. A normally operating circuit will have the expected amount of current flow.

There are two main types of ammeters and each requires a different hook-up into the circuit. Prior to using an ammeter in a circuit, make sure the meter is capable of handling the current in the circuit. Circuits like the starting motor circuit have very high amounts of current and only specially designed meters are able to withstand the current flow of these circuits. Often ammeters have ranges of measurement. Always select the proper range prior to connecting the meter.

One type of ammeter has two leads, a positive lead and a negative lead. In order for this type of ammeter to accurately read current flow, the meter must be connected in series to the circuit (**Figure 4-3**). The circuit should be disconnected from its power source, then opened by disconnecting a connector in the circuit and inserting the red lead in the connector half coming from the positive side of the battery and inserting the black lead into the other half. The circuit is now completed through the meter and, when activated, current will flow through the circuit and the meter, which will indicate the amount.

> ⚙⚙ **SERVICE TIP** Current flow with an inductive clamp has a very distinct advantage over old style meters that required opening the circuit to make a measurement. The inductive clamp does not require the circuit be disturbed in order to get a reading.

Inductive ammeters are not connected into series with the circuit. This type of meter utilizes an inductive pickup, which monitors the magnetic field formed by current flowing through a conductor (**Figure 4-4**). These meters normally have two leads, positive and negative, and an inductive clamp. The positive and negative leads are connected to their appropriate terminals of the multimeter. The inductive lead is clamped around a conductor in the circuit. When the circuit is activated, current flow can be read on the meter. Meters used to measure current flow in the starting and charging systems normally are the inductive type.

Classroom Manual
Chapter 4, page 109

Normally the positive lead is red and the negative lead is black.

Figure 4-3 An ammeter is always connected in series with the circuit. This allows circuit current to flow through the meter.

Figure 4-4 An inductive pickup for an ammeter.

Voltmeters

Electrical circuits and components cannot operate properly if a proper amount of voltage is not available to them. Not only must the battery be able to deliver the proper amount of voltage, but the proper amount of voltage must be available to operate the intended component. As current flows through a resistance, voltage is lost or dropped. In circuits where undesired resistance is present, the voltage available for the circuit is decreased. Because of these conditions, voltage is measured at the source, at the electrical loads, and across the loads and circuits.

Classroom Manual
Chapter 4, page 110

Connecting the positive lead to any point within a circuit and connecting the negative lead to a ground will measure the voltage at the point where the positive lead has been connected. If battery voltage is being measured, the positive lead is connected to the positive terminal of the battery and the negative lead to the negative post or to a clean spot on the common ground. The voltmeter will read the potential difference between the two points and display that difference on its scale. In the case of a 12-volt battery, the potential at the positive post is 12 volts and the potential at the negative is zero. Therefore, the meter will read 12 volts.

The positive lead can be connected anywhere within the circuit and it will measure the voltage present at that point. To measure the amount of available voltage, the negative lead can be connected to any point in the common ground circuit. To measure the amount of voltage drop across any wire or component of the circuit, the positive lead is connected to the battery side of the component and the negative lead is connected directly to the other side of the component. For example, to measure the voltage drop of a light bulb in a simple circuit, connect the positive lead to the battery or power lead at the bulb and the negative lead to the ground side of the bulb (**Figure 4-5**). The meter will read the amount of voltage drop across the bulb when the circuit is activated. If no other resistances are present in the circuit, the amount of voltage drop will equal the amount of source voltage. However, if there is a resistance present in the ground connection, the bulb will have a voltage drop that is less than the battery voltage. If the ground is bad, the voltage drop across the ground will equal the source voltage minus the amount of voltage dropped by the light. In order to measure the voltage drop and available voltage at various points within a circuit, the circuit must be activated (**Figure 4-6**).

Battery voltage is typically called *source voltage*.

Figure 4-5 To measure voltage drop, the voltmeter is connected in parallel or across the component being tested.

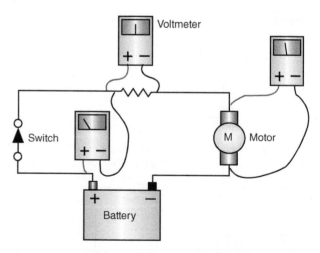

Figure 4-6 Voltmeter connections for measuring voltage drop across various parts of a circuit.

Ohmmeters

An ohmmeter is frequently used to test for circuit completeness or continuity. If a circuit is complete, the meter will display a low resistance, whereas if the circuit is open, a large amount of resistance and an infinite reading will be displayed.

Non-auto-ranging meters have a variety of scales to measure within. A technician chooses between scales of 1, 100, or 10,000 ohms. The anticipated resistance determines the scale to be selected.

To measure the resistance of a circuit or component, disconnect the power from the circuit or remove the component from the circuit (**Figure 4-7**). With the meter zeroed in the appropriate scale, connect one of the meter's leads to the power side of the component and the other lead to the ground side. Make sure you are not holding the metal leads while taking measurements. If you do this, you will become a parallel circuit with the component or circuit you are checking and you will get inaccurate results. Always hold the leads with the plastic handle. The meter will display the amount of resistance present between the two points. If the resistance is greater than the selected scale, the meter will show a reading of infinity (**Figure 4-8**). When this reading results, the technician should move up a scale on the meter. If subsequent scale changes result in continued readings of infinity, it can be assumed that there is no continuity between the two measured points.

Ohmmeters can also be used to compare the resistance of a component to the value it should have. Many electrical components have a specified resistance value, which is

Many technicians miss the K or M multipliers used with ohmmeter readings. K means the reading shown is multiplied by 1000. M means the reading should be multiplied by 1 million.

Classroom Manual
Chapter 4, page 112

An infinite reading simply means that the resistance is greater than what can be measured in that scale on the ohmmeter.

Figure 4-7 Using an ohmmeter to measure resistance. Note that the circuit's fuse has been removed.

Figure 4-8 A digital ohmmeter showing an infinite reading.

Digital multimeters may display "O.L" when there is an infinite measurement.

listed in the service information. This resistance value is important as it controls the amount of voltage drop and the amount of current that will flow in the circuit. If a component does not have the proper amount of resistance, the circuit will not operate properly.

Prior to testing a component or circuit with an ohmmeter, the service information should be checked for precautions regarding the impedance of the meter. Ohmmeters supply their own power to measure resistance and often electronic components can be damaged by causing excessive current to flow in them. Using an ohmmeter with the proper impedance will avoid damage to the components. If an impedance specification is not given in the manual, it can be safely assumed that any ohmmeter is appropriate for that circuit.

Test Lights

Often a test light is used to verify the presence of voltage at a particular point. Although the test light cannot give a reading of the amount of voltage present, it can clearly display the presence of voltage. Normally a test light has one wire lead. This is connected to ground. The positive lead is actually a piercing probe that is slipped into the circuit at the desired point to verify voltage. If voltage is present, the bulb in the test light will shine.

A self-powered test light is often used instead of an ohmmeter to test for continuity. A self-powered test light does not rely on the power of the circuit to light its bulb. Rather, it contains a battery for a power supply; when connected across a completed circuit, the bulb will light. This type of test light is connected to the circuit in the same manner as an ohmmeter.

Neither type of test light should be used to test electronic circuits or components. The use of test lights should be limited to checking for battery voltage at various points within an electrical circuit.

MULTIMETERS

A **multimeter** is a meter that combines the functions of a voltmeter, ammeter, and ohmmeter.

A multimeter is often referred to as a VOM (volt-ohm meter).

Impedance is best defined as the operating resistance of an electrical device.

Ohmmeters, ammeters, and voltmeters can be separate meters or can be combined in one meter called a **multimeter**. Most multimeters can measure DC voltage, DC current (amps), AC voltage, and resistance (ohms). Some meters can also measure frequency. Multimeters have either a needle-type analog display, or a numeric digital display. To use this type of meter, the desired function and the scale are selected prior to connecting the meter. Once the meter is set in a function, the **digital multimeter (DMM)** should be used as an individual function type meter.

A DMM with high **impedance** is normally required for testing electronic components and circuits. The high impedance prevents a surge of high current through the sensitive semiconductors when the tester is connected into a circuit. High current can destroy electronic circuits and components.

Using Meters

Meters can be used to test electrical components, such as switches, loads, and semiconductors. Switches can be tested for operation and for excessive resistance with a voltmeter, test light, or ohmmeter. To check the operation of a switch with a voltmeter or a test light, connect the meter's positive lead to the battery side of the switch. With the negative lead connected to a good ground, voltage should be measured at this point. Without closing the switch, move the positive lead to the other side of the switch. If the switch is open, no voltage will be present at that point. The amount of voltage present at this side of the switch should equal the amount on the other side when the switch is closed. If the voltage decreases, the switch is causing a voltage drop due to excessive resistance. If no voltage is present on the ground side of the switch with it closed, the switch is not functioning properly and should be replaced.

If a switch has been removed from the circuit, it can be tested with an ohmmeter or a self-powered test light. By connecting the leads across the switch connections, the action of the switch should open and close the circuit.

Classroom Manual
Chapter 4, page 115

Never test an electronic component with a self-powered test lamp, especially an air bag circuit.

Ohmmeters can be used to test semiconductors. Because semiconductors allow current flow only in certain directions, they can be tested by connecting an ohmmeter to their connections. Some digital meters have a feature called a diode test that allows for accurate testing of diodes. This feature should be used. By placing the leads of the meter on the semiconductor connections, continuity should be observed. Reversing the meter leads to the same connections should result in different readings of continuity. For example, a diode should show good continuity (low resistance) when the leads are connected to it and poor continuity (high resistance) when the leads are reversed.

In addition to meters and test lights, troubleshooting electrical problems may involve the use of fused jumper wires, circuit breakers, or short detection devices. A jumper wire is simply a length of wire, normally equipped with alligator clips on either end. These clips allow the wire to be easily and safely connected to various points within a circuit. Jumper wires are primarily used to bypass a particular point within a circuit and should contain an in-line fuse or circuit breaker to prevent damage to the circuit being tested and the wire itself.

ELECTRICAL PROBLEMS

Normally, the circuits in an automobile have a circuit protection device placed in the common path from the battery. This protection device is usually a fuse or a circuit breaker. They are designed to protect the wires and components from excessive current. When a great amount of current flows through the fuse or breaker, an element will burn out, causing the circuit to be opened and stopping current flow. This action prevents the high current from burning up the wires or components in the circuit.

Classroom Manual
Chapter 4, page 112

High current is caused by low resistance. A decrease in the amount of resistance is typically the result of a **short**. A short is best defined as an additional and unwanted path to ground. Most shorts, such as what occurs when a bare wire contacts the frame of the car, create an extremely low resistance parallel branch. A slow-turning motor can also cause low resistance and high current.

A short to ground can be present before the load in the circuit or internally within the load or component. A short can also connect two or more circuits together, causing additional parallel legs and uncontrolled operation of components. An example of a possible result from a wire-to-wire short would be the horn blowing each time the brake pedal is depressed. This could be caused by a wire-to-wire short between the horn and brake light circuits. Shorts are one of the three common types of electrical problems.

A circuit breaker is a protection device that resets itself after it has been tripped by high current. Circuit breakers open due to the heat of high current. While open, the breaker cools and closes again to complete the circuit. When a fuse is blown or burned out by high current, it must be replaced to reactivate the circuit. When diagnosing a circuit with a short, a completed circuit makes locating the fault easier. By inserting a circuit breaker in place of the fuse, the circuit can be activated and still protected.

A number of techniques and test instruments can be used to locate a short. The most common of these is called a short tester. This device actually includes a circuit breaker and a magnetic sensing device (**Figure 4-9**). The latter may be called a Gauss gauge or a

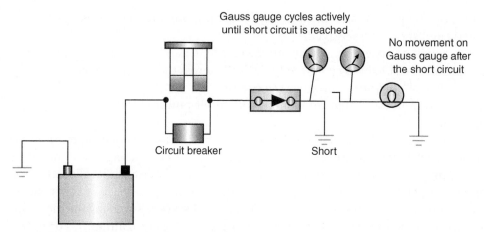

Figure 4-9 A typical way to find the location of a short.

compass. The gauge relies on the magnetic field formed by current flow to indicate the location of a short. The cycling of the circuit breaker is indicated by the sweeps of the detector's needle as it is moved along the length of the circuit. When the needle no longer sweeps in response to the breaker, the location of the short can be assumed to be before that point in the circuit. Carefully moving the gauge back through the circuit should locate the exact place of the short.

Another common electrical fault is the **open**. An open causes an incomplete circuit and can result from a broken or burned wire, a loose connection, or a faulty component. If a circuit is open, there is no current flow and the component will not operate. If there is an open in one leg of a parallel circuit, the remaining part of that parallel circuit will operate normally.

To determine the location of an open, a test light or voltmeter is used. By probing along the circuit, the completed path can be traced. The point at which voltage is no longer present is the point at which the open is. If the circuit is open, there will be no current flow. This means that if the open is after the load, source voltage will be measured after the load because there is no voltage drop in the circuit. Often a jumper wire is used to verify or isolate the problem area.

Excessive resistance at a connector (**Figure 4-10**), internally in a component, or within a wire is also a common electrical problem. High resistance will cause low current flow and the component will not be able to operate normally, if at all.

The source of high resistance is easily found by measuring the voltage drop across the circuit. It is important to remember that excessive resistance can occur anywhere in the circuit. The best way to identify the source of high resistance is to divide the affected circuit into three parts: the component, the ground circuit, and the power circuit. If the voltage drop across the component is equal to source voltage (**Figure 4-11**), there are no other resistances in the circuit. If the voltage drop is less than source voltage, check the voltage drop across the ground circuit. If the voltage drop is more than 0.1 volts, there is high resistance in that part of the circuit. The source can be located by separating that circuit into small measurable parts and then measuring each part. The part that has the excessive voltage drop across it is the source for the high-resistance problem. The power side of the circuit is checked in the same way.

Lab Scopes

Lab scope is the name given to a small (sometimes handheld) oscilloscope. An oscilloscope is actually a visual voltmeter. Lab scopes have become the diagnostic tool of choice for many good technicians. A scope allows you to see voltage changes over time

Figure 4-10 The various transmission connectors should be inspected and checked for high-resistance problems.

Figure 4-11 A simple light circuit with unwanted resistance. Notice the reduced voltage drop across the lamp and the reduced circuit current.

(**Figure 4-12**). Voltage is displayed across the screen of the scope as a waveform. By displaying the waveform, the scope shows the slightest changes in voltage. This is a valuable feature for a diagnostic tool. With a scope, precise measurement of voltage is possible. When measuring voltage with an analog voltmeter, the meter only displays the average values at the point being probed. Digital voltmeters simply sample the voltage several times each second and update the meter's reading at a particular rate. If the voltage is constant, good measurements can be made with both types of voltmeters. A scope will display any change in voltage as it occurs. This is especially important for diagnosing intermittent problems.

The screen of a lab scope is divided into small divisions of time and voltage (**Figure 4-13**). These divisions set up a grid pattern on the screen. The horizontal movement of the waveform represents time. Voltage is measured with the vertical position of the waveform. Since the scope displays voltage over time, the waveform moves from the

Special Tools

Lab scope
Assortment of jumper wires

Figure 4-12 Voltage and time axes on a scope's screen.

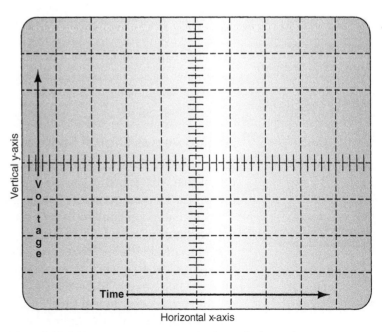

Figure 4-13 Grids on a scope screen serve as time and voltage references.

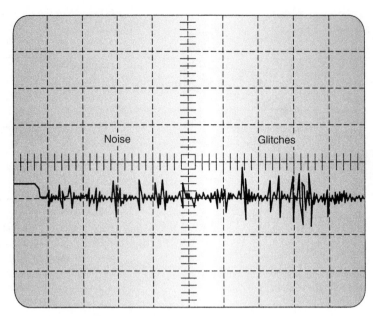

Figure 4-14 RFI noise and glitches may appear on the voltage signal.

left (the beginning of measured time) to the right (the end of measured time). The value of the divisions can be adjusted to improve the view of the voltage waveform. For example, the vertical scale can be adjusted so that each division represents 0.5 volts and the horizontal scale can be adjusted so that each division equals 0.005 seconds (5 milliseconds). This allows you to view small changes in voltage that occur in a very short period of time. The grid serves as a reference for measurements.

Since a scope displays actual voltage, it will display all electrical noises or disturbances that accompany the voltage signal (**Figure 4-14**). Noise is primarily caused by **radio frequency interference (RFI)**, which may come from the ignition system. RFI is an unwanted voltage signal that rides on a signal. This noise can cause intermittent problems with unpredictable results. The noise causes slight increases and decreases in the voltage. When a computer receives a voltage signal with noise, it will try to react to the minute changes. As a result, the computer responds to the noise rather than the voltage signal.

Electrical disturbances or **glitches** are momentary changes in the signal. These can be caused by intermittent shorts to ground, shorts to power, or opens in the circuit. These problems can occur for only a moment or may last for some time. A lab scope is handy for finding these and other causes of intermittent problems. By observing the voltage signal and wiggling or pulling a wiring harness, any looseness can be detected by a change in the voltage signal.

A **glitch** is best defined as abnormal movement of the waveform.

Analog versus Digital Scopes

Analog scopes show the actual activity of a circuit and are called live scopes. This simply means that what is taking place at the point being measured or probed is what you see on the screen. The vast majority of oscilloscopes used in automotive service are digital scopes.

A digital scope, commonly called a **digital storage oscilloscope (DSO)**, converts the voltage signal into digital information and stores it in its memory. To help in diagnostics, a technician can "freeze" the captured signal for close analysis. DSOs also have the ability to capture low-frequency signals. Low-frequency signals tend to flicker when displayed on an analog screen. To have a clean waveform on an analog scope, the signal must be repetitive and occur in real time.

DSO is a common acronym for **digital storage oscilloscope.**

Most DSOs have a sampling rate of one million samples per second. This is quick enough to serve as an excellent diagnostic tool. This fast sampling rate allows slight

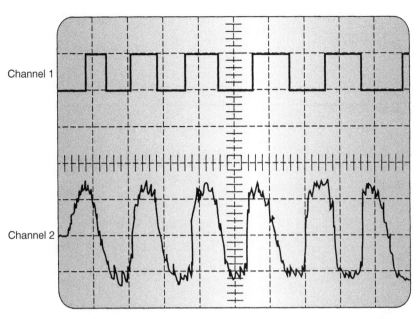

Figure 4-15 Some scopes can display two traces at one time; these are called dual-trace scopes.

changes in voltage to be observed. Slight and quick voltage changes cannot be observed on an analog scope.

A DSO uses an analog-to-digital (A/D) converter to digitize the input signal. Since digital signals are based on binary numbers, the trace appears slightly choppy when compared to an analog trace. But the voltage signal is sampled more often, which results in a more accurate waveform. The waveform is constantly being refreshed as the signal is pulled from the scope's memory. Remember, the sampling rate of a DSO can be as high as a million times per second.

Both an analog and a digital scope can be dual-trace scopes (**Figure 4-15**). This means they both have the capability of displaying two traces at one time. By watching two traces simultaneously, you can watch the cause and effect of a sensor, or compare a good or normal waveform to the one being displayed. Many scopes have four or more channels.

Waveforms

A waveform represents voltage over time. Any change in the **amplitude** of the trace indicates a change in the voltage. When the trace is a straight horizontal line, the voltage is constant (**Figure 4-16**). A diagonal line up or down represents a gradual increase or decrease in voltage. A sudden rise or fall in the trace indicates a sudden change in voltage.

Scopes can display AC and DC voltage either one at a time or simultaneously, as in the case of noise caused by RFI. Noise results from AC voltage riding on a DC voltage signal. The consistent change of polarity and amplitude of the AC signal causes slight changes in the DC voltage signal. A normal AC signal changes its polarity and amplitude over a period of time. The waveform created by AC voltage is typically called a **sine wave** (**Figure 4-17**). One complete sine wave shows the voltage moving from zero to its positive peak, then moving down through zero to its negative peak and returning to zero. If the rise and fall from positive and negative are the same, the wave is said to be **sinusoidal**. If the rise and fall are not the same, the wave is nonsinusoidal. Therefore, it can be said that all AC voltage waveforms are not sine waves.

One complete sine wave is a **cycle**. The number of cycles that occur per second is the frequency of the signal. Checking frequency or cycle time is one way of checking the operation of some electrical components. Input sensors are the most common

A captured signal is a signal stored in a DSO's memory.

Binary numbers are strings of zeros and ones, or ons and offs, which represent a numeric value.

The height of a waveform is called its **amplitude**.

A **sine wave** is a waveform of a single frequency alternating current.

The term **sinusoidal** means the wave is shaped like a sine wave.

A **cycle** is one set of changes in a signal that repeats itself several times.

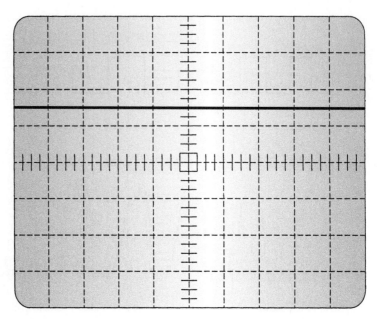

Figure 4-16 A constant voltage waveform.

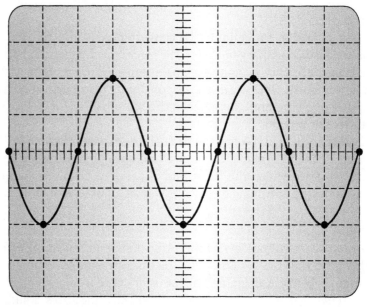

Figure 4-17 An AC voltage sine wave.

components that produce AC voltage. Permanent magnet voltage generators produce an AC voltage that can be checked on a scope (**Figure 4-18**). AC voltage waveforms should also be checked for noise and glitches, which may send false information to the computer.

DC voltage waveforms may appear as a straight line or a sloping line showing a change in voltage. Sometimes a DC voltage waveform will appear as a square wave or digital signal, which shows voltage making an immediate change (**Figure 4-19**). Square waves have straight vertical sides and a flat top. This type of wave represents voltage being applied (circuit being turned on), voltage being maintained (circuit remaining on), and no voltage applied (circuit is turned off). Of course, a DC voltage waveform may also show gradual voltage changes.

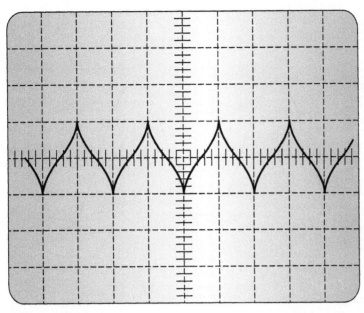

Figure 4-18 An AC voltage trace from a typical permanent magnet generator-type pickup or sensor.

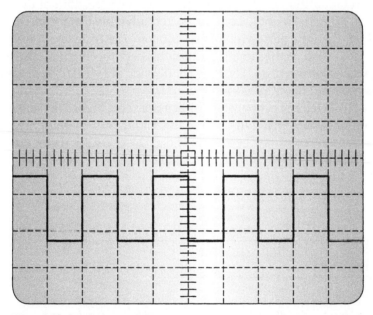

Figure 4-19 A typical square wave.

Scope Controls

Depending on the manufacturer and model of the scope, the type and number of its controls will vary. However, nearly all scopes have these: vertical (y-axis) adjustments, horizontal (x-axis) adjustments, and trigger adjustments.

The vertical adjustment actually controls the voltage displayed. The voltage setting of the scope is the voltage that will be shown per division (**Figure 4-20**). If the scope is set at 0.5 volts (500 millivolts), this means a 5-volt signal will need 10 divisions. Likewise, if the scope is set to 1 volt, 5 volts will need only 5 divisions. While using a scope, it is important to set the vertical so that voltage can be accurately read. Setting the voltage too low may cause the waveform to move off the screen, while setting it too high may cause the trace to be flat and unreadable. The vertical position control allows the vertical position of the trace to be moved anywhere on the screen.

Examples of permanent magnet generators are some crankshaft and camshaft sensors, ignition pickup units, vehicle speed sensors, and wheel speed sensors.

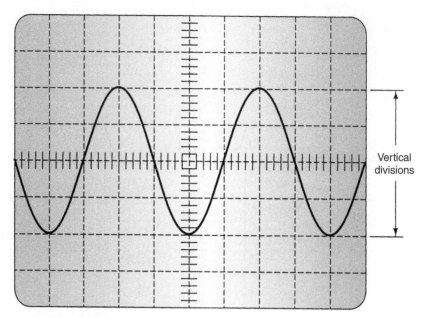

Figure 4-20 Vertical divisions represent voltage.

Some of the commonly found components powered by digital signals are fuel injectors, mixture-control solenoids, and digital EGR valves.

The horizontal position control allows the horizontal position of the trace to be set on the screen. The horizontal control is actually the time control of the trace (**Figure 4-21**). If the time per division is set too low, the complete trace may not show across the screen. Also, if the time per division is set too high, the trace may be too crowded for detailed observation. The time per division (TIME/DIV) can be set from very short periods of time (billionths of a second) to full seconds.

Trigger controls tell the scope when to begin a trace across the screen. Setting the trigger is important when trying to observe the timing of something. Proper triggering will allow the trace to repeatedly begin and end at the same points on the screen. There are typically numerous trigger controls on a scope. The AUTO setting will display a trace regardless of the time base.

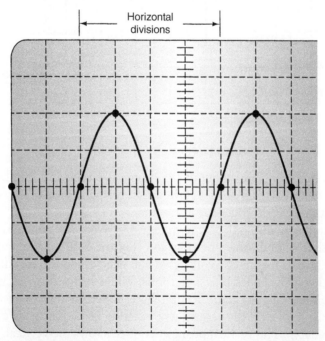

Figure 4-21 Horizontal divisions represent time.

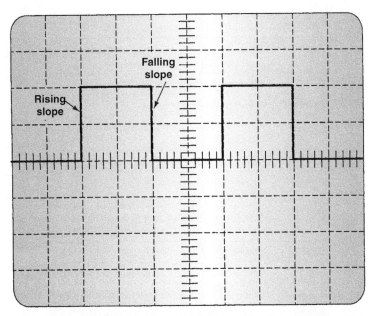

Figure 4-22 The trigger can be set to start the trace with a rise or fall of the voltage.

Slope and level controls are used to define the actual trigger voltage. The slope switch determines whether the trace will begin on a rising or falling of the voltage signal (**Figure 4-22**). The level control determines where the time base will be triggered according to a certain point on the slope.

A trigger source switch tells the scope which input signal to trigger on. This can be Channel 1, Channel 2, line voltage, or an external signal. External signal triggering is very useful when you want to observe a trace of a component that may be affected by the operation of another component. An example of this would be observing pressure control solenoid activity when changes in vehicle speed are made. The external trigger would be voltage change at the VSS. The displayed trace would be the cycling of the pressure control solenoid. Channel 1 and Channel 2 inputs are determined by the points of the circuit being probed. Some scopes have a switch that allows inputs from both channels to be observed at the same time or alternately.

SCAN TOOLS

A scan tool is one of the most important diagnostic tools available to a technician. The scan tool interfaces with the computer network on the vehicle to retrieve trouble codes, to see data from sensors and commands from the computer network as well as freeze frame data associated with stored trouble codes, and to take a "snapshot" of data at any time. Many scan tools can command outputs as well, such as shifting the transmission for diagnostic purposes. Many original equipment manufacturer scan tools are using laptop-based scan tools, to save the cost of developing hardware and concentrate on software. Many recent scan tools can be updated via the Internet. Some scan tools can program or flash modules when a unit is updated or replaced.

A computer will set a DTC when a voltage signal is entirely out of its normal range. If a signal is within its normal range but still not correct for the conditions, the computer will not set a DTC. However, there is still a problem that can cause a customer concern. To help identify the cause of the concern, the signals to and from the computer should be carefully observed. This is done with a scan tool. The tool is set to read **serial data**, which is often displayed in PID values (**Figure 4-23**).

A **scan tool** is actually a computer designed to communicate with the vehicle's computer.

The communications to and from the computer are commonly referred to as the system's **serial data**.

PID Name	Description of PID
EPC	Commanded Electronic Pressure Control Pressure—in psi
GEAR	Commanded Gear—not actual
LINEDSD	Commanded Line Pressure—in psi
OSS	Input from Output Shaft Speed Sensor—in rpm
RPM	Input from Engine Speed Sensor—in rpm
SSA	Commanded State of Shift Solenoid No. 1—ON or OFF
SSB	Commanded State of Shift Solenoid No. 2—ON or OFF
SS1F	Shift Solenoid No. 1 Circuit Fault—YES or NO
SS2F	Shift Solenoid No. 2 Circuit Fault—YES or NO
TCCACT	Slippage of Torque Converter Clutch—in rpm
TCCCMD	Commanded State of Torque Converter Clutch Solenoid—in %
TCCF	Torque Converter Clutch Solenoid Circuit Fault—YES or NO
TFT	Transmission Fluid Temperature—in voltage or degrees
TP	Throttle Position—in voltage
TR	Transmission Range Sensor—by position
TRANRAT	Actual Transmission Gear Ratio—by position
TSS	Input from Turbine Shaft Speed Sensor—in rpm

Figure 4-23 Common transmission PIDs for Ford products.

Although OBD II regulations standardized many DTCs for emission problems, transmission DTCs may not reflect the same problems across the different transmission models produced by the different manufacturers. There are some codes that reflect the same basic problem. For example, a P0705 code relates to a TR range sensor circuit problem. However, each manufacturer may have several other DTCs that describe particular failures, such as Ford's P0708 code, which says a particular circuit for the TR sensor is open. Always consult service information when deciphering a DTC. It is also important to note that if a problem does not increase emissions, the malfunction indicator lamp (MIL) light may not be illuminated even if a DTC has set.

GENERAL TRANSMISSION DIAGNOSIS

Classroom Manual
Chapter 4, page 123

A host of inputs and outputs are involved in shifting the transmission. Many of the same inputs are used for the engine and transmission control. The TCM and ECM share information to make decisions (**Figure 4-24**).

The controls of a transmission direct hydraulic flow by using solenoid-controlled valves. When used to control TCC operation, the solenoid opens a hydraulic circuit to the TCC spool valve, causing the spool valve to move and direct mainline pressure to apply the clutch. Electronically controlled shifting is accomplished in much the same way. Shifting occurs

Figure 4-24 The components involved in the operation of a typical late-model transmission system.

when a solenoid is either turned on or turned off (**Figure 4-25**). At least two shift solenoids are incorporated into the system and shifting takes place by controlling the solenoids. The desired gear is put into operation through a combination of on and off solenoids.

Several sensors and switches are used to inform the control computer of the current operating conditions. Most of these sensors are also used to calibrate engine performance. The computer then determines the appropriate shift time for maximum efficiency and best feel. The shift solenoids are controlled by the computer, which either supplies power to the solenoids or supplies a ground circuit. The techniques for diagnosing electronic transmissions are basically the same techniques used to diagnose TCC systems.

Figure 4-25 Shift solenoid action.

PRELIMINARY CHECKS

Some of the common problems that affect shift timing and quality, as well as the timing and quality of TCC engagement, are incorrect battery voltage, a blown fuse, poor connections, a defective TP sensor or VSS, defective solenoids, wires to the solenoids or sensors crossed, corrosion at an electrical terminal, or faulty installation of aftermarket accessories.

Electrical circuit problems, faulty electrical components can cause improper shift points.

Often computer-controlled transmissions will start off in the wrong gear. This can happen for several reasons, either because of internal transmission problems or external control system problems. Internal transmission problems can be faulty solenoids or stuck valves. External problems can be the result of a complete loss of power or ground to the control circuit or a fail-safe protection strategy initiated by the computer to protect itself or the transmission from an observed problem.

GM calls their preliminary diagnosis a "diagnostic circuit check."

Default gear. Transmissions usually have a default gear. The default gear is used when the transmission has an electrical problem. The transmission will start off in the default gear and stay in the default gear. The default gear allows the driver to "limp in" to a service center. Typically the default gear is simply the gear that is applied when the shift solenoids are off.

Basic System Checks

Troubleshooting a transmission's electronic control system is like troubleshooting any other electronic system; you need to make preliminary checks of the system before moving into specific checks.

The first step in a preliminary check is verifying system voltage. With a DMM, check the open-circuit voltage of the battery. Minimum voltage at the battery should be 12.6 volts. If the battery voltage is below this, recharge the battery. If the voltage is still low after recharging, replace it.

Continue by checking the condition of the battery cables. Conduct a voltage drop test across each cable. Make sure the system is activated when doing a voltage drop test. There should be no more than 0.1 volt dropped across the positive or the negative cable. If the voltage drop exceeds that amount, clean or replace the cables.

A visual inspection of the circuit should follow. Carefully look at the entire system and check for damaged or corroded wires, loose connections, and damaged connectors

Line F to "A" solenoid negative   Line EX to "A" solenoid positive   Line EY to "B" solenoid positive

Line G to "B" solenoid negative

Connector

Plugs

Line B to PWM solenoid negative

Line EZ to PWM solenoid positive

Line C to temperature switch for vehicles with traction control only

Line D to TCC solenoid negative   Line A to TCC solenoid positive

Figure 4-26 The position and condition of the transmission wiring harness and its connectors should be carefully checked.

(**Figure 4-26**). Pay particular attention to the connectors. Make sure there are no bent or broken terminals. If any are present, replace the connector. Also look inside each connector for signs of corrosion. If any is found, clean the wires and terminals. If the corrosion cannot be cleaned, replace the wires or the connector.

Continue your basic inspection by checking the fuse or fuses to the control module. To accurately check a fuse, either test it for continuity with an ohmmeter or check each side of the fuse for power when the circuit is activated.

CUSTOMER CARE Vehicles equipped with control computers may require a relearn procedure after the battery is disconnected. Many computers memorize and store vehicle operation patterns for optimum driveability and performance. When the vehicle's battery is disconnected, this memory is lost. The computer will use default data until new data from each key start is stored. Customers often complain of driveability problems during the relearn stage because the vehicle acts differently than it did before it was serviced. To reduce the possibility of complaints, the vehicle should be road tested and the correct relearn procedure followed.

⚠ **Caution**
Remember to disconnect power to the component before checking it with an ohmmeter. Failure to do so can result in the meter being destroyed.

If the system has no apparent problems, continue testing the system. The diagnostic procedure should include checking the system with a scan tool. The scan tool will display any trouble codes in the system. More important to the technician, scan tools will display serial data. The serial data stream allows you to monitor system sensor and actuator activity during operation. Comparing the test values to the manufacturer's specifications will greatly help in diagnostics.

It is possible that the data displayed by a scan tool is not the actual value. Most computer systems will disregard an input that is well out of range and rely on a default value held in its memory, as well as set a trouble code. Default values are hard to recognize and do little for diagnostics; this is why detailed testing with a DMM or lab scope can be valuable. These test instruments are also used to further test the system after a scan tool has identified a problem.

Road Test

Critical to proper diagnosis of the transmission and TCC control systems is a road test. The road test should be conducted with a scan tool connected to the vehicle to monitor engine and transmission operation.

During the road test, the vehicle should be driven in the normal manner. All pressure and gear changes should be noted. Also, the various computer inputs should be monitored and the readings recorded for future reference. Some scan tools have the capability of saving readings such as screenshots and snapshots to a removable SD card or thumb drive for later review. Critical information from the inputs includes engine speed, vehicle speed, manifold vacuum, operating gear, and the time it took to shift gears.

> ⚙ **SERVICE TIP** If the transmission will not upshift after you have serviced a Hyundai with a Mitsubishi KM-series transmission, check the connectors. It is possible to mistakenly connect the output connector to the input sensor connector and vice versa. The connectors are identical; only the wire colors are different, and these should be used to verify a correct connection.

Basic Transmission Circuits

As a comparison of cold-start shifting and schedule shifting, if the engine is cold, the transmission will shift later and converter clutch engagement will also be later. When the engine warms up, the transmission will shift more quickly and have an earlier converter clutch engagement.

To summarize how the circuitry of an electronically controlled transmission interrelates with a computerized engine control system, a summary of the transmission controls for a Ford 6R80 follows. This system is similar to the other systems used by Ford, as well as by other manufacturers. The transmission is controlled by the PCM of the engine control system when used with a gasoline engine, but uses a TCM under diesel applications. In either case the TCM/ECM or PCM receives inputs from both the engine and the transmission, which are shared over a CAN network with the BCM and ABS computers. Based on this information, the PCM can send signals to operate components in both the engine and transmission. There are many inputs to the PCM from the transmission. The transmission range (TR) sensor informs the PCM what gear has been selected. The vehicle speed sensor inputs the mph at which the vehicle is traveling. The turbine speed sensor is a Hall effect switch that sends signals to inform the PCM of the speed of the input to the transaxle. A second Hall effect sensor is used for the output shaft speed sensor so the gear ratio of the transmission can be accurately calculated and monitored for faults or slipping during shifts. The transmission oil temperature sensor monitors the temperature of the ATF. This signal determines whether the PCM should go to a cold-start shift cycle or shift according to its schedule.

The 6R80 has five shift solenoids, which are detailed in **Figure 4-27**. Note that shift solenoid E actually controls the application of shift solenoid D, which can either be off for engine braking in first gear, third, fourth, fifth or sixth gear, depending on gear selection.

There are seven solenoids in total in the 6R80; the electronic pressure control solenoid that regulates transmission operating pressure, a modulated solenoid that controls the converter clutch, and five shift solenoids that control fluid flow to the various holding and clutching devices.

Because the engine and transmission control systems share information, a common component may affect both systems. Nearly all computer control systems have some form of self-diagnostics. During this mode of operation, the computer will display a trouble code that represents a circuit or component it determines is faulty. Part of the basic procedure for diagnosing a transmission is to retrieve the trouble codes from the computer. This information will determine what step you should take next (**Figure 4-28**).

Driver Selected Gear	PCM Command	Solenoid A	Solenoid B	Solenoid C	Solenoid D	Solenoid E	Line Pressure Control Solenoid	TCC Solenoid
P	P	OFF	ON	OFF	OFF	OFF	Variable	OFF
N	N	OFF	OFF	OFF	OFF	OFF	Variable	OFF
R	R	OFF	OFF	OFF	OFF	OFF	Variable	OFF
D	D	OFF	OFF	OFF	OFF	OFF	Variable	OFF
	D1	ON	ON	OFF	ON	OFF	Variable	OFF
	D2	ON	ON	ON	ON	OFF	Variable	ON/OFF
	D3	ON	OFF	OFF	ON	ON	Variable	ON/OFF
	D4	ON	ON	OFF	OFF	ON	Variable	ON/OFF
	D5	OFF	OFF	OFF	OFF	ON	Variable	ON/OFF
	D6	OFF	ON	ON	OFF	ON	Variable	ON/OFF

Figure 4-27 6R80 Solenoid application chart. In Solenoid A, C, and the TCC solenoids the PCM controls pressure proportionally, meaning that low current produces a low pressure and high current produces a high pressure (directly proportional). The line pressure, B and D solenoids, with no current flow to the solenoid the pressure is high and goes down with increased current flow (inversely proportional). Solenoid E is either fully on or off. When E is off, it allows D to control the low-reverse clutch, when on it allows D to control oil to the overdrive clutch.

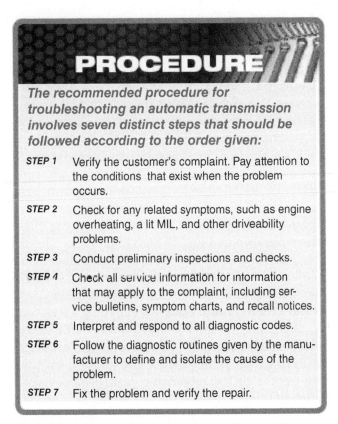

PROCEDURE

The recommended procedure for troubleshooting an automatic transmission involves seven distinct steps that should be followed according to the order given:

STEP 1   Verify the customer's complaint. Pay attention to the conditions that exist when the problem occurs.

STEP 2   Check for any related symptoms, such as engine overheating, a lit MIL, and other driveability problems.

STEP 3   Conduct preliminary inspections and checks.

STEP 4   Check all service information for information that may apply to the complaint, including service bulletins, symptom charts, and recall notices.

STEP 5   Interpret and respond to all diagnostic codes.

STEP 6   Follow the diagnostic routines given by the manufacturer to define and isolate the cause of the problem.

STEP 7   Fix the problem and verify the repair.

Figure 4-28 A flow chart showing the major steps that should be followed while diagnosing an EAT.

As you can readily tell, in this system, as well as the others, engine performance problems can cause the transmission to perform unsatisfactorily. Since the computer controls the outputs to the engine and the transmission based on input sensors common to both, it is very important that you perform basic engine checks before spending time diagnosing the transmission. This was important when diagnosing nonelectronically controlled transmissions and is even more important when diagnosing a transmission.

Classroom Manual
Chapter 4, page 131

On-Board Diagnostics (OBD)

On-board diagnostic capabilities are incorporated into a vehicle's computer to monitor virtually every component that can affect emission performance. Each component is checked by a diagnostic routine to verify that it is functioning properly. In addition to self-diagnostics, OBD provide a means to check the operation of the control systems.

The system continuously monitors operating conditions for possible system malfunctions. It compares system conditions against programmed parameters. If the conditions fall outside of these limits, it detects a malfunction and sets a trouble code that indicates the portion of the system that has the fault. It is very important that you refer to the service information when interpreting DTCs. When diagnosing a transmission concern, make sure there are no engine-related codes that could affect the operation of the transmission. If there is an engine problem, fix it before continuing with your diagnosis.

Troubleshooting Transmissions

The following steps provide a general outline for troubleshooting transmissions. There are slight variations in different years and with different models. Always refer to the manufacturer's information before beginning your diagnosis. Below is a brief explanation of the steps that should be followed:

1. **Interview the Customer**—Gather as much information as possible from the customer. Ask the customer to describe the driving conditions present when the problem appears. This should include weather, traffic, and speed. The initial steps during diagnosis should verify the customer's concern.

2. **Check the MIL**—The MIL should turn on when the ignition is turned on and the engine is not running. When the engine is started, the MIL should go off. If either of these does not occur, troubleshoot the lamp system before continuing. The MIL is basically an engine malfunction light, but if the TCM detects a problem that may affect emissions, it will send a request over the data bus to the PCM to turn on the MIL lamp.

3. **Visual Inspection**—Check all wires for frays, looseness, and damage. Also, check the wiring for burned or chafed spots, pinched wires, or contact with sharp edges or hot exhaust parts. Try not to wiggle the wires while doing this, a wiggle may correct an intermittent problem that may be hard to find later. Check the condition of the battery and all sensors and actuators for physical damage. Check all connections to sensors, actuators, control modules, and ground points. Also check all vacuum hoses for pinches, cuts, or disconnects. Correct any problems.

4. **Check DTC(s) and Freeze Frame Data**—When using a scan tool, an * (asterisk) next to the DTC often indicates there is stored freeze frame data associated with that DTC. During diagnostics, it is helpful to know the conditions present when the DTC was set, such as was the vehicle running or stopped, what was the engine's temperature, and was the air/fuel ratio rich or lean. Print or record all DTC and related information. If there was a no or poor communication DTC, solve that problem before continuing.

5. **Check Service History and Service Publications**—There may be a TSB or other service alert that may have the necessary repair information. Check these and follow the procedures outlined in them before continuing. Service history can give clues about cause of the problem, as the problem may be related to a recent repair.

6. **Clear DTC and Freeze Frame Data**—Using the scan tool, clear all DTCs and freeze frame data (**Figure 4-29**). Make sure you do this only after you have recorded all retrieved data.

7. **Check DTCs**—Repeat step 4. If there are no DTCs, check the status of the monitors' readiness and the pending codes on the scan tool. Do what is necessary to

The malfunction indicator lamp (MIL) is a warning lamp in a vehicle's instrument panel that lets the driver know when the vehicle's electronic control units detected a problem.

Figure 4-29 After all retrieved data are recorded, clear all DTCs and freeze frame data.

complete the necessary drive cycles before continuing. For many DTCs, the PCM will enter into the fail-safe mode. This means the PCM has substituted a value to allow the engine to run. Refer to the service information to determine if any DTCs indicate the fail-safe mode, if so, follow the appropriate diagnostic steps. Current DTCs indicate a problem that is present. Use the DTC chart to determine what was detected, the probable problem areas, and how to diagnose that DTC.

8. **Basic Inspection**—When the DTC is not confirmed in the previous check, use the problems symptoms chart in the service information to diagnose a problem when no code is displayed but the problem is still occurring.

9. **DTC Chart**—Use the DTC chart to determine what was detected, the probable problem areas, and how to diagnose that DTC (**Figure 4-30**).

10. **Check for Intermittent Problems**—If the cause of the problem has not yet been determined, proceed to check for an intermittent problem.

11. **Perform Repairs**—Once the cause of the concern has been identified, perform all required services.

12. **Repair Verification Test**—After repairs have been made, check your work by clearing all codes, checking the MIL, and rechecking for codes.

Diagnostic trouble codes (DTCs) are alphanumerical codes generated by the electronic control system to indicate a problem in a circuit or subsystem.

 Caution

Do not clear DTCs unless directed by a diagnostic procedure. Clearing DTCs will also clear valuable freeze frame and failure records data.

Malfunction Indicator Lamp Operation

The MIL is only used to inform the driver that the vehicle should be serviced soon. It also informs a technician that the computer has set a trouble code. OBD II systems continuously monitor the entire system, switches on a MIL if something goes wrong, and stores a fault code in the PCM when it detects a problem.

When the ignition is initially turned on, the MIL will momentarily flash ON then OFF and remain ON until the engine is running or the computer determines there are no DTCs stored in memory. The TCM monitors all transmission-related electrical components and typically sets a DTC if it detects a problem. However, not all set codes illuminate the MIL. The MIL is used to inform the driver of a serious problem (**Figure 4-31**). Therefore, DTCs can be set even though the MIL is not illuminated. Many of the diagnostic steps required under OBD II must be performed under specific operating conditions. The software in the TCM determines which tests will be conducted and when they will be run. Based on

DTC	COMPONENT	DESCRIPTION	CONDITION	SYMPTOM
P0705	Digital TR	Digital TR circuit failure.	This indicates an invalid pattern in TR-D. It is caused by a short to ground or an open in TR4, TR3A, TR2, and/or TR1 circuits. This DTC will not be set by an incorrectly adjusted TR sensor.	An increase in EPC pressure (harsh shifts). Defaults to (D) or D for all gear positions.
P0712	TFT, wiring, PCM	315°F (157°C) indicates the TFT circuit is grounded.	The voltage drop across the sensor exceeds the scale set for that temperature.	Firm shift feel.
P0713	TFT, wiring, PCM	–40°F (–40°C) indicates the TFT circuit is open.	The voltage drop across the sensor exceeds the scale set for that temperature.	Firm shift feel.
P0721	OSS	The OSS sensor signal is noisy.	The PCM has detected an erratic OSS signal.	Harsh shifts, abnormal shift schedule, no TCC engagement.
P0731	Shift Solenoid A (SSA), Shift Solenoid B (SSB), or internal parts	1st gear error.	No 1st gear.	Incorrect gear selection, depending on the failure mode and manual lever position. Shift errors may also be due to other internal transmission concerns (stuck valves, damaged friction material). Engine rpm could be higher or lower than expected.
P0732	SSA, SSB, or internal parts	2nd gear error.	No 2nd gear.	Incorrect gear selection, depending on the failure mode and manual lever position. Shift errors may also be due to other internal transmission concerns (stuck valves, damaged friction material). Engine rpm could be higher or lower than expected.
P0733	SSA, SSB, or internal parts	3rd gear error.	No 3rd gear.	Incorrect gear selection, depending on the failure mode and manual lever position. Shift errors may also be due to other internal transmission concerns (stuck valves, damaged friction material). Engine rpm could be higher or lower than expected.
P0734	SSA, SSB, or internal parts	4th gear error.	No 4th gear.	Incorrect gear selection, depending on the failure mode and manual lever position. Shift errors may also be due to other internal transmission concerns (stuck valves, damaged friction material). Engine rpm could be higher or lower than expected.
P0963	EPC solenoid	The EPC solenoid is shorted to the power circuit.	The voltage through the EPC solenoid is checked. An error will be noted if tolerance is exceeded.	Maximum EPC pressure, harsh engagements and shifts.
P1760	EPC, wiring, PCM	The EPC solenoid circuit is shorted or the output driver is faulty.	The PCM detected a loss of EPC during operation.	Unexpected reduction in engine torque.

Figure 4-30 A sample of transmission-related DTCs.

Figure 4-31 An example of an MIL that illuminates to warn the driver that transmission control has been switched to the default mode.

these tests, or monitors, the TCM will store different types of DTCs, as well as their history. The DTCs remain in memory until a certain number of trips or start cycles have completed and the problem has not re-occurred. The DTCs will also be removed when the battery is disconnected or cleared with the scan tool.

Data Link Connector

Standards require the **Data Link Connector (DLC)** for OBD II systems to be mounted in the passenger compartment out of sight of vehicle passengers. The standard DLC is a 16-pin connector. The same seven pins are used for the same information, regardless of the vehicle's make, model, and year. The connector is "D"-shaped and has guide keys that allow the scan tool to be only installed one way. Using a standard connector design and by designating the pins, it allows data retrieval with any scan tool designed for OBD II. Some European vehicles meet OBD II standards by providing the designated DLC along with their own connector for their own scan tool.

According to the OBD standards, the DLC must be easily accessible from the driver's seat (**Figure 4-32**). Therefore, they are located somewhere between the left end of the instrument panel and to the right of the center of the panel. The DLC cannot be hidden behind panels.

The DLC is designed only for scan tool use. Any generic scan tool can be connected to the DLC and can access the diagnostic data stream. The connector pins are arranged in two rows and are numbered consecutively. Seven of the 16 pins have been assigned by the OBD II standard. The remaining nine pins can be used by the individual manufacturers to meet their needs and desires.

Diagnostic Trouble Codes

Diagnostic trouble codes (DTCs) are designed to help technicians identify and locate problems in the transmission's control system.

Some OBD II codes are universal in that they are the same for all manufacturers and vehicle models. Other codes are manufacturer specific and must be interpreted through reference to the appropriate service information.

DTCs are set by the computer when a signal from a component or circuit is not within the normal operating range for the operating conditions. A code will also be set when there is no signal from an input circuit or to an output device. Transmission control modules also have the ability to conduct a self-test. During this test, the sensor circuits are checked and if a problem is found, a DTC is set.

Figure 4-32 Standard location for the DLC in an OBD II system.

There are basically two types of DTCs. A hard code is a DTC that represents a problem that is present at the time of retrieval. These are the codes that should be responded first, during diagnostics. Soft codes are those DTCs that are not currently present. These codes can be retrieved and represent an intermittent problem or a problem that existed but is no longer present.

An OBD II DTC is a five-character code with both letters and numbers (**Figure 4-33**). This is called the alphanumeric system. The first character of the code is a letter. This defines the system where the code was set. Currently there are four possible first character codes: "B" for body, "C" for chassis, "P" for powertrain, and "U" for undefined. The U-codes are designated for future use.

The second character is a number. This defines the code as being a mandated code or a special manufacturer code. A "P0XXX, P2XXX, P3XXX" code means that the fault is defined or mandated by OBD II. A P1 code means the code is manufacturer specific.

The third through fifth characters are numbers. These describe the fault. The third character tells you where the fault occurred. The remaining two characters describe the exact condition that set the code. The numbers are organized so that the various codes related to a particular sensor or system are grouped together.

DTCs that relate to transmission faults can be caused by engine or transmission input and/or output devices. These codes may seem to indicate a problem with an input or output circuit but may actually be caused by an internal transmission problem. Remember, codes are set by out-of-range values. Therefore, when the TCM is receiving a too low or

The SAE J2012 standards specify that all DTCs will have a five-digit alphanumeric numbering and lettering system. The following prefixes indicate the general area to which the DTC belongs:

1. P — powertrain
2. B — body
3. C — chassis

The first number in the DTC indicates who is responsible for the DTC definition.

1. 0 — SAE
2. 1 — manufacturer

The third digit in the DTC indicates the subgroup to which the DTC belongs. The possible subgroups are:

0 — Total system
1 — Fuel-air control
2 — Fuel-air control
3 — Ignition system misfire
4 — Auxiliary emission controls
5 — Idle speed control
6 — PCM and I/O
7 — Transmission
8 — Non-EEC powertrain

The fourth and fifth digits indicate the specific area where the trouble exists. Code P1711 has this interpretation:

P — Powertrain DTC
1 — Manufacturer-defined code
7 — Transmission subgroup
11 — Transmission oil temperature (TOT) sensor and related circuit

Figure 4-33 An explanation of the five digits in an OBD II trouble code.

high input signal, the cause is not necessarily a bad sensor. The sensor can be fine and a mechanical or hydraulic transmission problem is causing the abnormal signals. Not only can internal transmission problems cause codes to be set, but so can basic electrical problems. Problems such as loose connections, broken wires, corrosion, and poor grounds will affect the signals.

OBD II codes are quite definitive and can lead you to the cause of the problem. For example, there are five codes that relate to the transmission fluid temperature circuit. Code P0710 indicates there is a fault in the sensor's circuit, P0711 indicates the circuit is operating out of the acceptable range, P0712 indicates the circuit has lower than normal input signals, P0713 indicates the input signals are higher than normal, and P0714 indicates the circuit appears to have an intermittent problem.

If the TCM is unable to communicate with the PCM, there is a data bus problem. These problems will normally result in poor operation as well as the inability to retrieve DTCs from the TCM. The PCM constantly monitors the data bus and if it is unable to establish communication, it will order a data bus DTC.

Freeze Frame Data

One of the mandated capabilities of OBD II is the *freeze frame*. Although the regulations mandate just emission-related DTCs, manufacturers can choose to include this feature for other systems. With this feature, the PCM takes a screen shot of the activity of the various inputs and outputs at the time the PCM illuminated the MIL or set a DTC. The PCM uses this data for identification and comparison of similar operating conditions if the same problem occurs in the future. This feature is also valuable to technicians, especially when trying to identify the cause of an intermittent problem. The action of sensors and actuators when the code was set can be reviewed. This can be a great help in identifying the cause of a problem. The informations held in freeze frame are actual values; they have not been altered by the adaptive strategy of the PCM.

Once a DTC and the related freeze frame data are stored in memory, they will stay there even if other emission-related DTCs are set. This data can only be removed with a scan tool. When a scan tool is used to erase a DTC, it automatically erases all associated freeze frame data. The data are stored by priority, information related to misfire and fuel control have priority over other DTCs. Fuel system misfires will overwrite any other type of data except for other fuel system misfire data.

Test Modes

All OBD II systems have the same basic test modes and all are accessible with an OBD II scan tool. Always refer to the manufacturers' information when using these test modes for diagnosis.

Mode 1 is the **Parameter Identification (PID)** mode. This mode allows access to current emission-related data values of inputs and outputs, calculated values, and system status information. Some PID values are manufacturer specific; others are common to all vehicles. This information is referred to as serial data.

Mode 2 is the freeze frame data access mode. This mode permits access to emission-related data values from specific generic PIDs. The number of freeze frames that can be stored is limited. The freeze frame information updates if the condition recurs.

Mode 3 permits scan tools to obtain stored DTCs. The information is transmitted from the PCM to the scan tool following a Mode 3 request. The DTC, its descriptive text, or both can be displayed on the scan tool.

Mode 4 is the PCM reset mode. It allows the scan tool to clear all emission-related diagnostic information from its memory. When this mode is activated, all DTCs, freeze frame data, DTC history, monitoring test results, status of monitoring test results, and on-board test results are cleared and reset.

Mode 5 is the oxygen sensor monitoring test. This mode gives the actual oxygen sensor outputs during the test cycle. These are stored values, not current values that are retrieved in Mode 1. This information is used to determine the effectiveness of the catalytic converter.

Mode 6 is the output state mode (OTM) and can be used to identify potential problems in the noncontinuous monitored systems.

Mode 7 reports the test results for the continuous monitoring systems.

Mode 8 is the request for control of an on-board system test or component mode. It allows the technician to control the PCM, through the scan tool, to test a system. In some cases, the scan tool will only set the conditions for conducting a test and not actually conduct the test. An example of this is the EVAP leak test, which is done with other test equipment.

Mode 9 is the request for vehicle information mode. This mode reports the vehicle's identification number, calibration identification, and calibration verification. This information can be used to see if the most recent calibrations have been programmed into the PCM.

Mode 10 is called Permanent Codes. These codes cannot be reset by a scan tool or battery disconnection. Permanent Codes can only be erased by the ECM after the code has cleared by passing three consecutive trips where the fault is no longer present. This prevents a technician from clearing a code and forcing a monitor to pass.

Serial Data

Although the first steps in diagnosis include retrieving DTCs, there are problems that will not be evident by a code. These problems are solved with further testing, symptom charts, or pure logic. This logic must be based on an understanding of the transmission and its controls. It is possible to pinpoint the exact cause of a transmission problem: Monitor the serial data with a scan tool. The serial data stream allows you to monitor system activity during operation. Comparing the observed values to the manufacturer's specifications will greatly help in diagnostics.

To observe the serial data, enter the desired PID into the scan tool. Many PIDs are standard for all OBD II systems, however not all vehicles will support all PIDs and there are manufacturer-defined PIDs that are not part of the OBD-II standard. Manufacturers list and define the various applicable PIDs in their service information. This information should be compared to the observed data. Always refer to the service information and the scan tool's operating instructions for proper identification of normal or expected data. If an item is not within the normal values, record the difference and diagnose that particular item.

Most computer systems will disregard inputs that are well out of range and rely on a default value held in its memory. These default values are hard to recognize and do little for diagnostics; this is why detailed testing with a DMM or lab scope is preferred by many technicians. These test instruments are also used to further test the system after a scan tool has identified a problem.

CVI. Chrysler's use of CVI for adjusting line pressure to compensate for wear provides data concerning the transmission's wear. Chrysler was the first manufacturer to offer this type of information. The 41TE's TCM monitored the oil volume necessary to apply a clutch. Based on this, the TCM could adjust the flow rate of the oil by adjusting the pulse width to the shift solenoids. The TCM also monitored the input and output speed sensors to calculate the change rate of a shift and the gear ratio present after the shift. These two items make up the CVI value that can be displayed on a scan tool. The displayed values relate to both volume and time.

Watching the CVI (**Figure 4-34**) can help diagnose hard shifting or slippage during shifting problems. As a clutch begins to wear or an internal leaks starts, the CVI will rise which indicates a slightly longer shift transition. A low CVI number indicates a short shift,

Clutch	When Updated	Oil Temp.	Proper CVI (64 CVI = 1 cubic inch)
L/R	2–1 or 3–1 manual downshift	>110°F (43°C)	82–134
2C	3–2 kickdown	>110°F (43°C)	25–64
OD	2–3 upshift	>110°F (43°C)	30–64
4C	3–4 upshift	>110°F (43°C)	30–64
UD	4–3 part throttle kickdown	>110°F (43°C)	25–92
2–3 OD	After first 2 to 3 shift	>65°F (18°C)	Higher than OD

Figure 4-34 The standard CVI indexes for a Chrysler RFE transmission.

while a higher number indicates a longer shift. Abnormal CVI reading can be caused by other things besides clutch wear, such as a weak solenoid, faulty pressure regulator, bad TCM, or an internal oil leak. Other Chrysler transmissions have a similar system in their late-model units. These systems are called the Shift Time Adaptation system and the readout on the scan tool is given in torque or strength of the shift.

GM also has a similar system. The scan tool can display shift time and shift error data for every shift. Shift time is the actual time it took to shift; while shift error is the difference between the desired and actual shift times. When a positive value is displayed, the actual shift time took longer than the desired shift time. These values are displayed as TAP (Transmission Adaptive Pressure) cells. Along with these values is a code that signifies the throttle position when the less than ideal shifts occurred. These codes range from 4 to 16; 4 means light throttle, 8 means medium throttle, and 12 and higher results from a wide throttle opening. These values are displayed for each shift. When the TAP cells are within specifications and are very similar, the transmission is operating normally. If one or more gears have a higher value, the clutches or solenoids for those gears may be bad or worn.

Intermittent Faults

An intermittent fault is a fault that is not always present. It may not activate the MIL or cause a DTC to be set. By studying the system and the relationship of each component to another, you should be able to create a list of possible causes for the intermittent problem.

Most intermittent problems are caused by faulty connections or wiring. Refer to a wiring diagram for each of the suspected circuits or components to identify the connections and components in them. That entire circuit should be carefully inspected. Check for burnt or damaged wire insulation, damaged terminals at the connectors, corrosion at the connectors, loose connectors, loose wire terminals, and loose ground wire or straps.

The vehicle can also be taken for a test drive with the scan tool connected. The scan tool will monitor the activity of a circuit while the vehicle is being driven. This allows a look at a circuit's response to changing conditions. The snapshot or freeze frame feature stores the conditions and operating parameters at command or when the PCM sets a DTC. If the snapshot is taken when the intermittent problem is occurring, the problem will be easier to diagnose.

With a scan tool, actuators can be activated and their functionality tested. The results of their operation can be monitored. Also the outputs can be monitored as they respond to changes from the inputs. When an actuator is activated, watch the response on the scan tool. Also listen for the clicking of the relay that controls that output. If no clicking is heard, measure the voltage at the relay's control circuit, there should be a change of more than 4 volts when it is activated.

To locate the source of the problem, a voltmeter can be connected to the suspected circuit and the wiring harness wiggled (**Figure 4-35**). As a guideline for what voltage should be expected in the circuit, refer to the reference value table in the service information. If the voltage reading changes with the wiggles, the problem is in that circuit. The vehicle can also be taken for a test drive with the voltmeter connected. If the voltmeter

Shake slighty

Swing slighty

Figure 4-35 By slightly wiggling and moving the wires at a connector, the cause of an intermittent problem may be found.

readings become abnormal with changing operating conditions, the circuit being observed probably has the problem.

Service Information

After retrieving the DTCs, find the description of the DTC in the service information. There is typically more than one possible cause of the problem. The description of the DTC also leads to pinpoint tests. These are designed to guide you through a step-by-step procedure. To be effective, each step should be performed, in the order given, until the problem has been identified.

Make sure to check all available service information related to the DTCs. There may be a TSB related to the code. This may be a recommendation to reprogram the computer's software.

Often, accurately defining the problem and locating related information in TSBs and other materials can identify the cause of the problem. No manufacturer makes a perfect vehicle or transmission and as the manufacturer recognizes common occurrences of a problem, they will issue a statement regarding the fix of the problem. Check all available information before moving on in your diagnostics.

Correct diagnosis depends on correctly interpreting all data collected and performing all subsequent tests properly. The following information will help:

- Instructions for retrieving DTCs and obtaining freeze frames
- Instructions for diagnosing a no communication problem between the system and the scan tool
- Current technical bulletins for the vehicle
- A fail-safe chart that shows what strategy is taken when certain DTCs are set
- The operating manual for the scan tool
- A DTC chart with the codes and possible problem areas
- A parts locator for the system
- An electrical wiring diagram for the system
- Identification of the various terminals of the PCM
- DTC troubleshooting guide
- Component testing sequences

Reprogramming Control Modules. Often, there will be a TSB that recommends reprogramming of the computer. This is typically called *flash programming* the computer. When a computer is flashed, the old program is erased and a new one is written in. Reprogramming is often necessary when the manufacturer discovers a common concern that can be solved through changing the system's software. New programs are downloaded into the scan tool and then downloaded into the computer through a dedicated circuit.

Each type of scan tool has a different procedure for flash programming. Always follow the manufacturers' instructions. Many laptop-based scan tools require an interface device known as a J2534 interface tool. This interface is also known as a pass-through-programming tool. The interface tool allows the communication between the vehicle network and the laptop computer.

MN flash programming has been called *flashing, reflashing, and programming* by those in the field.

> **NOTE** Never connect a standard battery charger to a vehicle during the flash programming procedure. Computer damage can result from poorly filtered AC voltage from the charger. Before re-flashing, make sure the vehicle's battery is adequately charged or connect the specified charger or battery to the circuity.

Some scan tools are connected to a PC and the software is transferred from an OEM website. **Photo Sequence 6** shows a typical procedure for flashing a BCM.

Electronic Defaults

When the TCM detects a serious transmission problem, or a problem in its circuit (in some cases an engine control or data bus problem), it may switch to a default, limp-in, or fail-safe mode. Limp-in mode may also be initiated if the TCM loses its battery power feed. This **default mode** allows for limited driving capabilities and is designed to prevent further transmission damage while allowing the driver to drive with decreased power and efficiency to a service facility for repairs.

While diagnosing a problem in a transmission, always refer to the service information to identify the normal "default" operation of the transmission. You could spend time tracing the wrong problem by not recognizing that the transmission is operating in default.

The capabilities of a transmission while it is in limp-in mode depend on the extent of the fault, the manufacturer, and the model of transmission. When the TCM moves into the limp-in mode, a DTC is set and the transmission will only operate in this mode until the problem is corrected. Examples of operating characteristics during limp-in mode include the following:

- The transmission locks in third gear when the gear selector is in the DRIVE position or second gear when it is in a lower position.
- The transmission will remain in whatever gear it was in but will shift into third or second gear and stay there as soon as the vehicle slows down.
- The transmission will only use first and third gears while in the DRIVE position.
- The transmission will operate only in PARK, NEUTRAL, REVERSE, and second gears and will not upshift or downshift.

If the malfunction causes improper system operation, the computer may minimize the effects of the malfunction by using fail-safe action. During this, the computer will control a system based on programmed values instead of the input signals. This allows the system to operate on a limited basis instead of shutting down completely.

PHOTO SEQUENCE 6
Flashing a BCM

P6-1 Use the scan tool to retrieve the BCM's part number or the vehicle's VIN and record them.

P6-2 Connect the scan tool to a PC that can link the scan tool to the flash software. Some scan tools will connect directly to an Internet site.

P6-3 Enter the BCM part number in the appropriate field and select "Show Updates" on the menu.

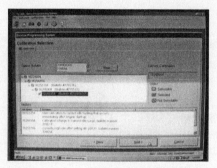

P6-4 Select the desired flash line.

P6-5 Hit the "next" button to begin downloading of the software.

P6-6 Monitor the progress of the downloading to the scan tool.

P6-7 Once downloading is complete, connect a battery pack or factory approved charger to the vehicle's battery. Turn it on and maintain about 13.5 volts at the battery.

P6-8 Connect the scan tool to the DLC and turn the scan tool on.

P6-9 Move through the menus on the scan tool until the desired flash screen is shown. Then follow all instructions given on the tool.

Symptom-Based Diagnosis

At times, no DTCs are set but a problem exists. To discover the cause of these concerns, the description of the problem or its symptoms should be used to determine what systems or components should be checked. Before diagnosing a problem based on its symptoms, make sure the following:

- The PCM and MIL are operating correctly.
- There are no stored DTCs.
- All data observed on the scan tool are within normal ranges.
- There is communication between the scan tool and the control system.
- There are no TSBs available for the current symptom.
- All of the grounds for the BCM/PCM are sound.
- All vehicle modifications are identified.
- The vehicle's tires are properly inflated and are the correct size.

There is typically a section in the service information dedicated to symptom-based diagnosis. Although a customer may describe a problem in nontechnical terms, you should summarize the concern to match one or more of the various symptoms listed by the manufacturer. Each potential problem area should be checked. It is important to realize that some problems may cause more than one symptom.

Sometimes the symptom will not match any of those described in the service information. This does not mean it is time to guess. It means it is time to clearly identify what is working right. By eliminating those circuits and components that are working correctly from the list of possible causes of the problem, you can identify what may be causing the problem and what should be tested further.

Communication Checks

Performing diagnostic checks on late-model vehicles should include a communications check. If the different control modules are not communicating with each other, there is no way to properly diagnose the systems. When a scan tool is installed, it will try to communicate with every module that could be in the vehicle. If an option is not there, the scan tool will display "No Comm" for that control module. That same message will appear if the module is present but not communicating. Therefore, always refer to the service information to identify what modules should be present before coming to any conclusions.

The system periodically checks itself for communication errors. The different buses send messages to each other immediately after it sends a message. The message between messages checks the integrity of the communication network. All of the modules in the network also receive a message within a specific time. If the message is not received, the control module will set a DTC stating it did not receive the message.

There are three types of DTCs used by CAN buses:

- **Loss of Communication**—Loss of communication (and Bus-off) DTCs are set when there is a problem with the communication between modules (**Figure 4-36**). This could be caused by bad connections, wiring, or the module.
  Note: In most cases, a lost communication DTC is set in modules other than the module with the communication problem.
- **Signal Error**—The control modules can run diagnostics on some input circuits to determine if they are operating normally. If a circuit fails the test, a DTC will set.
- **Internal Error**—The modules also run internal checks. If there is a problem, it will set an internal error DTC.

DTC	COMPONENT	DESCRIPTION	CONDITION	SYMPTOM
P1657	TCM	TCM communication link error.	Controller area network (CAN) link error detected by PCM.	No engagements, no adaptive or self-learning strategy. Limited fuel and spark.
U0073	TCM CAN	TCM communication link error.	CAN link error detected by PCM.	No engagements, no adaptive or self-learning strategy.
U0100	TCM CAN	TCM communication link error.	CAN link error detected by PCM.	No engagements, no adaptive or self-learning strategy.
U0101	TCM/PCM	TCM communication error.	PCM lost communication with the TCM.	• Engine driveability concerns. • Will turn on MIL and driver warning lamp.
U0121	ABS	TCM communication link error.	PCM/TCM has detected an error in the CAN wheel rpm information from the ABS system.	No engagements, no adaptive or self-learning strategy.
U0401	PCM/TCM	Invalid data received from the PCM or engine components.	Data received from the PCM or engine components is not correct for the vehicle operating conditions.	Transmission may enable limp-home strategies or increase pressures. Engine components and PCM may or may not set additional DTCs. P2544, fuel monitor error, ECT sensor failure, and MAF sensor failures may be present. MIL may illuminate.
U0415	Wheel speed sensor	TCM communication link error.	PCM/TCM has detected an error in the CAN wheel rpm information from the ABS system.	No adaptive or self-learning strategy.

Figure 4-36 Examples of some of the DTCs that reflect a problem with network communications.

Bus Wire Service. If the bus wire needs repair, due to an open, short, or high resistance, it must not be relocated or untwisted. The twisting serves an extremely important purpose (**Figure 4-37**). After a bus wire has been repaired by soldering, wrap that part of the wire with vinyl tape. Never run the repair wire in such a way as it bypasses the twisted sections. CAN bus wires are likely to be influenced by noise if you bypass the twisted wires.

Figure 4-37 CAN bus wires are twisted pairs. When making wire repairs, it is important to keep these wires twisted.

TESTING ELECTRONIC CIRCUITS AND COMPONENTS

Diagnosing a computer-controlled system is much more than accessing DTCs. You need to know what to test, when to test it, and how to test it. Most electronic circuits can be checked in the same way as other electrical circuits. However, only high-impedance meters should be used. Further testing may be required to define a particular DTC or when the problem does not set a DTC and when verifying that a component or circuit is faulty.

To pinpoint the exact cause of a transmission problem, you will need to use logic and basic electrical troubleshooting equipment, such as wiring diagrams, diagnostic charts, a scan tool, and basic electrical testing tools. The operation of some components can be monitored with the scan tool; however, additional tests are normally necessary. These tests include the following:

- **Ohmmeter Checks**—Most sensors and output devices can be checked with an ohmmeter.
- **Voltmeter Checks**—Many sensors, output devices, and their wiring can be diagnosed by checking the voltage to them, and in some cases, from them. Plus voltage drop tests are extremely valuable.
- **Lab Scope Checks**—The activity of sensors and actuators can be monitored with a lab scope or a graphing multimeter. By watching their activity, abnormal behavior can be observed.

⚡ WARNING **Before disconnecting any electronic component, be sure the ignition is turned off. Disconnecting components may cause high induced voltages and computer damage.**

Testing Sensors

To monitor conditions, the computer uses a variety of sensors. All sensors perform the same basic function. They detect a mechanical condition (movement or position), chemical state, or temperature and change them into electrical signals that can be used by the PCM.

A sensor is any device that sends an input signal to a computer.

If a DTC directs you to a faulty sensor or sensor circuit, or if you suspect a faulty sensor, it should be tested. Always follow the correct procedures when testing sensors and other electronic components. Also, make sure you have the correct specifications for each part. Sensors are tested with a DMM, scan tool, and/or lab scope or GMM.

If the preliminary tests pointed to a possible problem in an input circuit, the circuit should be tested. Make sure to check all suspect circuits for resistance problems; conduct voltage drop tests on those circuits. Often, the manufacturers will give specific testing procedures for specific sensors; always follow them.

Some sensors are simple on-off switches. Others are variable resistors that change resistance according to temperature changes. Some are voltage or frequency generators, while others send varying signals according to the rotational speed of a device. Knowing what they are measuring and how they respond to changes are the keys to accurately testing a sensor.

Testing Switches

Many different switches are used as inputs or control devices. Each of the switches serves a different purpose. However, all of these switches either complete or open an electrical circuit. By completing a circuit, these switches either provide a ground for the circuit or connect two wires together. To open the circuit, switches either disconnect the ground from the circuit or disconnect two wires.

Most of the switches are either mechanically or hydraulically controlled. The operation of these switches can be easily checked with an ohmmeter (**Figure 4-38**). With the meter

Figure 4-38 Checking a switch with an ohmmeter. The meter is placed in parallel with the component after the power is removed from the circuit.

connected across the switch's leads, there should be continuity or low resistance when the switch is closed and there should be infinite resistance across the switch when it is open. A test light can also be used. When the switch is closed, power should be present at both sides of the switch. When the switch is open, power should be present at only one side.

Most grounding switches react to some mechanical action to open or close. There are some, however, that respond to changes in pressure or temperature. An example of this type of switch is the power steering pressure switch. This switch informs the PCM when power steering pressures reach a particular point. When the power steering pressure exceeds that point, the PCM knows there is an additional load on the engine and will increase idle speed.

A switch can be easily tested with an ohmmeter. Disconnect the connector at the switch. Refer to the wiring diagram to identify the terminals at the switch if there are more than two. Connect the ohmmeter across the switch's terminals. Perform whatever action is necessary to open and close the switch.

Switches can also be checked with a voltmeter. The signal to the PCM from supply side switches should be 0 volts with the switch open and supply voltage when the switch is close. Using a voltmeter is preferred because it tests the circuit as well as the switch. If less than supply voltage is present with the switch closed, there is unwanted resistance in the circuit. Expect the opposite readings on a ground side switch.

TR Sensor. The Manual Lever Position or Transmission Range (TR) sensor is a switch that provides information to the computer as to what operating range has been selected by the driver. Based on that information, the computer determines the proper shift strategy for the transmission. The following is a list of problems that may result from a faulty TR Sensor:

- No upshifting
- Slipping out of gear
- High line pressure in transmissions equipped with a pressure control solenoid
- Delayed gear engagements
- Engine starts in other lever positions besides PARK and NEUTRAL

Checking the TR switch may be difficult as the switch assembly on late-model vehicles may be made up of more than one switch. The best way to check a TR switch is with a scan tool. The transmission range can be read off the scan tool directly, and/or the position of the several switch states can be checked at the same time (**Figure 4-39**).

Pressure Switch. The pressure switches (**Figure 4-40**) used in today's transmissions either complete or open an electrical circuit. Normally, open switches will have no continuity across the terminals until oil pressure is applied to it. Normally, closed switches will have continuity across the terminals until oil pressure is applied to it. Refer to the wiring diagram to determine the type of switch and test the switch with an ohmmeter. There should be continuity when the switch is closed and no continuity when the switch is open. Base your expected results on the type of switch you are testing.

Figure 4-40 To check a normally open pressure switch, apply air to the bottom oil passage and check for continuity across the terminals.

Figure 4-39 Checking a TR switch at the PCM and TR sensor connectors. The colors mark the different locations for placing the meter's leads to test the switch.

PCM #43 & TR #2

PCM #19 & TR #3

PCM #32 & TR #6

PCM #17 & TR #4

PCM #18 & TR #5

Pressure switches can be checked by applying air pressure to the part that would normally be exposed to oil pressure. When applying air pressure to these switches, check them for leaks. Although a malfunctioning electrical switch will probably not cause a shifting problem, it will, if it leaks. If the switch leaks off the applied pressure in a hydraulic circuit to a holding device, the holding member may not be able to function properly.

Temperature Switch. Temperature responding switches operate in the same way. When a particular temperature is reached, the switch opens. This type of switch is best measured by removing it, connecting it to an ohmmeter, and submerging it in heated water. A good temperature responding switch will open (have an infinite reading) when the water temperature reaches the specified amount. If the switch fails this test, it should be replaced.

Testing Temperature Sensors

Temperature sensors are designed to change resistance with changes in temperature. A temperature sensor is based on a thermistor. Some thermistors increase resistance with an increase of temperature. Others decrease the resistance as temperature increases. The PCM changes the operation of many systems based on temperature. Temperature sensor problems are often caused by wiring faults or loose or corroded connections. Nearly all temperature sensors are NTC thermistors. Typically, the PCM supplies a reference voltage of 5 volts to the sensor. That voltage is changed by the change of the sensor's resistance

⚠ **Caution**

Always disconnect an electrical switch or component from the circuit or from a power source before testing it with an ohmmeter. Failure to do so will damage the meter.

Figure 4-41 Many testers can show where to place the leads when testing an input or output.

An NTC thermistor is a Negative Temperature Coefficient thermistor and a PTC is a Positive Temperature Coefficient thermistor that reacts in the opposite way as an NTC.

and is fed back to the PCM. Based on the return voltage (voltage drop at the sensor), the PCM calculates the exact temperature. When the sensor is cold, its resistance is high, and the return voltage signal is also high. As the sensor warms up, its resistance drops and so does the voltage signal. Many testers are able to show where to place the probes of the tester to check sensors, such as an ECT (**Figure 4-41**).

Temperature sensor circuits should be tested for opens, shorts, and high resistance. Often, a DTC will be set. Scan tool data should also be looked at. If the observed temperature is the coldest possible, the circuit is open. If the observed temperature is the highest possible, the circuit has a short. Also, if the connector to the sensor is disconnected, the readings should drop to cold. High-resistance problems will cause the PCM to respond to a lower temperature than the actual temperature. This can be verified by using a good thermometer (infrared is best) to measure the temperature and compare it to the scan tool readings. There will be a slight difference between the two if the sensor circuit is working properly. Unwanted resistance in the circuit can cause delayed or poor shifting.

DMM Checks. Obviously, these sensors can be checked with an ohmmeter. To do so, remove the sensor. Use an infrared temperature probe to measure its temperature and measure the resistance across it. Compare your reading to the chart of normal resistances given in the service information (**Figure 4-42**). With the sensor installed, its terminals may be backprobed to connect a voltmeter to the sensor terminals. The sensor should provide the specified voltage drop at any temperature.

°F	°C	Resistance in ohms
−40 to −4	−40 to −20	967k–284k
−3 to 31	−19 to −1	284k–100k
32 to 68	0 to 20	100k–37k
69 to 104	21 to 40	37k–16k
105 to 158	41 to 70	16k–5k
159 to 194	71 to 90	5k–2.7k
195 to 230	91 to 110	2.7k–1.5k
231 to 266	111 to 130	1.5k–0.8k
267 to 302	131 to 150	0.8k–0.54k

Figure 4-42 An example of acceptable resistance readings for a fluid temperature sensor when it is at various temperatures.

The sensors can also be tested by removing them and placing them into a container of water with an ohmmeter connected across the sensor terminals. A thermometer is also placed in the water. When the water is heated, the sensor should have the specified resistance at any temperature. If the sensor does not have the specified resistance, replace the sensor.

Then, disconnect the wiring from the sensor to the computer, connect an ohmmeter from each sensor terminal to the appropriate computer terminal. Both sensor wires should indicate less resistance than specified by the manufacturer. If the wires have high resistance, the wires or wiring connectors must be repaired.

⚡ **WARNING** **Never apply an open flame to a temperature sensor for test purposes. This action will damage the sensor.**

Lab Scope Checks. Thermistor activity can be monitored with a lab scope. Connect the scope across to the output of the thermistor or temperature sensor. Run the engine and watch the waveform. As the temperature increases, there should be a smooth increase or decrease in voltage. Look for glitches in the signal. These can be caused by changes in resistance or an intermittent open.

Testing Pressure Sensors

Most pressure sensors are piezometric sensors (**Figure 4-43**). A silicon chip in the sensor flexes with changes in pressures. One side of the chip is exposed to a reference pressure, which is either a perfect vacuum or a calibrated pressure. The other side is the pressure that will be measured. As the chip flexes in response to pressure, its resistance changes. This changes the voltage signal sent to the PCM. The PCM looks at the change and calculates the pressure change. Some manufacturers have developed special testers to check for problems in a pressure switch.

The term *piezoresistive* represents the characteristic of something that changes resistances in relationship to changes in pressure.

DMM Checks. Normally, the PCM sends a voltage reference signal to a pressure sensor. With the ignition switch on, backprobe the reference wire and measure the voltage. If the reference wire does not have the specified voltage, check the reference voltage at the PCM. If the voltage is within specifications at the PCM, but low at the sensor, repair the wire. When this voltage is low at the PCM, check the voltage supply wires and ground wires for the PCM. If the wires are good, replace the computer.

Then connect the voltmeter from the sensor's ground wire to the battery ground. If the voltage drop across this circuit exceeds specifications, repair the ground wire from the sensor to the computer. To check the voltage signal of a pressure sensor, turn the ignition switch on and connect a voltmeter to the sensor signal wire. Operate the transmission in a way that should cause the monitored pressure to change. If there was no change in the voltage signal or if it was not what was expected, replace the sensor.

Figure 4-43 A piezoresistive sensor is made up of a silicon diaphragm sealed in a quartz plate.

Some pressure sensors use a diaphragm to sense pressure changes. As the diaphragm moves, the resistance of the sensor changes. To test these, remove the sensor and apply varying amounts of air pressure to the port below the diaphragm. Observe the resistance changes with an ohmmeter. Compare your findings to the specifications, if they are available.

Lab Scope Checks. To check a pressure sensor with a lab scope, connect the scope to the sensor's output and a good ground. Operate the transmission in a way that should cause the monitored pressure to change. If there was no change in the voltage signal or if it was not what was expected, replace the sensor. If the signal is erratic, the sensor or sensor wires are defective.

Testing Speed Sensors

Transmissions rely on electrical signals from a speed sensor to control shift timing (**Figure 4-44**). The use of this type of sensor negates the need for hydraulic signals from a governor. When this sensor fails or sends faulty readings, it can cause complaints that include no overdrive, no converter-clutch engagement, and no upshifts.

There are basically two types of speed sensors, permanent magnetic (PM) generator sensors and Hall effect switches. Identifying the type of sensor used in a particular application dictates how the sensor should be tested.

If the PCM does not receive a speed signal it will set a DTC, it may also set a code if the signal does not correlate with other inputs. Check the wiring and connectors at the sensor and the control modules. Make sure the connections are tight and not damaged.

Figure 4-44 The vehicle speed sensor will display a pulse each time a tooth of the rotor passes by the sensor.

DMM Checks. Speed sensors can be checked with an ohmmeter. Most manufacturers list a resistance specification. The resistance of the sensor is measured across the sensor's terminals (**Figure 4-45**). The typical range for a good sensor is 800–1400 ohms of resistance. If there is no continuity, the sensor is open and should be replaced. Reposition the leads of the meter so that one lead is on the sensor's case and other to a terminal. There should be no continuity in this position. If there is any measurable amount of resistance, the sensor is shorted.

A speed sensor can also be checked with the vehicle on a hoist. The vehicle should be positioned so the drive wheels are free to rotate. Backprobe the sensor's output wire and connect the voltmeter leads from this wire to ground. Select the 20 V AC scale on the meter. Then start the engine. Put the transmission in a forward gear and observe the meter. If the sensor's voltage is not 0.5 volts or more, replace the sensor. If the signal is correct, backprobe the sensor's terminal at the PCM and measure the voltage with the wheels rotating. If 0.5 volts is at this terminal, the trouble may be in the PCM.

⚠ WARNING **While conducting this procedure, it is possible to damage the CV joints if the drive wheels are allowed to dangle in the air. Place safety stands under the front suspension arms to maintain proper drive axle operating angles. Also, when running the vehicle on a lift make sure you stay clear of the wheels and other rotating parts. If you are caught by something powered by the engine, serious injury can result.**

When 0.5 volts is not available, turn the ignition off and disconnect the wire from the sensor to the PCM. Connect an ohmmeter across the wire. The meter should read 0 ohms. Repeat the test with the leads connected to the sensor's ground and the PCM ground terminal. This wire should also have 0 ohms. If the resistance of these wires is more than specified, repair the wires.

Lab Scope Checks. Magnetic pulse generators can be tested with a lab scope. Connect the scope's leads across the sensor's terminals. The expected pattern is an AC signal, which should be a perfect sine wave when the speed is constant (**Figure 4-46**). When the speed

Figure 4-45 Checking a speed sensor with an ohmmeter.

Ohmmeter

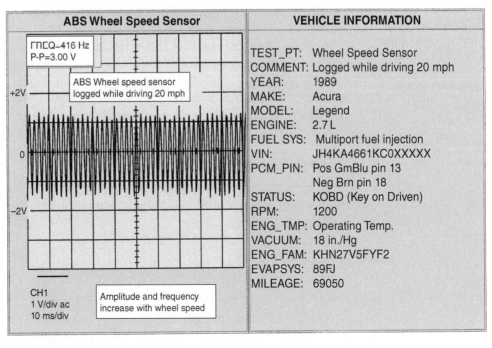

Figure 4-46 The waveform from a wheel speed sensor.

| ABS Wheel Speed Sensor | VEHICLE INFORMATION |

ⲀⲞⲢⲈQ–416 Hz
P-P=3.00 V

ABS Wheel speed sensor logged while driving 20 mph

+2V

0

−2V

CH1
1 V/div ac
10 ms/div

Amplitude and frequency increase with wheel speed

TEST_PT: Wheel Speed Sensor
COMMENT: Logged while driving 20 mph
YEAR: 1989
MAKE: Acura
MODEL: Legend
ENGINE: 2.7 L
FUEL SYS: Multiport fuel injection
VIN: JH4KA4661KC0XXXXX
PCM_PIN: Pos GmBlu pin 13
            Neg Brn pin 18
STATUS: KOBD (Key on Driven)
RPM: 1200
ENG_TMP: Operating Temp.
VACUUM: 18 in./Hg
ENG_FAM: KHN27V5FYF2
EVAPSYS: 89FJ
MILEAGE: 69050

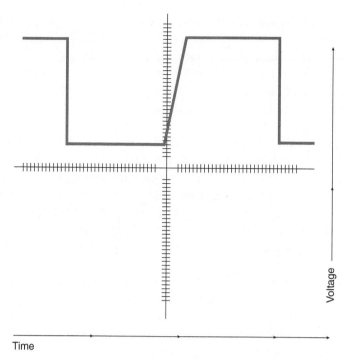

Voltage

Time

Figure 4-47 The waveform from a Hall effect switch with a bad transistor.

is changing, the AC signal should change in amplitude and frequency. If the readings are not steady and do not smoothly change with a change in speed, suspect a faulty connector, wiring harness, or sensor.

A Hall effect switch produces a voltage pulse when there is the presence of a magnetic field. The Hall effect is the consequence of moving current through a thin conducting that is exposed to a magnetic field; as a result of this, voltage is produced.

Hall Effect Sensors. To test a Hall effect switch, disconnect its wiring harness. Connect a low-voltage source across the positive and negative terminals of the Hall layer. Then connect a voltmeter across the negative and signal voltage terminals. Insert a metal feeler gauge between the Hall layer and the magnet. Make sure the feeler gauge is touching the Hall element. If the sensor is operating properly, the meter will read close to battery voltage. When the feeler gauge blade is removed, the voltage should decrease. On some units, the voltage will drop to near zero. Check the service information to see what voltage you should observe when inserting and removing the feeler gauge.

When observing a Hall effect sensor on a lab scope, pay attention to the downward and upward pulses. These should be straight. If they appear at an angle (**Figure 4-47**), this indicates the transistor is faulty causing the voltage to rise slowly. The waveform should be a clean and flat square wave. Any change from a normal trace means the sensor should be replaced.

Testing Throttle Position Sensors

A throttle position (TP) sensor sends information to the computer as to what position the throttle is in. If this sensor fails, the following problems can result:

- No upshifts
- Quick upshifts
- Delayed shifts
- Line-pressure problems with transmissions that have a line-pressure control solenoid
- Erratic converter clutch engagement

A basic TP sensor has three wires. One wire carries the 5-volt reference signal, another serves as the ground for the resistor, and the third is the signal wire. When the throttle plates are closed, the signal voltage will be around 0.6–0.9 volts. Vehicles that use throttle

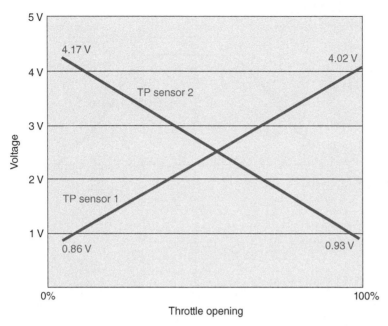

Figure 4-48 Dual TP sensors voltage ranges. (NOTE: There are several types of sensors, always use the chart for your specific vehicle.)

actuator control (TAC) usually have two TP sensors. On one sensor, as the throttle opens, there is less resistance and the voltage signal increases. At wide-open throttle the signal will be approximately 3.5–4.7 volts. On the second sensor, the voltage may start high and fall as the throttle opens (**Figure 4-48**). Some manufacturers use two sensors that rise at different voltage levels. The two TP sensors are used to check for accuracy. If both TP sensors do not read properly, a trouble code or a reduced power situation may result. All changes in throttle position should result in a change in voltage. Often, the connector terminals for the sensor are gold plated. The plating makes the connector more durable and corrosion resistant.

Scan Tool Checks. The action of a TP sensor can also be monitored on a scan tool. Compare the position, expressed in a percentage, to the voltage specifications for that throttle position.

A faulty TP sensor may not cause a DTC to be set. The diagnostic capabilities of the PCM must be able to determine if the sensor is working correctly or not. If it does not have this capability, it may not set a code. The PCM must be able to look at the input from the TP sensor and compare it to other inputs, such as engine speed, MAP inputs, and airflow. If the PCM determines that the TP signal does not reflect a true value based on engine speed and load inputs, it will set a code.

DMM Checks. With the ignition on, connect a voltmeter from the sensor signal wire to ground. Slowly open the throttle and observe the meter. The reading should increase smoothly and gradually. If the TP sensor does not have the specified voltage at a particular throttle opening or if the voltage signal is erratic, replace the sensor.

Connect the meter between the reference wire and the ground. Normally, the voltage should be 5 volts. If the reference wire is not supplying the specified voltage, check the voltage on this wire at the computer terminal. If the voltage is within specifications at the computer, but low at the sensor, repair the reference wire. When the voltage is low at the computer, check the voltage supply wires and ground wires on the computer. If these wires are satisfactory, replace the computer.

A TP sensor can be checked with an ohmmeter or a voltmeter. If checked with an ohmmeter, you should be able to watch the resistance of the sensor change as the throttle

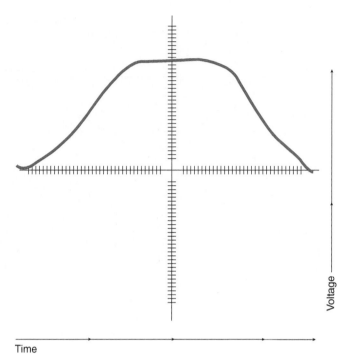

Time

Figure 4-49 The waveform from a good TP sensor as the throttle is opened and closed.

is opened and closed. Often, there will be a resistance specification given in the service information. Compare your reading to this.

Lab Scope Checks. Testing a TP sensor with a lab scope is a good way to watch the sweep of the resistor. The waveform is a DC signal that moves up as the voltage increases. Most potentiometers in computer systems are fed a reference voltage of 5 volts. Therefore, the voltage output of these sensors will range from 0.5 to 4.5 volts. The change in voltage should be smooth. Connect the scope to the sensor's output and a good ground and watch the trace as the throttle is opened and closed. The resulting trace should look smooth and clean, without any sharp breaks or spikes in the signal (**Figure 4-49**). A bad sensor will typically have a glitch (a downward spike) somewhere in the trace (**Figure 4-50**) or will not have a smooth transition from high to low. These glitches are an indication of an open or short in the sensor. Also look for sudden changes in voltage. These can be caused by changes in resistance or an intermittent open in the circuit or sensor.

Mass-Airflow Sensor

A mass-airflow sensor (MAF) is used to determine engine load by measuring the mass of the air being taken into the throttle body. When this sensor fails or sends faulty signals, the engine runs roughly and tends to stall as soon as you put the transmission into gear. The transmission may also not respond to changes in loads, which can lead to wrong gear operation.

The mass-airflow sensor is typically a wire, located in the intake air stream that receives a fixed voltage. The wire is designed so that it changes resistance in response to temperature changes. When the wire is hot, the resistance is high and less current flows through the wire. When the wire is cold, the resistance is low and larger amounts of current pass through it. The amount of air passing over the wire determines the amount of resistance the wire has. The computer monitors the amount of current flow and interprets the flow as the mass or volume of the air.

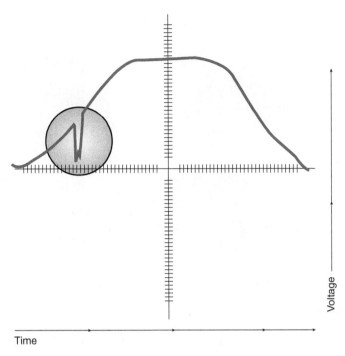

Voltage

Time

Figure 4-50 The waveform of a defective TP sensor. Notice the glitch while the throttle opens.

This sensor can be checked with a multimeter set to the Hz frequency range. Check the service information for specific values. Normally, at idle, 30 Hz is measured and the frequency will increase as the throttle opens.

A scan tool may also be used to test this sensor; most have a test mode that monitors MAF sensors. The output of some MAFs can be observed with a DMM; their output is variable DC voltage. While diagnosing these systems, keep in mind that cold air is denser than warm air.

DIAGNOSIS OF COMPUTER VOLTAGE SUPPLY AND GROUND WIRES

Basically, if the computer system's inputs and outputs and the engine and hydraulic and mechanical functions of the transmission are all good, but a problem exists, the computer or its circuit is bad. Replace the computer only as a last resort. Although there are DTCs that relate directly to the PCM or TCM, it may be difficult to identify a faulty computer. Since the computer reacts to the inputs it receives, abnormal input signals may result in abnormal computer operation. Also, most computers are designed to react to out-of-range inputs by ignoring the signals and using a fixed value held in its memory. This results in less than efficient operation.

SERVICE TIP Never replace a computer unless the ground wires and voltage supply wires are proven to be in satisfactory condition.

All PCMs cannot operate properly unless they have good ground connections and the correct voltage at the required terminals. A wiring diagram for the vehicle must be used for the following tests. Backprobe the battery terminal at the PCM and connect

Figure 4-51 Using a digital multimeter to check the PCM circuit.

a digital voltmeter from this terminal to ground (**Figure 4-51**). Always ground the black meter lead.

The voltage at this terminal should be 12 volts with the ignition switch off. If 12 volts are not available at this terminal, check the computer's fuse and related circuit. Turn the ignition on and connect the red meter lead to the other battery terminals at the PCM with the black lead still grounded. The voltage measured at these terminals should also be 12 volts. When the specified voltage is not available, conduct a voltage drop test to determine the location of the unwanted resistance in the voltage supply wires to these terminals. These terminals may be connected through fuses, fuse links, or relays.

If the computer is receiving battery voltage, check the computer's ground. The best way of doing this is to place one probe of a voltmeter on the case or ground connection of the computer and the other at a known good ground. With the system powered, the meter should read less than 0.1 volt. If the reading is higher than that, conduct voltage drop tests on the rest of the ground circuit.

Ground Circuits

Ground wires usually extend from the computer to a ground connection on the engine or battery. With the ignition on, connect a voltmeter from the battery ground to the computer ground. The voltage drop across the ground wires should be 30 millivolts or less. If the voltage is greater than that, repair the ground wires or connection.

Not only should the computer ground be checked, but so should the ground (and positive) connection at the battery. Checking the condition of the battery and its cables should always be part of the initial visual inspection before diagnosing an engine control system.

A good ground is especially critical for all reference voltage sensors. The problem here is not obvious until it is thought about. A bad ground will cause the reference voltage (normally 5 volts) to be higher than normal. As a result, the computer will be making decisions based on the wrong information. If the output signal is within the normal range for that sensor, the computer will not notice the wrong information and will not set a DTC.

Electrical Noise. Poor grounds can also allow EMI or noise to be present on the reference voltage signal. This noise causes small changes in the voltage going to the sensor. Therefore, the output signal from the sensor will also have these voltage changes. The best way to check for noise is to use a lab scope.

EMI is something that occurs when voltage changes in a nearby wire or component and influences the voltage in the wire or component that it is close to.

Connect the lab scope between the 5-volt reference signal wire at the sensor and its ground. The trace on the scope should be flat. If noise is present, move the scope's negative probe to a known good ground. If the noise disappears, the sensor's ground circuit is bad or has resistance. If the noise is still present, the voltage feed circuit is bad or there is EMI in the circuit from another source.

Circuit noise may be evident by a flickering MIL, a popping noise on the radio, or by an intermittent engine miss. The most common sources of noise are electric motors, relays and solenoids, AC generators, ignition systems, switches, and A/C compressor clutches. Typically, noise is the result of an electrical device being turned on and off.

Sometimes the source of the noise is a defective suppression device. Manufacturers include these devices to minimize or eliminate electrical noise. Some of the commonly used noise suppression devices are resistor-type secondary cables and spark plugs, shielded cables, capacitors, diodes, and resistors. Capacitors or chokes are used to control noise from a motor or generator. If the source of the noise is not a poor ground or a defective component, check the suppression devices.

DETAILED TESTING OF ACTUATORS

Once the TCM determines that a correction or adjustment must be made to the system, an output signal is sent to a control device or actuator. These actuators are as follows: solenoids, switches, relays, or motors; physically act or carry out the command sent by the TCM.

Actuators are electromechanical devices that convert an electrical current into mechanical action. This action can be used to open and close valves, engage or disengage gears, control vacuum to other components, or open and close switches. When the TCM receives an input signal indicating a change in one or more of the operating conditions, the TCM determines the best strategy for handling the conditions. The TCM then controls a set of actuators to achieve a desired effect or strategy goal. In order for the computer to control an actuator, it must rely on a component called an output driver.

The driver usually completes the ground circuit of the actuator. The ground can be applied steadily if the actuator must be activated for a selected amount of time. Or the ground can be pulsed to activate the actuator in pulses. Output drivers are transistors or groups of transistors that control the actuators. These drivers operate by the digital commands from the TCM. If an actuator cannot be controlled digitally, the output signal must pass through an A/D converter before flowing to the actuator.

Most systems allow for testing an actuator through a scan tool. Actuators that are duty cycled are more accurately diagnosed this way. Serial data can be used to diagnose outputs. The data should be compared against specifications to determine the condition of an actuator. Also, when an actuator is suspected as being faulty, make sure the inputs related to the control of that actuator are within normal range. Faulty inputs will cause an actuator to appear faulty.

Many systems have operating modes that can be accessed with a scan tool to control the operation of an output. Common names for this mode are the output state control (OSC) and output test mode (OTM). In this mode, an actuator can be enabled or disabled or the duty cycle or the movement of the actuator can be increased or decreased. While the actuator is being controlled, related PIDs are observed as an indication of how the system reacted to the changes. The actuators that can be controlled by this mode vary. Always refer to the service information to determine what can be checked and how it should be checked. Here is a summary of typical output controls and what happens on a late-model vehicle:

TCC PC Solenoid

- The TCM commands the TCC PC solenoid pressure to apply and release the TCC.
- When the ignition is on and engine off, there are no limits to this control. The solenoid remains on until commanded OFF.
- When the engine is running, the solenoid cannot be commanded OFF and when the transmission is in PARK, it cannot be commanded OFF for an extended time.

Shift Solenoids 1 and 2

- The TCM commands shift solenoids ON or OFF.
- When the ignition is on and the engine is off, the solenoid remains on until commanded OFF.
- When the engine is running, the transmission must be in PARK or NEUTRAL. Also there cannot be any active transmission DTCs.

Line PC Solenoid

- The scan tool can be used to request pressure in increments of 30 psi (207 kPa) from 0–270 psi (0–1862 kPa).
- When the ignition is on and the engine is off, there are no limits to the output control.
- When the engine is running and the transmission is in PARK or NEUTRAL, the engine's speed must be less than 1500 rpm. When the transmission is not in PARK or NEUTRAL, the TCM does not allow a pressure that may cause damage to the transmission. Also there cannot be any active transmission DTCs.

PC Solenoids 2, 3, 4, and 5

- The TCM commands the PC solenoids to apply or release the clutches.
- When the ignition is on and the engine is off, the solenoid remains on until commanded OFF.
- Also there cannot be any active transmission DTCs.

High Side Driver 2

- The TCM commands the driver to send voltage to the solenoid.
- When the ignition is on and the engine is off, there are no limits to this control.
- When the engine is running, the high side drivers cannot be commanded ON or OFF.

Testing with a DMM

Some actuators are easily tested with a voltmeter by checking the input voltage at the actuator. If there is the correct amount of input voltage, check the condition of the ground. If both of these are good, then the actuator is faulty.

When checking anything with an ohmmeter, logic can dictate good and bad readings. If the meter reads infinite, this means there is an open. Based on what you are measuring across, an open could be good or bad. The same is true for very low resistance readings. Across some things, this would indicate a short. For example, you do not want an infinite reading across the windings of a solenoid. You want low resistance. However, you want an infinite reading from one winding terminal to the case of the solenoid. If you have low resistance, the winding is shorted to the case.

Testing with a Lab Scope

When actuators are faulty, it is because they are electrically faulty or mechanically faulty. By observing the action of an actuator on a lab scope, you will be able to watch its electrical activity. Normally if there is a mechanical fault, this will affect its

Figure 4-52 A typical control signal for a pulse-width modulated solenoid.

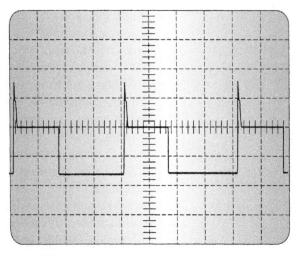

Figure 4-53 Typical control signal for a solenoid.

electrical activity as well. Therefore, you get a good sense of the actuator's condition by watching it on a lab scope.

Some actuators are controlled pulse-width modulated signals (**Figure 4-52**). These signals show a changing pulse width. These devices are controlled by varying the pulse width, signal frequency, and voltage levels. By watching the control signal, you can see the turning on and off of the solenoid (**Figure 4-53**). The voltage spikes are caused by the discharge of the coil in the solenoid.

Both waveforms should be checked for amplitude, time, and shape. You should also observe changes to the pulse width as operating conditions change. A bad waveform will have noise, glitches, or rounded corners. You should be able to see evidence that the actuator immediately turns off and on according to the commands of the computer.

Make sure you follow the instructions from the scope's manufacturer. If the scope is set wrong, the scope will not break. It just will not show you what you want to be shown. To help with understanding how to set the controls on a scope, keep the following things in mind: The vertical voltage scale must be adjusted in relation to the expected voltage signal; the horizontal time base or milliseconds per division must be adjusted so the waveform appears properly on the screen.

Minor adjustments of the trigger line may be necessary to position the waveform in the desired vertical position. Trigger slope indicates the direction in which the voltage signal is moving when it crosses the trigger line. A positive trigger slope means the voltage signal is moving upward as it crosses the trigger line, whereas a negative trigger slope indicates the voltage signal is moving downward when it crosses the trigger line.

Software packages, often programmed in a lab scope or GMM, are available to help you properly interpret scope patterns and set up the scope. These also contain an extensive waveform library that you can refer to and find what the normal waveform of a particular device should look like. The library also contains the waveforms caused by common problems. You can also add to the library by transferring waveforms to a PC from the lab scope. After the waveforms have been transferred, notes can be added to the file. The software may also include the theory of operation and diagnostic procedures for common inputs and outputs.

TESTING SOLENOIDS

Most scan tools have the capability of controlling shift points by manually controlling shift solenoids. If your scan tool does not have this feature, special testers are available to do this. This check operates the solenoids on command thereby bypassing the computer.

Doing this, you can observe the operation of the solenoids and the transmission. If the customer's complaint is still evident when the computer is bypassed, you know the problem is not the computer. If the problem is not evident, you know the problem is the computer and/or computer circuit.

Before continuing, however, you must first determine if the solenoids are case grounded and fed voltage by the computer or if they always have power applied to them and the computer merely supplies the ground. While looking in the service information to find this, also find the section that tells you which solenoids are on and which are off for each of the different gears.

The special solenoid testing tools connect into the solenoid assembly and allow the technician to switch gears by depressing or flicking a switch. This check can be conducted on all transmission solenoids except those that have a duty cycle. Check the service information to identify this type of solenoid. **Photo Sequence 7** shows a typical procedure for checking solenoids with solenoid tester.

PHOTO SEQUENCE 7
Checking the Condition of Solenoids with a Solenoid Tester

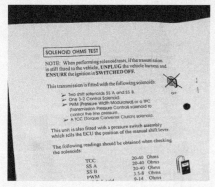

P7-1 In the service information, locate the wiring diagram for the solenoids and determine if the solenoids are power controlled or ground controlled. While looking at the service information, locate the section that tells you which solenoids are on and which are off for each of the different gears.

P7-2 The solenoid tester.

P7-3 Make sure the ignition switch is off and disconnect the car's wiring harness that leads to the solenoids.

P7-4 If the solenoids are controlled by their ground, connect a jumper wire with an in-line 20-amp fuse to the tester and a known good ground.

P7-5 Complete the connection of the tester to the transmission and then program it according to the tester's instructions.

P7-6 Start the engine and move the shift lever into Drive.

P7-7 Now turn on solenoid #1; the transmission should be in first gear. Pay attention to how it shifts.

P7-8 Increase your speed, and then turn on solenoid #2. The transmission should have immediately shifted into second gear. Pay attention to how it shifts.

P7-9 When third gear is selected, solenoid #1 will be turned off. Pay attention to how it shifts.

P7-10 Likewise, when you want fourth gear, solenoid #2 will be turned off. Pay attention to how it shifts.

P7-11 After checking all gears under light, half, and full throttle, return to the shop and summarize your findings.

P7-12 Disconnect the tester from the system and proceed to check solenoid circuits that seemed not to work correctly.

⚙ SERVICE TIP This type of tester is easily made. Simply secure a harness for the transmission you want to test. Connect the leads from the harness to simple switches. Follow the solenoid/gear pattern when doing this. To change gears, all you will need to do is turn off one switch and turn on the next.

If the transmission shifted fine with the movement of the switches, you know that the transmission is fine. Any shifting problem must be caused by something electrical. If the transmission did not respond to the switch movements, the problem is probably in the transmission.

At times, a solenoid valve will work fine during light throttle operation but may not exhaust enough fluid when pressure increases. To verify that the valve is not exhausting,

Terminal No. (symbol)	Color	Terminal Description
E6-6 (SL1+) - **E6-5** (SL1-)	Y - L	Shift solenoid valve SL1 signal
E6-4 (SL2+) - **E6-3** (SL2-)	BR - Y	Shift solenoid valve SL2 signal
E6-2 (SL3+) - **E6-1** (SL3-)	L - B	Shift solenoid valve SL3 signal
E7-7 (SL4+) - **E7-6** (SL4-)	B - W	Shift solenoid valve SL4 signal
E7-5 (SL5+) - **E7-4** (SL5-)	L - BR	Shift solenoid valve SL5 signal
E7-3 (SLT+) - **E7-2** (SLT-)	B - G	Shift solenoid valve SLT signal
E7-1 (SLU+) - **E7-8** (SLU-)	W - L/R	Shift solenoid valve SLU signal
E6-7 (SL) - **E5-7** (E1-)	L/R - BR	Shift solenoid valve SL signal
E7-17 (SR) - **E5-7** (E-)	G - GR	Shift solenoid valve SR signal

Figure 4-54 Solenoid-related pins and testing points at the TCM for a Toyota AA80E transmission.

Ohmmeter

Figure 4-55 Checking a solenoid for a short-to-ground.

activate the solenoid and then increase engine speed while pulling on the throttle cable. If the solenoid valve cannot exhaust, the transmission will downshift. Restricted solenoids should be suspected whenever the transmission shifts roughly or early under heavy loads or full throttle but shifts fine under light throttle.

Solenoids can be checked for circuit resistance and shorts-to-ground. This test can be conducted at the transmission case connector. By identifying the proper pins in the connector, individual solenoids can be checked with an ohmmeter (**Figure 4-54**). Remember, lower than normal resistance indicates a short, whereas higher than normal resistance indicates a problem of high resistance. If you get an infinite reading across the solenoid, the solenoid windings are open. The ohmmeter can also be used to check for shorts-to-ground. Simply connect one lead of the ohmmeter to one end of the solenoid windings and the other lead to a good ground (**Figure 4-55**). The reading should be infinite. If there is any measurable resistance, the winding is shorted to ground.

Solenoids can also be tested on a bench. Resistance values are typically given in the appropriate service information (**Figure 4-56**). A solenoid may be electrically fine but still may fail mechanically or hydraulically. A solenoid's check valve may fail to seat or the porting can be plugged. This is not an electrical problem, rather it could be caused by the magnetic field collecting metal particles in the ATF and clogging the port or check valve. These would cause erratic shifting, no shift conditions, wrong gear starts, no or limited passing (kickdown) gear, or binding shifts. When a solenoid affected in this way is activated, it will make a slow dull thud. A good solenoid tends to snap when activated.

Sump Temperature		Resistance in Ohms	
°F	°C	TCC PWM	Shift and PC solenoids
32	0	9.5	20.0
68	20	10.5	22.0
104	40	11.5	24.5

Figure 4-56 The resistance specifications for the solenoids in an Allison series 1000 transmission.

Figure 4-57 The connector at the solenoids in a typical Chrysler transmission.

Chrysler Solenoids

The typical Chrysler solenoid assembly consists of four solenoids, four ball valves, three pressure switches, and three resistors. The ball valves are operated by the solenoids and together they cause the transaxle to shift. The solenoids should be checked for proper operation. This can be done through the eight-pin connector at the assembly (**Figure 4-57**).

Using an ohmmeter with the harness disconnected from the assembly, check the resistance across pins 1 and 4, 2 and 4, and 3 and 4. The resistance should be between 270 and 330 ohms. With this test, you have checked the resistance for each of the switches. To check the action of these switches, apply air pressure to the feed holes for the overdrive, low-reverse, and overdrive pressure switches. With air applied, the resistance reading across each of the pin combinations should now be 0 ohms.

Each of the four solenoids can also be checked with an ohmmeter. Connect the meter across pins 4 and 5, 4 and 6, 4 and 7, and 4 and 8. The resistance across any of these combinations should be 1.5 ohms.

Ford Solenoids

The solenoids on the 6F50/55 are incorporated into a solenoid body that is attached to the main valve body located under the transmission pan (**Figure 4-58**). The solenoid valve body is only serviced as an assembly, and when it is replaced the PCM must be programmed for the new solenoid assembly

The shift solenoids can be checked using the scan tool, provided there are no codes set for the transmission range, turbine shaft speed, or output shaft speed sensors. The PCM will not allow the output state control from the scan tool if any of these faults are present. If the solenoids do not operate from the overrides on the scan tool, then the harness is checked for opens and shorts. If the harness checks out OK, then the solenoid resistance can be checked from the transmission harness connector, and finally the following test can be performed on the 6F50/55:

Shift solenoid electrical check—Current can be commanded to the individual shift solenoids with the scan tool if the transmission is in Park or Neutral. The amperage of the circuit read on the scan tool can indicate if the solenoid is open, shorted, or partially shorted.

Line pressure check—The transmission line pressure can be increased or decreased on demand with the scan tool. The actual pressure can be read with a gauge attached to the line pressure tap on the transmission. This checks the line pressure control circuit and solenoid. Once again, the transmission must be in Park or Neutral.

Commanded gear shifting—The transmission can be shifted by the scan tool to help diagnose shifting concerns. The transmission must be in Drive and being driven

Transmission case

Valve body   Solenoid body   Side cover

Figure 4-58 The transmission harness connector for a Ford E4OD. The pin numbers are needed for testing of the solenoids.

over 6 mph. The PCM will not allow a gear and speed combination that would damage the engine or transmission, such as shifting into 1st gear at 60 mph.

GM Solenoids

The solenoids of 6L80 are an integral part of the valve body and TCM, which is located in the transmission oil pan (**Figure 4-59**).

These solenoids are simple on/off devices. Because of this, they can be checked through normal diagnostic methods. However, some transmissions, such as the GM 4L80-E (**Figure 4-60**) controls line pressure according to engine load signals. A simple on/off shift solenoid cannot regulate the flow of oil in a metered manner. Therefore, the 4L80-E is equipped with a solenoid, which controls line pressure. This solenoid is a three-port spool valve, electronic pressure regulator that controls pressure based on current flow through its coil winding. The amount of current flow through the solenoid is

SOLENOID	ACROSS	the RESISTANCE should be:
SOLENOID A	E to F	20–40 ohms
SOLENOID B	E to G	20–40 ohms
PWM SOLENOID	E to B	10–15 ohms
ALL	A to D	20–40 ohms
ALL	A to D	2–15 ohms (reversed leads)

Figure 4-59 The transmission harness connector for a 4T60-E.

Figure 4-60 Testing the electrical components in a 4L80-E can be done through the transmission harness connector.

controlled by the PCM. As current through the pressure control solenoid increases, line pressure decreases.

Solenoids that are controlled by duty cycle or a constantly changing signal are more active than the simple on/off solenoids. Therefore, they are more likely to wear. Many transmission rebuilders recommend that this type of solenoid be automatically replaced during a transmission overhaul. Duty cycle solenoids are less prone to clogging because they are cycled fully on every 10 seconds. This helps to keep the regulator valve clean.

The action of a pressure control solenoid can be monitored on a DSO. Connect the DSO directly to the solenoid's circuit. The pattern should be compared to those given in the service information. A PWM solenoid is used to apply and release the TCC in such a manner that the TCC will apply and release smoothly under all conditions. The PWM solenoid in the 4L80-E uses a negative duty cycle and its activity can also be monitored on a scope.

Honda Solenoids

Honda's shift solenoids are normally closed. When the computer sends battery voltage to them, they open and allow fluid to flow through them. When the fluid flows through the

Figure 4-61 Identification of the terminals used to test Honda solenoids.

solenoid, it releases pressure that was preventing a shift valve from moving. When the pressure is exhausted, the valve is able to move and the transaxle shifts.

A Honda transmission can be shifted, without the computer, by jumping 12 volts to terminal 1 at the shift solenoids (**Figure 4-61**). Connect a good ground to terminal 2. On four-speed models, apply voltage to solenoid B to operate in first gear, apply voltage to both shift solenoids for second gear, to solenoid A for third, and to none of them for fourth gear. If the transaxle shifts normally when the solenoids are energized, the solenoids sensors or the controller are at fault and the mechanical and hydraulics of the transaxle are normal.

Disconnect the jumper wire, and check the valve at the fluid passage in the valve body. If the valve binds or moves slowly, or if the solenoid does not work, replace solenoids A and B.

To observe the action of the solenoids, connect a pressure gauge to the transaxle taps for the solenoids. Observe the pressure readings when the transaxle upshifts. If the pressure is greater than 3 psi when the solenoid valve is open, the valve has a restriction or the solenoid is not opening fully and not allowing a full release of pressure. If the valving is restricted, the solenoids should be replaced.

Use an ohmmeter to electrically check the solenoids. The resistance across a shift solenoid should be 12–24 ohms and 14–30 ohms across the lockup solenoid. If the resistance of the solenoids is not within these figures, they should be replaced.

> ⚙️ SERVICE TIP There are small filters located in the valve body to protect the solenoids. If these filters are plugged, they will cause a restriction to oil flow. The filters are mounted in a preformed rubber gasket and can be cleaned, however the rubber gaskets should be replaced on every rebuild.

Repairing the System

After identifying the cause of the problem, repairs should be made. When servicing or repairing OBD II circuits, the following guidelines are important:

■ Do not connect aftermarket accessories into an OBD II circuit.

■ Do not move or alter grounds from their original locations.

- Always replace a relay with an exact replacement. Damaged relays should be thrown away, not repaired.
- Make sure all connector locks are in good condition and are in place.
- After repairing connectors or connector terminals, make sure the terminals are properly retained and the connector is sealed.
- When installing a fastener for an electrical ground, be sure to tighten it to the specified torque.

After repairs have been made, the system should be rechecked to verify that the repair has taken care of the problem. This may involve a road test in order to verify that the complaint has been resolved. Record the fail records or freeze frame data taken before the repair. Use the scan tool to erase any DTCs. Then operate the vehicle within the conditions noted in the fail records or the freeze frame data. After driving the vehicle through a variety of conditions, recheck the DTCs. If the vehicle worked fine and no new codes were set, the repair has been verified.

Some control systems have ATF monitors that must be reset (sometimes with a scan tool) after service. Other systems may have relearn procedures that must be followed after the transmission has been overhauled.

Relearn Process

Vehicles equipped with engine or transmission computers may require a relearn procedure after the battery has been disconnected or after some repairs to the transmission and control system have been made. These include the replacement of the transmission, torque converter, clutches, and seals, as well as overhauling the valve body.

This is an important step because it tells the computer to look at new operating parameters. Based on these new values, the computer will readapt to a transmission that is functioning differently than the one before the repairs. When the vehicle's battery is disconnected, this memory is lost. The computer will use default data until new data from each key start is stored. As the computer memorizes vehicle operation for each new key start, driveability is restored. Vehicle computers may memorize vehicle's operation patterns for 40 or more key starts. Always refer to the service information as some transmissions have the ability to learn quickly when certain conditions are met.

Customers often complain of problems during the relearn stage because the vehicle acts differently than before it was serviced. Depending on the vehicle and how it is equipped, the following complaints may exist: harsh or poor shift quality, rough or unstable idle, hesitation or stumble, rich or lean running, and poor fuel economy. These complaints should disappear after a number of drive cycles. To reduce the possibility of complaints, after any service that requires that the battery be disconnected, the vehicle should be road tested. If a specific relearn procedure is not available, the following procedure may be used:

1. Set the parking brake and start the engine in P or N. Allow the engine to warm up to normal operating temperature or until the electric cooling fan cycles on.
2. Allow the vehicle to idle for about a minute in the N position, then move the gear selector into the D position and allow it to idle in gear for 1 minute.
3. Road test the vehicle. Accelerate at normal throttle openings (20–50 percent) until the vehicle shifts into top gear.
4. Then maintain a cruising speed with a light to medium throttle.
5. Decelerate to a stop, make sure you allow the transmission to downshift, and use the brakes to bring the vehicle to a stop.
6. If a driveability problem still exists, repeat the sequence.

⚠ Caution

Static electricity can destroy or render an electronic part useless. When handling any electronic part, do whatever is possible to reduce the chances of electrostatic buildup on your body and the inadvertent discharge to the electronic part.

Some systems have a "quick learn" process. As a result of this procedure, all previously learned parameters are set to a default value. This may initially result in poor shift quality, however as the vehicle is driven the system will relearn and adjust the transmission's operation to provide good shift quality. Before initiating the quick learn, certain conditions must be met:

- There should be no stored DTCs.
- The transmission should be shifted through all gear ranges with the engine running.
- The fluid should be at its appropriate level.
- The brake pedal should be depressed.
- The engine should be idling.

Once these conditions are met, the scan tool will display instructions to complete the procedure. After completing the process, the transmission should be placed in Park. Then the vehicle can be driven to test it and to allow the computer to learn. After the test drive, the system should be scanned for DTCs and for abnormal serial data.

Some manufacturers recommend a specific relearn procedure that is designed to establish good driveability during the relearn process. These procedures are especially important for all vehicles equipped with an electronically controlled converter and/or transmission. Always complete the procedure before returning the vehicle to the customer.

Chrysler

Chrysler uses a drive verification procedure for their 948RE 9HP48:

1. Reinstall any removed components.
2. Reconnect any disconnected electrical connectors.
3. With the scan tool, erase DTCs.
4. With the scan tool, display the transmission temperature.
5. Start the engine and allow the transmission fluid temperature to reach 43°C (110°F).
6. Check for transmission fluid leaks.
7. Check the transmission fluid level and adjust if necessary.

Road Test Procedure

1. Operate the vehicle in a manner that will allow 15–20 upshifts through all of the gears.
2. Allow these shifts to occur from 0 km/h (0 mph) to 72 km/h (45 mph) with a constant throttle opening of 20 to 25 degrees.
3. Operate the vehicle below 40 km/h (25 mph) and then perform a throttle kick-down maneuver to first gear. Do this several times from less than 40 km/h (25 mph).
4. With the scan tool, read TCM DTCs. If no transmission codes have set, then the vehicle repair is complete. If codes do set, then diagnose them according to the diagnostic charts for the trouble codes.

Ford

Ford Motor Company also specifies a relearn procedure for some of their transmissions and transaxles. All of the specific procedures begin with a fluid check and warming the ATF to normal temperatures. The drive relearn procedures are specific for the different transmissions.

The following procedure is for the 6R80 transmission:

1. Make sure to perform the adaptive learning drive cycle on a level road surface, with the engine and transmission at normal operating temperature. Make sure the transmission fluid is at the correct level.
2. Install a scan tool and record then clear the DTCs.

3. With the selector lever in D, moderately accelerate from stop to 80 km/h (50 mph), allowing the transmission to shift into sixth gear. Keep vehicle speed and throttle steady for a minimum of 15 seconds.
4. With the transmission in sixth gear and maintaining steady speed and throttle, lightly apply and release the brake to operate stop lamps. Maintain the speed and throttle for a minimum of 5 seconds.
5. Brake to a stop and remain stopped in D for a minimum of 20 seconds.
6. Repeat Steps 3 through 5 five times.

If the vehicle is equipped with a 4R70W transmission, road test the vehicle. With the gear selector in overdrive, moderately accelerate the vehicle to 50 mph for a minimum of 15 seconds. The transmission should be in fourth gear at the end of that time. Then hold the speed with a steady throttle and lightly apply and release the brake for about 5 seconds. Then stop and park the vehicle with the gear selector in the D position for a minimum of 20 seconds. Repeat this procedure five times.

CASE STUDY

We once had a Chevrolet pick-up truck come into the shop with the complaint of the transmission defaulting to second gear at times while driving. Compounding our concerns was the fact that the truck was a rebuilt salvaged vehicle. We decided to test drive the vehicle to see if we could duplicate the problem while driving. After driving with a scan tool most of the afternoon, the vehicle finally defaulted to second gear. Nothing leading to a cause was discovered from a snapshot taken when the problem occurred. During our visual inspection, we discovered that the wiring harness had been stripped out of its looms during the rebuild process. We were confident that the wiring had become damaged since it was in such a sorry state. After much searching, we were unable to find a cause visually. We continued to drive the vehicle and troubleshoot this very intermittent condition and finally determined that the transmission solenoids were losing power when the problem occurred. The problem was finally traced to the ignition switch in the steering column, but the related wiring in the area had never been disturbed. The lesson is that even though it appeared that the wiring would be the cause, we lost a lot of time chasing problems that did not exist. Always depend on the facts and logical diagnosis regardless of appearances!

ASE-STYLE REVIEW QUESTIONS

1. *Technician A* says a generic DTC is designated by a P1XXX, P2XXX, or P3XXX code.
   Technician B says manufacturer-specific codes are designated by P0XXX codes.
   Who is correct?
   A. A only
   B. B only
   C. Both A and B
   D. Neither A nor B

2. While checking a transmission range sensor:
   Technician A says typically the range sensor is actually made up of several switches.
   Technician B says a faulty TR sensor may allow the engine to start in other gear positions besides Park and Neutral.
   Who is correct?
   A. A only
   B. B only
   C. Both A and B
   D. Neither A nor B

3. *Technician A* says a transistor is used to measure transmission temperature.

   Technician B says the PCM changes the operation of many systems based on temperature.

   Who is correct?

   A. A only
   B. B only
   C. Both A and B
   D. Neither A nor B

4. *Technician A* says diagnosis of the transmission and TCC control with a road test is necessary.

   Technician B says the road test should be conducted with a scan tool connected to the vehicle to monitor engine and transmission operation.

   Who is correct?

   A. A only
   B. B only
   C. Both A and B
   D. Neither A nor B

5. While checking voltage drops:

   Technician A connects the voltmeter across the load with the circuit activated.

   Technician B connects the negative meter lead to ground and the positive lead to the component being tested.

   Who is correct?

   A. A only
   B. B only
   C. Both A and B
   D. Neither A nor B

6. *Technician A* says high current is caused by high resistance.

   Technician B says a short to ground can occur internally within a component.

   Who is correct?

   A. A only
   B. B only
   C. Both A and B
   D. Neither A nor B

7. *Technician A* says when used to control TCC operation, a solenoid opens a hydraulic circuit to the TCC spool valve.

   Technician B says this causes the spool valve to move and directs mainline pressure to apply the clutch.

   Who is correct?

   A. A only
   B. B only
   C. Both A and B
   D. Neither A nor B

8. While checking a switch:

   Technician A says a multimeter is a meter that combines the functions of a voltmeter, an ammeter, and an ohmmeter.

   Technician B says *impedance* is best defined as the operating resistance of an electrical device.

   Who is correct?

   A. A only
   B. B only
   C. Both A and B
   D. Neither A nor B

9. *Technician A* says an ohmmeter can be used to test semiconductors.

   Technician B says a low-impedance ohmmeter should be used on electronic circuits.

   Who is correct?

   A. A only
   B. B only
   C. Both A and B
   D. Neither A nor B

10. *Technician A* says the default gear is the normal gear selection.

    Technician B says the default gear is normally the gear selected by the TCM when the solenoids are off.

    Who is correct?

    A. A only
    B. B only
    C. Both A and B
    D. Neither A nor B

ASE CHALLENGE QUESTIONS

1. While discussing magnetic pulse generators:

   Technician A says magnetic pulse generators can be tested with a lab scope.

   Technician B says the expected pattern is a DC signal.

   Who is correct?

   A. A only
   B. B only
   C. Both A and B
   D. Neither A nor B

2. While testing temperature sensors:

   Technician A says typically the PCM supplies a reference voltage of 5 volts to the sensor.

   Technician B says if the sensor is at its highest value on the scan tool, the sensor or its circuit is shorted.

   Who is correct?

   A. A only
   B. B only
   C. Both A and B
   D. Neither A nor B

3. *Technician A* says vehicles equipped with engine or transmission computers may require a relearn procedure after repairs to the transmission and control system have been made.

   Technician B says that the customer may notice a performance difference in the transmission operation if the relearn procedure is not performed.

   Who is correct?

   A. A only
   B. B only
   C. Both A and B
   D. Neither A nor B

4. *Technician A* says the fluid pressure switches used in today's transmissions either complete or open an electrical circuit.

   Technician B says that a fluid pressure switch can also be checked with air.

   Who is correct?

   A. A only
   B. B only
   C. Both A and B
   D. Neither A nor B

5. While discussing diagnosing a transmission concern,

   Technician A says make sure there are no engine-related codes that could affect the operation of the transmission.

   Technician B says if there is an engine problem, fix it after diagnosing the transmission.

   Who is correct?

   A. A only
   B. B only
   C. Both A and B
   D. Neither A nor B

Name _____   **Date** _____

USING A DSO ON SENSORS AND SWITCHES

Upon completion of this job sheet, you should be able to connect a DSO and observe the activity of various sensors and switches.

ASE Education Foundation Correlation

This job sheet addresses the following **AST/MAST** tasks:

A.11   Diagnose electronic transmission/transaxle control systems using appropriate test equipment and service information. (**P-1**)

B.3   Inspect, test, adjust, repair, and/or replace electrical/electronic components and circuits including computers, solenoids, sensors, relays, terminals, connectors, switches, and harnesses; demonstrate understanding of relearn procedure. (**P-1**)

We Support

ASE | **Education Foundation**

Tools and Materials

- A vehicle with accessible sensors and switches
- Service information for the above vehicle
- Component locator manual for the above vehicle
- A DSO
- A DMM

Describe the Vehicle Being Worked On

Year _____ Make _____ Model _____

VIN _____

Procedure

1. Connect the DSO across the battery. Make sure the scope is properly set. Observe the trace on the scope. Is there evidence of noise? Explain.

   _____

   _____

   _____

2. Locate the A/C compressor clutch control wires. Start and run the engine. Connect the DMM to read available voltage. Observe the meter, then turn the compressor on. What happened on the meter?

   _____

   _____

   _____

   _____

3. Turn off the engine but keep the ignition on. Locate the TP sensor. List each wire and describe the purpose of each.

   _____

   _____

   _____

   _____

4. Connect the DMM to read reference voltage at the TP sensor. What do you read?

_____

_____

Now move the leads to read the output of the sensor. Starting with the throttle closed, slowly open the throttle until it is wide open. Watch the voltmeter while doing this. Describe your readings.

_____

_____

5. Now connect the DSO to read reference voltage at the TP sensor. What do you see on the trace?

_____

_____

Now move the leads to read the output of the sensor. Starting with the throttle closed, slowly open the throttle until it is wide open. Watch the trace while doing this. Describe your readings.

_____

_____

6. Now run the engine. Locate the oxygen sensor and identify the purpose of each wire going to it. Connect the DMM to read voltage generated by the sensor. (To do this, you may use an electrical connector for the O_2 sensor positioned away from the hot exhaust manifold.) Watch the meter and describe what happened.

_____

_____

_____

_____

7. Now connect the DSO to read voltage output from the sensor. Watch the trace and describe what happened.

_____

_____

_____

_____

8. Explain what you observed as the differences between testing with a DMM and a DSO.

_____

_____

_____

_____

Instructor's Response

_____

_____

_____

Name _____   Date _____

TESTING OF ELECTRICAL AND ELECTRONIC COMPONENTS

Upon completion of this job sheet, you will be able to inspect and test, adjust, repair, or replace transmission-related electrical and electronic components.

ASE Education Foundation Correlation

The job sheet addresses the following **AST/MAST** tasks:

A.11 Diagnose electronic transmission/transaxle control systems using appropriate test equipment and service information. (**P-1**)

B.3 Inspect, test, adjust, repair, and/or replace electrical/electronic components and circuits including computers, solenoids, sensors, relays, terminals, connectors, switches, and harnesses; demonstrate understanding of relearn procedure. (**P-1**)

We Support

ASE | Education Foundation

Tools and Materials

- Digital multimeter
- Service information
- Scan tool
- Test light
- Self-powered test light

Protective Clothing

Goggles or safety glasses with side shields

Describe the Vehicle Being Worked On

Year _____ Make _____ Model _____

VIN _____ Engine type and size _____

Transmission type and model _____

Procedure

1. Using service information and component locator for the vehicle, list all of the electrical and electronic components that are located on the transmission.

   _____

   _____

   _____

2. Carefully check all electrical wires and connectors for damage, looseness, and corrosion. Record your findings.

   _____

   _____

3. Check all ground cables and connections. Corroded battery terminals and/or broken or loose ground straps to the frame or engine block will cause problems. Record your findings.

   _____

   _____

4. Check the fuse or fuses to the control module. To accurately check a fuse, either test it for continuity with an ohmmeter or check each side of the fuse for power when the circuit is activated. Record your findings.

_____

_____

5. Make sure the battery static voltage is at least 12.6 volts, and that the alternator is charging normally. If a problem is found here, correct it before continuing to diagnose the system. Record your findings.

_____

_____

6. With the system on, measure the voltage drop across connectors and circuits. Record your findings.

_____

_____

7. Prepare the vehicle for a road test. Connect the scan tool to the DLC. Set the scan tool up to record a snapshot if the problem occurs while you are driving. **Have an assistant drive the vehicle while you operate the scan tool, do not drive the vehicle and operate the scan tool at the same time**. All pressure and gear changes should be noted. Also, the various computer inputs should be monitored and the readings recorded for future reference. Use the snapshot function of the scan tool to record the data present when the transmission appears to be malfunctioning. Summarize the results of the road test.

_____

_____

8. After completing the road test, use the scan tool to retrieve trouble codes and review the snapshot data recorded. Explain your findings.

_____

_____

_____

_____

9. Referring to the service information, describe what is indicated by the trouble codes (if applicable).

_____

_____

10. If a problem with shift points was noted during the road test, make sure the transmission is not operating in its default mode. Record your findings.

_____

_____

11. If a DTC or test sequence pointed to a problem in an input circuit, the input should be tested to determine the exact malfunction. Briefly describe the procedure outlined by the manufacturer.

_____

_____

_____

_____

_____

12. Describe the method of testing for the transmission range switch.

_____

_____

_____

_____

13. Describe the testing method for the transmission electronic pressure control (EPC) solenoid if equipped. Record your findings.

_____

_____

_____

14. Describe the operation and testing of the input and output shaft speed sensors. Does this transmission use both? Record your findings.

_____

_____

15. Pressure switches are used in transmissions to determine if a shift has occurred by sensing apply pressures to a clutch pack. Describe the operation and testing of the pressure switches on this transmission. Record your findings.

_____

_____

_____

_____

16. A type of sensor commonly used is a potentiometer. Rather than open and close a circuit, a potentiometer controls the circuit by varying its resistance in response to something. A TP sensor is a potentiometer that sends very low voltage back to the computer when the throttle plates are closed and increases the voltage as the throttle is opened. Most TP sensors receive a reference voltage of 5 volts. What the TP sensor sends back to the computer is determined by the position of the throttle. A TP sensor on most late-model vehicles is best checked by watching a scan tool or oscilloscope. You should be able to watch the resistance across the TP sensor change as the throttle is opened and closed. Compare your reading to the specification. Record your findings.

_____

_____

_____

17. With a voltmeter, you will be able to measure the reference voltage and the output voltage. Both of these should be within specified amounts. If the reference voltage is lower than normal, check the voltage drop across the reference voltage circuit from the computer to the TP sensor. Record your findings.

   _____

   _____

18. Vehicle speed sensors provide road speed information to the computer. Most VSS sensors are AC voltage generators and rely on a stationary magnet wrapped with a coil of wire. A rotating shaft fitted with iron teeth is located close to the sensor. Each time a tooth passes through the magnetic field an electrical pulse is present. By counting the number of teeth on the output shaft, you can determine how many pulses per revolution you will measure with a voltmeter set to AC volts. How many teeth are on the output shaft? How many pulses did you observe on the meter?

   _____

   _____

   _____

19. A mass-airflow sensor is used to determine engine load by measuring the mass of the air being taken into the throttle body. The mass-airflow sensor is a wire located in the intake air stream that receives a fixed voltage. The wire is designed so that it changes resistance in response to temperature changes. This sensor can be measured with a digital multimeter set to the Hz-frequency range. Check the service information for specific values. Summarize the specifications for this and record what you observed while checking this sensor.

   _____

   _____

   _____

20. A scan tool can also be used to test a mass-airflow sensor. While diagnosing these systems, keep in mind that cold air is denser than warm air. Record your findings.

   _____

   _____

21. Temperature sensors are designed to change resistance with changes in temperature. A temperature sensor is based on a thermistor. Some thermistors increase resistance with an increase in temperature. Others decrease the resistance as temperature increases. Obviously these sensors can be checked with an ohmmeter. To do so, remove the sensor. Then determine the temperature of the sensor and measure the resistance across it. Compare your reading with the chart of normal resistances given in the service information. Was the sensor you tested a PTC or NTC?

   _____

   _____

   _____

22. The controls of an electronically controlled transmission direct hydraulic flow through the use of solenoid valves. If you were unable to identify the cause of a transmission problem through the previous checks, you should continue your diagnostics by testing the solenoids. Before continuing, you must first determine if the solenoids are case grounded and fed voltage by the computer or if they always have power applied to them and the computer merely supplies the ground. While looking in the service information to find this, also find the area that tells you which solenoids are on and which are off for each of the different gears. Summarize how the solenoids in your transmission work and how they are switched for the various speed gears and other functions.

    _____

    _____

    _____

    _____

23. To begin this test you should connect a scan tool to the DLC that has the capability to control the transmission shifting.

    _____

    _____

    _____

24. Increase your speed, and then turn on the solenoid for second gear. The transmission should have immediately shifted into second gear. Record your findings.

    _____

    _____

25. When you want to shift into third, select the combination of solenoid off and on to provide that gear. Record your findings.

    _____

    _____

26. Do the same for fourth and fifth gears. Record your findings.

    _____

    _____

27. Solenoids can be checked for circuit resistance and shorts to ground. The test can be conducted at the transmission case connector. By identifying the proper pins in the connector, individual solenoids can be checked with an ohmmeter. Record your findings.

    _____

    _____

28. Solenoids can also be tested on a bench. Resistance values are typically given in the service information for each application. Summarize the resistance specifications and the results of your test.

    _____

    _____

29. A solenoid may be electrically fine and still fail mechanically or hydraulically. A solenoid's check valve may fail to seat or the porting can be plugged. To check for this, listen to the solenoid when it is activated. When a solenoid affected in this way is activated, it will make a slow dull thud. A good solenoid tends to snap when activated. Record your findings.

30. Describe how line pressure is controlled by your assigned transmission.

    _____

    _____

31. Describe a cause of high line pressure.

    _____

32. Describe a cause of low line pressure.

    _____

33. Using the scan tool, can you command line pressures? How could this help you in your diagnostics?

    _____

    _____

34. Use the scan tool to see if any "shift adapts" are being used for this transmission and record them here.

    _____

35. What do shift adapts indicate?

    _____

Problems Encountered

_____

_____

_____

Instructor's Comments

_____

_____

_____

CHAPTER 5

REBUILDING TRANSMISSIONS AND TRANSAXLES

Upon completion and review of this chapter, you should be able to:

- Diagnose noise and vibration problems and determine needed repairs.
- Remove and install a transmission/ transaxle assembly from a car or light truck.
- Disassemble, clean, and inspect a transmission/transaxle.
- Inspect and repair the bores, passages, bushings, vents, and mating surfaces of a transmission case.

- Inspect, repair, and replace extension housing and extension housing bushings and seals.
- Inspect and replace external seals and gaskets.
- Reassemble a transmission/transaxle after servicing it.

Basic Tools

Basic mechanic's tool set

Clean lint-free rags

Appropriate service information

Terms To Know

Crocus cloth

Dowels

DIAGNOSIS OF NOISE AND VIBRATION PROBLEMS

Abnormal noises and vibrations can be caused by faulty bearings, damaged gears, worn or damaged clutches and bands, or a bad oil pump, as well as by contaminated fluid or an improper fluid level. Torque converter and cracked flexplate problems can also be the cause of vibrations or noises (**Figure 5-1**).

A customer will often complain of a transmission noise that in reality is caused by something else in the driveline and not the transmission or torque converter. Bad CV or U-joints, wheel bearings, brake calipers, and dragging brake pads can generate noises that customers, and unfortunately some technicians, mistakenly blame on the transmission or torque converter. The entire driveline should be checked before assuming the noise is transmission related.

Most vibration problems are caused by an unbalanced torque converter assembly, a poorly mounted torque converter, or a faulty engine or transmission mount, or a defective output shaft. The key to determining the cause of the vibration is to pay particular attention to the vibration in relation to engine and vehicle speed. If the vibration changes with a change in engine speed, the cause of the problem is most probably the torque converter, flexplate or engine related. If the vibration changes with vehicle speed, the cause is probably the wheels, tires, driveshaft, output shaft or the driveline connected to it. The latter type of problem can be a bad extension housing bushing or universal joint, which would become worse at higher speeds.

This chapter covers the mechanical aspects of diagnosis, disassembly, inspection, service, and assembly of an automatic transmission. The hydraulic components are detailed in Chapter 8.

Refer to a good manual transmission textbook for procedures on diagnosing the entire driveline.

Problem	Probable Cause(s)
Ratcheting noise	The return spring for the parking pawl is damaged, weak, or misassembled
Engine speed sensitive whine	Torque converter is faulty Faulty pump
Popping noise	Pump cavitation—bubbles in the ATF Damaged fluid filter or filter seal
Buzz or high-frequency rattle Whine or growl	Cooling system problem Stretched drive chain Broken teeth on drive and/or driven sprockets Nicked or scored drive and/or driven sprocket bearing surfaces Pitted or damaged bearing surfaces
Final drive hum	Worn final drive gear assembly Worn or pitted differential gears Damaged or worn differential gear thrust washers
Noise in forward gears	Worn or damaged final drive gears
Noise in specific gears	Worn or damaged components pertaining to that gear
Vibration	Torque converter is out of balance Torque converter is faulty Misaligned transmission or engine Output shaft bushing is worn or damaged Input shaft is out of balance The input shaft bushing is worn or damaged

Figure 5-1 A basic noise and vibration diagnostic chart.

Begin your diagnosis by determining if the cause of the problem is the driveline or the transmission. To do this, put the transmission in gear and apply the foot brakes. If the noise is no longer evident, the problem must be in the driveline or the output of the transmission. If the noise is still present, the problem must be in the transmission, engine, flexplate, or torque converter.

Noise problems are also best diagnosed by paying attention to the speed, operating gear, and the conditions at which the noise occurs. If the noise is engine speed related and is present in all gears, including Park and Neutral, the oil pump is the most probable source because it rotates whenever the engine is running.

However, if the noise is engine speed related and is present in all gears except Park and Neutral, the most probable sources of the noise are those parts that rotate in all gears, such as the drive chain, input shaft, and torque converter. The noise from a torque converter is typically a whining noise that is not present when the transmission is in Park or Neutral. It is important to note that a vibration will not normally accompany the noise if it is caused by an internal part of the transmission. This is because those parts do not have enough mass to set up a vibration.

Noises that only occur when a particular gear is operating must be related to those parts responsible for providing that gear, such as a brake or clutch. Often, the exact cause

Classroom Manual
Chapter 5, page 158

of noise and vibration problems can only be identified through a careful inspection of a disassembled transmission.

Regardless of the exact fault in the transmission that is causing a vibration or noise, the transmission will need to come out and the entire unit will have to be checked and carefully inspected. Proper diagnosis prior to disassembling the transmission will identify the specific areas that should be carefully looked at and can prevent unnecessary transmission removal and teardown.

TRANSMISSION/TRANSAXLE REMOVAL

⚠️ **WARNING** **Be sure to wear safety glasses or goggles when working under the vehicle and when handling ATF. ATF, as well as rust and dirt, can cause serious damage to your eyes.**

RWD Vehicles

Removing the transmission from a rear-wheel-drive (RWD) car is generally more straightforward than removing one from a front-wheel-drive (FWD) model, as there is typically one crossmember, one driveshaft, and easy access to cables, wiring, cooler lines, and bellhousing bolts. Transmissions in FWD cars, because of their limited space, can be more difficult to remove because you may need to disassemble or remove large assemblies such as engine cradles, suspension components, brake components, splash shields, or other pieces that would not usually affect RWD transmission removal.

The following is a list of components typically removed or disconnected while removing an automatic transmission from a RWD vehicle. This list is arranged in a suggested order of events. Some vehicles require more than this, others require less.

Battery ground cable
Transmission oil pan
Torque converter access plate
Transmission dipstick tube (if applicable)
Transmission cooler lines
Vacuum hose to modulator
Electrical connectors to solenoids
Electrical connectors to sensors
Gear selector linkage
Transmission range sensor
Starter motor (as required)
Exhaust heat shields
Electrical connectors to oxygen sensors
Exhaust pipes and catalytic converters
Driveshaft
Torque converter to flexplate bolts or nuts
Transmission mounts
Crossmember
Bellhousing to engine bolts

Safe removal of transmissions requires the purchase or fabrication of tools to help support the engine in the chassis and to lift and carry the transmission away from the vehicle.

🔧 **Special Tools**
Drain pan
Transmission jack
Engine support and/
or heavy-duty chain

⚙️ **SERVICE TIP** If you plug the cooler line fittings on the housing and the lines themselves, you will prevent the frustration and possible danger of having ATF drip down your neck or into your eyes while you are removing the transmission.

The exact procedure for removing a transmission will vary with each year, make, and model of vehicle. Always refer to the service information for the procedure you should follow. Normally the procedure begins with placing the vehicle on a hoist so that you can easily work under the vehicle and under the hood.

Once the vehicle is in position, disconnect the negative battery cable and place it away from the battery. Disconnect and remove any transmission linkages connected to the engine. Also remove the ATF dipstick, if the vehicle is equipped with one.

Then raise the vehicle and disconnect the parts of the exhaust system that may interfere with transmission removal. Disconnect all electrical connections at the transmission (**Figure 5-2**). Make sure you place them away from the transmission so that they are not damaged during transmission removal or installation. Disconnect and remove the starter motor.

Remove the torque converter inspection plate or dust cover (**Figure 5-3**). Some transmissions may require the removal of the starter motor for access to the transmission torque converter bolts. Place an index mark on the converter and the flexplate to ensure that the two will be properly mated during installation. Using a flywheel turning tool, rotate the flywheel until some of the converter-to-flexplate bolts are exposed. Loosen and remove the bolts. Then rotate the flywheel until more bolts are accessible. Remove them and continue the process until all the mounting bolts are removed. Once the bolts are removed, slide the converter back into the transmission.

Place a drain pan under the transmission and drain the fluid from the transmission. Once the fluid is out, place the oil pan back onto the transmission and keep it in place with three or four bolts. Move the drain pan under the rear of the transmission. Then remove the driveshaft.

Disconnect the shift and other linkages from the transmission. Disconnect the cooler lines from the transmission (**Figure 5-4**). Cap the ends of the lines to prevent leakage and dirt from entering the cooling system. Disconnect the brackets for the transmission vent and dipstick tube (if equipped). Then remove the vent and dipstick tubes (**Figure 5-5**).

Place a transmission jack under the transmission and secure the transmission to it. The use of a transmission jack allows for easier access to parts hidden by crossmembers or hidden in the space between the transmission and the vehicle's floor pan. Using a chain or other holding fixture, secure the transmission to the jack's pad. Using two chains in an "X" pattern around the transmission works well to secure the transmission to the jack. If the transmission begins to slip on the jack while you are removing it, never try to catch it. Let it fall.

Keyway and lock

Figure 5-2 Be sure to examine the electrical connectors before attempting to separate them. Most have locks that must be unlocked.

Torque converter inspection plate

Figure 5-3 Location of a typical torque converter (and flexplate) inspection cover and bolts.

Figure 5-4 Disconnect the fluid cooler lines at the transmission.

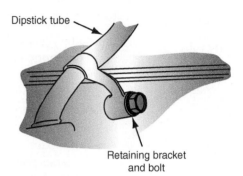

Dipstick tube

Retaining bracket
and bolt

Figure 5-5 Disconnect the brackets for the transmission vent and dipstick tube and then remove the vent and dipstick tubes.

Remove the transmission mounting bolts. Then remove the crossmember at the transmission. After the mounts are free from the transmission, lower the transmission slightly so you can easily access the top transmission-to-engine bolts. Loosen and remove these bolts. Then remove the remaining transmission-to-engine bolts.

Move the transmission away from the engine until the unit is clear of the alignment dowels at the rear of the engine. Then slowly lower the transmission (**Figure 5-6**). Make sure the converter hub and any associated shafts have a clear path while you are lowering the transmission. Once the transmission is out of the vehicle, carefully move it to the work area and mount it to a stand.

> ⚠ **Caution**
>
> If the vehicle is a hybrid vehicle, make sure you follow all precautions and procedures for disconnecting the high-voltage circuit before proceeding to remove the transmission.

Frame   Lift pad

Lift arm

Figure 5-6 Move the transmission away from the engine until the unit is clear of the alignment dowels at the rear of the engine and the converter hub has a clear path while lowering the transmission.

FWD Vehicles

Obviously the procedure for removing a transmission from a FWD vehicle is different. Because all the components are confined to a very small area, the exact items that need to be removed or disconnected vary from vehicle to vehicle. The following is a list of components that are typically removed or disconnected. This list is arranged in a suggested order of events.

Battery ground cable
Underhood electrical connectors to transaxle
Front wheels
Electrical connectors at the wheel brake units
Brake calipers
Steering knuckles
Drive axles
Transmission oil pan
Transaxle to engine brackets
Torque converter access plate
Torque converter drain plug
Transmission dipstick tube (if equipped)
Transmission cooler lines
Electrical connectors to solenoids
Electrical connectors to sensors
Gear selector linkage
Transmission range sensor
Starter motor
Exhaust heat shields
Electrical connectors to oxygen sensors
Exhaust pipes and catalytic converters
Torque converter to flexplate bolts or nuts
Transaxle to engine mount
Crossmember
Bellhousing to engine bolts

> **⚙ SERVICE TIP** Some technicians use digital cameras or cell phone video to help recall the locations of underhood items by taking pictures before work is started. This technique can be valuable considering how complex the underhood systems of cars have become.

⚠ **Caution**

Always begin the transmission removal procedure by disconnecting the battery ground cable. This is a safety precaution to help avoid any electrical surprises when removing starters or wiring harnesses. It is also possible to send voltage spikes, which may kill the PCM if wiring is disconnected when the battery is still connected.

On some vehicles, the recommended procedure may include removing the engine with the transaxle. Always refer to the appropriate service information before proceeding to remove the transaxle. You may waste much time and energy if you don't check the service information first.

> **CUSTOMER CARE** Because much of your time will be spent working under the hood, make sure you use fender covers and take extra care not to damage the body panels or finish of the vehicle.

Begin removal by placing the vehicle on a lift. However, before raising the vehicle, look around the engine bay to see if any interference will occur between the firewall and engine components, such as fans and fan shrouds, fuel lines, exhaust systems, or

Figure 5-7 A typical engine support fixture for a FWD vehicle.

electrical components, when the transaxle is removed. If any causes for interference are found, these problems should be corrected before continuing. Also, any bellhousing bolts, wiring, or cables accessible from above should be removed. Before raising a FWD vehicle to begin transaxle removal, a support fixture should be attached to the engine (**Figure 5-7**).

Raise the vehicle to a comfortable height and remove all but the three or four corner bolts of the transmission pan, depending on the shape of the pan. Place a drain pan large enough to catch all the oil under the transaxle. Carefully remove bolts from one side of the pan. Back off the bolt or bolts on the other side just enough to allow the pan to drop slightly as you pry it loose. Be careful, as some pans will come loose without being pried and if you loosen the pan too much, you may have a large mess to clean up! When the fluid stops draining, replace the pan with a minimum of bolts, taking care not to lose the remaining bolts.

To remove FWD driveshafts, you must first loosen the large nut that retains the outer CV joint, which is splined on the shaft to the hub.

Now raise the vehicle and remove the front wheels. Tap the splined CV joint shaft with a soft-faced hammer to see if it is loose. Most will come loose with a few taps. Many Ford FWD cars use an interference fit spline at the hub. You will need a special puller for this type of CV joint. The tool pushes the shaft out and on installation pulls the shaft back into the hub.

⚠ **Caution**

It is recommended that the large nut that retains the outer CV joint be loosened with the vehicle on the floor and the brakes applied to reduce possible damage to the CV joints and wheel bearings.

 SERVICE TIP To control the mess, some technicians disconnect a cooler line at the radiator and place the end of the line in a drain pan. The engine is then cranked and fluid pumped into the drain pan. This can be done to move most of the fluid out of the transmission. Do not crank the engine for more than 30 seconds at one time.

Figure 5-8 The knuckle assembly is moved out of the way to allow the stub shaft to be pulled out of the hub.

The lower ball joint must now be separated from the steering knuckle. The ball joint will be either bolted to the lower control arm or held into the knuckle with a pinch bolt. Once the ball joint is loose, the control arm can be pulled down and the knuckle can be pushed outward to allow the splined CV joint shaft to slide out of the hub (**Figure 5-8**). The inboard joint can then be pried out or it will slide out. Some transaxles have retaining clips that must be removed before the inner joint can be removed (**Figure 5-9**).

On some cars, the inner CV joints have flange-type mountings. These must be unbolted for removal of the shafts. In some cases, the flange-mounted driveshafts may be left attached to the wheel and hub assembly and unbolted only at the transmission flange. The free end of the shafts should be supported and placed out of the way. Doing this will greatly decrease the amount of time needed to remove and install the transmission.

Figure 5-9 Location of the various switches, connectors, and levers on a typical transaxle.

At this time, you should study the underbody layout to aid reinstallation. This is another time when use of a digital camera will help your memory when it comes time to reinstall the transmission. In many shops, the technician repairing the transmission is not the same person who removes and reinstalls it. If the installer does not remember the exact location of all the underbody and engine bay components, or the correct adjustments of control cables, the transmission may not operate properly when it is reinstalled. Improper installation can ruin a transmission or cause drivability, noise, and vibration problems.

Now the shift linkages, vacuum hoses, electrical connections, and control cables should be disconnected (Figure 5-9). The exhaust system may also need to be lowered or partially removed. The inspection cover between the transaxle and engine should be removed to allow access to the torque converter bolts (sometimes the starter must be removed to access the flywheel bolts). There will be three to six bolts or nuts securing the converter to the flexplate, depending on the application.

Mark the position of the converter to the flexplate to help maintain balance or runout. It will be necessary to rotate the crankshaft to remove the converter bolts. This can be done by using a long ratchet and socket on the front crankshaft bolt or by using a flywheel-turning tool, if space permits.

Carefully remove the cooler lines by holding the case fitting with one wrench and loosening the line nut with a line wrench. Doing this assures that you will not twist the steel lines, which would damage them and restrict their flow. Some transmission lines are connected to the transmission with quick-disconnects, some of which may require the use of special tools.

With a transmission jack supporting the transmission, remove the transaxle mounts. The mounts may be at the top, side, or bottom of the unit. Some mounts consist of two parts (**Figure 5-10**). Make sure you remove both parts as well as all transaxle mounts.

Now remove the starter. The starter wiring may be left connected but you will need to hang the starter with heavy mechanic's wire to avoid damage to the cables. You can also remove the starter from the vehicle to get it totally out of the way.

Now pull the transaxle away from the engine. It may be necessary to use a pry bar between the transaxle and engine block to separate the two units. Make sure the converter comes out with the transaxle. This prevents bending the input shaft, damaging the oil pump, or distorting the drive hub. After separating the transaxle from the engine, retain the torque converter in the bellhousing. This can be done simply by bolting a small combination wrench to a bellhousing bolt hole across the outer edge of the converter.

Figure 5-10 Remove the bolts from the transaxle insulator, then lower the transaxle and remove the nuts and bolts to remove the insulator bracket.

 Caution

To prevent accidental damage, pay attention to the proper location of all mounting bolts and hardware when the transmission is removed. For example, on some torque converters, using longer than normal fly wheel-to-converter bolts can dimple the torque converter clutch friction surface, causing the converter clutch to chatter.

Caution

Never allow the starter to hang by the wires attached to it. The weight of the starter can damage the wires or, worse, break the wire and allow the motor to fall, possibly on you or someone else. Always securely support the starter and position it out of the way after you have unbolted it from the engine.

Caution

Never force the torque converter back into the oil pump if it has slipped out. Turn the converter while pushing back on it until it aligns with the transmission shafts and goes back into place.

Alignment pins are commonly referred to as dowels.

⚠ Caution

Dowels or locating pins are used to keep the engine and transmission in perfect alignment. Torque converter housing bolts alone cannot maintain perfect alignment. Perfect alignment is necessary to prevent flexplate breakage.

Classroom Manual
Chapter 5, page 169

⚠ Caution

Never use an impact wrench on torque converter bolts. Impact wrenches can drive the bolts through the cover, which will warp the inside surface and prevent proper clutch apply or may damage the clutch pressure plate.

Installation

Transmission installation is generally a reverse of the removal procedure. Care must be taken to avoid destroying the new or rebuilt transmission during installation. A quick check of the following list will greatly simplify your installation and reduce the chances of destroying something during installation:

- Make sure the block alignment pins (dowels) are in the appropriate bores and are in good shape and that the alignment holes in the bellhousing are not damaged (**Figure 5-11**).
- Make sure the pilot hole in the crankshaft is smooth and not out-of-round. This will allow the converter to move in and out on the flexplate.
- Make sure the pilot hub of the converter is smooth and cover it with a light coating of chassis lubricant to prevent chafing or rust.
- Make sure the converter's drive hub is smooth and coat it with trans gel or petroleum jelly.
- Secure all wiring harnesses out of the way to prevent their being pinched between the bellhousing and engine block. If the wires get pinched, not only will there be a large electrical short, but you may also destroy the car's computer.
- Flush out the oil cooler lines and the cooler itself to remove any material that could damage the transmission. It is recommended that torque converters be replaced, as it is not possible to tell how much the unit has been damaged, and it is nearly impossible to remove all contamination from inside.
- Always perform an endplay check and check the overall height before reinstalling a torque converter or installing a fresh unit out of the box.
- Pour one quart of the recommended fluid into the converter before mounting the converter to the transmission. This will assure that all parts in the converter have some lubrication before start-up.

Slide the converter into the transmission, making sure that all drives are engaged. Double-check this by using the height dimension you measured during transmission removal. In older transmissions, the torque converter must engage into the input shaft stator splines and the oil pump, later models also have those or a direct driveshaft or an oil pump driveshaft.

Secure the converter in the transmission as you did during removal. Then transfer the transmission to the jack and move it under the car. Raise the transmission to get close alignment to the engine block. If the transmission has a full-circle bellhousing, you will need to align the converter drive studs or bolt pads before you push the transmission up against the block, as this is difficult to do once the transmission is against the engine block. Make sure you set the torque converter onto the flexplate in alignment with the marks made during removal. These will provide proper balance for the unit. Once the converter

Dowel pins

Figure 5-11 The dowel pins may remain in the engine or the transmission case when the transmission is removed. In either case, they should be inspected.

is aligned, be sure the block dowels line up with the bellhousing, then push the transmission against the engine block. Check to be sure that nothing has been caught between the block and the bellhousing. Start two bellhousing bolts across from each other and slowly tighten them. Check for free rotation of the torque converter while doing this. Never use the bolts to pull the transmission against the engine block. Then, install the rest of the bolts and torque them to specifications.

On FWD cars, if the engine is held in place with a support bar, the transmission jack may be removed at this time. If the car has a split cradle-type subframe, it should be installed now.

On RWD vehicles, it may be necessary to leave the transmission jack in place while components such as exhaust crossovers or frame crossmembers are installed. Do not connect the gearshift linkage until the transmission is mounted to the crossmember and the transmission jack is removed. Once the rear transmission mount is attached to the crossmember, the jack may be removed.

Install the converter drive bolts (or nuts). You should notice while installing these fasteners that the converter has some noticeable fore and aft movement. This amount varies with different transmissions, but it is generally less than 1/2 inch. This is normal and necessary, as it allows the converter to move on the flexplate and also allows for a noninterference fit at the oil pump drive gear. If there were no movement, premature pump failure would result.

The cooler lines should now be installed. Make sure the lines are connected to the bores they were originally in. Then, tighten the line fittings, making sure you do not twist or distort the line. The remaining components—starter, electrical connections, dipstick tube (if equipped), and so on—can now be installed.

While connecting the manual shift linkage (**Figure 5-12**), pay attention to the condition of the plastic bushings. If these are worn or missing, hard or inaccurate linkage movement will result. Replace the bushings if needed.

Some other areas that require special attention during installation are the grounding straps and any rubber tubes. These are often overlooked during transmission removal and installation and can cause problems if not checked. All rubber tubes are suspect, as they are exposed to heat, which causes them to crack. New parts should be used if necessary. The ground straps must be in good condition and free of corrosion, as these provide an electrical ground path to the body of the car during operation. Failure to clean or attach these straps or cables can also result in poor signals to the PCM, voltage spikes that can damage the PCM, electrolysis through the fluid that welds internal transmission components, or even fused manual linkage cables as the current looks for a path to flow through.

On RWD vehicles, the driveshaft must be installed using the marks you made during removal (**Figure 5-13**). Be sure to coat the slip yoke with ATF before sliding it into the

Caution

After the transmission has been installed, make sure everything is properly realigned. Repositioning the transmission an inch or less can have a big effect on the manual shift linkage adjustment.

Figure 5-12 Connect the gear selector rod or cable. Make sure the plastic bushings are in good shape and that they are properly seated.

Figure 5-13 The index marks on the driveshaft flange yoke and the pinion flange ensure the driveshaft is installed at the correct phase.

extension housing. The driveshafts of FWD cars are installed in the reverse manner in which they were removed and the related components (strut lower ball joint bolts, and so on) should be installed just as they were before they were removed. Use a new nut on the outer CV joint stud, if the nut is of self-locking type. The torque of this nut is critical. Torquing the nut should be done with the car's tires on the ground and with the brakes held. Air-type impact tools should not be used, as they can damage the wheel bearing or hub.

Connect the battery cables. Then add about half of the total quantity of the proper ATF to the transmission. This amount varies, but an average amount is four quarts. Some technicians will also connect a pressure gauge to the transmission and take pressure readings during the initial operation of the fresh transmission. Some imported-transmission manufacturers do require that this be done because line pressures must be set after start-up. Connecting the pressure gauges will also give you an indication of whether there is sufficient oil pressure to road test the car.

Put the vehicle on a lift, then start the engine and apply the brakes. Move the shift selector through all of the gear ranges. Then place the selector into Park. Check the fluid level and add ATF until you reach the "add" or "cold" mark. Many late-model transmissions are not equipped with a dipstick or fill tube. On these transmissions it will be necessary to follow the specific procedure indicated in the service information. Then, inspect the transmission for any signs of leakage or loose bolts. If a pressure gauge was attached to the transmission and the pressure reading was fine, disconnect the gauge after you shut down the engine.

> **⚙ SERVICE TIP** Make sure that any bolts you replace are the same length and quality of those you are replacing.

> **⚙ SERVICE TIP** Check the service information before taking a transmission on its initial road test. Computer-controlled automatic transmissions require that a "Learning Procedure" be followed, which includes various driving conditions. Since you need to road test the transmission and need to teach the transmission, you may as well do both at the same time.

Road test the vehicle to check the operation of the transmission, as well as anything else that may have been affected by your work. Make any necessary adjustments. The road test should cover at least 20 miles in order to completely warm up the transmission. Recheck and adjust the fluid level. It should be between the "add" and "full" marks on the dipstick. Again, follow the service information procedures for those transmissions without a dipstick tube. DO NOT OVERFILL! After the road test, visually inspect the transmission again for signs of leakage. Also carefully look around the engine and transmission to see if any wires, hoses, or cables are disconnected or positioned in an undesirable spot.

This sequence covers the general disassembly of an automatic transmission. Complete overhaul of common transmissions is covered in Chapter 10 of this manual.

DISASSEMBLY

The disassembly of a transmission and a transaxle are similar; therefore, the disassembly, inspection, and reassembly guidelines can be grouped together.

Once the transmission has been removed from the vehicle, certain things should be checked before beginning to disassemble the transmission. Measure the depth at which the converter fits into the transmission case (**Figure 5-14**). To do this, hold a straightedge across the bellhousing and measure into the pilot hub or a drive pad on the converter. Record this dimension and save it for use when installing the converter.

Also inspect the flexplate for warpage and cracks (**Figure 5-15**). Pay attention to the condition of the teeth on the starter ring gear. Also check the flexplate for excessive runout and elongated bolt holes. Replace the flexplate if there is evidence of damage. Also inspect the converter-attaching bolts. Replace any worn bolts with suitable equivalents. While inspecting the flexplate, check the starter ring gear for looseness and damage. Pay special attention to the welds that secure the ring gear to the plate. It is common for these and the areas around them to crack.

Check the flexplate for cracks at the crankshaft mounting bolt holes. The best way to do this is to remove the plate and hold it up toward a bright light. If there are cracks, the light will shine through the cracks. If there is any flexplate damage, the flexplate should be replaced. Before reinstalling the flexplate, check service information to make sure the attaching bolts are either reusable or should be replaced. Also make sure the flexplate is installed in the correct direction and that all spacers are in place.

Check the drive hub of the torque converter. It should be smooth and show no signs of wear. Pay particular attention to the area the seal rides on. If the hub is worn, the torque converter should be replaced and the oil pump drive should be carefully inspected for scoring or other damage.

Before disassembling the transmission, check out the causes of any leakage. If there is evidence of leakage or leakage is the reason for the teardown, determine the path of the leakage before cleaning the area around the seals. At times, the leakage may be from sources other than the seal. Leakage could be from worn gaskets, loose bolts, cracked housings, or loose line connections.

Inspect the outside sealing area of the seal to see if it is wet or dry. If it is wet, see whether the oil is running out or if it is merely a lubricating film. Check both the inner and outer parts of the seals for wet oil, which means leakage.

A transmission overhaul normally includes replacing all rubber and paper gaskets, frictional materials, and worn bushings; replacing any damaged transmission components; cleaning the planetary gearsets and the valve body; making correct endplay adjustments; and flushing the cooler lines or replacing the cooler. The torque converter is generally replaced as well.

The term teardown is often used to describe the process of disassembling a transmission.

Classroom Manual
Chapter 5, page 176

Figure 5-14 After the transmission and engine have been separated, measure and record the depth of the converter into the housing. This measurement will be used during reassembly to ensure that the torque converter and oil pump are properly assembled.

Figure 5-15 Carefully inspect the flexplate and torque converter mounting hardware.

When removing a seal, inspect the sealing surface, or lips, before cleaning it. Look for signs of unusual wear, warping, cuts and gouges, or particles embedded in the seal. On spring-loaded lip seals, make sure the spring is seated around the lip, and that the lip was not damaged when first installed. If the seal's lip is hardened, this was probably caused by heat from either the shaft or the fluid.

If the seal is damaged, check all shafts for roughness, especially at seal contact areas. Look for deep scratches or nicks that could have damaged the seal. Determine if a shaft spline, keyway, or burred end could have caused a nick or cut in the seal lip during installation. Inspect the bore into which the seal was fitted. Look for nicks and gouges that could create a path of oil leakage. A coarsely machined bore can allow oil to seep out through a spiral path. Sharp corners at the bore edges could have scored the metal case of the seal when it was installed. These scores can make a path for oil leakage.

Cleaning and Inspection

Before disassembling the automatic transmission, care should be taken to clean away any dirt, undercoating, grease, or road grime on the outside of the case. This ensures that dirt will not enter the transmission during disassembly. Once the transmission is clean outside, you may begin the disassembly.

> ⚙ **SERVICE TIP** Record the model and serial number of the unit you are working on. You will need this information to purchase the correct parts during rebuild.

When cleaning automatic transmission parts, avoid the use of solvents, degreasers, and detergents that can decompose the friction composites used in a transmission. It is best not to attempt to clean the friction members, as this will damage the parts. Use compressed air to dry components; don't wipe down parts with a rag. The lint from a rag can easily destroy a transmission after it has been rebuilt.

There are many different ways to clean automatic transmission parts. Some rebuilding shops use a parts washing machine, which does an excellent job in a small amount of time, to thoroughly clean a transmission case, converter housing, and extension housing. These parts washers use hot water and a special detergent that are sprayed onto the parts as they rotate inside the cleaner. The key to good cleaning with these machines is to use a very small amount of soap with very hot water. Many rebuilders simply clean the parts in a mineral spirits tank, or an aqueous cleaner with which the parts are brushed and then hand cleaned. No matter what type of cleaning procedure is followed, the transmission and parts should be rinsed with water, then thoroughly dried with compressed air before reassembly.

⚠ **WARNING** Always wear safety goggles when using compressed air to dry something. The air pressure can easily move dirt, metal, or other debris around the work area. If these get into your eye, they can cause permanent damage.

Special Tools
Solvent
Scrub brush
Compressed air
  nozzle

> ⚙ **SERVICE TIP** It is best to mount the transmission on a bench mount made especially for working on transmissions, (**Figure 5-16**). These mounts are available from many tool suppliers.

After the case is clean, remove the torque converter and carefully inspect it for damage. Check the converter hub for grooves caused by hardened seals. Also check the bushing contact area. If the bushing contact area of the converter is damaged, you will have to replace the converter and pump bushing. Also, inspect the pump for any wear. To remove the converter, slowly rotate it as you pull it from the transmission, and have a drain pan

Figure 5-16 A typical transaxle holding fixture.

handy to catch the fluid. It should come right out without binding. This is a good time to check the input shaft splines, stator support splines, and the converter's pump drive hub for any wear or damage. Converters with direct driveshafts should be checked to be sure that no excessive play is present at the drive splines of the shaft or the converter. If any play is found in the converter, the converter or the shaft must be replaced.

Classroom Manual
Chapter 5, page 161

Position the transmission to perform an endplay check. The transmission endplay checks can provide the technician with information about the condition of internal bearings and seals, as well as clues to possible causes of improper operation found during the road test. These measurements will also determine the thickness of the thrust washer(s) during reassembly. Thrust-washer thickness sets the endplay of various components. Excessive endplay allows clutch drums to move back and forth too much, causing the transmission case to wear. If there is insufficient clearance, rapid wear may occur due to inadequate oil clearances. Assembled endplay measurements should be between minimum and maximum specifications, but preferably at the low end of the specifications. **Photo Sequence 8** covers a typical procedure for checking endplay.

Position the transmission so that the shaft centerline is vertical. This allows the weight of the internal components to load the shafts toward the rear of the transmission. Most GM transmissions can be checked by mounting a dial indicator to read the input shaft movement. Zero the dial indicator and lift upward on the input shaft. Most Chrysler and some Ford transmissions require removing the oil sump pan and valve body and prying the input shell upward. Chrysler and Ford transmissions may be measured horizontally (**Figure 5-17**), but you may want to center the output shaft with a slip yoke for more accurate readings.

The 5R110W and 6L80 are examples of transmissions that use a direct driveshaft.

Input shaft

Dial indicator

Figure 5-17 The endplay of the input shaft should be checked during disassembly and reassembly.

PHOTO SEQUENCE 8
Measuring Input Shaft Thrust Play (Endplay)

P8-1 Position the transmission so the input shaft is facing up.

P8-2 Mount a dial indicator to the transmission case so the indicator's plunger can move with the input shaft as the shaft is pulled up and pushed down.

P8-3 Lightly pull up on the shaft, then push it down until it stops. Then zero the indicator.

P8-4 Pull the shaft up until it stops and then read the indicator. This is the total amount of endplay.

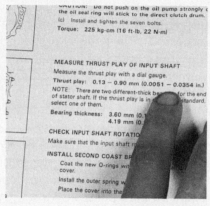

P8-5 Locate the endplay specifications in the service information.

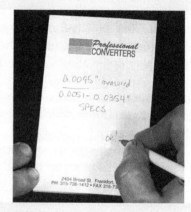

P8-6 Compare your measurements to the specifications.

SERVICE TIP Record endplay measurements before disassembly and during reassembly.

Input shaft endplay is measured with a dial indicator. The dial indicator should be solidly clamped to the bellhousing and the plunger positioned so that it is centered on the end of the input shaft. Move the input shaft in and out of the case and note the reading on the indicator. Compare this reading with the specifications for that transmission. If the endplay is incorrect, it should be corrected during assembly by installing a thrust washer, snap ring, or spacer of correct thickness between the input shaft and the output shaft. Some transmissions require additional endplay checks during disassembly. The manufacturer's recommended procedure for checking endplay should always be followed.

Transfer shaft
gear

Dial
indicator

Figure 5-18 Some transaxles require additional endplay checks during assembly and disassembly.

Some transaxles may require more than one check. As most transaxles have their shaft centerlines to one side, the transaxle may need to be partially disassembled in order to gain access to the shafts.

It is possible that an output or final drive endplay check should be made on either a transmission or transaxle (**Figure 5-18**). The purpose of these settings is to provide long gear life and to eliminate engagement "clunk." All measurements taken during this phase of teardown should be kept during overhaul and used as a guide during the rebuild. Having this information gives the technician places to look for worn parts during the teardown and helps obtain the necessary shims to correct undesired endplay. If the endplay is excessive, this may indicate that the thrust washers or bearings are worn or the sealing rings and grooves are worn. During rebuild, endplay settings should be kept to the minimum allowable amount. Once the endplay checks have been made and recorded, disassembly can begin.

 Special Tools

Dial indicator
    mounting fixture
Dial indicator
Transmission stand

> ⚙⚙ SERVICE TIP When working on automatic transmissions, there is no such thing as being too clean. All dirt, grease, and other materials should be cleaned off any parts that are going to be reused.

A clean work area, clean tools, and an orderly sequence during teardown will help you avoid lost time and frustration during reassembly. As you remove the various assemblies from the transmission, place them on the workbench in the order that they were removed. The correct position of thrust washers, snap rings, and even bolts and nuts is of great importance and should be noted. Unless the transmission suffered a major breakdown or was overheated, the internal components of the transmission will be somewhat clean due to the nature of the fluid. A light cleaning with mineral spirits or a simple blowing off with low-pressure air is all that is necessary for cleaning most parts for inspection.

Each subassembly should be completely disassembled, and the parts thoroughly cleaned and checked for signs of excessive wear, scoring, overheating, chipping, or cracking. If there is any question as to the condition of a part, replace it.

Special Tools

Transmission mount
Mineral spirits
Snap ring pliers
Dial indicator
Assortment of special
    tools as needed for
    particular
    transmission

Basic Disassembly

The following is a general procedure for disassembling a transmission. Always refer to the manufacturer's recommended procedure for the specific transmission.

With the transmission mounted on a fixture, remove the oil pan. Carefully inspect the oil pan for types of foreign matter. An analysis of the type of matter may help you determine what parts need to be carefully examined. Some filters do an excellent job of cleaning the pan of debris. If necessary, take the filter apart and see what kind of debris is present.

Remove the bellhousing from the transmission case, if it is not part of the case. Unscrew and remove all externally mounted solenoids and switches, such as the input and output shaft speed sensors. Remove the O-rings or seals for each after the part is removed.

Many transaxles have a valve body cover at the side of the unit (**Figure 5-19**). Remove the retaining bolts and then remove the cover. Be careful not to bend the cover's flange while removing it. With the cover removed, a wiring harness to various solenoids and sensors may be exposed. Many transmissions, such as the 6L80, have their TCM located on the valve body itself. In these cases, the solenoids are integral to the TCM. Disconnect the connectors of the harness before attempting to remove the valve body.

Unbolt and remove the solenoids from the valve body (**Figure 5-20**). In some transmissions the solenoids are part of a single unit, along with the TCM, remove this unit. Whenever disconnecting any electrical part, make sure to inspect the wiring and connectors.

Unbolt the valve body. Most valve bodies are assemblies of two or more valve units with separators and gaskets (**Figure 5-21**). Carefully separate the total assembly. Make sure you note the location of all check balls before removing them. Also, the length of the bolts may vary; therefore, keep track of where each bolt came from.

Remove the manual valve from the valve body to prevent the valve from dropping out. Back off the servo piston stem locknut and snugly tighten the piston stem to prevent the front clutch drum from dropping out when removing the front pump.

Using the correct puller, remove the front pump from the case (**Figure 5-22**). Lay this assembly to the side for further inspection.

 Caution

When removing the valve body, the steel check balls may fall out. Be prepared for this to happen, and take steps so that you don't lose them. Better yet, try to remove the valve body so the balls will remain in place while the valve body is separated from the case.

Figure 5-19 If the transaxle has a side valve body cover, remove it to gain access to the valve body.

Figure 5-20 Remove the solenoids from the valve body. Keep track of their exact location.

Figure 5-21 This valve body is comprised of several separate assemblies.

Pullers

Figure 5-22 The proper tool should be used to pull the oil pump from the transmission housing.

Remove the front clutch thrust washer and bearing race. Now back off the front brake band servo piston stem to release the band. Remove the brake band strut and front brake band. The drum and band may be removed together.

Remove the front and rear clutch assemblies (**Figure 5-23**). Note the positions of the front pump thrust washers and rear clutch thrust washer, if the transmission has them.

Remove the rear clutch hub, front planetary carrier, and connecting shell. Note the positions of the thrust bearings and the front planetary carrier and thrust washer.

Figure 5-23 When removing clutch assemblies, be careful not to lose track of the placement of the thrust washers.

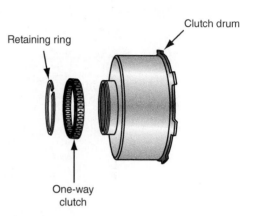

Figure 5-24 One-way clutches are normally retained by a snap ring.

Remove the output shaft snap ring. It will often be easier to remove the snap ring if the carrier is removed first. Remove the carrier snap ring and remove the carrier. Now remove the output shaft snap ring.

Remove the rear connecting drum from the housing. Then using a screwdriver, remove the large retaining snap ring of the rear brake assembly (**Figure 5-24**). Tilt the extension housing upward and remove the rear brake assembly.

Then unbolt and remove the extension housing. Be careful not to lose the parking pawl, spring, and retaining washer. Pull out the output shaft. Remove the inner race of the one-way clutch, the thrust washer, piston return spring, and thrust ring. Using compressed air, remove the rear brake piston and front servo.

⚡ **WARNING Cover the servo with a rag to prevent ATF from blowing into your face and the servo piston from popping into your face.**

Band servos and accumulators are pistons with seals in a bore held in position by springs and retaining rings. Remove the retaining rings and pull the assembly from its bore for cleaning (**Figure 5-25**). Check the condition of the piston and springs. Cast iron seal rings may not need replacement, but elastomer seals should always be replaced.

Figure 5-25 All servos and accumulators should be removed from their bores in the case and the bores carefully inspected.

Once the transmission has been disassembled into its various subassemblies, each subassembly should be disassembled, cleaned, inspected, and reassembled.

TRANSMISSION CASE SERVICE

The transmission case should be thoroughly cleaned and all passages blown out. Make sure all electrical components have been removed from the case before cleaning it. After the case has been cleaned, all bushings, fluid passages, bolt threads, and clutch plate splines should be checked. The passages can be checked for restrictions and leaks by applying compressed air to each one. If the air comes out the other end, there is no restriction. To check for leaks, plug off one end of the passage and apply air to the other. If pressure builds in that passage, there are probably no leaks in it.

Modern transmission cases are made of aluminum, primarily to save weight. Aluminum is a soft material that is susceptible to corrosion and can be deformed, scratched, cracked, or scored easily. Special attention should be given to the following areas: clutch, oil pump, servo, and accumulator bores. All bores should be smooth to avoid scratching or tearing the seals. The servo piston could also hang up in a bore that is deeply scored. Check the fit of the servo piston in the bore without the seal, if possible, to be sure it has free travel. There should be no tight spots or binding over the whole range of travel. Any deep scratches or gouges that cause binding of the piston will require case replacement.

Case-mounted accumulator bores are checked the same way as servo bores. The oil pump bore at the front of the case should be free of any scratches that would keep the O-ring from sealing the outer diameter of the pump to the front of the case. Case-mounted hydraulic clutch bores are prone to the same problems as servo bores. Look for any scratches or gouges in the sealing area that would affect the rubber seals. It is possible to damage these areas during disassembly, so be careful with tools used during overhaul.

Sealing surfaces of the case should be inspected for surface roughness, nicks, or scratches where the seals ride (**Figure 5-26**). Any problems found in servo bores or clutch drum bores can cause pressure leakage in the affected circuit. Imperfections in steel or

The inspection, service, and testing of all hydraulic components are detailed in Chapter 7 of this manual.

Classroom Manual
Chapter 5, page 154

Figure 5-26 All sealing surfaces and bores of the transmission case should be carefully inspected for cracks, grooves, and scratches.

Crocus cloth is a very fine polishing paper. It is designed to remove very little metal, so it is safe to use on critical surfaces. Never use regular sandpaper, as it will remove too much metal.

The oil passages in the case are commonly called worm tracks.

cast iron parts can usually be polished out with crocus cloth. Care should be taken not to disturb the original shape of the bore. Under no circumstances should sandpaper be used. Sandpaper will leave too deep a scratch in the surface. Use the crocus cloth inside clutch drums to remove the polished marks left by the cast iron sealing rings. This will help the new rings rotate with the drum as designed. As a rule, all sealing rings, either cast iron or Teflon, are replaced during overhaul, as this gives the desired sealing surface required for proper operation.

Passages in the case guide the flow of fluid through the case. Although not common, porosity in this area can cause cross-tracking of one circuit to another. This can cause bind-up (two gears at once) or a slow bleed-off of pressure in the affected circuit, which can lead to slow burnout of a clutch or band. If this is suspected, try filling the circuit with solvent and watching to see if the solvent disappears or leaks away. If the solvent goes down, you will have to check each part of the circuit to find where the leak is. Be sure to check that all necessary check balls were in position during disassembly.

Check the valve body mounting area for warpage with a straightedge and feeler gauge. This should be done in several locations so that any crossover from one worm track to another is evident. If there is a slight burr or high spot, it can be removed by flat-filing the surface.

A long straightedge should be laid across the lower flange of the case to check for distortion. Any warpage found here may result in circuit leakage, causing any number of hydraulically related problems. Case warpage should be less than 0.002 inch. Cases with bolted-in center supports should be checked with the support bolted in place. Cases are being made lighter and often will distort during service. This causes a pressure loss to some circuits, causing band or clutch burnouts.

Be sure to check all bellhousing bolt holes and dowel pins. Cracks around the bolt holes indicate that the case bolts were tightened with the case out of alignment with the engine block. The case should be replaced if the following problems are present: broken worm tracking, cracked case at the oil pump to case flange, cracked case at clutch housing pressure cavity, ears broken off the bellhousing, or oil pan flange broken off the case. Although it is possible to weld the aluminum case, it is not possible to determine if the repair will hold. A transmission case is very thin and welding may distort the case.

If any of the bolts that were removed during disassembly have aluminum on the threads, the thread bore is damaged and should be repaired. Thread repair entails installing a thread insert, which serves as new threads for the bolt, or re-tapping the bore. After the threads have been repaired, make sure you thoroughly clean the case to remove all metal filings.

The small screens found during teardown should be inspected for foreign material. These screens are used to prevent valve hang-up at the pressure regulator and they must be in place. Most screens can be removed easily. Care should be taken when cleaning, because some cleaning solvents will destroy the plastic screens. Low air pressure (approximately 30 psi) can be used to blow the screens out in a reverse direction.

Bushings (**Figure 5-27**) in a transmission case are normally found in the rear of the case and require the same inspection and replacement techniques as other bushings in the transmission. Always be sure that the oil passage to a pressure-fed bushing or bearing is open and free of dirt and foreign material. It does no good to replace a bushing without checking to be sure it has good oil flow.

Vents are located in the pump body or transmission case and provide for equalization of pressures in the transmission. These vents can be checked by blowing low-pressure air through them, squirting solvent or brake cleaning spray through them, or by pushing a small-diameter wire through the vent passage. A clean, open passage is all you need to verify proper operation.

Classroom Manual
Chapter 5, page 170

LEGEND

1. Bushing, stator shaft (front)
2. Bushing, oil pump body
3. Bushing, reverse input clutch (front)
4. Bushing, reverse input clutch (rear)
5. Bushing, stator shaft (rear)
6. Bushing, input sun gear (front)

7. Bushing, input sun gear (rear)
8. Bushing, reaction carrier shaft (front)
9. Bushing, reaction gear
10. Bushing, reaction carrier shaft (rear)
11. Bushing, case
12. Bushing, case extension

Figure 5-27 Bushings are used throughout a transmission.

SERVICE TIP The oil pump gears may seize to the pump plate in a 5R55E if the bellhousing is warped around the bushing bore. Be extra careful when inspecting this bellhousing as this fault can destroy a transmission rebuild.

EXTENSION HOUSING

Check the extension housing (**Figure 5-28**) for cracks, especially around the case mounting surface and the pad that attaches to the transmission mount. Using a straightedge and feeler gauge set, check the flatness of the mating surface. Lay the straightedge across the

Seal

Extension
housing

Figure 5-28 The extension housing should be checked for defects and flatness. Always replace its seals and gaskets during reassembly.

surface and attempt to insert feeler gauges of various sizes between the bottom of the straightedge and the surface. If a 0.002-inch gauge fits under the straightedge, the surface may need to be resurfaced or the housing replaced. Minor problems may be corrected by filing down the surface. However, filing should only be done when it is necessary. To correctly file the surface, select the largest single-cut file available. Place the file across one end of the surface and pull or draw the file across the surface to the opposite end. Lift the file off the surface and place it back at its original position, then draw the file to the other end. Repeat this process until the surface is corrected. Never move the file from side to side.

Carefully inspect all threaded and non-threaded bores in the housing. All damaged threaded bores should be repaired by running a tap through the bore or by installing threaded inserts. If any condition exists that cannot be adequately repaired, the case should be replaced.

At the rear of the extension housing is the slip yoke bushing. This bushing will normally wear to one side due to loads imparted on it during operation. Oil-feed holes at this bushing must be checked to make sure oil can get to this bushing. Always make sure this bushing is aligned correctly during replacement or premature failure can result.

Before installing the rear extension housing, assemble the parking pawl pin, washer, spring, pawl, and any other assembly that is enclosed by the extension housing. Be sure they are assembled properly. Install the housing and tighten the bolts to specifications.

Classroom Manual
Chapter 5, page 174

GASKETS

Gaskets are used to create a seal between two flat surfaces. This seal must be able to prevent fluid leaks while undergoing changes in pressure and temperature. Whenever a transmission is overhauled, new gaskets should be used throughout the transmission. The only exception to this is the transmission oil pan gasket, which can be reused if it is in good condition. If the oil pan gasket is damaged or deformed in any way, do not reuse it. Most rebuilders don't reuse oil pan gaskets; they always replace them. The risk of the gasket not sealing is not worth the money saved. Some oil pan and side cover gaskets are intended to be reused, however. It is safe to reuse these gaskets, provided they are not damaged. Check with the service information to be sure.

While disassembling a transmission, keep all of the old gaskets. These will not be reinstalled, but may be used to select the correct gaskets for reassembly. Often transmission rebuilding kits come with a few different gaskets for the same part. This is particularly true of valve body gaskets. To help select the correct gasket for the transmission and the part, compare the old gasket with the new ones. Obviously, the new gasket that matches the old one is the gasket that should be used.

When installing a gasket, make sure both surfaces are clean and flat. Any imperfections in a sealing surface should be corrected before installing a new gasket. These imperfections will prevent the gasket from providing a good seal. To remove minor scratches from a sealing surface, use a fine flat file. Make sure you don't remove more metal than is necessary. Also make sure the surface is flat after you have filed away the scratches.

To ensure flatness, spread a piece of medium-grade (approximately 300 grit) sandpaper on a flat surface. Then set the sealing surface of the object on the paper. Move the object in a figure 8 pattern around the surface of the sandpaper. Apply an even amount of pressure on the object while doing this. When the sandpaper leaves a mark on all parts of the surface, the sealing surface is flat. Stop sanding it and clean it off.

Oil pans are typically made of stamped steel. The thin steel tends to become distorted around the attaching bolt holes. These distortions can prevent the pan from fitting tightly against the transmission case. To flatten the pan, place the mounting flange of the pan on a block of wood and flatten one area at a time with a ball-peen hammer.

Transmission gaskets should not be installed with any type of liquid adhesive or sealant, unless specifically noted by the manufacturer. If any sealer gets into the valve body, severe damage can result. Also, sealant can clog the oil filter. If a gasket is difficult to install, a thin coating of transmission assembly lube can be used to hold the gasket in place while assembling the parts.

One type of gasket that presents unique installation problems is the cork-type gasket. These gaskets tend to change shape and size with changes in humidity. If a cork-type gasket is slightly larger than it should be, soak it in water and lay it flat on a warm surface. Allow it to dry before installing it. If the gasket is slightly smaller than required, soak it in warm water prior to installation.

Whenever you are using a cork-type gasket to create a seal between two parts, make sure you properly tighten the two surfaces together. Tighten the attaching bolts or nuts, in a staggered pattern, to the specified torque so that the gasket material is evenly squeezed between the two surfaces. If too much torque is applied, the gasket may split.

SEALS

Four types of seals are used in automatic transmissions: O-ring, square-cut, lip, and sealing rings (**Figure 5-29**). These seals are designed to stop fluid from leaking out of the transmission. Internal oil seals prevent the loss of hydraulic pressure in the transmissions control and apply circuits. The latter ensures proper shifting of the transmission.

O-ring and square-cut seals are used to seal nonrotating parts (**Figure 5-30**). These seals can provide a good seal at high-pressure points within the transmission. Common points that are sealed with these seals are oil pumps, servos, and clutch pistons.

When installing a new O-ring or square-cut seal, coat the entire surface of the seal with assembly lube. Make sure you don't stretch or distort the seal while you work it into its holding groove. Some stretching may be necessary to work the seal over a shaft or fitting, but do not stretch it more than is needed. After a square-cut seal is installed, double-check to make sure it is not twisted. The flat surface of the seal should be parallel with the bore. If it is not, fluid will easily leak past the seal.

Lip seals are commonly used around rotating shafts and apply pistons. Lip seals that are used to seal a shaft typically have a metal flange around their outside diameter. The shaft rides on the lip seal at the inside diameter of the seal assembly. The rigid outer diameter provides a mounting point for the lip seal and is pressed into a bore. Once pressed into the bore, the outer diameter of the seal prevents fluid from leaking into the bore while the inner lip seal prevents leakage past the shaft.

Piston lip seals are set into a machined groove on the piston. These lip seals are not housed in a rigid metal flange. They are designed to be flexible and provide a seal while the piston moves up and down. While the piston moves, the lip flexes up and down. Any distortion or damage to the seal will allow fluid to escape. If fluid escapes from around a piston, the piston will not move with the force and speed that it should.

The most important thing when installing a lip seal is to make sure the lip is facing in the correct direction. The lip should always be aimed toward the source of pressurized fluid. If installed backward, fluid under pressure will easily leak past the seal. Also remember to make sure the surfaces to be sealed are clean and not damaged.

Teflon or metal sealing rings are commonly used to seal servo pistons, oil pump covers, and shafts. These rings may be designed to provide for a seal, but they may also be designed to allow a controlled amount of fluid leakage. Leakage may be allowed to lubricate some shaft bushings in the transmission. Sealing rings are either solid or cut rings. Cut sealing rings are one of three designs: open-end, butt-end, or locking-end.

The square cut seal is commonly called a lathe-cut seal.

Most seal problems after a rebuild are caused by incorrect installation. Never rush through the installation process and always use the correct tool for the job.

Classroom Manual
Chapter 5, page 175

LEGEND

2	Ring, turbine shaft front oil seal
5	Seal, oil pump
13	Seal, output shaft
15	Seal, case extension
20	Seal assembly, prop shaft front slip yoke oil
57	Seal, manual 2–1 band servo piston
66	Seal, low and reverse servo piston
67	Ring, low and reverse accumulator piston outer oil seal
69	Ring, low and reverse accumulator piston inner oil seal
201	Seal assembly, torque converter oil
219	Ring, oil seal, overrun clutch housing
404	Seal, 3rd clutch accumulator piston outer
406	Seal, 3rd clutch accumulator piston inner
501	Ring, turbine shaft rear oil seal
503	Ring, turbine shaft intermediate oil seal
505	Piston assembly, overrun clutch
527	Seal, 4th clutch piston inner
531	Seal, 4th clutch piston outer

606	Piston, forward clutch
619	Piston, direct clutch
622	Seal, direct clutch piston intermediate
637	Seal, intermediate clutch piston inner
638	Seal, intermediate clutch piston outer
639	Ring, direct clutch housing oil seal
685	Seal assembly, forward clutch piston

Figure 5-29 Location of the various gaskets and seals used in today's transmissions.

Figure 5-30 O-ring and square-cut seals are used to seal nonrotating parts, such as the front pump.

Solid sealing rings are commonly used in late-model transmissions. These rings are made of a Teflon-based material and are never reused. To remove them, pry them out of their groove and carefully cut through the seal. Installing a new seal requires two tools: an installation tool and a resizing tool. It is important to note that the seals used in a particular transmission or in different transmissions may be a different size. This means two tools are required for each of the different sizes unless you have a universal tool set.

To install a Teflon seal, place the installation tool over the hub or shaft. Warm the new seal in hot water to soften the material, this will make installation easier. Lubricate the new seal and tool. Slide the seal over the tool and seat it into its groove on the hub or shaft. The installation tool will stretch the seal uniformly and just enough to install it. Due to the stretch, the resizing tool must be used to ensure a proper fit. The resizing tool is tube or cylinder with a long chamfer or bevel. The tool is designed to compress the seal into its original size after installation. To resize the seal, lubricate it and the resizing tool. Then carefully work the tool over the seal. While doing this, make sure the seal is sitting in its groove.

Open-end sealing rings fit loosely into a machined groove. The ends of the rings do not touch when they are installed. This type of ring is typically removed and installed with a pair of snap ring pliers. The ring should be expanded just enough to move it off or onto the shaft.

Butt-end sealing rings are designed so that their ends butt or touch each other once the seal is in place. This type of seal can be removed with a small screwdriver. The blade of the screwdriver is used to work the ring out of its groove. To install this type of ring, use a pair of snap ring pliers and expand the ring to move it into position.

Locking-end rings may have hooked ends that connect or ends that are cut at an angle (these are called scarf-cut seals) to hold the ends together. These seals are removed and installed in the same way as butt-end rings. After these rings are installed, make sure the ends are properly positioned and touching.

REASSEMBLY AND TESTING

Before proceeding with the final assembly of all components, it is important to verify that the case, housing, and parts are clean and free from dust, dirt, and foreign matter (use an air gun). Have a tray available with clean ATF for lubricating parts. Also have ready a jar or tube of Vaseline for securing washers during installation.

Coat all parts with the proper type of ATF. Soak bands and clutches in the fluid for at least 15 minutes before installing them. All new seals and rings should have been installed before beginning final assembly.

Seal Installation

All seals should be checked in their own bores. They should be slightly smaller or larger (±3 percent) than their groove or bore. If a seal is not the proper size, find one that is. Do not assume that because a particular seal came with the overhaul kit, it is the correct one.

Never install a seal when it is dry. The seal should slide into position and allow the part it seals to slide into it. A dry seal is easily damaged during installation. Coat all seals with transmission assembly lube. Following are some guidelines for installing a seal in an automatic transmission:

1. Install only genuine seals recommended by the manufacturer of the transmission.
2. Use only the proper fluids as stated in the appropriate service information.
3. Keep the seals and fluids clean and free of dirt.
4. Before installing seals, clean the shaft or bore area. Carefully inspect these areas for damage. File or stone away any burrs or bad nicks and polish the surfaces with a fine crocus cloth. Then clean the area to remove the metal particles. In dynamic applications, the sliding surface for the seal should have a mirror finish for best operation.

Locking-end sealing rings may be called locking rings if they are made of metal, or scarf-cut rings if they are made of Teflon.

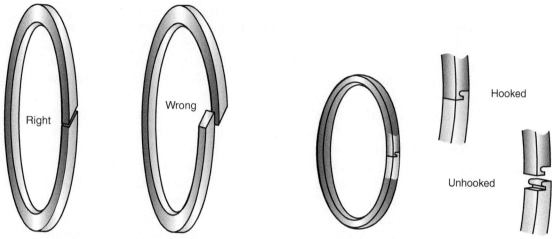

Figure 5-31 Installation of Teflon oil seal rings.

Figure 5-32 Installation of cast iron oil seal rings.

Automatic transmission parts distributors package parts kits to fit the level of service to be performed. Two common ones are the resealing and reconditioning kits. The resealing kit includes all the necessary parts, such as gaskets and cast iron, Teflon, rubber, and metal rings and seals, to reseal a transaxle or transmission. A transmission reconditioning kit includes all the necessary seals, gaskets, filters, bushings, clutch friction and steel discs, and brake bands. Do not assume that installing an overhaul kit will fix any transmission problems. You still must find the "root cause" of any problem to prevent a returning or disappointed customer.

> ⚙ **SERVICE TIP** When tightening any fastener that directly or indirectly involves a rotating shaft or other part, rotate that part during and after tightening to ensure that the part does not bind.

5. Lubricate the seal, especially any lips, with transmission assembly lube to ease installation.

6. All metal sealing rings should be checked for proper fit. Since these rings seal on their outer diameter, the seal should be inserted in its bore and should feel tight there. If the seal has some form of locking ends, these should be interlocked prior to trying the seal in its bore (**Figures 5-31** and **5-32**). The fit of the sealing rings in their shaft groove should also be checked. If the ring can move laterally in the groove, the groove is worn and this will cause internal fluid leaks. To check the side clearance of the ring, place the ring into its groove and measure the clearance between the ring and the groove with a feeler gauge. Typically, the clearance should not exceed 0.003 inch.

7. While checking the clearance, look for nicks in the grooves and for evidence of groove taper or stepping. If the grooves are tapered or stepped, the shaft will need to be replaced. If there are burrs or nicks in the grooves, they can be filed away.

8. Any distorted or undersized sealing rings should be replaced.

9. Always use the correct driver when installing a seal (**Figure 5-33**) and be careful not to damage the seal during installation. When possible, press the seal into position. This usually prevents the garter spring from moving out of position during installation.

Transmission Reassembly

Prior to reassembly, check service information for the order and procedure for installing the different components and units. Also, be sure to examine all thrust washers carefully and coat them with petroleum jelly before placing them in the housing. Install the thrust ring, piston return spring, thrust washer, and one-way clutch inner race into the case. Align and start the bolts into the inner race from the rear of the case. Torque the bolts to specifications.

Figure 5-33 Using a special driver to install an oil seal. Note how the tool prevents damage to the seal.

 Special Tools

Appropriate set of
    seal removers and
    installers
Fresh ATF
Feeler gauge
Torque wrench

Classroom Manual
Chapter 5, page 175

Many Teflon seals may look different than the ones shown in these figures and they may require the application of heat for installation. Always refer to the service information when installing seals and other transmission components.

⚠ **Caution**

Check and verify that the return spring is centered onto the race before tightening.

Lubricate and install the rear piston into the case. After determining the correct number of friction and steel plates, install the steel dished plate first. Make sure the steel disc is facing in the correct direction, then install the steel and friction plates, the retaining plate, and snap ring (**Figure 5-34**).

Using a suitable blowgun with a rubber tapered tip, air check the clutch pack operation. After the rear brake has been completely assembled, measure the clearance between the snap ring and the retainer plate. Select the proper thickness of selective snap ring or retaining plate that will give the correct ring-to-plate clearance if the measurement does not meet the specified limits.

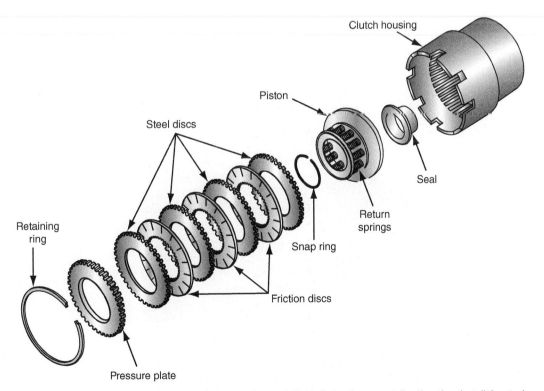

Figure 5-34 When assembling a clutch pack, be sure the steel disc is facing the correct direction, then install the steel and friction plates, the retaining plate, and snap ring.

> ⚙ **SERVICE TIP** The ends of snap rings are slightly tapered. When installing snap rings, make sure the taper faces up or toward the outside of the shaft. This will provide a good gripping surface for the snap ring pliers during installation and for removal the next time the transmission is taken apart.

On some models, the output shaft, bearing, and appropriate gauging shims are placed into the transmission housing. The output shaft washer and bolt are then installed. While holding the output shaft and gear assembly, torque the output shaft nut to specifications. Then install a dial indicator and check the travel of the output shaft as it is pushed and pulled. Remove the gauging shims and install the correct sizes of service shims (**Figure 5-35**), output shaft gear, washer, and nut. Torque the output shaft nut to specifications. Using a pound-inch torque wrench, check the turning torque of the output shaft and compare this reading to specifications.

Place the small thrust washer on the pilot end of the transaxle output shaft. Then place the rear clutch assembly, front clutch drum, turbine shaft, and thrust washer into the housing. Locate and align the rear clutch over its hub. Gently move the rear clutch and turbine shaft around, rotating the assembly to engage the teeth of the friction discs with the rear clutch hub. Align the direct clutch assembly over the front clutch hub. Move the input shaft back and forth, rotating it so the front clutch friction discs engage with the front clutch hub.

Position the thrust washer to the back of the rear planetary carrier. Install the rear planetary carrier and thrust washer into the housing to engage the rear planetary ring gear. Install the front thrust washer and the drive shell assembly, engaging the common sun gear with the planetary pinions in the rear planetary carrier. Assemble the front planetary gear assembly into the front planetary ring gear. Make sure the planetary pinion gear shafts are securely locked to the planetary carrier.

Install the one-way sprag into the one-way clutch outer race with the arrow on the sprag facing the front of the transmission. Some overrunning clutches do not have an

Figure 5-35 A differential bearing race selective shim. Measuring rotational torque will determine the correct size shim to use. If the measurement is less than the specification, a thicker selective shim should be used; likewise, if the measurement is greater than the specification, use a thinner shim.

Align marks identifying
nonthreaded hole in
cam and case

Overrunning
clutch
assembly

Figure 5-36 For proper installation, an overrunning
clutch should be indexed to the case prior to removal.

arrow to use for guidance. For these, the manufacturers recommend that an index mark be made on the clutch and the transmission case (**Figure 5-36**). This index mark should be made at the non-threaded bore in the clutch. The bores for the mounting of the cam are slightly countersunk on one side. The clutch needs to be installed with this side facing rearward or toward the piston retainer. Install the connecting drum and sprag by rotating the drum clockwise, using a slight pressure and wobbling to align the plates with hub and sprag assembly. The connecting drum should now be free to rotate clockwise only. This check will verify that the sprag is correctly installed and operative. Now install the rear internal gear and the shaft's snap ring.

⚙⚙ **SERVICE TIP** This snap ring may be thinner than the clutch drum snap ring, so be sure you are using the correct size. Should you have insufficient space to install the snap ring into the drum groove, pull the connecting drum forward as far as possible. This will give you sufficient groove clearance to install the drum snap ring.

Secure the thrust bearing with petroleum jelly and install the rear planet carrier and the snap ring.

Assemble the front and rear clutch drum assemblies together and lay them flat on the bench. Be sure the rear hub thrust bearing is properly seated, then measure from the face of the front clutch drum to the top of the thrust bearing. Install the thrust washer and pump front bearing race to the pump. Then measure from the pump shaft (bearing race included) to the race of the thrust washer. Normally, the difference in measurements should be about 0.02 inch. If the thrust washer is not within the limits, replace it with one of the correct thickness.

Total endplay should now be checked. Set the transmission case on end, front end up. Be sure the thrust bearings are secure with assembly lube. Pick up the complete front clutch assembly and install it into the case. Be sure all parts are seated before proceeding with the measurement. Using a dial indicator or caliper, measure the distance from the rear hub thrust bearing to the case. Next measure the pump with the front bearing race and gasket installed. Tolerance should fall within specifications. If the difference between the measurements is not within tolerance, select the proper size front bearing race. If it is necessary to change the front bearing race, be sure to change the front clutch thrust washers the same amount.

Figure 5-37 Install the brake band and strut and finger-tighten the band servo piston stem just enough to keep the band and strut snug from falling out.

Figure 5-38 When installing the oil pump, make sure it is in the proper position. Also, tighten the bolts in the correct order and to the correct specification.

Install the brake band servo. Use extreme care not to damage the O-rings. Lubricate around the seals. Then install and torque the retainer bolts to specifications. Loosen the piston stem. Install the brake band and strut (**Figure 5-37**) and finger-tighten the band servo piston stem just enough to prevent the band and strut from falling out. Do not adjust the band at this time. Air check for proper performance.

Place some assembly lube in two or three spots around the oil pump gasket and position it on the transaxle housing. Next align the pump and install it with care (**Figure 5-38**). Tighten the pump, attaching bolts to specifications in the specified order. After the bolts are tight, check the rotation of the input shaft. If the shaft does not rotate, disassemble the transmission to locate the misplaced thrust washer.

Before installing the bellhousing, check the bolt hole alignment. Install the bellhousing and torque the retaining bolts to specifications.

If bands are used in the transmission, check the fit of the bands, servo operation and necessary clearances according to service information.

Before proceeding with the installation of the valve body assembly, it is good practice to perform a final air check of all assembled components. This will ensure that you have not overlooked the tightening of any bolts or damaged any seals during assembly.

Assemble the parking pawl assembly. Place the assembly into its position and install the extension housing with a new gasket, then tighten the attaching bolts to the proper specifications.

On transaxles, the differential assembly should be disassembled, cleaned, inspected, and reassembled. After it has been reassembled, measure its endplay with gauging shims. Then select a shim thick enough to correct the endplay. After installing the proper shims, measure the differential turning torque with a pound-inch torque wrench. Increase or decrease the shim thickness to provide the correct turning torque.

Install the valve body. Be sure the manual valve is in alignment with the selector pin. Tighten the valve body attaching bolts to the specified torque.

Before installing the solenoids, check to verify that they are operating properly. Connect a solenoid checker to each solenoid. When the solenoids are activated, you should hear the solenoid click on. On some transmissions the integrity of a solenoid can be checked with an ohmmeter. If the solenoid is good, install it. If the solenoid does not check out, replace it.

 Caution

Be certain that the proper length bolts are installed in the related depth holes. The lengths of the bolts vary.

⚙️ **SERVICE TIP** It is a good idea to partially prefill all torque converters before installing them in the vehicle.

Before installing the oil pan, check the alignment and operation of the control lever and parking pawl engagement. Make a final check to be sure all bolts are installed in the valve body. Install the oil pan with a new gasket. Torque the bolts to specifications.

Lubricate the oil pump's lip seal and the converter neck before installing the converter. Also put a coat of petroleum jelly on the seal's garter spring. This will prevent the spring from being knocked loose while installing the torque converter. Install the converter, making sure that the converter is properly meshed with the oil pump drive gear.

The transmission is now ready for installation into the vehicle. Use the reverse of the removal procedures. Never bolt the converter onto the flexplate and then try to install the transmission onto the converter. Remember to follow the correct fluid filling procedure and check service information for transmission relearn instructions.

CASE STUDY

A student at the community college was taking an automatic transmission course. He had rebuilt his transmission, but seemed to be a little concerned about his work. He was having a hard time getting the input shaft to turn after bolting the transmission pump into the case. Another student told him that "the engine will be able to turn it, no problem." Overhearing the conversation, I asked if I could help. The student was able to turn the input shaft, but only with the help of a pair of pliers. I looked at his worksheet and noticed that the entry for the endplay measurement before disassembly was blank. He had not checked the endplay before disassembly. We proceeded to check to see if it was within specification, and there was no endplay in his assembly. After the student took the assembly apart a fourth time, he found that he had switched two thrust bearings and caused the problem with endplay. As frustrated as he was, he was happy to find the problem. Can you imagine what would have happened if he had installed the transmission and "let the engine turn it"? The amount of heat generated would have been tremendous, and would have destroyed the transmission in a short time. Two lessons here: Always do needed endplay readings, before and after teardown and it is an effective way to check for proper assembly!

ASE-STYLE REVIEW QUESTIONS

1. *Technician A* says before tearing down a transmission, you should inspect the debris trapped in the fluid filter.
   Technician B says installing an overhaul kit will take care of all the problems with a transmission.
   Who is correct?
   A. A only
   B. B only
   C. Both A and B
   D. Neither A nor B

2. While discussing endplay checks:
   Technician A says these checks should be taken before the transmission is disassembled.
   Technician B says these checks should be taken after the transmission is reassembled.
   Who is correct?
   A. A only
   B. B only
   C. Both A and B
   D. Neither A nor B

3. *Technician A* says abnormal noises from a transmission will never be caused by faulty clutches or bands.

   Technician B says abnormal noises from a transmission are typically caused by a faulty torque converter.

   Who is correct?

   A. A only   C. Both A and B

   B. B only   D. Neither A nor B

4. An output shaft or final drive endplay check should be made on either a transmission or transaxle.

   Technician A says this is done to provide long gear life and to eliminate engagement "clunk."

   Technician B says endplay should be measured during teardown, which helps obtain the shims necessary to correct undesired endplay.

   Who is correct?

   A. A only   C. Both A and B

   B. B only   D. Neither A nor B

5. *Technician A* says when removing a seal, inspect the sealing surface or lips.

   Technician B says this check should be done before cleaning the seal.

   Who is correct?

   A. A only   C. Both A and B

   B. B only   D. Neither A nor B

6. *Technician A* says vibration problems can be caused by an unbalanced torque converter.

   Technician B says vibration problems can be caused by a faulty output shaft.

   Who is correct?

   A. A only   C. Both A and B

   B. B only   D. Neither A nor B

7. *Technician A* says once the transmission is removed and before it is disassembled, the outside of the transmission should be cleaned.

   Technician B says no matter what type of cleaning procedure is followed, the transmission and parts should be rinsed with water and then thoroughly dried with compressed air before reassembly.

   Who is correct?

   A. A only   C. Both A and B

   B. B only   D. Neither A nor B

8. While checking a transmission's vent:

   Technician A applies a vacuum to it and says that if the vent leaks, it should be replaced.

   Technician B runs ATF through the vent and says that if the vent cannot hold fluid, it must be replaced.

   Who is correct?

   A. A only   C. Both A and B

   B. B only   D. Neither A nor B

9. *Technician A* says seals should be kept clean and free of dirt, before and during installation.

   Technician B says most seals should be installed dry.

   Who is correct?

   A. A only   C. Both A and B

   B. B only   D. Neither A nor B

10. While discussing transmission lip seals:

    Technician A says the most important thing when installing a lip seal is to make sure the lip is facing in the correct direction.

    Technician B says the lip should always be aimed away from the source of pressurized fluid.

    Who is correct?

    A. A only   C. Both A and B

    B. B only   D. Neither A nor B

ASE CHALLENGE QUESTIONS

1. *Technician A* says input shaft endplay is measured with a micrometer.

   Technician B says a dial indicator should be solidly clamped to the bellhousing and the plunger positioned so that it is centered on the end of the input shaft.

   Who is correct?

   A. A only   C. Both A and B

   B. B only   D. Neither A nor B

2. *Technician A* says all metal sealing rings should be checked for proper fit.

   Technician B says since these rings seal on their outer diameter, the seal should be inserted in its bore and should feel tight there.

   Who is correct?

   A. A only   C. Both A and B

   B. B only   D. Neither A nor B

3. *Technician A* says to check all bellhousing bolt holes and dowel pins in the transmission case for cracking.

Technician B says cracks around the bolt holes indicate that the case bolts were tightened with the case in alignment with the engine block.

Who is correct?

A. A only

B. B only

C. Both A and B

D. Neither A nor B

4. *Technician A* says at the rear of the extension housing is the slip yoke bushing.

Technician B says this bushing will normally wear to one side due to loads imparted on it during operation.

Who is correct?

A. A only

B. B only

C. Both A and B

D. Neither A nor B

5. When removing a front-wheel-drive transmission,

Technician A says to remove the bolts and cables that are accessible from the engine bay before beginning work under the vehicle.

Technician B says it is usually necessary to install an engine support fixture before removing a front-wheel-drive transmission.

Who is correct?

A. A only

B. B only

C. Both A and B

D. Neither A nor B

Name _____   Date _____

REMOVING A TRANSMISSION OR TRANSAXLE

Upon completion of this job sheet, you should be able to describe the procedures that must be followed in order to remove a transaxle or transmission from a vehicle.

ASE Education Foundation Correlation

This job sheet addresses the following **AST/MAST** task:

C.1   Remove and reinstall transmission/transaxle and torque converter; inspect engine core plugs, rear crankshaft seal, dowel pins, dowel pin holes, and mating surfaces. **(P-2)**

We Support
ASE | Education Foundation

Tools and Materials

- Vehicle with automatic transmission
- Hoist
- Transmission jack
- Engine support fixture
- Drain pan
- Droplight or good flashlight
- Service information

Describe the Vehicle Being Worked On

Year _____ Make _____ VIN _____

Model _____

Model and type of transmission _____

Procedure   **Task Completed**

This job sheet is designed to allow you to take a good look at what would be involved in removing a transmission from an assigned vehicle.

1. Refer to service information and look for any precautions or special instructions that relate to the removal of a transmission from this vehicle. Describe them here.

   _____

   _____

2. Before beginning to remove the transmission from this vehicle, what is the first thing that should be disconnected? Why?

   _____

   _____

3. According to service information, identify what should be removed or disconnected from above before raising the vehicle on the hoist. List those items here.

   _____

   _____

   _____

   _____

   _____

4. Raise the vehicle to a comfortable working height.   ☐

5. Carefully examine the area around the transmission and identify everything that should be disconnected or removed before unbolting the transmission from the engine. List those items here.

_____

_____

_____

_____

_____

Problems Encountered

_____

_____

_____

Instructor's Comments

_____

_____

_____

Name _____   Date _____

DISASSEMBLY AND INSPECTION OF A TRANSMISSION

Upon completion of this job sheet, you should be able to disassemble and inspect the major components of a transmission.

ASE Education Foundation Correlation

This job sheet addresses the following **MAST** task:

C.6   Disassemble, clean, and inspect transmission/transaxle. **(P-2)**

We Support

ASE | Education Foundation

Tools and Materials

- Automatic transmission or transaxle on a bench
- Compressed air and air nozzle
- Supply of clean solvent
- Lint-free shop towels
- Service information

Describe the Transmission Being Worked On

Model and type of transmission _____

Year _____ Make _____ VIN _____

Model _____

Procedure

Task Completed

1. Disassemble the transmission into major units. Set each unit aside until this job sheet refers to it. Describe any problems you encountered while disassembling the transmission. Be sure to follow the procedures given in the service information while taking the transmission apart.

   _____

   _____

   _____

   _____

2. Clean the planetary gearset in clean solvent and allow it to air dry. If compressed air is used to help the drying process, firmly hold the pinion gears to prevent the bearings from moving.   ☐

3. Clean all thrust washers, thrust bearings, and bushings in clean solvent and allow them to air dry.   ☐

4. Place one member of the planetary gearset on the sun gear. Rotate the gearset slowly. Do the same for the other parts of the gearset. Describe the feel of the gears' rotation.

   _____

   _____

   _____

5. Inspect each member of the gearset for damaged or worn gear teeth. Describe their condition.

   _____

   _____

6. Inspect the splines of each gearset member and describe their condition.

_____

_____

_____

7. Remove any buildup of material or dirt that may be present between the teeth of the   ☐
   gears.

8. Look at the carrier assemblies and check for cracks or other damage. Describe your
   findings.

_____

_____

_____

9. Inspect the entire gearset for signs of discoloration. Describe your findings and explain
   what is indicated by them.

_____

_____

_____

10. Check the output shaft and its splines for wear, cracking, or other damage. Describe
    their condition.

_____

_____

_____

11. Carefully inspect the driving shells and drive lugs for wear and damage. Describe their
    condition.

_____

_____

12. Summarize the condition of the planetary gearsets and shaft.

_____

_____

_____

13. Check the thrust washers and bearings for damage, excessive wear, and distortion.
    Describe their condition.

_____

_____

_____

14. Check all of the bushings for signs of wear, scoring, and other damage. Describe their condition.

_____

_____

_____

15. Check the shafts that ride in the bushings. Describe their condition.

_____

_____

_____

16. Summarize the condition of the thrust washers, bearings, and their associated shafts.

_____

_____

_____

Problems Encountered

_____

_____

_____

Instructor's Comments

_____

_____

_____

Name _____   Date _____

SERVICING ACCUMULATORS AND SERVOS

Upon completion of this job sheet, you will be able to inspect the servo's and accumulator's bore, piston, seals, pin, spring, and retainers.

ASE Education Foundation Correlation ――――――――――――――――――――――

This job sheet addresses the following **MAST** task:

C.8   Inspect servo and accumulator bores, pistons, seals, pins, springs, and retainers; determine needed action. (**P-2**)

We Support

ASE | **Education Foundation**

Tools and Materials

- Snap ring pliers
- Small file
- Crocus cloth
- Knife

Protective Clothing

Goggles or safety glasses with side shields

Describe the Vehicle Being Worked On

Year _____ Make _____ Model _____

VIN _____ Engine type and size _____

Transmission type and model _____

Procedure

Task Completed

1. The servo and accumulator assemblies may be serviceable with the transmission in the vehicle. Others require complete disassembly of the transmission. Before disassembling a servo or any other component, carefully inspect the area to determine the exact cause of the leakage. Do this before cleaning the area around the seal. Look at the path of the fluid leakage and identify other possible sources. These sources could be worn gaskets, loose bolts, cracked housings, or loose line connections. Describe your findings.

   _____

   _____

2. Inspect the outside area of the seal. If it is wet, determine if oil is leaking out or if it is merely a lubricating film of oil. Describe your findings. Describe your findings.

   _____

   _____

3. Remove the servo and continue to look for causes of the leak. Check both the inner and outer parts of the seal for wet oil, which means leakage. Describe your findings.

   _____

   _____

4. When removing the seal, inspect the sealing surface, or lips. Look for unusual wear, warping, cuts and gouges, or particles embedded in the seal. Describe your findings.

   _____

   _____

5. Remove the retaining rings and pull the assembly from the bore for cleaning. Carefully check the bore and report its condition here.

_____

_____

6. Check the condition of the piston and springs. Cast iron seal rings may not need replacement, but rubber and elastomer seals should always be replaced. Describe your findings.

_____

_____

7. Begin the disassembling of an accumulator by removing the accumulator plate snap ring. ☐

8. Remove the accumulator plate, the spring and accumulator pistons. If rubber seal rings are installed on the piston, replace them whenever you are servicing the accumulator. ☐

9. Lubricate the new accumulator piston ring and carefully install it on the piston. ☐

10. Lubricate the accumulator cylinder walls and install the accumulator piston and spring. ☐

11. Install the accumulator plate and retaining snap ring. ☐

12. A servo is serviced in a similar fashion. The servo's piston, spring, piston rod, and guide should be cleaned and dried. ☐

13. Check the servo piston for cracks, burrs, scores, and wear. Servo pistons should be carefully checked for cracks and their fit on the guide pins. Describe your findings.

_____

_____

14. Check the piston's bore and seal groove for defects or damage. Describe your findings.

_____

_____

15. Inspect the servo or accumulator spring for possible cracks. Also check where the spring rests against the case or piston. Describe your findings.

_____

_____

16. Inspect the servo cylinder for scores or other damage. Describe your findings.

_____

_____

17. Move the piston rod through the piston rod guide and check for freedom of movement. Describe your findings.

_____

_____

Task Completed

18. Check the band servo components for wear and scoring. Describe your findings.

_____

_____

19. When reassembling the servo, lubricate the seal ring with ATF and carefully install it on the piston rod. ☐

20. Lubricate and install the piston rod guide with its snap ring into the servo piston. ☐

21. Install the servo piston assembly, return spring, and piston guide into the servo cylinder. ☐

22. Lubricate and install the new lip seal. ☐

Problems Encountered

_____

_____

_____

Instructor's Comments

_____

_____

_____

Name _____   Date _____

REASSEMBLING A TRANSMISSION/TRANSAXLE

Upon completion of this job sheet, you should be able to properly measure and set endplay and preload during the assembly of a transmission or transaxle.

ASE Education Foundation Correlation ———————————————

This job sheet addresses the following **MAST** tasks:

C.9   Assemble transmission/transaxle. (**P-1**)

C.11   Measure transmission/transaxle end play or preload; determine
            needed action. (**P-1**)

We Support

Education Foundation

Tools and Materials

- Basic hand tools
- Special tools for the transmission
- Clean ATF in a tray
- Petroleum jelly
- Air nozzle
- Micrometer
- Dial indicator

Describe the Transmission That Was Assigned to You and the Vehicle It Came Out of

Year _____ Make _____ Model _____

VIN _____ Engine type and size _____

Model and type of transmission _____

Procedure

Task Completed

1. Before proceeding with the final assembly of all components, it is important to verify that the case, housing, and parts are clean and free from dust, dirt, and foreign matter. All new seals and rings should have been installed before beginning final assembly. Coat the seals with assembly lube or transmission fluid to ease installation and assure proper lubrication.

2. Coat all parts with the proper type of ATF. Soak bands and clutches in the fluid for at least 15 minutes before installing them.   ☐

3. Examine all thrust washers carefully and coat them with petroleum jelly before placing them in the housing.   ☐

4. Install the thrust ring, piston return spring, thrust washer, and one-way clutch inner race into the case. Align and start the bolts into the inner race from the rear of the case. Torque the bolts to specifications. What are the specifications?

   _____

   _____

5. Lubricate and install the rear piston into the case.   ☐

6. After determining the correct number of friction and steel plates, install the steel dished plate first, then the steel and friction plates, and finally the retaining plate and snap ring. How many steel plates did you have?

   _____

7. Using a suitable blowgun with a rubber tapered tip, air check the rear brake operation. What were the results?

_____

_____

8. After the rear brake has been completely assembled, measure the clearance between the snap ring and the retainer plate. Select the proper thickness of retaining plate that will give the correct ring to plate clearance if the measurement does not meet the specified limits. What were the results?

_____

_____

9. On some models, the output shaft, bearing, and appropriate gauging shims are placed into the transmission housing. The output shaft washer and bolt are then installed. While holding the output shaft and gear assembly, torque the output shaft nut to specifications. What are the specifications?

_____

10. Install a dial indicator and check the travel of the output shaft as it is pushed and pulled. What were the results?

_____

_____

11. Remove the gauging shims and install the correct sizes of service shims, output shaft gear, washer, and nut.   ☐

12. Torque the output shaft nut to specifications. Using a pound-inch torque wrench, check the turning torque of the output shaft and compare this reading to specifications. What were the results?

_____

_____

13. Place the small thrust washer on the pilot end of the transaxle output shaft.   ☐

14. Place the rear clutch assembly, front clutch drum, turbine shaft, and thrust washer into the housing.   ☐

15. Locate and align the rear clutch over its hub. Gently move the rear clutch and turbine shaft around, rotating the assembly to engage the teeth of the friction discs with the rear clutch hub. Align the direct clutch assembly over the front clutch hub. Move the input shaft back and forth, rotating it so the front clutch friction discs engage with the front clutch hub.   ☐

16. Position the thrust washer to the back of the rear planetary carrier.   ☐

17. Install the rear planetary carrier and thrust washer into the housing to engage the rear planetary ring gear.   ☐

18. Install the front thrust washer and the drive shell assembly, engaging the common sun gear with the planetary pinions in the rear planetary carrier.   ☐

19. Assemble the front planetary gear assembly into the front planetary ring gear. Make sure the planetary pinion gear shafts are securely locked to the planetary carrier.   ☐

20. Install the one-way sprag into the one-way clutch outer race with the arrow on the sprag facing the front of the transmission.   ☐

21. Install the connecting drum with sprag by rotating the drum clockwise using a slight pressure and wobbling to align the plates with hub and sprag assembly. The connecting drum should now be free to rotate clockwise only. This check will verify that the sprag is correctly installed and operative. What were the results?

_____

_____

22. Install the rear internal gear and the shaft's snap ring.   ☐

23. Secure the thrust bearing with petroleum jelly and install the rear planet carrier and the snap ring.   ☐

24. Assemble the front and rear clutch drum assemblies together and lay them flat on the bench.   ☐

25. Make sure the rear hub thrust bearing is properly seated, then measure from the face of the front clutch drum to the top of the thrust bearing. What did you measure, and how does it compare to the specifications?

_____

_____

_____

26. Install the thrust washer and pump front bearing race to the pump.   ☐

27. Measure from the pump shaft (bearing race included) to the race of the thrust washer. If the thrust washer is not within the limits, replace it with one of the correct thickness. What were the results?

_____

_____

28. Total endplay should now be checked. Set the transmission case on end, front end up.   ☐

29. Make sure the thrust bearings are secure with petroleum jelly. Pick up the complete front clutch assembly and install it into the case. Be sure all parts are seated before proceeding with the measurement. Using a dial indicator or caliper, measure the distance from the rear hub thrust bearing to the case. What were the results?

_____

_____

30. Measure the pump with the front bearing race and gasket installed. Tolerance should fall within specifications. If the difference between the measurements is not within tolerance, select the proper size front bearing race. If it is necessary to change the front bearing race, be sure to change the front clutch thrust washers the same amount. What were the results?

_____

_____

_____

Task Completed

31. Install the brake band servo. Use extreme care so as not to damage the O-rings. ☐
    Lubricate around the seals.

32. Install and torque the retainer bolts to specifications. What are the specifications?

    _____

33. Loosen the piston stem. ☐

34. Install the brake band strut and finger-tighten the band servo piston stem just enough ☐
    to keep the band and strut snug and prevent them from falling out. Do not adjust the
    band at this time.

35. Air check for proper performance. What were the results? ☐

36. Place some petroleum jelly in two or three spots around the oil pump gasket and ☐
    position it on the transaxle housing.

37. Align the pump and install the pump with care. ☐

38. Tighten the pump attaching bolts to specifications in the specified order. What are the
    specifications and the specified order?

    _____

39. Check the rotation of the input shaft. If the shaft does not rotate, disassemble the
    transmission to locate the misplaced thrust washer. What were the results?

    _____

    _____

40. Install the bellhousing and torque the retaining bolts to specifications. What are the ☐
    specifications?

41. Before proceeding with the installation of the valve body assembly, it is good practice
    to perform a final air check of all assembled components. This will ensure that you
    have not overlooked the tightening of any bolts or damaged any seals during assembly.
    What are the specifications?

    _____

42. Assemble the parking pawl assembly. Place the assembly into its position and install
    the extension housing with a new gasket, then tighten the attaching bolts to the proper
    specifications. What are the specifications?

    _____

43. On transaxles, the differential assembly should be disassembled, cleaned, inspected,
    and reassembled. After it has been reassembled, measure its endplay with gauging
    shims. Then select a shim thick enough to correct the endplay. What are the
    specifications?

    _____

44. After installing the proper shims, measure the differential turning torque with a
    pound-inch torque wrench. What were your results?

    _____

45. Install the valve body. Be sure the manual valve is in alignment with the selector
    pin. Tighten the valve body attaching bolts to the specified torque. What are the
    specifications?

    _____

46. Before installing the solenoids, check to verify that they are operating properly. Connect the solenoid to a 12-volt source and ground the other terminal. What happened?

_____

_____

_____

47. Before installing the oil pan, check the alignment and operation of the control lever and parking pawl engagement. Make a final check to be sure all bolts are installed in the valve body. ☐

48. Install the oil pan with a new gasket. Torque the bolts to specifications. What are the specifications?

_____

49. Lubricate the oil pump's lip seal and the converter neck before installing the converter. ☐

50. Install the converter, making sure that the converter is properly meshed with the oil pump drive gear. ☐

51. The transmission is now ready for installation in the vehicle. Use the reverse of the removal procedures. Remember to follow proper fluid filling procedures. ☐

Problems Encountered

_____

_____

_____

Instructor's Comments

_____

_____

_____

CHAPTER 6

TORQUE CONVERTER AND OIL PUMP SERVICE

Upon completion and review of this chapter, you should be able to:

- Diagnose torque converter problems and determine needed repairs.
- Perform a stall test and determine needed repairs.
- Perform converter clutch system tests and determine needed repairs.
- Diagnose electrically controlled torque converter clutches.
- Inspect converter flexplate, converter attaching bolts, converter pilot, and converter pump drive surfaces.
- Measure torque converter endplay and check for interference.
- Check the stator clutch.
- Check a torque converter and transmission cooling system for contamination.
- Inspect, leak test, flush, and replace cooler, lines, and fittings.
- Inspect, measure, and replace an oil pump assembly and related components.

Basic Tools

Basic mechanic's tool set

DMM

Appropriate service information

Terms To Know

Ballooning

Heat exchanger

Mineral spirits

GENERAL DIAGNOSTICS

Many transmission problems are related to the operation of the torque converter. Normally torque converter problems will cause abnormal noises, poor acceleration in all gears, normal acceleration but poor high-speed performance, or transmission overheating.

To test the operation of the torque converter, many technicians perform a stall test. The stall test checks the holding capacity of the converter's stator overrunning clutch assembly, as well as the clutches and bands in the transmission.

However, torque converter problems can often be identified by the symptoms, and therefore the need for conducting a stall test is minimized. If the vehicle lacks power when it is pulling away from a stop or when passing, it has a restricted exhaust or the torque converter's one-way stator clutch is slipping. To determine which of these problems is causing the power loss, test for a restricted exhaust first.

Exhaust Restriction Test

The easiest way to test for an exhaust restriction is to use a vacuum gauge (**Figure 6-1**). Connect the vacuum gauge to a source of engine manifold vacuum. Observe the vacuum reading with the engine at idle. The manifold vacuum should be approximately 18 in Hg. Quickly open the throttle and observe the vacuum reading. Then quickly release the throttle to allow it to close. The vacuum reading should drop to 0 at first and then quickly

Many engine problems can cause a vehicle to behave as if the torque converter is malfunctioning, or actually cause the converter clutch to either lockup early or not at all. Clogged fuel injectors or ignition misfires can be misinterpreted as torque converter complaints. Bad vacuum lines, EGR valves, or engine speed or coolant sensors can prevent the clutch from locking up at the proper time.

Vacuum
gauge

Throttle
body

Figure 6-1 To check for an exhaust restriction,
connect a vacuum gauge to the engine and observe
the readings while the engine speed is raised and
quickly lowered.

increase to about 1.5 to 2 in Hg. After releasing the throttle back to idle, the vacuum
should recover quickly to the first reading, approximately 18 in Hg. If the vacuum reading
is low at first but then drops to 0 before even reaching wide-open throttle, the exhaust is
restricted. Common causes for exhaust restrictions are a plugged catalytic converter or a
collapsed exhaust pipe.

Another way to determine if the exhaust is restricted is with a pressure gauge. Insert the
gauge into the exhaust manifold's bore for the oxygen sensor. Then, start the engine. Bring
the engine to 2000 rpm and observe the gauge. If the exhaust is not restricted, the reading
will be less than 1.25 psi. A very restricted exhaust will have a reading of over 2.75 psi.

If there is no evidence of a restricted exhaust from these tests, it can be assumed that
the torque converter's stator clutch is slipping and not allowing any torque multiplication
to take place in the converter. To repair this problem, the torque converter should be
replaced.

Other Common Problems. If the engine's speed flares up during acceleration in Drive
and does not have normal acceleration, the clutches or bands in the transmission are
slipping. This symptom is similar to the slipping of a clutch in a manual transmission.
Often, this problem is mistakenly blamed on the torque converter.

The torque converter is often blamed as the cause for problems simply based on the
customer's complaint. Complaints of thumping or grinding noises are often blamed on
the converter when they are really caused by bad thrust washers or damaged gears and
bearings in the transmission. This type of noise can also be caused by non-transmission
parts, such as bad CV joints and wheel bearings.

Also, many engine problems can cause a vehicle to act as if it has a torque converter
problem. This is especially true of converter clutches, which may engage early or not at
all. Engine problems can cause the converter clutch to lockup early or not at all. Common
engine problems that can cause torque converter like symptoms include the following:
clogged fuel injectors and ignition problems, vacuum lines, EGR valves, or speed
sensors.

A faulty engine mount can also feel like a bad torque converter, allowing vibrations
that seem to be related to the torque converter. These vibrations can also be caused by an
out of balance torque converter. If the torque converter is not properly balanced, the
vibration will increase at higher speeds and while the transmission is in Park, since
the converter rotates at engine speed.

Special Tools

Vacuum gauge
Various tee fittings
Various lengths of
  hoses

Common exhaust
restrictions are a
plugged catalytic
converter or a
collapsed exhaust
pipe.

Sources for engine
manifold vacuum are
found in the intake
manifold, below the
throttle plate.

TESTING CONVERTER CLUTCHES

All transmissions are equipped with a torque converter clutch (TCC). Most converter clutches are controlled by the PCM or TCM (**Figure 6-2**). The computer turns on the converter clutch solenoid, which opens a valve and allows hydraulic pressure to engage the clutch. When the computer turns the solenoid off, the clutch disengages. PCM or TCMs can modulate the application of the torque converter clutch to minimize complaints of harsh engagement, either through a pulse-width modulated or variable-force solenoids, which allows the PCM or TCM to apply and release the converter in steps, which helps prevent unwanted sensations of converter operation to the driver.

Classroom Manual
Chapter 6, page 186

One of the trickiest parts of diagnosing a converter clutch problem is differentiating a normal-acting converter clutch from an abnormal-acting one. You should pay attention to the action of all converter clutches, whether or not they are suspected of having a problem. By knowing what a normal clutch feels like, it is easier to feel abnormal clutch activity. A malfunctioning converter clutch can cause a wide variety of driveability problems. Normally, the application of the clutch should feel like a smooth engagement into another gear. It should not feel harsh, nor should there be any noises related to the application of the clutch.

Classroom Manual
Chapter 6, page 188

To properly diagnose converter clutch problems, you must know when they should engage and disengage and understand the function of the various controls involved with the system. Although the actual controls for a converter clutch vary with the different manufacturers and models of transmissions, they all will have certain operating conditions that must be met before the clutch can be engaged.

Care should be taken during diagnostics because abnormal clutch action can be caused by engine, electrical, clutch, or torque converter problems.

Before the converter clutch is applied, the vehicle must be traveling at or above a certain speed. The vehicle speed sensor sends this speed information to the computer. Also, the converter clutch will not engage when the engine is cold; therefore, a coolant temperature sensor provides the computer with information regarding engine temperature. During sudden deceleration or acceleration, the clutch should be disengaged. One of the

Figure 6-2 The TCM controls the application pressure for the TCC.

Figure 6-3 Proper adjustment of the brake light switch is essential for proper operation of a torque converter clutch.

Special Tools

Assistant to drive vehicle
Scan tool

Torque converter clutch shudder is sometimes called chatter or a stick-slip condition.

sensors used to tell the computer when these driving modes are present is the accelerator pedal position sensor (APPS). Transmissions use a gear switch to inform the computer of the transmission gear range. The computer will then allow clutch engagement if other operating conditions are satisfactory. A brake switch or brake pedal position sensor is also used in clutch circuits or as a direct input to the TCM to disengage the clutch when the brakes are applied (**Figure 6-3**). These key sensors—the VSS, ECT, APPS, transmission gear switches, and brake switch—should be checked for operation with the scan tool as part of your diagnosis of converter problems.

Diagnosis of a converter clutch circuit should be conducted in the same way as any other computer system. The computer will recognize problems within the system and store trouble codes that reflect the problem area of the circuit. The codes can be retrieved and displayed by a scan tool. An important thing to remember is that if a problem with a vehicle does not affect emissions the MIL does not have to be displayed.

Engagement Quality

The vehicle should be taken on a road test to verify the customer's concern and to gather more information about the vehicle. Prior to operating the vehicle, make sure you have as much detail about the concern as possible. This should include the driving conditions at which the concern is most noticeable. If a DTC was set, check for information possibly captured as a freeze frame. A freeze frame allows duplication of the conditions that were present when the code was set.

During the road test, determine if the TCC engages and disengages. Have your assistant drive the vehicle in conditions listed in the service information. Watch the scan tool to determine when the TCC applies. If it does neither, detailed testing of the control system is required. To help, if the clutch does not apply there should also be a DTC stored. The engagement of the clutch should be smooth. Also pay attention to any noise or

Step	Findings	Remedy
1. Does the TCC engage and disengage?	Yes	Go to step #2
	No	Go to symptoms chart—diagnose the system
2. Describe the TCC shudder during upshifts	Light	Go to step #3
	Medium	Go to step #3
	Heavy	Go to symptoms chart—is not TCC related
3. Is the vibration or shudder vehicle speed related and not gear related?	Yes	Go to symptoms chart—is not TCC related
	No	Go to step #4
4. Is the vibration or shudder engine speed related and not gear related?	Yes	Go to symptoms chart—is not TCC related
	No	Go to step #5
5. Is the vibration or shudder occur in coast, cruise, or reverse gear?	Yes	Go to symptoms chart—is not TCC related
	No	Go to step #6
6. Does the vibration or shudder occur during long periods of light braking?	Yes	Go to symptoms chart—is not TCC related
	No	Go to step #7
7. Did the vibration or shudder only occur in step #2?	Yes	There is a probable TCC problem—diagnose the system
	No	Go to symptoms chart—is not TCC related

Figure 6-4 An evaluation form to use while doing a road test to check the torque converter.

vibration that may take place during shifting. If the vibration only occurs during the gear changes that should cause the TCC to engage or disengage, there is evidence of TCC shudder. If the vibration is present at other times, the cause is probably something other than the TCC (**Figure 6-4**). If your scan tool has the capability, turn the TCC on and off with the scan tool while observing the shift quality. Also, check the misfire counters on the scan tool as the problem occurs. A misfire when the clutch applies can be misdiagnosed as a TCC problem. Other areas to check with the scan tool are fuel trims that may be skewed rich or lean. A fuel injector cleaning may be needed if the fuel trims are compensating for a lean mixture condition. Also, do not rule out items such as a cracked or carbon-tracked spark plug. A cracked or carbon-tracked plug and/or a bad coil will cause a tip-in shudder at around 50 mph on many vehicles. The vehicle may need to be driven until the shudder occurs, then an oscilloscope or a scan tool may be used to find the misfire in the shop. An ignition misfire produces a rather sharp but small jerking sensation when the TCC (torque converter clutch) is first engaged and the tip of the foot is used to lightly accelerate (tip-in acceleration). Under harder acceleration or higher speeds the problem may not be felt. At the lowest speed the TCC engages, the fuel mixture is typically as lean as it can get; under these conditions the ignition misfire will occur. The TCC adds additional load to the engine already in a lean condition, so if the ignition system is not up to specifications due to coil, plug wire or boot, or spark plug problems a misfire will occur.

If the clutch prematurely engages or is not being applied by full pressure, a shudder or vibration results from the rapid grabbing and slipping of the clutch. The clutch begins to engage, then slips, because it cannot hold the engine's torque and become fully engaged. The torque capacity of the clutch is determined by the oil pressure applied to the clutch and the condition of the frictional surfaces of the clutch assembly.

If the shudder, sometimes called a stick-slip condition, is only noticeable during the engagement of the clutch, the problem is typically the converter. When the shudder is evident after the engagement of the clutch, the engine, transmission, or another component of the driveline may be the cause. *Tip-in shudder* is a commonly used term to describe the condition where the clutch releases and reapplies under a light load. This is a clutch control problem.

The ability of a converter clutch to hold torque is its torque capacity.

Classroom Manual
Chapter 6, page 197

Nearly all torque converters are welded-together units and therefore cannot be disassembled for inspection or repair. The primary objective when servicing a torque converter should be to check the serviceability of the unit.

When clutch apply pressure is low and the clutch cannot firmly engage fully, shudder will occur. A faulty clutch solenoid or its return spring may cause this. The valve controlled by the solenoid is normally held in position by a coil-type return spring. If the spring loses tension, the clutch will be able to prematurely engage. Because insufficient pressure is available to hold the clutch, shudder occurs as the clutch begins to grab and then slips. If the solenoid valve and/or return spring is faulty, they should be replaced, as should the torque converter. Turbine shaft seals can also cause shudder if the seals are worn or damaged because full pressure may not be able to build for TCC to apply.

An out-of-round torque converter prevents full clutch engagement, which will cause shudder, as will contaminated clutch friction material. The friction material can become contaminated by metal particles circulating through the torque converter and collecting on the clutch. Broken or worn clutch dampener springs will also cause shudder.

Besides replacing the torque converter and/or replacing other components of the system, one possible correction for shudder is the use of an ATF with friction modifiers. Sometimes the torque converter shudder can be remedied by changing the filter and fluid. The oil pan can be inspected at that time also. Excessive amounts of wear particles would indicate a more serious problem, requiring a rebuild with torque converter replacement. Be sure to use the specified fluid and check for the latest technical bulletins related to that problem. Some rebuilders may recommend that an oil additive be added to the ATF. The additive is designed to improve or alter the friction capabilities of regular ATF.

CUSTOMER CARE After replacing a converter because of clutch problems, be sure to explain to the customer that other engine or transmission problems may become more evident with the new converter clutch. Because the old converter didn't fully engage, the fluid in the converter may have dampened many of the vibrations from the engine and transmission. The new converter will now connect the engine to the transmission so engine and transmission vibrations can be mechanically transmitted through the drivetrain.

Special Tools

Short piece of hose
Graduated container
OSHA-approved air nozzle

Torque Converter-Related Cooler Problems

Besides causing transmission overheating problems, the vehicle may stall when the transmission is shifted into reverse gear. The cause of this problem may be plugged transmission cooler lines, or the cooler itself may be plugged. Fluid normally flows from the torque converter through the transmission cooler. If the cooler passages are blocked, fluid is unable to exhaust from the torque converter and the converter clutch piston remains engaged. When the clutch is engaged, there is no vortex flow in the converter and therefore little torque multiplication takes place in the converter.

To verify that the transmission cooler is plugged, disconnect the cooler return line from the radiator or cooler (**Figure 6-5**). Connect a short piece of hose to the outlet of the cooler and allow the other end of the hose to rest inside an empty container. Start the engine and measure the amount of fluid that flows into the container after 20 seconds. Normally, 1 quart of fluid should flow into the container. If less than that filled the container, a plugged cooler is indicated.

To correct a plugged transmission cooler, disconnect the cooler lines at the transmission and the radiator. Blow air through the cooler, one end at a time, then through the cooler lines. The air will clear large pieces of debris from the transmission cooler. Always use low air pressure of no more than 50 psi. Higher pressures may damage the cooler. If there is little air flow through the cooler, the radiator or external cooler must be removed

Classroom Manual
Chapter 6, page 195

Figure 6-5 To verify if a fluid cooler is plugged, disconnect the return fluid line and observe the flow of fluid in and out of the cooler.

and flushed or replaced. A good radiator shop can replace the cooler without a total radiator replacement, provided the core section is still in good condition. If the transmission cooler is external, it makes sense to replace it when a cooler or contaminated fluid problem is encountered.

TCC is a commonly used acronym for torque converter clutch.

GENERAL CONVERTER CONTROL DIAGNOSTICS

All testing of TCC controls should begin with a basic inspection of the engine and transmission. Too often, technicians skip this basic inspection and become frustrated during diagnostics because of conflicting test results. The basic inspection should include the following:

Classroom Manual
Chapter 6, page 197

1. A road test to verify the complaint and further define the problem.
2. A careful inspection of the engine and transmission.
3. A check of the PCM for codes.
4. A check of the mechanical condition of the engine, the output of ignition system, and the operation of the fuel system.
5. A check of the entire intake system for vacuum leaks.

> ⚙ SERVICE TIP It is important to realize that some oil coolers cannot be flushed effectively. Check the service information for any notations regarding this.

When inspecting wires, look for burnt spots, bare wires, and damaged or pinched wires. Make sure the wiring harness to the electronic control unit has a tight and clean connection. Also, check the source voltage at the battery before beginning any detailed tests. If the voltage is too low or too high, the system cannot function properly.

AUTHOR'S NOTE Engine problems can cause transmission and torque converter problems. The opposite is also true; transmission problems can cause engine problems. For example, if the TCC does not disengage, the engine will stall when the vehicle is at a stop.

SERVICE TIP Electronic converter and transmission controls are capable of displaying DTCs. However, the basic inspection is important because the computer may not display every problem, or even illuminate the MIL, unless the failure is emission related.

ELECTRONIC CONVERTER CLUTCH CONTROL DIAGNOSTICS

The TCC is controlled by the PCM or TCM. The computer turns on the converter clutch solenoid (**Figure 6-6**), which opens a valve and allows fluid pressure to engage the clutch. When the computer turns the solenoid off, the clutch disengages.

A faulty clutch solenoid or its return spring may cause low apply pressure. The valve controlled by the solenoid is normally held in position by a coil-type return spring. If the spring loses tension, the clutch will engage too soon. Because there is insufficient pressure to hold the clutch, shudder occurs as the clutch begins to grab and then slip. If the solenoid valve and/or return spring are faulty, they should be replaced, as should the torque converter. If the TCC fails to release, it can cause the engine to jerk and stall when the vehicle is stopping. Confirming the correct application of the TCC is detailed in **Photo Sequence 9**.

A malfunctioning converter clutch can cause a wide variety of driveability problems. Normally, the application of the clutch should feel like a smooth engagement into another gear. It should not feel harsh, nor should there be any noises related to the application of the clutch.

Figure 6-6 A typical torque converter clutch solenoid.

P9-1 Technician starts by connecting the scan tool.

P9-2 The scan tool is set up for the vehicle. If the technician suspects a problem with the TCC or its apply, now is a good time to check for trouble codes for both the engine and transmission.

P9-3 The technician brings up the following data set: rpm, vehicle speed, TCC apply, TCC pressure control solenoid, and TCC slip (all parameters may not be available on all vehicles). The data set may have to be selected manually if not grouped together.

P9-4 A second technician is going to drive the vehicle, while the first technician is going to operate the scan tool. NEVER drive and try to operate a scan tool by yourself.

P9-5 Always operate the vehicle in an area free of heavy traffic. Heavy traffic makes the steady speed driving necessary for TCC application hard to maintain.

P9-6 Now that the vehicle is at highway speeds with a steady throttle, see if the TCC shows applied. NOTE: The TCC may apply on many vehicles in gear ratios as low as second gear, with a significant amount of slip, but at highway speeds there should be less than around 100 rpm slip.

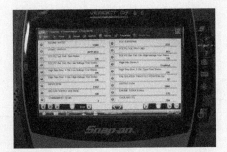

P9-7 With the TCC commanded on, tap the brake pedal. The TCC should release and then apply almost immediately. The rpm should momentarily go up and then back down.

P9-8 If there was a problem with the TCC that was duplicated during the test drive, the technician should consult the service information to determine the best course of action.

Electrical Checks

Each manufacturer has specific test sequences for testing an electronic converter clutch. In all cases, the system is first taken on a road test and then checked for DTCs. Normally vehicles equipped with an OBD-II system, will display TCC-related DTCs along with the DTCs of other systems. Generally speaking, the TCM/PCM has the capability to check the TCC circuit by watching for engine load and rpm information as the clutch is applied and released, as well as checking the electrical circuits for shorts or opens by the comprehensive component monitor.

To properly diagnose converter clutch problems, you must know when the TCC should engage and disengage and understand the function of the various components involved with the system (**Figure 6-7**). Although the actual controls for a TCC vary with the different models of transmissions, they all have certain operating conditions that must be met before the clutch can be engaged.

The TCM monitors vehicle speed, gear range, and throttle position to determine when to apply the TCC. Specific conditions must be present (these depend on the model of vehicle and transmission) before the TCM will apply the TCC solenoid. If one of those changes is no longer within the specified parameters, then the TCC will be disengaged. Normally all of the following conditions must be met:

- The transmission is operating in the specified gear(s).
- The vehicle is traveling above the minimum speed.
- The TP sensor senses a throttle opening that is greater than idle.
- The brake light switch is not on (brakes are not applied).

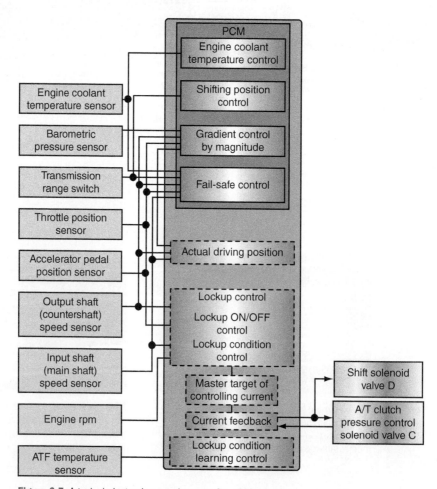

Figure 6-7 A typical electronic control system for a TCC.

- The temperature of the engine coolant or ATF is above a predetermined level.
- The vehicle's speed does not drop 6 or more mph below the set cruise control speed.
- No knocking is detected by the knock sensors.

Checking TCC Operation

The following test verifies that the TCC is working correctly.

1. Check for DTCs. Keep the scan tool connected.
2. Bring up the PIDs for engine rpm and the TCC apply command or status.
3. Drive the vehicle, with the gear selector in the D position, at highway speeds for about 15 minutes to bring the engine to normal operating temperature.
4. Using the scan tool, make sure that the ATF temperature is within specification. Normally this is about 110°F (43°C).
5. Once normal operating temperature is reached, drive at a constant speed of about 50 mph (80 km/h).
6. Lightly depress the throttle and pay attention to changes in engine speed. If there is a sudden increase in engine speed, the TCC was not locked, even if it was commanded on.
7. If the TCC was locked, resume the cruising speed.
8. Tap brake pedal with your left foot and observe the speed of the engine. Engine speed should increase and the TCC should show released when the brake pedal is tapped and decrease about 5 seconds after pedal is released.
9. Reduce speed to about 30 mph (48 km/h) and maintain that speed.
10. Fully release the throttle pedal and then depress the throttle to partially open it again. Observe the scan tool parameters TCC released, TCC applied. Likewise, engine speed will increase when the TCC is released and decrease when the TCC is applied. If this does not occur, further diagnosis is necessary.

Always make sure that you take an assistant along to read the scan tool. Do not drive and operate the scan tool.

TCC Diagnostics

Diagnosis of a TCC circuit should be conducted in the same way as any other computer system. Make sure you understand the customer's concern and verify the problem by operating the vehicle. Before conducting any test on the system, make sure the fluid in the transmission is at the correct level and is in good condition. Also conduct a good inspection of the vehicle. This should include a gear selector linkage check and a look at the vehicle to see if any modifications were made to it that may affect the operation of the TCC. Also make sure to check for a TSB that may relate to the concern. If there is such a bulletin, follow the given test and repair procedure before proceeding.

With the engine and transmission at normal operating temperature, connect the scan tool to the vehicle. Retrieve and record all DTCs. The DTCs are set when the computer recognizes problems within the system. The stored DTCs reflect the problem area of the circuit.

It is important that you retrieve all codes and deal with them in the proper sequence. Use the DTC chart in the service information to define the code and identify the possible causes. All engine-related DTCs should be examined first and repairs made to the faulty system. Then attention should be paid to the transmission- and torque converter–related codes (**Figure 6-8**). Pinpoint tests are conducted to diagnose the exact cause of the DTCs. It is also important that the system be checked for DTCs after repairs have been made.

It is important to note that before the TCM will set a DTC, a transmission fault must occur during four consecutive driving cycles. A torque converter DTC will be set when the fault occurs during five consecutive driving cycles. In some transmissions, the TCC system has its own monitor, similar to those found with emission-related systems. The monitor is continuously run and will set a DTC if a fault occurs during two driving cycles.

DTC	Description	Condition	Symptom
P0741	TCC slippage detected	The PCM noticed an excessive amount of slippage during normal operation.	TCC slippage, erratic operation, or no TCC operation. Flashing transmission MIL.
P0743	TCC solenoid circuit failure during on-board diagnostic	The TCC solenoid circuit does not provide the specified voltage drop across the solenoid. The circuit is open or shorted or the PCM driver failed during the diagnostic test.	With a short circuit, the engine will stall in second gear at low idle speeds with brake applied. With an open circuit, the TCC will not engage. This may cause the MIL to flash.
P0740	TCC electrical failure	The TCC solenoid circuit does not provide the specified voltage drop across the solenoid. The circuit is open or shorted or the PCM driver failed during the diagnostic test.	With a short circuit, the engine will stall in second gear at low idle speeds with brake applied. With an open circuit, the TCC will not engage. This may cause the MIL to flash.
P1740	TCC malfunction	A mechanical failure of the solenoid was detected.	If the solenoid is stuck in the ON position, the engine will stall in second gear at low idle speeds with brake applied. If the solenoid stays in the OFF position, the TCC will not engage. This may cause the MIL to flash.
P1742	TCC solenoid failed on	The TCC solenoid failed, which was caused by an electrical, mechanical, or hydraulic problem.	The transmission will exhibit harsh shifts.
P2758	TCC solenoid circuit failure, stuck ON.	The TCC solenoid circuit does not provide the specified voltage drop across the solenoid. The circuit is open or shorted or the PCM driver failed during the diagnostic test.	The TCC will never engage and there will be no adaptive or self-learning strategies.

Figure 6-8 A sample of TCC-related DTCs.

Make sure the vehicle has completed its necessary driving cycles before responding to a no-code situation. Before driving the vehicle through its required drive cycle, record and erase all DTCs. A typical driving cycle includes the following steps:

1. Warm engine to its normal operating temperature.
2. Check the ATF level and correct it as necessary.
3. With transmission in its normal Drive position, moderately accelerate to 50 mph (80 km/h).
4. Maintain that speed for a minimum of 15 seconds.
5. Lightly depress and release the brake pedal.
6. Resume the speed and maintain it for at least 5 seconds.
7. Bring the vehicle to a stop and keep it there for a minimum of 20 seconds.
8. Repeat Steps 3 through 7 at least five times.

TCC Monitors

The TCM determines the lockup status of the torque converter by comparing the engine speed to the turbine speed. The TCM then calculates the actual gear the transmission is in by comparing the turbine speed to the counter or output gear speed. When the TCM sees the conditions are suitable, it will do a quick check of the operation of the TCC, along with other transmission components. It checks the operation of the TCC by applying voltage to the appropriate solenoid to cause a lockup condition. When there is a difference of 70 rpm or more between engine speed and turbine speed and the TCM commands lockup, the TCM will set a code. In most cases, it will also turn on the MIL. Take into account systems that have a variable converter clutch apply system (always check with the specific service information for the vehicle concerned).

Although the monitor runs continuously, it will not run if certain codes are set. These codes typically relate to CAN communications, the electronic throttle control system, certain engine sensors, and the circuits for the transmission's solenoids. The monitor will also not run if the engine is not running, the vehicle is not moving more than 15 mph (25 km/h), the transmission is in low gear, the engine's temperature is lower than 104°F (40°C), or the temperature of the ATF is lower than 14°F (−10°C).

Other Scan Tool Checks

Most systems have the provision for operating outputs from the scan tool. This allows you to control the operation and main components of the transmission. The scan tool also displays PID information, along with the output control functions. With the output control, you can engage and disengage the TCC. The output control allows a technician to command three separate modes:

- ON—Commands the TCC solenoid On (100 percent duty cycle) to lock the TCC
- OFF—Commands the TCC solenoid Off (0 percent duty cycle) to unlock the TCC
- XXX—Cancels control of the TCC (returns to normal operation)

Most output control systems have a drive mode that allows you to control the TCC during a road test. During the drive mode, certain conditions must be met before the TCC solenoid can be commanded ON or OFF. When these conditions are not met, the scan tool will display an error message and the control function is ended. Examples of these conditions are the following:

To turn OFF the solenoid the following should be checked:

- The engine must be running.
- The gear selector is in its highest range.
- The vehicle must be traveling at a speed of at least 2 mph (3 km/h).

To command the solenoid ON the following should be checked:

- The engine must be running.
- The gear selector is in its highest range.
- The vehicle must be traveling at a speed of at least 2 mph (3 km/h).
- The ATF temperature must be between 60 and 240°F (6 and 116°C).
- There are no excessive loads on the engine.
- The engine's speed is higher than 1500 rpm.

Intermittent Problems

When there is an intermittent problem, using pending or soft codes may allow you to isolate the fault. These codes along with the customer's description of the problem and the road test results can lead you to the most probable problem area.

One way to identify the cause of an intermittent problem is by conducting a wiggle test with the scan tool connected to the vehicle. On the scan tool, select the TCC and TCCF PIDs. These will display the activity of the solenoid. Then with the scan tool, command the TCC solenoid ON and OFF, while monitoring the voltage and state of the solenoid on the scan tool (ON and OFF). When the solenoid changes from ON or OFF, it should click. Now wiggle the wiring and connectors at the transmission and observe the scan tool. If the state of voltage changes, there is a short or an open in the wiring or that connector.

Solenoid Checks

A common way to assess the condition of a solenoid is to check its resistance. Normally this is done by connecting an ohmmeter across the terminals of the solenoid (**Figure 6-9**). The specified resistance is typically listed for room temperature. This means the solenoid should not be checked if it is cold or hot. Most solenoids should have low resistance. The solenoid should also be checked for short-to-ground. This is done by connecting one of the meter's leads to a terminal at the solenoid and the other to a known good ground. Observe the meter. Repeat this test at the other solenoid terminal. If there is no short-to-ground, the readings will be close to infinite. If the solenoid is shorted, it should be replaced.

Figure 6-9 Checking the resistance of a TCC solenoid.

No Code Diagnosis

While testing any system, if no codes or codes indicating the system is operating normally are retrieved, conduct a thorough inspection of the system. Visually inspect all of the wires and connectors in the circuit. Make sure they are not loose, disconnected, or corroded. Also, refer to the symptom diagnostic chart given by the manufacturers.

If the clutch does not engage, make sure the converter clutch circuit has power. Check the circuit's fuse, make sure it is not blown. Check for voltage at the solenoid. If it is available, make sure the circuit's ground is good. If there is power available and the ground is good, check the voltage drop across the solenoid. The voltage drop should be very close to the source voltage. If less than that is measured, check the voltage drop across the power and ground sides of the circuit. If the voltage drop testing results in good results, remove the solenoid and test it with an ohmmeter. If the solenoid checks out fine with the ohmmeter, suspect clutch material, dirt, or other material plugging up the solenoid valve passages. If blockage is found, attempt to flush the valve with clean ATF. If the solenoid has a filter assembly (**Figure 6-10**), replace the filter after cleaning the fluid passages. If the blockage cannot be removed, replace the solenoid.

If the clutch engages at the wrong time, a sensor or switch in the circuit is probably the cause. If clutch engagement occurs at the wrong speed, check all speed related sensors. A faulty temperature sensor may cause the clutch not to engage. If the sensor is not reading the correct temperature, the PCM may never realize the temperature is suitable for engagement. Checking the appropriate sensors can be done with a scan tool, DMM, and/or lab scope. A check of the sensors is normally part of the manufacturer's system check.

Figure 6-10 A torque converter clutch solenoid with a replaceable filter.

To verify that a component is faulty, test it with the appropriate meter. All faulty wires and connectors should be repaired or replaced. TCC solenoids are not re-buildable and are replaced when they are faulty.

Mechanical Problems

Mechanical problems can cause the TCC not to operate properly. The causes of these problems typically result in the replacement of the torque converter with a new or rebuilt unit. It is important to keep these causes in mind when all electrical tests prove that system to be fine.

If the TCC does not apply, the cause can be fluid leaking from the converter or past its seals. This problem can also be caused by a severely worn TCC disc or friction material that has fallen off the clutch disc. These problems can be verified by a careful inspection of the torque converter.

If the TCC stays applied and tends to stall the engine at a stop, the problem may be a damaged piston plate or the plate that is stuck to the torque converter cover. These problems can be caused by overheating of the converter. Look at the converter for signs of heat to verify this problem. This problem can also be caused by the lack of end clearance inside the converter. In either case, the torque converter should be replaced.

Classroom Manual
Chapter 6, page 202

> ⚙ SERVICE TIP Stall testing is not recommended on many late-model transmissions. This test places extreme stress on the transmission and should only be conducted if recommended by the manufacturer.

STALL TESTING THE TORQUE CONVERTER

Two methods are commonly used to check the operation of the stator's one-way clutch: the stall test and the bench testing. Bench testing a converter will be covered later in this chapter. For now, we are still trying to determine if the converter is the cause of the customer's complaint. To bench test a converter, it must be removed from the vehicle and there is no need to do that unless we know the converter is at fault.

⚡ **WARNING** Make sure no one is around the engine or in front of the vehicle while a stall test is being conducted. A lot of stress is put on the engine, transmission, and brakes during the test. If something lets go, somebody can be seriously hurt.

⚡ **Caution**

If a stall test is not correctly conducted, the converter and/or transmission can be damaged.

To conduct a stall test, connect a tachometer to the engine and position it so that it can be easily read from the driver's seat (**Figure 6-11**). Set the parking brake, raise the hood, and place blocks in front of the vehicle's non-driving tires. Conduct the test outdoors if possible, especially if it is a cold day. If the test is conducted indoors, place a large fan in front of the vehicle to keep the engine cool. With the engine running, press and hold the brake pedal. Then move the gear selector to the "Drive" position and press the throttle pedal to the floor. Hold the throttle down for 2 seconds, note the tachometer reading, and immediately let off the throttle pedal and allow the engine to idle. Compare the measured stall speed to specifications (**Figure 6-12**).

If the torque converter and transmission are functioning properly, the engine will reach a specific rpm, which is the stall speed. If the tachometer indicates a speed above or below specifications, a possible problem exists in the transmission or torque converter. If a torque converter is suspected of being faulty, it should be removed and the one-way clutch checked on the bench.

1. Never conduct a stall test if there is an engine problem.
2. Check the fluid levels in the engine and transmission before conducting the test.
3. The engine should be at normal operating temperature during the test.

Trouble	Probable Cause
Stall rpm high in D₄, 2, 1 & R	Low fluid level or oil pump output Clogged oil strainer Pressure-regulator valve stuck closed Slipping clutch
Stall rpm high in R	Slippage of 4th clutch
Stall rpm high in 2	Slippage of 2nd clutch
Stall rpm high in D₄	Slippage of 1st clutch or 1st gear one-way clutch
Stall rpm low in D₄, 2, 1 & R	Engine output low Torque converter one-way clutch slipping

Figure 6-11 Before conducting a stall test, chock the wheels and place the tachometer in a position where it can be easily seen from the driver's seat.

Engine Liter	Transaxle Type	Converter Diameter	Stall rpm
1.7	A-404	9½ in. (241 mm)	2300–2500
2.2	A-413	9½ in. (241 mm)	2200–2410
2.6	A-470	9½ in. (241 mm)	2400–2630

Figure 6-12 A typical torque converter stall speed chart.

4. Never hold the throttle wide open for more than 5 seconds during the test.
5. Do not perform the test in more than two gear ranges without driving the vehicle a few miles to allow the engine and transmission to cool down.
6. After the test, allow the engine to idle for a few minutes to cool the transmission fluid before shutting off the ignition.

Special Tools
Tachometer
Large fan
Wheel chocks

If the stall speed is below the specifications, a restricted exhaust or slipping stator clutch is indicated. If the stator's one-way clutch is not holding, ATF leaving the turbine of the converter works against the rotation of the impeller and slows down the engine. With both of these problems, the vehicle would exhibit poor acceleration. This is caused by either a lack of power from the engine or no torque multiplication occurring in the converter. If the stall speed is only slightly below normal, the engine is probably not producing enough power and should be diagnosed and repaired.

If the stall speed is above specifications, the bands or clutches in the transmission may be slipping and not holding properly.

If the vehicle has poor acceleration but had good results from the stall test, suspect a seized one-way clutch. Excessively hot ATF in the transmission is a good indication that the clutch is seized. However, other problems can cause these same symptoms; therefore, be careful during your diagnosis.

A normal stall test will generate a lot of noise, most of which is normal. However, if you hear any metallic noises during the test, diagnose the source of those noises. Operate the vehicle, at low speeds, on a hoist with the drive wheels free to rotate. If the noises are still present, the source of the noise is probably the torque converter.

> **SERVICE TIP** Several other items influence the results of a stall test. These include the condition of the TCC, the condition of one-way clutches, the holding ability of clutches and bands, and the condition of the engine.

Visual Inspection

If there was a noise coming from the torque converter during the stall test, visually inspect the converter before pulling the transmission out and testing or replacing the converter. Remove the torque converter access cover on the transmission (**Figure 6-13**) and rotate the engine with a remote starter button. While the torque converter is rotating with the engine, check to make sure the torque converter bolts are not loose and are not contacting the bellhousing. Also, observe the action of the converter as it is spinning. If the converter wobbles, it may be due to a damaged flexplate or converter.

> ⚡ WARNING **Before rotating the engine with a remote starter button, make sure the engine's ignition or fuel injection system is disabled. Also keep all parts of your body away from the rotating flexplate. Serious injury can result from being careless during this check.**

Check the converter's balance weights to make sure they are still firmly attached to the unit. Carefully inspect the flexplate for evidence of cracking or other damage. Also check the condition of the starter ring gear, the teeth of which should not be damaged. The gear should be firmly attached to the flexplate.

Check the torque converter for **ballooning**. If excessive pressure is able to build up inside the converter, the converter will expand, or balloon. A stuck converter check valve typically causes this. If the converter is ballooned, it should be replaced and the cause of the problem also repaired.

Ballooning makes the torque converter look like it has been blown up like a balloon; it is caused by excessive pressure in the converter.

Figure 6-13 To inspect a torque converter while it is still in the vehicle, remove the access cover.

A ballooned torque converter will also damage other driveline components. If the converter is ballooned toward the rear of the unit, the transmission's oil pump is most likely bad. If it is ballooned toward the front, the crankshaft's thrust bearings are worn or damaged.

Converter Fluid. A stall test works well to circulate fluid inside the converter, therefore a small sample of the fluid can give a good indication of the condition of the converter. Pour a small amount of fluid from the converter onto a white absorbent cloth or paper towel or pour it through a fine filter. Examine and smell the fluid. Now examine the paper towel. Look for any signs of residue. If particles were found in the fluid or if there is any indication of engine coolant in the fluid, the converter should be replaced. If the fluid was not contaminated, empty the converter of all fluid.

External Measurements

Certain parts of the torque converter should be measured for wear. These also should be carefully inspected for signs of damage. The measurements are taken with the appropriately sized micrometer. Begin by measuring the distance between the two flats on the hub. Then measure the outside diameter (OD) of the hub (**Figure 6-14**). The latter measurement

Figure 6-14 Measure the torque converter's hub and compare it to specifications. (A) The OD between the flats of the hub and (B) the OD of the hub at the bushing wear area.

Figure 6-15 Measure the OD of the torque converter's pilot and compare it to specifications.

determines if the hub is worn where it rides in its bushing. Most specifications list a minimum dimension for these diameters. If either measurement of the hub is less than the minimum, the converter should be replaced.

The OD of the pilot on the cover of the torque converter should also be measured (**Figure 6-15**). The pilot is also a potential wear area. If the diameter is less than the minimum given in the specifications, replace the converter.

BENCH TESTING THE TORQUE CONVERTER

Further inspection of the torque converter requires that it be removed from the vehicle. These inspections (**Figure 6-16**) should also be done anytime you have the torque converter out of a vehicle, especially if the oil pump is damaged or if the customer's complaint appears to be related to the torque converter.

When removing the transmission, pay particular attention to the condition of the flexplate and converter mounting bolts. Problems that appear to be torque converter problems may be caused by a damaged flexplate or a loose torque converter.

Figure 6-16 Some of the areas of a torque converter that need to be carefully checked.

Once the converter is separated from the transmission, inspect the drive studs or lugs used to attach the converter to the flexplate. These are designed to hold the converter firmly to the flexplate and to keep the converter in line with the engine. Damaged studs or lugs will cause runout, misalignment, and vibration problems, and can result in damage to the bushings of the oil pump. If the threads are damaged slightly, they can be cleaned up with a thread file, tap, or die set. However, if they are badly damaged, the converter should be replaced. The converter should also be replaced if the studs or lugs are loose or damaged. Also check the shoulder area around the lugs and studs for cracked welds or other damage. If any damage is found, the converter should be replaced. An exception to this is when the internal threads of a drive lug are damaged. These can often be repaired by tapping the threads or by installing a threaded insert. Also inspect the converter attaching bolts or nuts and replace them if they are damaged.

Check the pilot of the converter for wear and other damage. Also check the area around the pilot for cracks. In either case, the converter should be replaced. If the area around the pilot has dimples and looks like it has contacted the flywheel bolts, the converter has ballooned. If the converter has ballooned, it should be replaced and the cause of the ballooning corrected.

Inspect the flexplate for signs of damage, warpage, and cracks. Check the condition of the teeth on the starter ring gear. Replace the flexplate if there is any evidence of damage.

Check the drive hub of the torque converter; it should be smooth and not show any signs of wear. The easiest way to check the smoothness of the hub is to run a fingernail across the surface. If any irregularities are felt, the hub should be smoothed with crocus cloth or the converter replaced. If the hub is worn, carefully inspect the oil pump drive and replace the torque converter. Light scratches, burrs, nicks, or scoring marks on the hub surface can be polished with fine crocus cloth. Be careful not to allow dirt to enter into the converter while polishing the hub. Use a rag to cover the opening in the hub. Dirt and dust that enter the converter through this opening can cause the converter to wear rapidly. After polishing, clean the hub with solvent and a clean, lint-free rag. If the hub has deep scratches or other major imperfections, the converter should be replaced.

Since misalignment can cause many different problems, it is important that you check for excessive runout of the flexplate and the converter's hub. Check flexplate runout with a dial indicator. Mount the indicator onto the engine block and set its plunger on the flexplate (**Figure 6-17**). Rotate the flexplate and observe the readings on the indicator. Typically, the maximum allowable runout is 0.03 in.

To check the runout of the converter's hub, mount the converter onto the flexplate. Tighten the retaining bolts to the specified torque. Position a dial indicator on the engine block. Set the plunger of the indicator on the hub (**Figure 6-18**). Rotate the torque

Special Tools
Vee-blocks
Dial indicator and
   mounting bracket

Figure 6-17 Setup for checking flexplate runout.

Figure 6-18 Setup for checking converter hub runout.

Figure 6-19 The pump driveshaft should be inspected in the areas that ride on the bushings and bearings. The shaft seals should always be replaced.

Figure 6-20 Typical stator support and oil flow in a torque converter.

converter and observe the readings on the indicator. Again, the typical maximum allowable runout is about 0.03 in. If the runout is excessive and there is a normal amount of flexplate runout, remove the torque converter and set it at a new position on the flexplate. Check the runout again. If the runout is now within specifications, mark its position on the flexplate. If repositioning the converter on the flexplate does not correct hub runout, the converter should be replaced.

Because of the weight of torque converters, the above check is not always reliable. To get an accurate check of torque converter runout, mount the converter on two vee-blocks. Set the dial indicator so it can read lateral runout of the converter housing. To do this, place the indicator's plunger against the top of the converter housing's flywheel mounting surface. Rotate the housing and observe the runout. To check radial runout, place the dial indicator on the outside end of the pump hub. Rotate the housing and observe the runout. Then move the indicator on the hub and toward the housing. Again rotate and measure the runout.

In general, a torque converter should be replaced if there is an excessive amount of metal in the transmission fluid. The debris is very difficult to remove from inside the torque converter. The entire unit should also be replaced if it has signs of overheating, fluid leakage around its seams or welds, worn drive stud shoulders, loose or damaged drive studs, a heavily grooved hub, or excessive hub runout. Of course, the torque converter should be bench tested before you decide to reuse it.

Remote Driveshafts. Transaxles that don't have their oil pump driven directly by the torque converter use a driveshaft (**Figure 6-19**) that fits into a support bushing inside the converter's hub (**Figure 6-20**). This bushing should be checked for wear. To do this, measure the inside diameter of the bushing and the outside diameter of the driveshaft. The difference between the two is the amount of clearance. This measurement should be compared to factory specifications. Normally the maximum allowable clearance is 0.004 inch. Excessive clearance can cause oil leaks. If the clearance is excessive, the bushing should be replaced.

Stator One-Way Clutch Diagnosis

The operation of the stator one-way clutch inside the torque converter is critical to overall effectiveness of the torque converter. If a problem occurs in this clutch assembly, the clutch will either fail to lock when rotated in either direction or fail to unlock when rotated

Torque wrench

Splined into stator
clutch inner race

Converter clutch
torquing tool

Figure 6-21 Checking one-way clutch with special tool fixture.

Classroom Manual
Chapter 6, page 193

in either direction. Although these problems are similar, they affect efficiency at opposite ends of the engine's operating speeds. However, in either case, fuel economy will be affected.

When the stator clutch does not lock, there is a disruption in vortex flow and a loss of torque multiplication in the torque converter. A vehicle with this problem will have sluggish low-speed performance but will perform normally at higher speeds when the stator is supposed to freewheel.

A vehicle with a constantly locked stator will have good low-speed and poor high-speed performance. Torque multiplication will always occur, as will speed reduction. A vehicle with a constantly locked stator will show signs of overheating. If you suspect a locked stator, check for a bluish tint on the hub of the converter. This discoloration typically results from overheating. It is normal for some blue to be evident at the spot where the hub was welded to the housing. If the most of the hub is blue, the converter has overheated.

To check the stator's one-way clutch with the converter on a bench, insert a finger into the splined inner race of the clutch. Attempt to turn the inner race in both directions. You should be able to turn the race freely in one direction and feel lockup in the opposite direction. If the clutch rotates freely in both directions or if the clutch is locked in both directions, the converter should be replaced.

Because this check does not put a load on the clutch assembly, it does not totally check the unit. Therefore, some manufacturers, such as Ford, recommend the use of a special tool set that holds the inner race and exerts a measurable amount of torque on the outer race, thereby allowing the technician to observe the action of the clutch while under load (**Figure 6-21**).

Internal Interference Checks

Internal converter parts hitting each other or hitting the housing may also cause noises. To check for any interference between the stator and turbine, place the converter face down on a bench (**Figure 6-22**). Then install the oil pump assembly. Make sure the oil pump drive engages with the oil pump. Insert the input shaft into the hub of the turbine. Hold the oil pump and converter stationary, and then rotate the turbine shaft in both directions. If the shaft does not move freely or makes noise, the converter must be replaced.

Figure 6-22 Checking stator-to-turbine interference.

Figure 6-23 Checking stator-to-impeller interference.

To check for any interference between the stator and the impeller, place the transmission's oil pump on a bench and fit the converter over the stator support splines (**Figure 6-23**). Rotate the converter until the hub engages with the oil pump drive. Then hold the pump stationary and rotate the converter in a counterclockwise direction. If the converter does not freely rotate or makes a scraping noise during rotation, the converter must be replaced. **Photo Sequence 10** covers the procedure for checking internal interference and endplay.

The input shaft may be referred to as the turbine shaft on some transaxles.

Endplay Check

The special tools required to check the internal endplay of a torque converter are typically part of the essential tool kit recommended by each manufacturer. However, these specialty tools can be individually purchased through specialty tool companies. Basically, the special tools are a holding tool and a dial indicator with a holding fixture. The holding tool is inserted into the hub of the converter and, once bottomed, is tightened in place. This locks the tool into the splines of the turbine. The dial indicator is fixed onto the hub. The amount indicated on the dial indicator, as the tool is lifted up, is the amount of endplay inside the converter. If this amount exceeds specifications, replace the converter.

Special Tools

Manufacturer-specified Holding tool

Dial indicator and mounting bracket

Converter Leakage Tests

If the initial visual inspection suggested that the converter has a leak, special test equipment can be used to determine if the converter is leaking. This equipment uses compressed air to pressurize the converter. Leaks are found in much the same way as tire leaks are; that is, the converter is submerged in water and the trail of air bubbles lead the technician to the source of leakage.

PHOTO SEQUENCE 10
Internal Interference and Endplay Checks

P10-1 To check for any interference between the stator and turbine, place the converter, face down, on a bench.

P10-2 Then install the oil pump assembly. Make sure the oil pump drive engages with the oil pump.

P10-3 Insert the input shaft into the hub of the turbine.

P10-4 Hold the oil pump and converter stationary, then rotate the turbine shaft in both directions. If the shaft does not move freely and/or makes noise, the converter must be replaced.

P10-5 To check for interference between the stator and the impeller, place the transmission's oil pump on a bench and fit the converter over the stator support splines.

P10-6 Rotate the converter until the hub engages with the oil pump drive. Then hold the pump stationary and rotate the converter in a counterclockwise direction. If the converter does not rotate freely or makes a scraping noise during rotation, the converter must be replaced.

P10-7 Special tools are required to check the internal endplay of a torque converter. These include a holding tool and a dial indicator with a holding fixture.

P10-8 Insert the holding tool into the hub of the converter and, once bottomed, tighten it in place. This locks the tool into the splines of the turbine.

P10-9 Fasten the dial indicator to the hub.

P10-10 Set the plunger of the dial indicator so it can read the movement of the tool.

P10-11 Zero the dial on the indicator.

P10-12 Lift up on the tool and observe the reading on the indicator. This is the amount of endplay inside the converter. Compare your reading to specifications.

Oil Pump Seal

If the backside of the torque converter is wet, it is very likely that the transmission pump seal is bad. There are many possible causes for leaks at the seal. These must also be looked at and corrected when the new seal is installed (**Figure 6-24**). These are some of the many causes of pump seal leakage:

- The seal has a missing or damaged spring.
- The lip of the seal is cut or damaged.
- The pump bore is worn, scratched, or damaged.
- The pump seal bore is off the centerline of the crankshaft.
- The pump drain hole is too small or is restricted.
- The pump bushing is worn, loose, or missing.
- The hub seal area is worn, rusted, or pitted.
- The hub has excessive runout.
- The hub is cracked or machined roughly.
- The hub is of incorrect length.

Figure 6-24 The converter/pump seal must be installed with the proper driver in order to seal.

- The hub or converter is out of balance.
- There is excessive crankshaft end clearance (should be no more than 0.006 inch).
- The crankshaft pilot bore is worn.
- The crankshaft pilot sleeve is missing.
- The flexplate spacer is broken or missing.
- The flexplate is broken or has excessive runout.
- The converter housing is loose.
- The dowels in the converter housing are worn or missing.
- The replacement seal has incorrect elastomer or is missing an auxiliary lip.
- The inside or outside diameter of the seal is incorrect.
- The elastomer of the seal is too thick and is restricting the drain hole.
- The helix direction of the seal is incorrect.
- The seal was installed upside down.
- The seal was damaged during installation by the use of a defective or wrong driver.
- The seal was damaged by careless installation of the torque converter.

Summary of Checks

If certain conditions are found while inspecting a torque converter, the converter must be replaced. Specifically, the converter should be replaced:

1. If there is a stator clutch failure.
2. If there is internal interference.
3. If the transmission's front pump is badly damaged.
4. If the converter hub is severely damaged or scored.
5. If there are signs of external fluid leaks.
6. If the drive studs or lugs are damaged or loose.
7. If there are signs of overheating.
8. If heavy amounts of metal were found in the fluid.
9. If the fluid was contaminated with coolant.
10. If any damage is evidence that the converter is no longer balanced.

Although some specialty shops will rebuild a converter, nearly all technicians replace the converter when any of the above exists. This is especially true of converters with a clutch. Rebuilding a converter is for the automatic transmission specialist and is not a normal task for an automobile technician. Because this procedure requires special equipment and knowledge, do not attempt to repair a faulty converter.

Starter Ring Gear Replacement

The starter ring gear is most often part of the flexplate. Therefore, whenever the teeth of the ring gear are damaged, the entire flexplate is replaced. There are some transmissions that are equipped with a torque converter fitted with a ring gear around the outer circumference of the torque converter cover. The ring gear is welded to the front of the converter cover and gear replacement involves breaking or cutting the welds to remove the old gear, then replacing the gear and welding it back onto the converter. This procedure is typically not recommended for torque converters fitted with a clutch, as the heat from welding can destroy the frictional surfaces inside the converter. In these cases, if the ring gear is damaged, the entire converter should be replaced.

SERVICE TIP A torque converter can collect a lot of debris after a major transmission problem, and no amount of soaking and flushing will remove all of the dirt from the converter. A new converter should be installed.

CLEANING

Flushing with solvent is not recommended for modern converters . It is recommended that the converter be replaced, rather than flushed or cleaned. If the converter will be used again and had no signs of being contaminated, pour about 2 quarts (1.9 liters) of clean transmission fluid into the converter. Swirl the fluid around the inside of the converter, then drain the fluid out. Keep in mind that the converter is generally replaced with most overhauls, due to the fact the condition of the internals of the converter cannot be inspected visually.

⚠️ **WARNING Never use water-based cleaners or mineral spirits to clean or flush the torque converter. Use only clean automatic transmission fluid designated for that transmission.**

Flushing removes all debris and sludge from the inside of the converter. Flushing is either done by a machine or by hand. Some transmission shops use a torque converter flushing machine that pumps a solvent through the converter as it is rotated by the machine. This method keeps the fluid moving inside the converter. The moving fluid is quite efficient at picking up any dirt present. The dirty solvent is pumped out of the converter and new solvent added during the flushing procedure.

Since any amount of dirt can destroy a transmission, most rebuilders replace the torque converter during a transmission overhaul. There is no accurate way of knowing if dirt remains in the converter after cleaning and flushing it. The cleaning process may loosen up the debris, which will break down and contaminate the fluid once the torque converter starts spinning and gets hot.

Some transmission shops may cut the torque converter shell in half, then clean the parts, examine them for wear, and replace any that are worn or broken. The shell is then welded back together. This is a job only for shops that are equipped to do this job right.

After the converter has been flushed, disconnect it from the machine. If the converter has a drain plug, invert and drain the complete assembly. Converters without drain holes or plugs can be drilled with a 1/8-inch drill bit between the top end of the impeller fin dimples. This hole will act as an air bleed to maximize flushing. After flushing and draining the converter, the bleeder hole is sealed with a closed-end pop rivet covered with sealant. Some technicians use a MIG welder to seal the hole rather than use a rivet. However, this is not recommended because the heat can destroy the torque converter clutch.

🔧 **Special Tool**
Torque converter flushing machine

Classroom Manual
Chapter 6, page 193

TORQUE CONVERTER REPLACEMENT

Extra care should always be taken when replacing a torque converter. Size and fit are not the only important variables. Nor does size alone determine the stall speed of the converter. Even if the converter has exactly the same stall speed, it may have a different torque ratio and should not be used. Always check and double-check the part or model number of the torque converter you are removing and compare it to the one you are going to install. Converters are typically identified by a sticker or a number code stamped into the converter housing.

⚙️ **SERVICE TIP** Many OBD II systems have an operational mode that allows for converter clutch break-in. The purpose of this mode is to allow the system to readjust to a new clutch in the torque converter. Whenever a torque converter is replaced in a vehicle equipped with OBD II, set the system into the clutch break-in mode. Doing this will allow the system to realize things have changed and will allow the converter clutch to work more efficiently.

A = 21.2/23.2 mm (.83/.91 in.)

Figure 6-25 Before removing a torque converter, check its installed depth in the bellhousing.

Minimum 10 mm (7/16 in.)

Straightedge

Torque converter

Figure 6-26 Checking proper installation of a torque converter in a transmission.

When replacing a torque converter, never use an impact wrench on the torque converter bolts. Impact wrenches can drive the bolts through the cover, which will warp the inside surface and cause the converter clutch to stay applied at all times, or it will damage the clutch's pressure plate.

Always perform an endplay check and check the depth of the torque converter in the bellhousing (**Figure 6-25**) before reinstalling a torque converter or installing a new unit.

Correct installation of a torque converter requires that the converter's hub be fully engaged with the transmission. To help in seating the torque converter into the transmission, push on the converter while rotating it. Care should be taken not to damage the pump seal while installing the converter. To verify full engagement, place a straightedge across the bellhousing or transaxle flange. Then measure from the straightedge to the pilot (**Figure 6-26**). Compare your readings to those specified by the manufacturer. If your reading is less than the specifications, the torque converter is not properly engaged in the transmission.

Converter Balance

When replacing a torque converter, make sure the manufacturer or rebuilder balanced it. If the converter is not balanced, noise and vibrations can result, as well as damage to the transmission.

If reusing a torque converter, make sure it is positioned in its original position on the flexplate and make sure the mounting bolts or nuts are tightened to specifications. Sometimes manufacturers install a slightly unbalanced torque converter in a precise position on the flexplate. This is done to offset or counter engine vibrations, and you have no way of knowing if this is what they did.

When installing a known balanced converter, it can be positioned anywhere on the flexplate as long as it is fully seated against it and the mounting bolts or nuts are properly torqued.

OIL PUMP SERVICE

The oil pump (**Figure 6-27**) of an automatic transmission should be carefully inspected during any transmission overhaul and especially when low line pressure was measured during a pressure test. Carefully remove the oil pump assembly (**Figure 6-28**). Some transmissions require the use of a special puller to remove the oil pump from the transmission case (**Figure 6-29**). Never pry the pump out; it is easy to damage the case if this is done.

Classroom Manual
Chapter 6, page 209

Figure 6-27 A typical gear-type oil pump.

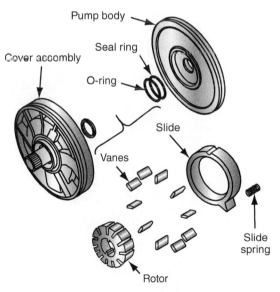

Figure 6-28 While removing a pump assembly, make sure to keep track of the position of all of the parts.

Figure 6-29 Using a puller to remove an oil pump from a transmission case.

Special Tools

Machinist bluing
Paint stick
Feeler gauge set
RTV

⚠ WARNING Some transmissions are fitted with oil control floats located between the oil pump and the gearsets. These floats can be easily damaged during pump removal and installation. The floats close the opening between the oil pan and the gearsets to prevent the chamber that contains the gears from overfilling with fluid, which can cause fluid aeration. The float can be easily dislodged and become stuck in one position.

Disassembly

Before disassembling the pump, mark the alignment of the gears. If acceptably worn gears are reinstalled in a position other than their wear pattern, excessive noise will result. When the pump halves have been separated, look at the relationship between the inner and outer gears. Most gears will have a mark on them indicating the top side of the gear (**Figure 6-30**).

If no mark is present, you should use a nondestructive type of marker to be sure you install the gears in the same position they were originally in (**Figure 6-31**). This ensures that the converter drive hub will mate correctly with the inner gear.

Another way of ensuring the proper position of the gears is to observe the existing wear pattern on the gears while they are being removed and make a note of this for use during reassembly. If the inner gear is installed upside down, there will not be enough free movement, which can result in one or more of the following problems: broken inner gear, broken flexplate, pump cover scoring, or broken transmission case ears. The proper procedure for disassembling an oil pump is shown in **Photo Sequence 11**. This procedure is given as an example; always refer to the procedures for the specific transmission you are working on.

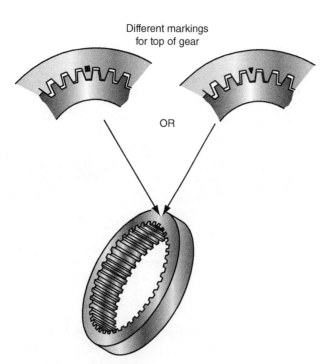

Figure 6-30 Identification marks on oil pump gearsets.

Figure 6-31 Marking the location of the outer and inner gears so that they can be properly meshed during reassembly.

PHOTO SEQUENCE 11
Proper Procedure for Disassembling an Oil Pump

P11-1 Remove the front pump bearing race, front clutch thrust washer, gasket, and O-ring. Inspect the pump bodies, pump shaft, and ring groove areas.

P11-2 Unbolt and separate the pump bodies.

P11-3 Mark the gears with machinist bluing ink or paint before removing them, so that the gears will remain in the same relationship during reassembly.

P11-4 Inspect the gears and all internal surfaces for defects and visible wear.

P11-5 Measure the distance between the outer gear and the pump housing. Clearance should be 0.003–0.006 in. or 0.08–0.15 mm. Replace the pump if the measurements exceed specifications.

P11-6 Measure the distance between the outer gear teeth and the crescent. Clearance should be 0.006–0.007 in. or 0.015–0.18 mm. Replace the pump if the clearance exceeds 0.008 in.

P11-7 Place the pump flat on the bench and, with a feeler gauge and straightedge, measure between the gears and the pump cover. The clearance should be no more than 0.002 in. (0.05 mm). Replace the pump if the measurement exceeds specifications.

P11-8 Put the halves back together.

P11-9 Torque the securing bolts to specifications. Replace all O-rings and gaskets.

Inspection

Inspect the pump bore for scoring on both its bottom and sides. A converter that has a tight fit at the pilot hub may hold the converter drive hub and inner gear too far into the pump, causing cover scoring. A front pump bushing that has too much clearance may allow the gears to run off center, causing them to wear into the crescent or the sides of the pump body.

> **SERVICE TIP** Some service informations will call for a maximum front bushing clearance of 0.004 inch; however, a dimension of less than that will avoid wear, as well as keep the pressure losses at the bushing to a minimum.

The stator shaft should be inspected for looseness in the pump cover (**Figure 6-32**). The shaft's splines and bushing should also be carefully looked at. If the splines are distorted, the shaft and the pump cover should be replaced. Because the bushings control oil flow through the converter and cooler, their fit must be checked (**Figure 6-33**). Bushings must be tight inside the shaft and provide the input shaft with a 0.0005- to 0.003-inch clearance.

Inspect the gears and pump parts for deep nicks, burrs, or scratches. Examine the pump housing for abnormal wear patterns. The fit of each gear into the pump body, as well as the centering effect of the front bushing, controls oil pressure loss from the high-pressure side of the pump to the low-pressure input side. Scoring or body wear will greatly reduce this sealing capability.

Gear- and rotor-type pumps are positive-displacement pumps.

On positive-displacement pumps, use a feeler gauge to measure the clearance between the outer gear and the pump pocket in the pump housing (**Figure 6-34**). If possible, also check the clearance between the outer pump gear teeth and the crescent (**Figure 6-35**), and between the inner gear teeth and the crescent. Compare these measurements to the specifications. Use a straightedge and feeler gauge to check gear side clearance (**Figure 6-36**) and compare the clearance to the specifications. If the clearance is excessive, replace the pump. Normally, the maximum clearance is 0.0015 inch (0.038 mm).

> **SERVICE TIP** All pumps, valve bodies, and cases should be checked for warpage and should be flat-filed to take off any high spots and burrs prior to reassembly.

Figure 6-32 Checking the play between the splines of the oil pump and the input shaft.

Input shaft

Front pump and stator support shaft

Converter assembly

Figure 6-33 Location of a stator shaft bushing.

Reaction shaft support

Hub

Bushing

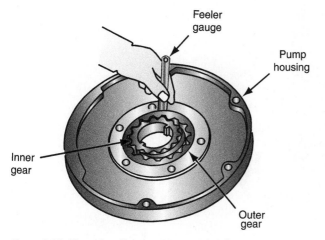

Figure 6-34 Measuring oil pump gear-to-pocket clearance.

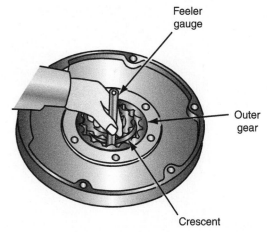

Figure 6-35 Measuring the clearance between the outer gear and the crescent.

Figure 6-36 Measuring the side clearance of the pump's gears.

Variable-Displacement pumps

Variable-displacement vane-type pumps require different measuring procedures. However, the inner pump rotor to converter drive hub fit is checked in the same way as described for the other pumps. The pump rotor, vanes, and slide are originally selected for size during assembly at the factory (**Figure 6-37**). Changing any of these parts during overhaul can destroy this sizing and possibly the body of the pump. You must maintain the original sizing if any parts are found to need replacement. These parts are available in select sizes for just this reason.

> **SERVICE TIP** Some early pump vanes cracked as a result of excessive load. These pumps did not have enough vanes to handle the hydraulic loads encountered during operation. Updated pumps with more vanes cured the problem.

Rotor Selection		Vane Selection	
Thickness (mm)	Thickness (in.)	Thickness (mm)	Thickness (in.)
17.593–17.963	0.7068–0.7072	17.943–17.961	0.7064–0.7071
17.963–17.973	0.7072–0.7076	17.961–17.979	0.7071–0.7078
17.973–17.983	0.7076–0.7080	17.979–17.997	0.7078–0.7085
17.983–17.993	0.7080–0.7084		
17.993–18.003	0.7084–0.7088		

Slide Selection	
Thickness (mm)	Thickness (in.)
17.983–17.993	0.7080–0.7084
17.993–18.003	0.7084–0.7088
18.003–18.013	0.7088–0.7092
18.013–18.023	0.7092–0.7096

Figure 6-37 Vane-type oil pump measurements and selection chart.

The vanes are subject to edge wear, as well as cracking and subsequent breakage. The outer edge of the vanes should be rounded with no flattening (**Figure 6-38**). These pumps have an aluminum body and cover halves; therefore, any scoring indicates that they should be replaced.

Inspect the reaction shaft's seal rings. If the rings are made of cast iron, check them for nicks, burrs, scuffing, or uneven wear patterns; replace them if they are damaged (**Figure 6-39**). Make sure the rings are able to rotate in their grooves. Check the clearance between the reaction shaft support ring groove and the seal ring. If the seal rings are the Teflon full-circle type, cut them out, and use the required special tool to replace them.

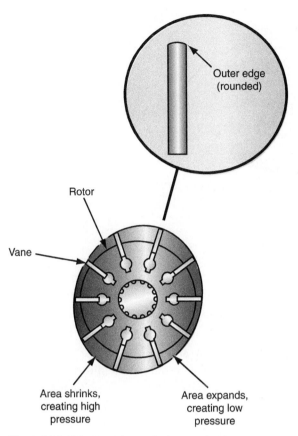

Outer edge
(rounded)

Rotor

Vane

Area shrinks,
creating high
pressure

Area expands,
creating low
pressure

Figure 6-38 Placement of vanes in the pump's rotor.

Seal
rings

Support
hub

Figure 6-39 Location of oil seal rings on
a typical oil pump assembly.

The outer area of most pumps utilizes a rubber seal (**Figure 6-40**). Check the fit of the new seal by making sure the seal sticks out a bit from the groove in the pump. If it does not, it will leak. The seal at the front of the pump is always replaced during overhaul. Most of these seals are the metal-clad lip seal type. Care must be taken to avoid damage to the seating area when removing the old seal.

Check the area behind the seal to be sure the drainback hole is open to the sump. If this hole is clogged, the new seal may blow out. The drainback hole relieves pressure behind the seal. A loose-fitting converter drive hub bushing can also cause the front pump seal to blow out.

Oil pump
assembly

Oil ring

Figure 6-40 Outer oil pump seal.

Figure 6-41 The wear plate's thickness should be measured.

Wear Plate Inspection. Some oil pumps have a wear plate. These are common in vane-type pumps. The plate should be carefully inspected for damage and distortion. The plate should not be scored, nicked, or have grooves cut into it. The thickness of the plate at the wear area should be measured and compared to specifications (**Figure 6-41**). If the plate is damaged or excessively worn, it should be replaced.

Oil Pump Body Bushing. The bushing of the oil pump should be carefully checked. It should be replaced if it is damaged or worn. The best way to determine if it is worn is to measure its inside diameter (**Figure 6-42**). If the measurement exceeds specifications, the bushing should be replaced.

Special tools are required to remove the old bushing and install a new one. To replace the bushing, remove all gears from the pump's body. Support the body so both ends of the bushing's bore are away from the bench (**Figure 6-43**). Install the removal tool into the bore and press or drive the bushing out of the bore. Clean the bore. Before installing the new bushing, apply a coat of Loctite® #620, or equivalent, on the inside of the bore. Then press or drive the bushing into the bore.

Removal/installer tool

Pump body

Bushing

Figure 6-42 Checking the oil pump's bushing for wear.

Figure 6-43 Special tools are required to remove and install the oil pump's bushing.

Figure 6-44 Oil pump bolts must be torqued in the sequence recommended by the manufacturer.

Reassembly

The pump seal can be installed with a hammer and seal driver. Apply a very thin coating of RTV sealant around the outside surface of the seal case, when installing the seal. Place some transmission fluid in the pocket of the pump housing and install the gears into the housing according to their alignment marks. Align and install the reaction shaft support and tighten the bolts to the specified torque. Make sure the pump is not binding after you have tightened it by using the torque converter to rotate the pump.

Place a layer of clean ATF in two or three spots around the oil pump gasket and position it on the transmission case. Tighten the pump, attaching bolts to specifications in the specified order (**Figure 6-44**). After the bolts are tight, check the rotation of the input shaft. If the shaft does not rotate, disassemble the transmission to locate the misplaced thrust washer or misaligned friction plate.

TRANSMISSION COOLER SERVICE

Vehicles equipped with an automatic transmission can have an internal or external transmission fluid cooler, or both (**Figure 6-45**). The basic operation of either type of cooler is that of a **heat exchanger**, meaning that heat from the fluid is transferred to something else, such as a liquid or air. Hot ATF is sent from the transmission to the cooler, where some of its heat is removed, then the cooled ATF returns to the transmission.

A **heat exchanger** may also be called an intercooler.

Figure 6-45 Typical transmission coolers.

Internal-type coolers are located inside the engine's radiator. Heated ATF travels from the torque converter to a connection at the radiator. Inside the radiator is a small internal cooler, which is sealed from the liquid in the radiator. ATF flows through this cooler and its heat is transferred to the liquid in the radiator. The ATF then flows out of a radiator connection, back to the transmission.

CUSTOMER CARE This type of cooler is less efficient as engine temperatures increase; therefore, you should recommend that an external cooler be installed in addition to the internal one if the vehicle is used in ways that require the engine to operate at high temperatures, such as carrying heavy loads or pulling trailers.

Special Tools

Catch pan
Various lengths of hoses
Graduated container
OSHA-approved air nozzle

⚡ **Caution**

Do not apply an air pressure of more than 50 psi to the cooler. Extreme pressures can cause the cooler to explode.

⚡ **Caution**

A clogged cooler will result in a subsequent failure of a new or rebuilt transmission.

External coolers are mounted outside the engine's radiator, normally just in front of it. Air flowing through the cooler removes heat from the fluid before it is returned to the transmission.

The engine's cooling system is the key to efficient transmission fluid cooling. If anything affects engine cooling, it will also affect ATF cooling. The engine's cooling system should be carefully inspected whenever there is evidence of ATF overheating or a transmission cooling problem.

If the problem is the transmission cooler, examine it for signs of leakage. Check the dipstick for evidence of coolant mixing with the ATF. Milky fluid indicates that engine coolant is leaking into and mixing with the ATF because of a leak in the cooler. At times, the presence of ATF in the radiator will be noticeable when the radiator cap is removed, as ATF will tend to float to the top of the coolant. A leaking transmission cooler core can be verified with a leak test.

External cooler leaks are evident by the traces or film buildup of ATF around the source of the leak and typically requires little time to determine the source of the leakage. However, internal coolers present a little bit more difficulty and a leak test should be conducted. To do this, place a catch pan under the fittings that connect the cooler lines to the radiator (**Figure 6-46**). Then, disconnect (**Figure 6-47**) and plug both cooler lines from the transmission (**Figure 6-48**). Tightly plug one end of the cooler and apply compressed air (50–75 psi) into the open end of the cooler.

⚡ WARNING **Be sure the radiator is cool to the touch. Hot coolant may spray on you if the system is hot. The hot coolant can cause serious injuries. Wearing safety goggles is a good precaution when working around hot liquids.**

⚡ WARNING **Wear safety goggles when working with compressed air.**

Figure 6-46 Cooler line fittings may use a retaining ring to lock the fitting to the line. These rings should never be reused and must be properly installed.

Inlet connector

Outlet connector (plug)

Figure 6-47 Before testing the cooling system, identify which is the in line and which is the out line.

Figure 6-48 Typical routing of transmission oil cooler lines for a transaxle.

Carefully check the coolant through the top of the radiator; if there is a leak in the transmission cooler, bubbles will be apparent in the coolant. If it leaks, the cooler must be replaced.

The inability of the transmission fluid to cool can also be caused by a plugged or restricted fluid cooling system. If the fluid cannot circulate, it cannot cool. To verify that the transmission cooler is plugged, disconnect the cooler return line from the radiator or cooler (**Figure 6-49**). Connect a short piece of hose to the outlet of the cooler and allow the other end of the hose to rest container. Start the engine and measure the amount of fluid that flows into the container after 20 seconds. Normally, 1 quart of fluid should flow into the container. If less than that filled the container, a plugged cooler is indicated.

A tube-type transmission cooler can be cleaned by using cleaning solvent or mineral spirits and compressed air. A fin-type cooler, however, cannot be cleaned in this way. Therefore, the normal procedure includes replacing the radiator (which includes the cooler). In both cases the cooler lines should be flushed to remove any debris.

If the cooler is plugged, disconnect the cooler lines at the transmission. Apply compressed air through one port of the cooler, then to the other port. The air should blow any

Figure 6-49 Setup for flushing a transmission's cooling system.

debris out of the cooler. Always use low air pressure, no more than 50 psi. Higher pressures may damage the cooler. If little air passes through the cooler, the cooler is severely plugged and may need to be replaced.

If the vehicle has two coolers, pull the inlet and outlet lines from the side of the transmission and, using a hand pump, pump **mineral spirits** into the inlet hose until clear fluid pours out of the outlet hose. To remove the mineral spirits from the cooler, release some compressed air into the inlet port, then pump some ATF into the cooler.

The transmission's cooling system can be flushed. This will remove nearly all of the debris, leaving a clean, dry system. To do this, a transmission cooler flusher is required. The flusher's container is filled with biodegradable flushing fluid. Compressed air pushes the fluid through the entire cooling system of the transmission. The system is then cleaned with forced hot water. When the discharge from the system is cleared, compressed air is expelled through the system until no moisture is evident in the discharge.

Check the condition of the cooler lines from their beginning to their ends. A line that has been accidentally damaged while the transmission has been serviced will reduce oil flow through the cooler and shorten the life of the transmission. If the steel cooler lines need to be replaced, use only double-wrapped and brazed steel tubing. Never use copper or aluminum tubing to replace steel tubing. The steel tubing should be double-flared and installed with the correct fittings.

> Mineral spirits can be purchased at most hardware stores.

> **Mineral spirits** is an alcohol-based liquid that leaves no surface residue after it dries.

SERVICE TIP A transmission cooler is a good hiding place for debris when there has been a major transmission failure. Blowing out the cooler and lines will not always remove all of the dirt. Install a temporary auxiliary oil filter in the cooler lines to catch any leftover debris immediately after rebuilding a transmission.

CASE STUDY

A customer brought her late-model Chevy Silverado to a shop. Her concern was the truck sometimes stalled when it came to a stop. The technician verified the concern and began diagnosis with a check of the engine. He found nothing unusual. He then checked the TSBs available for the truck and found several. Most were related to a stuck torque converter clutch solenoid; however, there were other possible causes. He began by unplugging the TCC solenoid to see if the problem was still there. If the solenoid is stuck and disconnected, the concern should disappear which is what it did. The technician knew the solenoid could be the cause but the concern could also be caused by dirt in the valve body or a bad signal from the PCM.

He raised the vehicle on a lift with the driving wheels off the ground. He checked the ground and power feed to the solenoid and found it to be within specifications. He then connected a test light across terminals A and D at the transmission. With the engine running while the transmission is in DRIVE, he observed the test light. The vehicle was then slowly accelerated to about 60 mph (97 km/h). The test light should have illuminated if the PCM was sending a signal to the solenoid. In this case, the test light did light.

While the vehicle was moving at this speed, he applied the brake pedal. This should cause the TCC to disengage, it did not. He therefore moved to check the brake pedal switch and its circuit. During the initial inspection, he found a potential source of the concern. A connector in the wiring harness near the brake switch was damaged. The connection was very loose. To verify that this was the cause of the problem, he taped the connector tightly together. He then repeated his tests and found the TCC disengaging when the brake pedal was pressed down. He then replaced the connector and reconnected everything he had disconnected for his tests. During a road test, he accelerated and stopped several times. Never did the truck stall. His conclusion was the poor connection caused an intermittent problem that would sometimes not turn off the TCC solenoid.

ASE-STYLE REVIEW QUESTIONS

1. *Technician A* says a ballooned torque converter can cause damage to the oil pump.
   Technician B says a ballooned torque converter is caused by excessive pressure in the torque converter.
   Who is correct?
   A. A only
   B. B only
   C. Both A and B
   D. Neither A nor B

2. While checking torque converter endplay:
   Technician A says the torque converter must be installed in the transmission.
   Technician B says the torque converter endplay is corrected by installing a thrust washer between the oil pump and the direct or front clutch.
   Who is correct?
   A. A only
   B. B only
   C. Both A and B
   D. Neither A nor B

3. While servicing a variable-displacement vane-type pump:
   Technician A says the pump rotor, vanes, and slide have selective sizes and may destroy the pump if the correct ones are not used.
   Technician B says the outer edge of the vanes should be flat.
   Who is correct?
   A. A only
   B. B only
   C. Both A and B
   D. Neither A nor B

4. *Technician A* says the gears in a gear-type oil pump should be replaced if the outer edges of the teeth are worn flat.
   Technician B says all parts of gear-type pumps are selectively sized.
   Who is correct?
   A. A only
   B. B only
   C. Both A and B
   D. Neither A nor B

5. *Technician A* says a cooler located in the radiator can be replaced by a radiator shop if necessary.
   Technician B says if an external cooler is plugged it may make sense to replace it.
   Who is correct?
   A. A only
   B. B only
   C. Both A and B
   D. Neither A nor B

6. *Technician A* says that if the stall speed of a torque converter is below specifications, a restricted exhaust or slipping stator clutch is indicated.
   Technician B says that if the stall speed is above specifications, the bands or clutches in the transmission may not be holding properly.
   Who is correct?
   A. A only
   B. B only
   C. Both A and B
   D. Neither A nor B

7. *Technician A* says torque converter clutch control problems are always caused by electrical malfunctions.
   Technician B says nearly all converter clutches are engaged through the application of hydraulic pressure on the clutch.
   Who is correct?
   A. A only
   B. B only
   C. Both A and B
   D. Neither A nor B

8. While checking the stall speed of a torque converter:
   Technician A maintains full-throttle power until the engine stalls.
   Technician B manually engages the TCC before running the test.
   Who is correct?
   A. A only
   B. B only
   C. Both A and B
   D. Neither A nor B

9. *Technician A* says a converter that has a tight fit at the pilot hub could hold the converter drive hub and inner gear too far into the pump, causing cover scoring.
   Technician B says a front pump bushing that has too much clearance may allow the gears to run off center, causing them to wear into the crescent or the sides of the pump body.
   Who is correct?
   A. A only
   B. B only
   C. Both A and B
   D. Neither A nor B

10. *Technician A* says a seized one-way stator clutch will cause the vehicle to have good low-speed operation but poor high-speed performance.
    Technician B says a freewheeling or non-locking one-way stator clutch will cause the vehicle to have poor acceleration.
    Who is correct?
    A. A only
    B. B only
    C. Both A and B
    D. Neither A nor B

ASE CHALLENGE QUESTIONS

1. *Technician A* says a lower-than-specified stall speed may be caused by a faulty stator one-way clutch.

   Technician B says a higher-than-specified stall speed may be caused by faulty clutch packs.

   Who is correct?

   A. A only
   B. B only
   C. Both A and B
   D. Neither A nor B

2. *Technician A* says a shudder after the converter clutch engages could be caused by a cracked spark plug.

   Technician B says a transmission fluid change may fix a TCC shudder.

   Who is correct?

   A. A only
   B. B only
   C. Both A and B
   D. Neither A nor B

3. The backside of a torque converter is found to be wet.

   Technician A says it may be caused by excessive torque converter hub runout.

   Technician B says it can be caused by insufficient input shaft endplay.

   Who is correct?

   A. A only
   B. B only
   C. Both A and B
   D. Neither A nor B

4. While discussing TCC electrical diagnosis:

   Technician A says the TCM/PCM has the capability to check the TCC circuit by watching for engine load and rpm information as the clutch is applied and released.

   Technician B says the electrical circuits are monitored for shorts or opens by the comprehensive component monitor.

   Who is correct?

   A. A only
   B. B only
   C. Both A and B
   D. Neither A nor B

5. *Technician A* says a plugged transmission cooler or cooler lines may cause the vehicle to stall when the transmission is shifted into a forward gear.

   Technician B says a plugged transmission cooler or cooler lines may overheat the transmission.

   Who is correct?

   A. A only
   B. B only
   C. Both A and B
   D. Neither A nor B

Name _____   Date _____

ROAD TEST A VEHICLE TO CHECK THE OPERATION OF A TORQUE CONVERTER CLUTCH

Upon completion of this job sheet, you should be able to road test a vehicle with an automatic transmission to verify and analyze the operation of a torque converter clutch.

ASE Education Foundation Correlation

This job sheet addresses the following **AST** task:

A.7   Perform lock-up converter system tests; determine needed action. (**P-3**)

This job sheet addresses the following **MAST** task:

A.9   Perform lock-up converter system tests; determine needed action. (**P-3**)

We Support
ASE | Education Foundation

Tools and Materials

- A vehicle with an automatic transmission and lockup converter
- Service information
- Pad of paper and pencil

Describe the Vehicle Being Worked On

Year _____ Make _____ Model _____

VIN _____ Engine type and size _____

Model and type of transmission _____

Procedure

Task Completed

1. Park the vehicle on a level surface.   ☐

2. Describe the condition of the fluid (color, condition, smell).

   _____
   _____

3. What is indicated by the fluid's condition?

   _____
   _____

4. In the appropriate service information, find the chart that shows the conditions that must be present in order for the converter clutch to be engaged. Describe these conditions.

   _____
   _____
   _____
   _____

5. Allow the engine and transmission to cool by letting the vehicle sit.   ☐

6. Take the vehicle for a test drive. Bring the vehicle to normal lockup speed in the appropriate gear and observe the action of the clutch. Record the speed and feel of the clutch engagement.

_____

_____

_____

7. Was the engagement of the clutch normal for the engine temperature and vehicle speed? Why or why not?

_____

_____

8. Now drive the car through the following conditions and note the action of the torque converter clutch.

Manual upshifting of the transmission during acceleration

Torque converter clutch action: _____

Part-throttle downshifting

Torque converter clutch action: _____

Wide-open throttle downshifting

Torque converter clutch action: _____

Cruising at highway speed, then lightly applying the brakes

Torque converter clutch action: _____

Cruising at highway speed, then letting completely off the throttle but not applying the brakes

Torque converter clutch action: _____

9. Summary of torque converter clutch operation:

_____

_____

_____

_____

Problems Encountered

_____

_____

_____

Instructor's Comments

_____

_____

_____

Name _____   Date _____

VISUAL INSPECTION OF A TORQUE CONVERTER

Upon completion of this job sheet, you will be able to conduct a visual inspection of a torque converter.

ASE Education Foundation Correlation ————————————————————————

This job sheet addresses the following **AST/MAST** task:

C.3   Inspect converter flex (drive) plate, converter attaching bolts, converter pilot, converter pump drive surfaces, converter end play, and crankshaft pilot bore. **(P-2)**

We Support

ASE | Education Foundation

Tools and Materials

- Clean white paper towels
- Droplight or good flashlight
- Hoist
- Service information
- Remote starter switch
- Special tools for the torque converter
- Dial indicator

Protective Clothing

- Goggles or safety glasses with side shields

Describe the Vehicle Being Worked On

Year _____ Make _____ Model _____

VIN _____ Engine type and size _____

Model and type of transmission _____

Procedure	**Task Completed**

1. Park the vehicle on a level surface.   ☐

2. Describe the condition of the fluid (color, condition, and smell).

_____

_____

3. What is indicated by the fluid's condition?

_____

_____

4. Connect a remote starter switch to the vehicle and disable the ignition. Explain how you did this.

_____

_____

5. Raise the vehicle on a hoist to a comfortable working height.   ☐

6. Put on eye protection.   ☐

7. Remove the torque converter access cover or shield.   ☐

8. Inspect the converter through the access hole. Use the remote starter switch to observe the entire converter. Describe your findings below.

_____

_____

_____

9. Carefully look for signs of fluid leakage around the torque converter and the access cover or shield. Describe your findings below and state what could be causing the leak.

_____

_____

_____

10. Look at the heads of all torque converter bolts. Do they show signs of contact with other parts?

_____

_____

_____

Out-of-Vehicle Torque Converter Checks

1. Detailed diagnosis of a torque converter takes place with the torque converter removed from the vehicle. Inspect the drive studs or lugs used to attach the converter to the flexplate. Summarize your findings.

_____

_____

_____

2. Check the threads of the lugs and studs. If they are damaged slightly, they can be cleaned up with a thread file, tap, or die set. However, if they are badly damaged, the converter should be replaced. The converter should also be replaced if the studs or lugs are loose or damaged. Carefully check the weld spots that position the studs and lugs on the converter. If the welds are cracked, replace the converter. Summarize your findings.

_____

_____

_____

3. Check the hub of the converter for scoring, nicks, excessive wear, burrs, and signs of discoloration. If polishing cannot remove the defects in the hub, the converter should be replaced. Summarize your findings.

_____

_____

_____

4. Carefully check the converter's flexplate for damage and cracks. Also check the ring gear for looseness. Summarize your findings.

_____

_____

_____

5. Noises may be caused by internal converter parts hitting each other or hitting the housing. To check for any interference between the stator and turbine, place the converter, face down, on a bench.

_____

_____

_____

6. Then install the oil pump assembly. Make sure the oil pump drive engages with the oil pump. ☐

7. Insert the input shaft into the hub of the turbine. ☐

8. Hold the oil pump and converter stationary, then rotate the turbine shaft in both directions. If the shaft does not move freely and/or makes noise, the converter must be replaced. Summarize your findings.

_____

_____

_____

9. To check for any interference between the stator and the impeller, place the transmission's oil pump on a bench and fit the converter over the stator support splines. Rotate the converter until the hub engages with the oil pump drive. Then hold the pump stationary and rotate the converter in a counterclockwise direction. If the converter does not freely rotate or makes a scraping noise during rotation, the converter must be replaced. Summarize your findings.

_____

_____

_____

10. Check endplay with the proper holding tool and a dial indicator with a holding fixture. Insert the holding tool into the hub of the converter and, once bottomed, tighten it in place. ☐

11. This locks the tool into the splines of the turbine while the dial indicator is fixed onto the hub. The amount indicated on the dial indicator as the tool is lifted up is the amount of endplay inside the converter. If this amount exceeds specifications, replace the converter. Summarize your findings.

_____

_____

_____

12. If the initial visual inspection suggests that the converter has a leak, special test equipment can be used to determine if the converter is leaking. This equipment uses compressed air to pressurize the converter. Leaks are found in much the same way as tire leaks are, that is, the converter is submerged in water and the trail of air bubbles leads the technician to the source of leakage. This test equipment can only be used to check converters with a drain plug and therefore is somewhat limited on current applications. However, some manufacturers list a procedure for installing a drain plug or draining the converter in their service information. Summarize your findings.

_____

_____

_____

Problems Encountered

_____

_____

_____

Instructor's Comments

_____

_____

_____

Name _____   Date _____

LOCKUP CONVERTER DIAGNOSIS

Upon completion of this job sheet, you will be able to test the operation of a lockup converter.

ASE Education Foundation Correlation ——————————————————————————

This job sheet addresses the following **AST** task:

A.7   Perform lock-up converter system tests; determine needed action. (**P-3**)

This job sheet addresses the following **MAST** task:

A.9   Perform lock-up converter system tests; determine needed action. (**P-3**)

We Support
ASE | Education Foundation

Tools and Materials

- Service information
- Scan tool
- DMM
- Goggles or safety glasses with side shields

Describe the Vehicle Being Worked On:

Year _____ Make _____ Model _____

VIN _____ Engine type and size _____

NOTE: Do not attempt to test drive a vehicle while operating a scan tool. Always take an assistant to drive while the technician observes the scan tool.

Procedure   Task Completed

1. As a preliminary measure, make visual checks of the transmission for any causes of a shudder, such as damaged or worn engine mounts. Record your findings here.

    _____

    _____

2. Does the engine seem to be running okay? Is there any evidence of an engine misfire or rough running?

    _____

    _____

3. Check the transmission fluid level and condition. Record your findings here.

    _____

    _____

4. Connect the scan tool and retrieve all DTCs. What are they?

    _____

    _____

5. If an engine-related problem is suspected, it should be corrected before moving on to   ☐
    the torque converter tests.

6. Operate the vehicle with the scan tool connected. Drive it at 40 mph. ☐

7. With the scan tool, engage the lockup converter's solenoid. Watch the tachometer while making the change. What did you observe? Engine speed should drop when the converter locks.

_____

_____

8. Disengage the clutch and observe the engine's speed. What did you observe?

_____

_____

9. The action of the lockup converter can also be checked with the vehicle stopped and ☐ the engine running. Place the shifter into the "Drive" position.

10. With the scan tool, engage the lockup solenoid. Describe what happens.

_____

_____

11. If there was no change in engine speed, what is evident?

_____

_____

12. The lockup solenoid can also be checked with the ignition on but the engine off. Engage the solenoid with the scan tool and listen. Did you hear a click?

_____

_____

13. If a problem with the solenoid is suspected, remove it from the transmission for ☐ testing.

14. Refer to the service information and locate the resistance checks for the solenoid. What are the specifications?

_____

_____

15. Measure the resistance of the solenoid across the test points specified. What were your results?

_____

_____

16. What can you conclude from these tests?

_____

_____

Problems Encountered

_____

_____

_____

Instructor's Comments

_____

_____

_____

Name _____   **Date** _____

CONDUCT A STALL TEST

Upon completion of this job sheet, you should be able to conduct a stall test.

ASE Education Foundation Correlation ——————————————

This job sheet addresses the following **AST** task:

A.6   Perform stall test; determine needed action. (**P-2**)

This job sheet addresses the following **MAST** task:

A.8   Perform stall test; determine needed action. (**P-2**)

We Support
⚙ ASE | Education Foundation

Tools and Materials

- Vehicle with an automatic transmission
- Tachometer
- Hoist
- Stethoscope
- Droplight or good flashlight
- Service information

Describe the Vehicle Being Worked On

Year _____ Make _____ Model _____

VIN _____ Engine type and size _____

Model and type of transmission _____

Procedure   **Task Completed**

1. Park the vehicle on a level surface.   ☐

2. Describe the level and condition of the fluid (color, condition, and smell).

   _____

   _____

3. What is indicated by the fluid's condition?

   _____

   _____

4. Connect a tachometer to the engine. If the vehicle has one in the instrument panel, there is no need to connect another one. Explain how you connected the tachometer.

   _____

   _____

5. Place the tachometer on the inside of the vehicle so that it can be easily read.

6. Mark the face of the tachometer with a grease pencil at the recommended maximum rpm for this test. What is the maximum rpm? _____

7. Check the engine's coolant level.   ☐

8. Block the front wheels and set the parking brake.   ☐

9. Place the transmission in Park and allow the engine and transmission to warm.   ☐

10. Place the gear selector into the gear that is recommended for this test. What is the recommended gear? _____

11. Press the throttle pedal to the floor with your right foot and firmly press the brake pedal with your left. Hold the brake pedal down.   ☐

12. When the tachometer needle stops rising, note the engine speed and let off the throttle. What was the highest rpm during the test? _____

13. Place the gearshift into Neutral and allow the engine to run at 1000 rpm for at least 1 minute. What is the purpose of doing this?

    _____

    _____

14. If noise is heard from the transmission during the stall test, raise the vehicle on a hoist.   ☐

15. Use the stethoscope to determine if the noise is from the transmission or the torque converter.   ☐

16. What are your conclusions from this test?

    _____

    _____

    _____

    _____

Problems Encountered

_____

_____

_____

Instructor's Comments

_____

_____

_____

Name _____   **Date** _____

AUTOMATIC TRANSMISSION OIL PUMP SERVICE

Upon completion of this job sheet, you should be able to inspect, measure, and replace the components of an oil pump assembly.

ASE Education Foundation Correlation

This job sheet addresses the following **MAST** task:

C.10   Inspect, measure, and reseal oil pump assembly and components. **(P-2)**

We Support
ASE | Education Foundation

Tools and Materials

- Machinist dye
- Feeler gauge

Describe the Transmission Being Worked On and the Vehicle the Transmission Is From

Year _____ Make _____ Model _____

VIN _____ Engine type and size _____

Model and type of transmission _____

NOTE: Each transmission is going to have a unique procedure. Consult service information for the specific transmission you are servicing.

Procedure

Task Completed

1. With the oil pump removed from the transmission, remove the front pump bearing race, front clutch thrust washer, gasket, and O-ring. Inspect the pump bodies, pump shaft, and ring groove areas. Record your findings.

    _____

    _____

    _____

2. Mount the pump on a stand and unbolt and separate the pump bodies.   ☐

3. Mark the gears with machinist bluing ink or paint before removing them so that the gears will remain in the same relationship during reassembly. What did you use to mark them and where did you mark them?

    _____

    _____

4. Inspect the gears and all internal surfaces for defects and visible wear. Record your findings.

    _____

    _____

5. With the pump mounted on the stand, use a feeler gauge to measure between the outer gear and the crescent in the pump housing. What tolerance do the specifications call for? What did you measure?

    _____

    _____

6. Measure between the outer gear teeth and the crescent. What tolerance do the specifications call for? What did you measure?

_____

_____

7. Place the pump flat on the bench and, using a feeler gauge and straightedge, measure between the gears and the pump cover. What tolerance do the specifications call for? What did you measure?

_____

_____

8. Measure the clearance between the O-ring and the ring groove. What tolerance do the specifications call for? What did you measure?

_____

_____

9. Using the stand to center the pump, torque the securing bolts to specifications. What is the specified torque?

_____

_____

10. Replace all O-rings and gaskets.

_____

_____

Problems Encountered

_____

_____

_____

Instructor's Comments

_____

_____

_____

CHAPTER 7

GENERAL HYDRAULIC SYSTEM SERVICE

Upon completion and review of this chapter, you should be able to:

- Remove and install the valve body from common transmissions.
- Check valve body mating surfaces.
- Inspect and measure valve body bores, springs, sleeves, retainers, brackets, check balls, and screens.
- Inspect and replace valve body spacers and gaskets.
- Check and/or adjust valve body bolt torque.

Terms To Know

Auxiliary transmission fluid pump (ATFP)

Manual shaft

Solenoid/pressure switch

Transmission fluid pressure (TFP) switch

INTRODUCTION

The hydraulic control center for a transmission is the valve body. All electronic automatic transmissions have electronic controls that modify the action of clutches. These too have a central control center, called the TCM or PCM. But, the actual control of the hydraulics takes place in the valve body. Therefore, the valve body is an extremely important part of the transmission and faults in it can cause a wide variety of problems.

A valve body can be a single unit or an assembly of many subassemblies. When it is comprised of several parts, the assembly may be made up of a main valve body, regulator valve body, secondary valve body, and accumulator body. The valve body can be mounted in several different locations, the common of which is to the side of the torque converter housing on below the gears.

The oil pump is driven by the torque converter. Fluid flows through the regulator valve to maintain a specified pressure through the main valve body to the manual valve, which directs pressure to each of the clutches. The actual application of the clutches is controlled by the action of the shift solenoids.

The main valve body typically contains the manual valve, torque converter check valve, shift solenoids, some shift valves, shift pressure control valves, oil relief valve, lockup shift valve, and lubrication control and check valves. The main valve body basically controls the hydraulic pressure going to the rest of the hydraulic control system.

Mainline pressure is the regulated pressure directly from the pump.

Hydraulic-Style Pressure Regulator

The pressure-regulator valve section is made up of the pressure-regulator valve, cooler check valve, lockup control valve, and some accumulators. The pressure-regulator valve keeps the hydraulic pressure from the ATF pump to the hydraulic control system constant. While doing this, the pressure-regulator valve is maintaining the correct level of line pressure. Excess pressure exhausts through the relief valve. Normal pressure moves through the transmission and through the manual shift valve to the various control valves. It also sends fluid to the torque converter. Line pressure will change according to need. This may be accomplished by the regulator valve. Some systems have a stator shaft arm that can work against a spring cap in the regulator valve. During acceleration or heavy loads, the torque demand causes the arm to move against the regulator's spring cap. This action allows line pressure to increase. To check the operation of the hydraulic pressure regulator valve, it is necessary to do a line pressure check with a pressure gauge.

Electronic Pressure Control Solenoid

Modern transmissions and transaxles use an electronic pressure control (EPC) solenoid installed in the valve body, which is PCM/TCM controlled (**Figure 7-1**). This solenoid is also called a line pressure control solenoid by some manufacturers. The EPC allows very precise control of transmission pressures via the PCM/TCM. The PCM/TCM knows how much line pressure is needed because of the inputs from the engine control sensors, such as the MAP and TP sensors. Line pressure is controlled by Pulse width Modulation (PWM) of the solenoid by the PCM/TCM. The current to the EPC is used to vary the line pressure by PWM of the solenoid (**Figure 7-2**). The more current that is used at the solenoid, the lower the pressure. This ensures that if the solenoid should fail, line pressure will be at its highest level. The transmission will shift hard, but high pressure to the clutch packs is better than low pressure. Low pressure would allow the clutches to slip and possibly destroy the clutches.

Electronic pressure
control solenoid

Figure 7-1 Line pressure control solenoid located in the valve body.

Figure 7-2 Electronic pressure control solenoid. Pressure increases as current decreases.

Diagnosis of a P2716 Open or Shorted Electronic Pressure Control Solenoid

A typical pressure-control solenoid electrical diagnosis would involve disconnecting the transmission electrical connector and measuring the resistance of the solenoid and wiring with a multimeter (**Figure 7-3**). If the resistance is not as specified, there is a problem with the solenoid or wiring inside the transmission case. If the resistance is correct, it is still important to check for the wiring being shorted to the case. This is done by checking each lead from the solenoid for grounding to the transmission case (**Figure 7-4**). In this measurement, there should be at least 10 K ohms resistance between the solenoid wiring and ground. If there is no open, high resistance, or grounding in the wiring at the transmission case, then the transmission harness is reconnected to the transmission and the PCM is disconnected. The harness from the transmission to the PCM is checked for opens (**Figure 7-5**) or grounds to the body (**Figure 7-6**). If opens or grounds are found, then the harness and especially the connectors need to be inspected for damage. If no damage is found, then the PCM is suspect, but replace the PCM only after you are sure that there are no external causes.

Checking the Solenoid

The PWM solenoid can be checked out of the vehicle. First, perform a simple resistance check, and then check by powering it with 12 volts with an incandescent bulb, such as an 1156 bulb in series (**Figure 7-7**). The turn signal bulb is used to provide a load and reduce

Figure 7-3 A multimeter should measure approximately 4.9 Ohms at the transmission connector at room temperature on this vehicle.

Figure 7-4 Checking the solenoid wiring for a short to ground inside the transmission case. There should be at least 10K Ohms resistance on each measurement.

Figure 7-5 Transmission harness is checked for opens or high resistance.

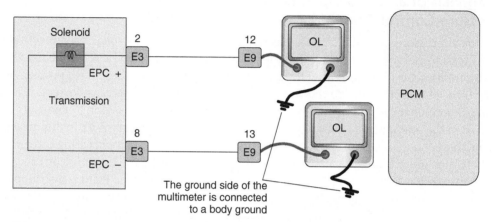

Figure 7-6 Checking the wiring harness for shorts to ground. Both these readings should be at least 10k ohms.

Figure 7-7 Checking the pressure control solenoid off the vehicle.

the amperage to the solenoid. Never connect a low-resistance solenoid directly to 12 volts, as it may be destroyed. Remember that this solenoid is pulsed rapidly, not connected to constant battery voltage on the vehicle. Almost any solenoid can be checked on or off the car. A solenoid with a higher resistance may not need the bulb in series to limit amperage. Check in the manufacturer's service information to be sure.

Figure 7-8 An auxiliary transmission fluid pump (ATFP).

Auxiliary Transmission Fluid Pump

Some transmissions, especially those made for hybrid vehicles, have an auxiliary fluid pump for the transmission. With hybrids, the auxiliary pump ensures that the transmission has fluid flow when the system is in stop/start. Since the engine is off during those times the transmission's pump is not providing fluid flow. Therefore, the transmission will not be ready to move the vehicle once the engine is restarted. The auxiliary pump provides fluid flow during those conditions.

The **auxiliary transmission fluid pump (ATFP)** is normally mounted on the outside of the torque converter housing (**Figure 7-8**). When the vehicle comes to a stop and the engine turns off while the ignition is still on, the pump immediately begins to supply fluid to the regulator valve. The ATFP runs until the engine restarts. A typical ATFP is a three-phase brushless DC motor. The TCM has total control of the pump. If the transmission cannot move the vehicle after the engine restarts, check the ATFP motor.

ATFP Motor Test. To check the motor, disconnect the wiring connector at the ATFP. Place the positive lead of an ohmmeter to one terminal at the connector on the pump. Connect the other meter lead to a good ground. You should receive an infinite reading. Do this at each terminal in the connector. If there is continuity between any of the terminals and ground, the motor should be replaced.

Replacement of ATFP. Normally, the replacement of the motor is straightforward. Disconnect electrical connector at the pump. Carefully inspect the connector, make sure it is free of dirt, oil, and rust. Remove the pump's retaining bolts and lift off the pump. After the pump has been removed, the feed and suction pipes, and the dowel pins can be removed. Inspect all parts and replace them if they show signs of damage or wear. Inside the feed pipe there may be a check ball, shake the pipe to see if the ball can move freely. If it cannot, attempt to clean the inside of the pipe. Then, clean the mounting surface for the pump.

To install a new pump, fit new O-rings on the suction pipe and install the pipe into the converter housing with the filter end toward the outside of the housing. Then fit new O-rings on the feed pipe and install the large end of the pipe into the housing. Carefully install the dowel pins into the housing. Then fit a new O-ring into its groove in the pump assembly and install the pump onto the housing. Tighten the retaining bolts to specifications and securely reattach the electrical connector.

Feed Pipes

Some transmissions have fluid feed pipes that move fluid from one point of the valve body to another. At times, it is necessary to remove the pipes to gain access to the valve body.

Figure 7-9 A feed pipe assembly.

The pipes can be retained in the valve body in a number of different ways. Some are retained by plate that is held in the housing by a snap ring. Others have flanges that are bolted to various locations on the valve body (**Figure 7-9**). Some others are merely pressed into bores in the valve body. To remove the pipes, the snap rings or bolts must be removed. Then the pipes are lifted away. If the pipes are pressed into a bore, they should be carefully pried out of the bore. It is important to be careful when removing them, make sure you do not bend or damage them. Also, make sure you do not damage their sealing surface on the valve body.

In most cases, the pipe's seals or O-rings should be replaced when the pipes are reinstalled. Some feed pipes have reusable seals. These are glued to the pipes by the manufacturer. If these seals are damaged, the entire pipe should be replaced.

Wires and Control Module

Often, there is a wiring harness or control module that must be removed before the valve body can be removed. Some of these harnesses are molded together and are removed as a unit. Others are individual wires that must be disconnected from the various components.

Molded leads are secured by their fit into the various connections (**Figure 7-10**). While removing the harness, inspect it for damage. Make sure all terminals are secure with their molded connector shell. Also, when reinstalling it, make sure all connections are secure and tight.

Some wiring harnesses are comprised of several groups of wires and connectors that go to individual components, such as the EPC solenoid, shift solenoids, TCC solenoid, and switches (**Figure 7-11**). To remove this type of harness, it must be disconnected for component. In some cases, the harness is secured to the valve body by retaining bolts or clips. Some have plastic locks at the point where the harness passes into the valve body that must be released. It is important that the harness and the various connectors are not damaged during removal or installation.

Control Module. Some transmissions have a single assembly that contains all of the solenoids, the TCM, and a lead frame assembly. The lead frame connects to the individual components, therefore fewer electrical connections are needed for these transmissions, eliminating many potential problems. The assembly bolts directly to the lower and upper valve body assemblies (**Figure 7-12**). The assembly connects to an engine harness with a 16-pin connector through a pass-through sleeve in the valve body cover. Often, to remove the assembly, some part of the shift linkage must be removed. This is then followed by disconnecting the electrical connectors.

> **⚡ Basic Tools**
>
> Basic hand tools
> Cleaning pan
> OSHA-approved air
>   nozzle
> Cleaning solution

⚠ WARNING **Be careful when handling the assembly. If it is jarred or dropped, damage to its internal components can result.**

Figure 7-10 A molded wiring harness that is secured to the valve body by its connections.

Figure 7-11 A wiring harness with connectors for the individual outputs and inputs located on the valve body.

Figure 7-12 A TCM/solenoid assembly mounted on a valve body.

Always replace the filter when reinstalling the assembly. Once the filter has been removed and reinstalled, it may be contaminated which could cause damage to the transmission.

VALVE BODY REMOVAL

If the pressure test indicated that there is a problem associated with the valves in the valve body, a thorough disassembly, cleaning in fresh solvent, careful inspection, and the freeing up and polishing of the valves may correct the problem. Sticking valves and sluggish valve movements are caused by poor maintenance, the use of the wrong type of fluid, and/or overheating of the transmission. The valve body of most transmissions can be serviced when the transmission is in the vehicle; however, it is typically serviced when the transmission has been removed for other repairs.

Typically, to remove a valve body from a transmission while it is still in the vehicle, begin by draining the fluid and removing the oil pan. Then disconnect the manual valve. Carefully remove the detent spring and screw assemblies. Loosen and remove the valve body screws (**Figure 7-13**). Before lowering the valve body and separating the assembly, hold the assembly with the valve body on the bottom and the transfer and separator plates on top. Holding the assembly in this way will reduce the chances of dropping the steel check balls that are located in the valve body (**Figure 7-14**). Lower the valve body and note where these check balls are located in the valve body; remove them and set them aside, along with the various screws.

Classroom Manual
Chapter 7, page 243

> ⚙ **SERVICE TIP** To avoid having to spend hours crawling on the floor of the shop looking for lost parts, place your hand or fingers over spring-loaded valves or plugs when removing them from the valve body.

On transmissions with a horizontal valve body, you can leave the separator plate in the transmission to keep the check balls above the spacer from falling out. The check balls in the valve body will not fall out if the valve body is lowered carefully. Then the spacer plate can be carefully removed, keeping it level as it is lowered, and the check balls on the plate will remain in their proper position.

Figure 7-13 When removing a valve body from the transmission, remove only the bolts that are necessary to lower the valve body. Also be careful not to lose any springs or check balls.

Figure 7-14 Location of the check balls in a typical valve body.

Removal of a valve body from a transmission on a bench is similar. Make sure you have the transmission positioned so that the valve body can be lowered in such a way that you can keep the check balls in place.

There are many different designs and configurations used in today's transmissions. Each of them requires specific steps for removal and installation. It is extremely important that you refer to the service information for these specifics. To illustrate these variables, valve body removal and installation procedures for some common transmissions follow, as well as a description of the components that are included in the valve body assembly.

Chrysler Transaxles

Chrysler's 41TE transaxle is found in many of their cars and minivans. The valve body assembly of these units is comprised of the valve body, a transfer plate, and a separator plate. The valve body contains many different valves and check balls (**Figure 7-15**). All of these control fluid flow to the TCC, the **solenoid/pressure switch** assembly, and the various hydraulic apply devices.

To remove the valve body, move the manual valve lever clockwise into low gear. Loosen and remove the valve body retaining bolts. Then, using a screwdriver, push the park rod rollers out of the guide bracket. Next, remove the valve body from the housing. Carefully pull the valve body assembly from the case. The **manual shaft** is attached to the valve body and must be carefully pulled from its bore.

Once removed, the valve body can be disassembled. This begins with the removal of the TR sensor assembly (**Figure 7-16**). The accumulator retaining plate and accumulator are then removed.

The transfer plate can now be unbolted and separated from the valve body. While separating these units, pay attention to the check balls. It is very possible that a loose check ball will drop out.

Remove the oil screen and overdrive clutch check valve from the separator plate. Then remove the separator plate (**Figure 7-17**). Then remove the thermal valve and check balls from the valve body. Keep track of their location and count the balls as they are removed. Disassemble, inspect, and clean the valve body components as needed.

To install the valve body, guide the park rod rollers into the guide bracket while positioning the valve body. Install the valve body-to-case mounting bolts and tighten them to specifications.

⚠ Caution

When assembling parts of a transmission, especially a valve body, make sure you do the following:

Always use new gaskets and seals.

▪ All fasteners must be tightened to torque specifications.

▪ Each part should be lubricated with clean transmission fluid.

▪ Be careful not to nick or damage any part or sealing surface while disassembling, cleaning, or assembling the parts.

The **solenoid/ pressure switch** is a single unit that consists of various solenoids and pressure switches.

The **manual shaft** is connected to the gearshift linkage.

Figure 7-15 The various components in the valve body of a Chrysler 45RFE.

Figure 7-16 The main connector, TR sensor, and other associated parts in this Chrysler valve body.

Figure 7-17 A valve body with its separator plate.

Ford Transmissions

The 4R70E transmission is commonly found in Ford pickups and SUVs. Removing the valve body from this transmission is a rather simple process. The valve body is bolted directly to the transmission case and few items need to be removed to remove it. The electrical connectors to the various solenoids mounted to or around the valve body need to be disconnected. Once they are, the wiring harness can be removed by disconnecting it from its main connector. After the harness is removed, the shift and TCC solenoids can be unbolted and removed. Then remove the retaining bolt for the manual valve detent lever spring and then the spring (**Figure 7-18**). Now unbolt the valve body from the transmission case. Remove and discard the pump outlet screen.

Remove the separator plate and discard the gaskets. Note the location of the check balls (**Figure 7-19**) and remove them before cleaning the valve body. Once the valve body is serviced, install the check balls and the separator plate. Install a new pump outlet screen and gaskets during reassembly.

Figure 7-18 The detent lever spring for the manual valve.

Figure 7-19 Location of the check balls in a 4R70W transmission.

To ensure proper alignment of the valve body, this transmission has two alignment bolts. Position the valve body gasket and valve body onto the case by using these alignment bolts as a guide. Loosely install all of the valve body retaining bolts. Then install the manual valve detent lever spring and tighten it in place. Before doing this, make sure the drive pin of the detent lever is correctly positioned in the manual valve.

Now tighten the valve body retaining bolts according to the specified sequence and to the correct torque. Install the shift and TCC solenoids and reconnect the wiring harness to the main connector and to the solenoids.

Six-Speed Transaxle. This transaxle is used in the Fusion-based front-wheel-drive (FWD) vehicles from Ford Motor Company. The main valve body can be serviced with the transaxle in the vehicle, as well as with the unit removed from the vehicle. To gain access to the valve body, the bolts retaining the oil pan should be removed and discarded. With the pan removed, clean the sealing surfaces of the pan and housing, being careful not to damage the surfaces. Then disconnect the wiring harness from the solenoids and speed sensors (**Figure 7-20**) and remove the harness. Then unbolt the lock plate for the temperature sensor and remove the sensor. Mounted in a corner of the valve body is the suction cover. This must be removed before removing the attaching bolts for the valve body. The valve body bolts should be loosened in a staggered sequence. Once those bolts are removed, the valve body can be removed. While removing the valve body, the manual control lever may fall out of its bore, so remove the valve body carefully.

Installation of the valve follows the opposite sequence from removal. Take extra care to keep the harness away from the mounting surface for the valve body. If the harness is caught between the valve body and housing, it can be damaged. All retaining bolts must be torqued to specifications and all harnesses and connectors must be securely in place. Now, the oil pan must be sealed with a thin application of sealer and new bolts tightened to specifications.

General Motors' Transmissions

GM's 4L60-E transmission is used in a wide variety of rear-wheel-drive (RWD) vehicles. Like the valve body in a Ford transmission, the valve body of a 4L60 is rather straightforward. Although the procedures are similar, the 4L60 requires a different sequence and, due to its design, additional steps. **Photo Sequence 12** covers the procedure for removing and installing a valve body when the 4L60 is still in the vehicle.

Figure 7-20 The various solenoids and switches located on the main valve body of a Ford six-speed transaxle.

PHOTO SEQUENCE 12
Removing and Installing a Valve Body

P12-1 Position the vehicle on a lift and loosen the transmission pan to drain the fluid.

P12-2 After the fluid has drained, remove the pan and the transmission filter.

P12-3 Allow the fluid to drip off the transmission, then disconnect the electrical connections to the various switches and solenoids mounted to or around the valve body.

P12-4 Remove the retaining clip for the PWM solenoid, and then do the same for the shift solenoids.

P12-5 Remove the TCC solenoid.

P12-6 Remove the retaining bolts or clips for the electrical harness. Move the harness off to the side of the transmission.

P12-7 Now remove the pressure switch retaining bolts and switch.

P12-8 Now unbolt and remove the transmission fluid pressure switch assembly. Carefully inspect this assembly for any damage or debris.

P12-9 Unbolt and remove the manual detent spring.

PHOTO SEQUENCE 12 (CONTINUED)

P12-10 Loosen and remove the remaining valve body mounting bolts.

P12-11 Lower the valve body just enough to disconnect the manual valve's linkage, then lower the valve body away from the transmission and remove it.

P12-12 Unbolt and remove the accumulator assembly.

P12-13 Before reinstalling the valve body, check the fit of the new gasket to the transmission.

P12-14 Reinstall the accumulator.

P12-15 Position the valve body on the case and loosely install some of the mounting bolts. Then, install the manual detent spring.

P12-16 Install the transmission fluid pressure switch assembly. Do not tighten the mounting bolts.

P12-17 Position the electrical harness and loosely install the retaining bolts.

P12-18 Install the TCC solenoid but do not tighten the mounting bolts.

PHOTO SEQUENCE 12 (CONTINUED)

P12-19 Now tighten all of the valve body mounting bolts according to the specified order and to the specified torque.

P12-20 Install the various solenoids and switches to the valve body. Then tighten all retaining bolts to the specified torque.

P12-21 Reconnect the electrical connectors to the various solenoids and switches.

P12-22 Install a new fluid filter.

P12-23 Install the transmission pan with a new gasket and tighten the retaining bolts to specifications.

P12-24 Fill the transmission with fluid, following the correct procedure and using the correct fluid.

The removal process begins with disconnecting the electrical connections to the various switches and solenoids mounted to or around the valve body (**Figure 7-21**). The wiring harness cannot be removed until the TCC solenoid has been removed. The order in which the solenoids are removed is important. Some retaining bolts and brackets are covered by the solenoids. Remove the PWM solenoid to gain access to the retaining bolts for the TCC solenoid. Once the TCC solenoid is removed, the electrical harness retaining bolts can be removed.

Remove the pressure switch retaining bolts and switch. Then unbolt and remove the **transmission fluid pressure (TFP) switch** assembly (**Figure 7-22**). This unit needs to be carefully inspected and checked for any damage or debris.

The manual detent spring can now be unbolted and removed. Once this is off, the remaining valve body mounting bolts can be removed. Once loosened, the valve body should be slightly lowered and the manual valve's linkage disconnected. This will allow for removal of the valve body.

The **transmission fluid pressure (TFP) switch** assembly contains five different pressure switches and is connected to five different hydraulic circuits.

Figure 7-21 The various solenoids and switches located on or near the valve body of a 4T45E transaxle.

Figure 7-22 A typical transmission fluid pressure (TFP) switch assembly.

Installation is done in the reverse order of removal. Again, it is important that the solenoids be installed according to the sequence. All of the valve body mounting bolts should be loosely installed as components are installed, until all of the bolts are in their proper location. Then the bolts should be tightened in the specified order and to the specified torque.

Honda Transaxles

The valve body assembly in this unique transaxle is comprised of a main valve body, the regulator valve body, the servo valve body, and the accumulator body (**Figure 7-23**). The main valve body contains the pump gears and the manual, modulator, shift, servo control, TCC control, and cooler check valves. The shift valves are controlled by fluid flow directed by one of the three shift solenoids.

The regulator valve body contains the regulator, TCC timing, and relief valves. The servo valve body contains additional shift valves, two forward speed accumulators, and the servo valve. The accumulator valve body contains two forward speed accumulators and a lubrication check valve.

Transmission fluid from the regulator passes through the manual valve to the different shift and control valves. Oil feed pipes and internal passages are used to move the fluid to the various clutches in the transmission.

Servicing this valve body requires transaxle removal and disassembly. To remove this valve body, the ATF feed pipes must be removed first. Then the ATF strainer and servo detent plate should be unbolted and removed. Now press down on the accumulator cover while loosening its mounting bolts. Loosen the bolts in a staggered pattern. The cover is spring-loaded and the threads in the servo valve body will strip if the cover is not evenly removed.

Now the servo valve body, servo separator plate, accumulator valve body, and regulator valve body can be removed. After these are removed, the stator shaft and shaft stop

Figure 7-23 The valve body assembly for a late-model Honda transaxle.

should be removed. Then unhook the detent spring from the detent arm and remove the detent arm shaft, detent arm, and control shaft.

Note the location of the cooler check ball and spring and remove them. Once these are removed, the main valve body can be unbolted and removed. While removing the valve body, be prepared to catch the TCC control valve and spring. These are held in position by the valve body.

Remove the pump gears, noting which side of the gears faces up. Now the main separator plate and its dowels can be removed. Clean and inspect all parts. Installation of the valve body assembly follows the reverse order of the removal procedure.

Toyota Transmissions

The valve body in most Toyota transmissions is comprised of an upper valve body, a lower valve body, a manual valve body, various covers, and separator plates. The individual hydraulic circuits extend to the housing of the transmission and may be connected by oil tubes.

To remove the valve body from an AA80E, the electrical connections to the various solenoids must be disconnected. The manual shaft detent assembly must be unbolted and removed, along with its spring and cover. The valve body can then be unbolted. Once the valve body is free, disconnect the shift lever from the manual valve. Then remove the transmission wire from the case. To do this, the retainer's bolt is removed and the harness pulled out (**Figure 7-24**).

In an U250E transaxle, the wiring harness must be disconnected and removed, along with the ATF temperature sensor. The oil strainer (filter) must be unbolted and removed with its O-ring. Then the valve body can be removed.

Retainer   Main harness

Figure 7-24 To remove the transmission wire in an AA80E transmission, unbolt the retainer and pull the harness out of the case.

VALVE BODY SERVICE

CUSTOMER CARE If while you are disassembling and inspecting the valve body you discover signs of poor transmission maintenance or abuse, make note of it. When you next talk to the customer, explain what you found and what the resulting problem was. Make sure you explain this in an understanding and inoffensive way.

SERVICE TIP Another trick is to use the cardboard sheet included in every gasket set. Fold this sheet in 1-inch pleats like an accordion and lay it on your bench with the slick side of the cardboard facing up. Then follow these steps:

1. Take the valves and springs out of the valve body and lay them in the different grooves in the sequence they were removed.
2. Clean the valve body castings.
3. Clean the valves and springs. Do not put them back on the cardboard; rather, put them directly into their bores in the valve body.

If previous tests suggest a problem with only one or two valves, start your inspection at those valves. Doing this will not only save you time but will also reduce the chance of something being misplaced or ruined during a total disassembly. If the transmission had heavily contaminated fluid, the entire valve body should be inspected and cleaned, or replaced.

Disassembly

A valve body contains many valves. These valves are typically held in their bores by a plug or cover plate. These must be removed to gain access to the valves. The cover plates are bolted or screwed to the valve body (**Figure 7-25**). Plugs can be held in place in a number of different ways. All of them can be released by pressing the plug slightly into the bore.

Begin disassembly by removing the manual shift valve from the valve body (**Figure 7-26**). Then, remove the pressure regulator retaining screws while keeping one hand around the spring retainer and adjusting screw bracket. Remove the pressure

Figure 7-25 Examples of the covers used to retain valves and their springs in the bores of a valve body.

Figure 7-26 Remove the manual shift valve. Often, there is a retaining pin that must be driven out to release the valve.

regulator valve. Then remove all of the valves and springs from the valve body. Make sure that you keep all springs and other parts with their associated valves.

It is important to keep track of the placement and position of the springs on each valve. Doing this will make reassembly easy and will ensure you are doing things correctly. One way of doing this is to draw a diagram indicating where the valve was and the position of the spring on the valve. Many technicians use a digital camera or the camera on a cell phone to capture the spring–valve relationship. It is also wise to photograph the placement of all check balls during disassembly.

Cleaning and Inspection

Remove the check balls and springs. Note their exact location and count them as they are removed. Compare your count to the number given in the service information. This will ensure that all have been removed and some will not be lost during cleaning. Never use a magnet to remove the check balls. The balls may become magnetized and tend to collect particles in the fluid. Be sure to do something that will help you remember where the balls were, such as photographing or drawing their location.

Valve Body, Plates, and Gaskets. Check the separator plate (spacer) for scratches or other damage (**Figure 7-27**). Scratches or score marks can cause oil to bypass correct oil passages and result in system malfunction. If the plate is defective in any way, it must be replaced. Check the oil passages in the upper and lower valve bodies for varnish deposits, scratches, or other damage that could restrict the movement of the valves. Make sure all fluid drain back openings are clear and have no varnish or dirt buildup. Check all of the threaded holes and related bolts and screws for damaged threads, and then replace as needed.

If the valve body has an oil strainer, make sure it is clean. This can be checked by pouring clean ATF through the strainer and observing the fluid flow out of the strainer. If there is poor flow, the strainer should be replaced.

With a straightedge laid across the sealing surface of the valve body, use a feeler gauge to check its flatness. If it is slightly warped, it can be flat filed. Be very careful when doing this. Keep the file flat and always file in one direction. If the surface is warped beyond repair, the valve body must be replaced. Warpage can produce a cross leak, which is a leak from one track to another. These can cause unwanted, partial engagement of a clutch or band. Proper torque of the valve body is critical for a successful repair. An over-torqued valve body can cause the valves to hang up in their bores.

Valve body separator plates are often called spacers.

In the figure labels:
1–2 shift valve governor plug
End cover
2–3 shift valve governor plug
End cover

Check surface
for scratches

Piston bores

Oil passages

Figure 7-27 Carefully check the bores, passages, and surface of the separator plate. It must be replaced if it is damaged.

> ⚙ **SERVICE TIP** Be sure to inspect the shift solenoid O-rings and the TCC solenoid O-rings for damage during disassembly. If they are damaged, attempt to identify the cause of the damage.

After the valves, springs, and electrical components have been removed, soak the valve body and separator and transfer plates in the parts cleaner, then wash the parts off with water. Thoroughly clean all parts and make sure all passages within the valve body are clear and free of debris. Carefully blow-dry each part with compressed air. Never wipe the parts with a rag or paper towel. Lint from either will collect in the valve body passages and cause shifting problems. Make sure to clean the worm tracks in the valve body and case.

⚡ **WARNING Wear safety glasses when drying parts with compressed air. The overspray of solvent and other cleaners can damage your eyes and lead to blindness.**

Valves. Continue disassembly by removing the valves and springs (**Figure 7-28**). The valves and springs can be retained in many different ways. Some have a cover plate that is bolted or screwed to the valve body. These may retain one or more valves. Others have a plug or sleeve retained by a pin, plate, or key at the outside of each bore. The plug can be removed by pushing it in slightly and removing the pin, plate, or key.

Examine each valve for nicks, burrs, and scratches. If the valves are coated with varnish, they must be soaked in an appropriate cleaner. If the valve lands have worn areas, this means the valve has been rubbing in its bore. The edges of the spools should be damage-free but should not be sharp. To identify a bent valve, install it into its bore and rotate it. If it binds or drags while it is being rotated, it is bent.

Make sure each valve properly fits into its respective bore. To do this, hold the valve body vertically and install an unlubricated valve into its bore. Let the valve fall of its own weight into the valve body until the valve stops (**Figure 7-29**). Then, place your finger over the valve bore and turn the valve body over. The valve should again drop by its own

Figure 7-28 Typical location of some of the valves and springs in a valve body.

Figure 7-29 Check the fit of the valve in its bore by dropping it into its bore.

weight. If the valve moves freely under these conditions, it will operate freely with fluid pressure. Repeat this test on all valves.

SERVICE TIP A good aluminum valve in an unworn valve body bore may not drop in its bore if it is wet, the fluid may cause the valve to stick in the bore. Always check these without a coating of fluid.

If steel valves do not move freely, the problem may be corrected by polishing the valve lands. If the valves are made of aluminum, then the valve body should be replaced. To polish a valve, use a polishing (Arkansas) stone or crocus cloth. It is helpful to let the cloth soak in ATF before using it to polish a valve. Evenly rotate the valve on the polishing material, make sure this does not round the edges of the valve. Polish the valves only enough to ensure free movement in the bore.

After the valve is polished, it must be thoroughly cleaned to remove all of the cleaning and abrasive materials. After the valve has been re-cleaned, it should be tested in its bore again.

> ⚙ SERVICE TIP Nearly all valve bodies are aluminum and therefore little polishing is required to remove the burrs or other defects.

Valve Bores. If a valve is smooth but sticks in its bore, carefully check the bore for dirt or nicks. Minor imperfections can be addressed; however, serious damage requires the replacement of the valve body.

To remove minor imperfections, do the following steps:

1. Soak a sheet of #600-grit abrasive paper in ATF for about 30 minutes.
2. Roll up half of the abrasive paper and insert it into the bore of the sticking valve (**Figure 7-30**).
3. Twist the roll of paper so it unrolls and expands to the size of the bore.
4. Then polish the bore by twisting it while moving it in and out of the bore.
5. Remove the paper.
6. Then clean the entire valve body in solvent and dry it with compressed air.

Place the valve into its bore and check its free movement. If the valve still cannot move freely in its bore, the valve body should be replaced. Individual valve body parts are usually not available.

> ⚙ SERVICE TIP Make sure to inspect and clean the manual valve. It is often easy to overlook this valve because it is often removed early in the disassembly process and it is not reinstalled until the valve body is installed in the transmission case.

Figure 7-30 Roll up some fine sandpaper and move it in the bore to remove any minor imperfections.

Valve to Bore Tolerances. One of the most common problems today involves excessive valve body bore wear due to the constant shuffling of the valves by a duty-cycle solenoid. This wear causes excessive valve to bore clearance. The correct clearance allows the valves to slide back and forth while preventing fluid leakage. The clearance also allows for maintaining a thin film of oil for the valve to ride on. The typical valve to bore clearance is 0.0005–0.0016 inch (0.0127–0.040 mm). If the bore is worn, or if there is excessive clearance, the entire valve body should be replaced.

Springs. Check each spring for signs of distortion. If any spring is damaged, the valve body should be replaced. A good way to check the spring is to lay them on their side and roll them. If they roll true, they are not distorted. If they wobble, they are distorted.

Reassembly

After the valve body has been cleaned, it should be allowed to air dry. Once dried, it should be dipped into a pan of clean ATF. While soaking the valve body in ATF, locate the installation specifications in the service information.

> ⚙ **SERVICE TIP** When assembling a valve body, always have your notes, photograph, and the correct service information close for quick reference. Many valves and springs look similar; however, each has its own bore and purpose.

Lubricate all parts with clean ATF. Then, install the valves and associated springs into their bores. It is important that you place the valve-retaining plugs or caps in the correct bore and in the correct direction (**Figure 7-31**). Once the cap is in position, carefully depress it with a small screwdriver and install the retaining clip (**Figure 7-32**).

Install all check balls and springs in their correct location. If you have any doubts as to where they should be placed, refer to the service information. Count the check balls as you install them to make sure you have inserted all of them.

Before beginning to install the valve body, check the new valve body gasket to make sure it is the correct one by comparing it to the old gasket. If the gasket appears to be the correct one, lay it over the separator plate and hold it up to a light, make sure no oil holes

Figure 7-31 The plugs used to retain the valves in a valve body have designated bores and must face the correct direction.

Figure 7-32 Once the cap is in position, carefully depress it with a small screwdriver and install the retaining clip.

Separator gasket

Separator

Figure 7-33 Always check a new valve body gasket by laying it over the separator plate and making sure it does not block any holes. If the gasket is correct, use clean ATF to hold it in place during assembly.

are blocked (**Figure 7-33**). Also, check to make sure the gasket seals off the worm tracks and will not allow the fluid to go where it should not go. Then install the bolts to hold valve body sections together and the valve body to the case. Tighten the bolts to the torque specifications to prevent valve body warpage and possible leak-over. Over-torqueing can also cause the bores to distort which would not allow the valves to move freely once the valve body is tightened to the transmission case.

Many transmissions use bolts of various lengths to secure the valve body to the case. It is important that the correct length bolt is used in each bore. It is so important that service information lists the size and location of the mounting bolts (**Figure 7-34**).

⚠ **Caution**

The primary cause of valve sticking is the over-tightening of the valve body bolts. Always be careful when handling the valve bodies because they are very precise components.

Each numbered bolt location corresponds to a specific bolt size and length, as indicated below:

1. M6 × 1.0 × 65.0
2. M6 × 1.0 × 54.4
3. M6 × 1.0 × 47.5
4. M6 × 1.0 × 35.0

5. M8 × 1.25 × 20.0
6. M6 × 1.0 × 12.0
7. M6 × 1.0 × 18.0

Figure 7-34 When installing the valve body mounting bolts, make sure you use the correct bolt size and length in the correct location. Refer to the service information for guidance.

CASE STUDY

A transmission technician was investigating a complaint from a customer who indicated that his vehicle had a very hard downshift at times when going around corners. The technician drove the vehicle with the customer since the complaint was intermittent. The technician let the customer drive, and the situation occurred. The technician made a note that the transmission did make a hard downshift under a load going uphill and slowing down to take a corner. The technician was able to determine that the downshift in question was from third to second gears. The technician was happy that he took the time to go with the customer to know exactly when and where the complaint occurred. The technician looked for technical bulletins when he arrived back at the shop and found that a problem in the valve body had been identified with a harsh 3–2 downshift. The bulletin called for a cleaning of the shift valve and replacement of the return spring, while checking for freedom of movement in the valve. The technician did not see any apparent problems, but performed the recommended procedure anyway. The technician was a bit wary that the transmission would be fixed when he test drove the vehicle, but to his relief the transmission performed flawlessly. The customer took the vehicle the same day. The customer also called the service manager the next day and expressed his gratitude to the technician for resolving his problem. It seemed that the customer had been to two other shops where the problem could not be duplicated.

ASE-STYLE REVIEW QUESTIONS

1. *Technician A* says electrical components from the valve body must not be soaked in parts cleaner.
   Technician B says a lint-free rag is a must when wiping down valves.
   Who is correct?
   A. A only
   B. B only
   C. Both A and B
   D. Neither A nor B

2. *Technician A* says a scratched separator plate must be replaced.
   Technician B says make sure all drain back openings in the spacer plate are clear.
   Who is correct?
   A. A only
   B. B only
   C. Both A and B
   D. Neither A nor B

3. *Technician A* says over-torqeuing the hold-down bolts of the valve body can cause the valves to stick in their bores.
   Technician B says flat-filing the surface of the valve body will allow the valve body to seal properly and prevent cross leaks.
   Who is correct?
   A. A only
   B. B only
   C. Both A and B
   D. Neither A nor B

4. While assembling a valve body after cleaning it:
   Technician A lubricates all parts with clean ATF.
   Technician B says it is important that the valve-retaining plugs or caps be placed in the correct bore and in the correct direction.
   Who is correct?
   A. A only
   B. B only
   C. Both A and B
   D. Neither A nor B

5. *Technician A* says never use a magnet to remove the check balls.
   Technician B says be sure to do something that will help you remember where the balls were, such as photographing or drawing their location.
   Who is correct?
   A. A only
   B. B only
   C. Both A and B
   D. Neither A nor B

6. When inspecting valves from the valve body:
   Technician A says the edges of the spools should be damage-free but should not be sharp.
   Technician B says to identify a bent valve, install it into its bore and rotate it. If it binds or drags while it is being rotated, it is bent.
   Who is correct?
   A. A only
   B. B only
   C. Both A and B
   D. Neither A nor B

7. While inspecting a valve body:

   Technician A says deep scratches in a valve's bore can be removed with #600-grit abrasive paper.

   Technician B says one of the most common transmission problems is excessive valve body bore wear due to the constant shuffling of the valves by a duty-cycle solenoid.

   Who is correct?

   A. A only   C. Both A and B

   B. B only   D. Neither A nor B

8. *Technician A* says that if diagnosis suggests a problem with only one or two valves, visual inspection of the valve body should start at those valves.

   Technician B says that if the transmission has heavily contaminated fluid, the entire valve body should be inspected and cleaned, or replaced.

   Who is correct?

   A. A only   C. Both A and B

   B. B only   D. Neither A nor B

9. While discussing some control modules:

   Technician A says some transmissions have a single assembly that contains all of the solenoids, the TCM, and a lead frame assembly.

   Technician B says the lead frame connects to the individual components, therefore fewer electrical connections are needed for these transmissions, eliminating many potential problems.

   Who is correct?

   A. A only   C. Both A and B

   B. B only   D. Neither A nor B

10. *Technician A* says that if a valve cannot be cleaned well enough to move freely in its bore, the valve body should be replaced.

    Technician B says that if there is even the slightest bit of damage or varnish buildup in a valve's bore, the entire valve body must be replaced.

    Who is correct?

    A. A only   C. Both A and B

    B. B only   D. Neither A nor B

ASE CHALLENGE QUESTIONS

1. While discussing removal of the valve body on an AA80E Toyota transmission:

   Technician A says the electrical connections to the various solenoids must be disconnected first.

   Technician B says the next step is to disconnect the shift lever from the manual valve.

   Who is correct?

   A. A only   C. Both A and B

   B. B only   D. Neither A nor B

2. Which of the following is not a common cause for sticking valves and sluggish valve movements?

   A. Over-torqeuing valve body bolts.

   B. A faulty pressure control solenoid.

   C. The use of the wrong type of fluid.

   D. Overheating of the transmission.

3. *Technician A* says the valve body of all transmissions and transaxles can be serviced with the transmission still in the vehicle, providing that surrounding components are removed first.

   Technician B says the Ford 4R70E transmission valve body has two alignment bolts for proper alignment of the valve body.

   Who is correct?

   A. A only   C. Both A and B

   B. B only   D. Neither A nor B

4. *Technician A* says oil feed pipes are used to move the fluid to the various clutches and apply devices in a transmission.

   Technician B says shift and TCC solenoids are often mounted directly to the valve body.

   Who is correct?

   A. A only   C. Both A and B

   B. B only   D. Neither A nor B

5. *Technician A* says valve body springs are distorted if they roll straight on the workbench when removed from the valve body instead of wobbling.

   Technician B says problems rarely result from excessive clearance between the valve and its bore.

   Who is correct?

   A. A only   C. Both A and B

   B. B only   D. Neither A nor B

Name _____   **Date** _____

VALVE BODY SERVICE

Upon completion of this job sheet, you should be able to disassemble, clean, inspect, and reassemble a valve body.

ASE Education Foundation Correlation

This job sheet addresses the following **MAST** task:

C.7   Inspect, measure, clean, and replace valve body (includes surfaces, bores, springs, valves, sleeves, retainers, brackets, check valves/balls, screens, spacers, and gaskets). **(P-2)**

We Support

ASE | Education Foundation

Tools and Materials

- Automatic transmission or transaxle on a bench
- Compressed air and air nozzle
- Inch-pound torque wrench
- Measuring calipers
- Supply of clean solvent
- Lint-free shop towels
- Service Information

Describe the Transmission Being Worked On

Model and type of transmission _____

Year _____ Make _____ VIN _____

Model _____

Procedure

Task Completed

1. Remove the valve body attaching screws from the transmission. Start at the outside bolts and work toward the center if the service information does not give specific instructions.   ☐

2. Remove the valve body assembly and place it in a container of clean solvent.   ☐

3. Remove the end plates and covers from the assembly.   ☐

4. In the space below, draw a simplified view of the valve body. Note the location of all the valves, check balls, and springs.   ☐

5. Begin to disassemble the unit. Lay all the parts on a clean surface in the order in which they were disassembled. This will help during the reassembly of the valve body. Be careful with the check balls; they may not be all the same size. Some transmissions use slightly larger or smaller check balls in circuits. How many check balls are in the valve body? Are they all the same size?

   _____

   _____

   _____

6. Remove all the gaskets from the assembly. Place them aside. Do not throw them away. They will be needed for comparison when installing new gaskets.   ☐

7. Thoroughly clean the main body and plates of the assembly. Do not dry the valve body with anything that would leave lint. Allow the unit to air dry or dry it off with compressed air.   ☐

Task Completed

8. Inspect the valve body for damage and cracks. Also check the flatness of the body and the plates. Describe your findings.

_____

_____

_____

_____

9. What are the valves made of? _____

10. Check all the valves for free movement in their bores. Also check the valves for wear, scoring, and signs of sticking. Describe your findings and recommendations.

_____

_____

_____

_____

11. Correctly reinstall the valves and springs in their bores.   ☐

12. Replace the end plates or covers. Install the retaining screws by hand, and then tighten   ☐
    them to the specific torque. The specified torque is _____.

13. Install the check balls into their proper location.   ☐

14. Compare the new gaskets with the old and install the correct new ones on the valve   ☐
    body.

15. Align the spring-loaded check balls and all other parts as needed.   ☐

16. Place the gasket on top of the transfer plate. Align the gasket holes with the transfer   ☐
    plate and the bores in the valve body.

17. Position the valve body on the transmission. Align the parking and other internal   ☐
    linkages. Then install the retaining screws by hand.

18. Tighten the screws to the specified torque. Tighten the screws in the center first, then
    work to the outside of the valve body. The specified torque is _____.

19. Summarize this job sheet.

_____

_____

_____

Problems Encountered

_____

_____

_____

Instructor's Comments

_____

_____

_____

CHAPTER 8
GEAR AND SHAFT SERVICE

Upon completion and review of this chapter, you should be able to:

- Inspect, measure, and replace thrust washers and bearings.
- Inspect and replace bushings.
- Inspect and measure a planetary gear assembly and replace parts as necessary.
- Inspect and replace shafts.
- Remove and install the various types of shaft seals.

- Inspect, repair, or replace transaxle drive chains, sprockets, gears, bearings, and bushings.
- Inspect and replace parking pawl, shaft, spring, and retainer.
- Inspect, measure, repair, adjust, or replace transaxle final drive components.
- Inspect and service the transmission housing and cases.

 Basic Tools

Basic mechanic's tool set

Clean lint-free rags

Appropriate service information

Clean ATF

Terms To Know

Etching
Selective thrust washers
Thrust load

INTRODUCTION

The overall reliability of a transmission depends on the structural integrity of the shafts and gears. Therefore these parts, as well as all thrust washers, bearings, and bushings (**Figure 8-1**), should be carefully inspected and replaced if damaged or worn.

THRUST WASHERS, BUSHINGS, AND BEARINGS

A thrust washer is designed to support a **thrust load** and keep parts from rubbing together, preventing premature wear on parts such as planetary gearsets (**Figure 8-2**) and transfer and final drive assemblies. **Selective thrust washers** come in various thicknesses to take up clearances and adjust end play at many different locations. Some transmissions have a selective thrust washer between the front and the rear multiple-disc assemblies (**Figure 8-3**). Thrust washers and thrust bearings are used wherever rotating parts must have their end play maintained. To control the end play of nonrotating parts, selective shims or spacers are used.

Thrust load is a load placed in parallel with the center of an axis.

Thrust washers are often referred to as thrust plates.

Figure 8-1 Location of thrust washers and bearings in a typical transaxle.

Many transmissions are fitted with thrust bearings, often called Torrington bearings (**Figure 8-4**). These thrust washers have small roller or needle bearings placed into them. The bearings reduce the rotational friction between the two surfaces they separate. Thrust bearings should be carefully inspected for distortion, cracks, and wear. If there is any evidence of damage, the bearing should be replaced.

Most flat thrust washers and bearings are not selective and have a fixed thickness. Typically, the selective types are numbered or colored by the manufacturer for easy

Planetary
gearset

Thrust
washer

Figure 8-2 The purpose of a thrust washer is to
support a thrust load and keep parts from rubbing
together, such as planetary gearsets.

Selective
thrust washer

Sun gear
driving shell

Output
shell

Figure 8-3 A selective thrust washer between the front
and rear multiple-friction disc assemblies.

Figure 8-4 Thrust bearings are often used in transfer shaft or countershaft assemblies.

Color Code	Depth	Thickness
Green	1.485–1.503 in. (37.706–38.184 mm)	0.050–0.054 in. (1.270–1.372 mm)
Yellow	1.504–1.521 in. (38.185–38.641 mm)	0.068–0.072 in. (1.727–1.829 mm)
Natural	1.522–1.538 in. (38.642–39.073 mm)	0.085–0.089 in. (2.159–2.261 mm)
Red	1.539–1.555 in. (39.074–39.505 mm)	0.102–0.106 in. (2.591–2.692 mm)
Blue	1.556–1.581 in. (39.506–40.165 mm)	0.119–0.123 in. (3.023–3.124 mm)

Figure 8-5 An example of the color code used to identify selective thrust washers.

identification (**Figure 8-5**). Thrust washers should be inspected for scoring, flaking, and
wear. Flat thrust washers should also be checked for broken or weak tabs (**Figure 8-6**).
These tabs are critical for holding the washer in place. On metal thrust washers, the tabs
may appear cracked at the bend; however, this is a normal appearance due to the characteristics of the materials used to manufacture them.

The use of the name
selective thrust washer
means the thrust
washer for this
application is available
in different thicknesses.
The correct thrust
washer must be
selected to provide the
correct end play or
clearance.

Figure 8-6 A thrust washer with tabs. These tabs must be carefully inspected for cracks or other damage.

Plastic thrust washers are typically color-coded to denote their size. These washers will not show wear unless they are damaged. The only way to check their wear is to measure the thickness with a micrometer and compare them to a new part. All damaged and worn thrust washers and bearings should be replaced.

Proper thrust washer thicknesses are important to the operation of an automatic transmission. After following the recommended procedures for checking the end play of various components, refer to the manufacturer's chart for the proper thrust plate thickness for each application.

Use transmission assembly lubricant to hold thrust washers in place during assembly. This will keep them from falling out of place, which will affect end play.

Bearings

All bearings should be checked for roughness before and after cleaning. Carefully examine the inner and outer races, and the rollers, needles (**Figure 8-7**), or balls for cracks, pitting, **etching**, or signs of overheating.

Sprag and roller clutches should be inspected in the same way as bearings. Check their operation by attempting to rotate them in both directions (**Figure 8-8**). If working properly, they will allow rotation in one direction only. If the unit was assembled with the sprags or rollers installed backward, the unit will spin freely in both directions.

The smooth surfaces of the race should be very smooth and without defects. The roller clutch springs should be checked for signs of overheating and wear. Sprag units should be checked for loose sprags.

Bushings/Sleeves

Bushings and Sleeves should be inspected for pitting and scoring. Always check the depth that bushings are installed to and the direction of the oil groove, if so equipped, before

---

Thrust washers are not included in an overhaul kit because they are selective fit. Even if the existing thrust washers are of correct size, they should be replaced with new ones if they are pitted or scored.

⚠ Caution

Under no circumstances should you use white lube or chassis lube. These greases will not mix in with the fluid and can plug up orifices, passages, and even hold check balls off their seats.

Etching is a discoloration or removal of some material caused by corrosion or some other chemical reaction.

Overdrive carrier pinion

Needle bearing

Figure 8-7 Needle bearings are often located inside the small pinion gears. Coating the inside of the gear with TransJel before the needle bearings are positioned will help to keep them in place.

Lock   Free

Hold

Figure 8-8 To check the action of a one-way clutch, hold the inner race and rotate the clutch in both directions. The clutch should rotate smoothly in one direction and lock in the other.

you remove them. Many bushings that are used in the planetary gearing and output shaft areas have oiling holes in them. Be sure to line these up correctly during installation or you may block off oil delivery and destroy the geartrain. If any damage is evident on the bushing, it should be replaced.

Bushing wear can be checked visually as well as by observing the lateral movement of the shaft that fits into the bushing. Any noticeable lateral movement indicates wear and the bushing should be replaced. The amount of clearance between the shaft and the bushing can be checked with a wire-type feeler gauge. Insert the wire between the shaft and the bushing; if the gap is greater than the maximum allowable, the bushing should be replaced. Normally, bushings must fit the shafts they ride on with about a 0.0015- to 0.003-inch clearance. You can check this fit by measuring the inside diameter of the bushing and the outside diameter of the shaft with a dial caliper or micrometer (**Figure 8-9**). This is a critical fit throughout the transmission and especially at the converter drive hub, where 0.004 inch is the desired fit. Excessive clearance will allow TIC fluid to drain back.

Most bushings or sleeves are press-fit into a bore or on a shaft. To remove one, drive it out of the bore with a properly sized bushing tool. Some bushings can be removed with a slide hammer fitted with an expanding or threaded fixture that grips the bushing. Another way to remove bushings is to carefully cut one side of the bushing and collapse it. Once collapsed, the bushing can be easily removed with a pair of pliers. Small-bore bushings that are located in areas where it is difficult to use a bushing tool can be removed by tapping the inside bore of the bushing with threads that match a selected bolt that fits into the bushing. After the bushing has been tapped, insert the bolt and use a slide hammer to pull the bolt and bushing out of its bore.

Whenever possible, all new bushings should be installed with the proper bushing driver. The use of these tools prevents damage to the bushing and allows for proper seating of the bushing into its bore (**Figure 8-10**).

Special Tools

Wire-type feeler gauge set
Micrometer
Bushing pullers and drivers
Hydraulic press

Figure 8-9 The inside diameter of bushings should be measured for wear with a telescoping gauge and a micrometer.

Figure 8-10 Typical bushing/sleeve removal and installation tools.

PLANETARY GEAR ASSEMBLIES

The purpose of inspection at this point is to try to eliminate the possibility of putting noise into a newly rebuilt unit. This would result in a costly setback, so close inspection of the planetary gearset is a must. All planetary gear teeth should be inspected for chips or stripped teeth. Any gear that is mounted to a splined shaft needs to be checked for mutilated or shifted splines. Helical gears have many advantages over straight-cut gears, such as providing low operating noise, but you must check the end play of the individual gears during your inspection. The helical cut of the gears makes them thrust to one side during operation. This can put a lot of load on the thrust washers and may wear them beyond specification. Checking these was discussed earlier, but particular attention should be given to the planetary carriers.

Look first for obvious problems like blackened gears or pinion shafts. These conditions indicate severe overloading and require that the carrier be replaced. Occasionally the pinion gear and shaft assembly can be replaced individually. When looking at the gears themselves, a bluish condition can be a normal condition, as this is part of a heat-treating process used during manufacture. Check the planetary pinion gears for loose bearings. Check each gear individually by rolling it on its shaft to feel for roughness or binding of the needle bearings. Wiggle the gear to be sure it is not loose on the shaft. Looseness will cause the gear to whine when it is loaded. Also, inspect the gear teeth for chips or imperfections, as these will also cause whine.

Check the gear teeth around the inside of the front planetary ring gear. Check the fit between the front planetary carrier to the output shaft splines. Remove the snap ring and thrust washer from the front planetary ring gear (**Figure 8-11**). Examine the thrust washer and the outer splines of the front drum for burrs and distortion. The rear clutch friction

Classroom Manual
Chapter 8, page 275

Special Tools
Feeler gauge set
Micrometer

Annulus gear
and support

Front planetary
gear assembly

#5 thrust
washer

Figure 8-11 The planetary assembly, its thrust washer, and annulus gear should be inspected for signs of abnormal wear.

Figure 8-12 The clearance between the pinion gears and the planetary carrier should be checked and compared to specifications.

discs must be able to slide on these splines during engagement and disengagement. With the snap ring removed, the front planetary carrier can be removed from the ring gear. Check the planetary carrier gears for end play by placing a feeler gauge between the planetary carrier and the planetary pinion gear (**Figure 8-12**). Compare the end play to specifications. On some Ravigneaux units, the clearance at both ends of the long pinion gears must also be checked and compared to specifications.

Check the splines of the sun gear. Sun gears should have their inner bushings inspected for looseness on their respective shafts. Also check the fit of the sun shell to the sun gear (**Figure 8-13**). The shell can crack where the gear mates with the shell. The sun shell should also be checked for a bell-mouthed condition where it is tabbed to the clutch drum. Any variation from a true round should be considered junk and should not be used. Look at the tabs and check for the best fit into the clutch drum slots. This involves trial fitting the shell and drum at all the possible combinations and marking the point where they fit the tightest. A snug fit here will eliminate bell mouthing due to excess play at the tabs. It can also reduce engagement noise in reverse, second, and fourth gears. This excess play allows the sun shell tabs to strike the clutch drum tabs as the transmission shifts from first to second or when the transmission is shifted into reverse.

The gear carrier should have no cracks or other defects. Replace any abnormal or worn parts. Check the thrust bearings for excessive wear, and if required, correct the

Input drum

Snap ring
(input drum to
rear sun gear)

Rear
sun gear

Thrust washer

Figure 8-13 The fit of all drums onto the splines of their mating shafts should be checked.

Figure 8-14 Measure the inside diameter of all bushings to determine wear.

input shaft thrust clearance by using a washer with the correct thickness. Determine the correct thickness by measuring the thickness of the existing thrust washer and comparing it to the measured end play. Now move the gear back and forth to check its end play. Some shop manuals will give a range for this check, but if none can be found you can figure about 0.007 to 0.025 inch as an average amount. All the pinions should have about the same end play.

Also inspect all bushings that may be inserted into the planetary gearsets. These are commonly found in sun gears (**Figure 8-14**). Measure the inside diameter and compare that dimension to specifications. If the diameter exceeds those specifications, replace the gear assembly.

SHAFTS

Carefully examine the area on all shafts that rides in a bushing, bearing, or seal. Check the entire length of the shaft for signs of overheating and other damage. If the shaft has gears, the teeth of the gears should be inspected for damage and breaks. Also inspect the splines for wear, cracks, or other damage (**Figure 8-15**). A quick way to determine spline wear is to fit the mating splines and check for lateral movement.

Classroom Manual
Chapter 8, page 257

Ring grooves

Splines

Bearing surface

Figure 8-15 All shafts, including their splines and ring grooves, should be carefully inspected for wear or other damage.

Shafts are checked for scoring in the areas where they ride in bushings. As the shaft is a much harder material than the bushings, any scoring on the shaft would indicate a lack of lubrication at that point. The affected bushing should appear worn into the backing metal. Because shaft-to-bearing fit is critical to correct oil travel throughout the transmission, a scored shaft should be replaced. Lubricating oil is carried through most shafts; therefore, an internal inspection for debris is necessary. A blocked oil delivery hole can starve a bushing, resulting in a scored shaft. The internal oil passage of a shaft may not be able to be visually inspected and only observation during cleaning will give an indication of the openness of the passage. Washing the shaft passage out with a solvent and possibly running a piece of small-diameter wire through the passage will dislodge most particles. Be sure to check that the ball that closes off the end of the shaft, if the shaft is so equipped, is securely in place. A missing ball could be the cause of burned planetary gears and scored shafts due to a loss of oil pressure. Any shaft that has an internal bushing should be inspected, as described earlier. Replace all defective parts as necessary.

Shafts should be checked for wear in the ring groove area. This is especially critical if the groove accommodates a metal ring. Make sure there is no step wear in the groove and that the sides and bottom are square. Also make sure the groove is not too wide for the ring. If a 0.005-inch feeler gauge would fit in the groove with the ring in place, the groove is worn and the shaft should be replaced.

Input and output shafts can be solid, drilled, or tubular. The solid and drilled shafts are supported by bushings, so the bushing journals of the shafts should be free of noticeable wear at these points. Small scratches can be removed with 320-grit emery paper. Grooved or scored shafts require replacement. The splines should not show any sign of waviness along their length. Check drilled shafts to be sure the drilled portion is open and free of any foreign material. Wash out the shaft with solvent and run a small-diameter wire through the shaft to dislodge any particles. After running the wire through the opening, wash out the shaft once more and blow it out with compressed air.

If the shaft has a check ball in the turbine shaft (**Figure 8-16**), be certain the ball seats in the correct direction. This particular check ball controls oil flow direction to the converter. Some shafts have a ball pressed into one end to block off one end of the shaft. This is used to hold oil in the shaft so the oil is diverted through holes in the side of the shaft. These holes supply oil to bushings, one-way clutches, and planetary gears. If the ball does not fully block the end of the shaft, oil pressure can be lost, causing failure of these components. Some shafts may be used to support another shaft (**Figure 8-17**), as in the GM 4L30. The output shaft uses the rear of the input shaft to center and support itself. The small bushing found in the front end of the output shaft should always be replaced on this transmission during rebuild. If the input shaft pilot is worn or scored, a replacement shaft will be necessary.

All hubs, drums, and shells should be carefully examined for wear and damage. Especially look for nicked or scored band application surfaces on drums, worn or damaged lug grooves in clutch drums, worn splines, and burned or scored thrust surfaces. Minor scoring or burrs on band application surfaces can be removed by lightly polishing the surface with a 600-grit crocus cloth. Any part that is heavily scored or scratched should be replaced.

> **Special Tools**
> Feeler gauge set
> 320-grit emery paper
> Cleaning solvent
> OSHA-approved air nozzle

Figure 8-16 If the turbine shaft is fitted with a check ball, make sure it is able to seat and unseat.

Figure 8-17 Some transmission shafts support another shaft through the bushings fitted to the inside diameter of one shaft. These bushings should be carefully inspected.

PARKING PAWL

CUSTOMER CARE Always remind your customers that they should not rely totally on the park gear selector position when parking the vehicle. The parking brake should be set in addition to fully placing the selector into Park. This is especially of concern when the engine is running. Many accidents have happened because the transmission slipped out of Park while the engine was running.

The parking pawl assembly (**Figure 8-18**) can be inspected after the transmission is disassembled or, on some transmissions, while the transmission is still in the vehicle. Examine the engagement lug on the pawl; make sure it is not rounded off. If the lug is worn, it may allow the pawl to slip out or not fully engage in the parking gear. Most parking pawls pivot on a pin; this also needs to be checked to make sure there is no excessive looseness at this point. The spring that pulls the pawl away from the parking gear must also be checked to make sure it can hold the pawl firmly in place. Also check the position and seating of the spring to make sure it will remain in that position during operation.

The pushrod or operating shaft (**Figure 8-19**) must provide the correct amount of travel to engage the pawl to the gear. Make sure the shaft is not bent or that the pivot hole in the internal shift linkage is not worn oblong. Also, make sure the bushing or sleeve that supports the manual shaft is in good condition.

Any components found unsuitable should be replaced. It should be noted that the components that make up the parking lock system are the only parts holding the vehicle in place when parked. If they do not function correctly, the car may roll or even drop into reverse when the engine is running, causing an accident or injury. Replace any questionable or damaged parts.

⚡ WARNING **A careful inspection of the parking pawl assembly is essential to avoid possible injury, death, and/or lawsuits.**

Before installing the rear extension housing, assemble the parking pin, washer, spring, and pawl. Be sure they are assembled properly. Then install the housing and tighten the bolts to specifications.

Figure 8-18 Each part of the parking pawl assembly should be carefully inspected.

Figure 8-19 The pushrod or operating shaft for the parking pawl must provide the correct amount of travel to engage the pawl to the gear. Make sure that the shaft is not bent and that the pivot hole in the internal shift linkage is not worn oblong.

DRIVE CHAINS

Classroom Manual
Chapter 8, page 272

The drive chains used in some transaxles should be inspected for side play and stretch. These checks are made during disassembly and should be repeated as a double-check during reassembly. Chain deflection is measured between the centers of the two sprockets. Typically very little deflection is allowed.

Deflect the chain inward on one side until it is tight (**Figure 8-20**). Mark the housing at the point of maximum deflection. Then deflect the chain outward on the same side until it is tight (**Figure 8-21**). Again mark the housing in line with the outside edge of the chain at the point of maximum deflection. Measure the distance between the two marks. If this distance exceeds specifications, replace the drive chain.

 Special Tools

Paint stick
Machinist's rule
Bushing pullers and
  drivers
Hydraulic press

Check each link of the chain by pushing and pulling the links away from the pin that holds them together. All of the links should move very little and each move the same amount. It is important to realize that a chain is only as strong as its weakest link. Check each link carefully.

Be sure to check for an identification mark on the chain during disassembly. These can be painted or dark-colored links, and may indicate either the top or the bottom of the chain, so be sure you remember which side was up.

Chain deflection is commonly referred to as chain slack.

The sprockets should be inspected for tooth wear and for wear at the point where they ride. If the chain is found to be too slack, it may have worn the sprockets in the same manner that engine timing gears wear when the timing chain stretches. A slightly polished appearance on the face of the gears is normal.

Bearings and Bushings

The bearings and bushings used on the sprockets need to be checked for damage. The radial needle thrust bearings must be checked for any deterioration of the needles and cage. The running surface in the sprocket must also be checked, as the needles may pound into the gear's surface during abusive operation. The bushings should be checked for any signs of scoring, flaking, or wear. Replace any defective parts.

Typically, the bearings and bushings are removed with a puller (**Figure 8-22**) and installed with a driver and a press (**Figure 8-23**).

The removal and installation of the chain drive assembly of some transaxles requires that the sprockets be spread slightly apart (**Figure 8-24**). The key to doing this is to spread the sprockets just the right amount. If they are spread too far, they will not be easy to install or remove.

Figure 8-20 While measuring the slack of the drive chain, outwardly deflect the chain, and make a mark to that point of deflection.

Figure 8-21 Continue measuring the slack of the drive chain by deflecting the chain inwardly. Make the point of deflection. The distance between the outward and inward marks is the amount of chain slack.

Figure 8-22 A special tool used with a slide hammer to remove bearings and bushings in a chain drive assembly.

Figure 8-23 The bearings and bushings in a chain drive assembly should be installed with a driver and a hydraulic press.

Figure 8-24 To remove and install some drive chains, they must be slightly spread apart with a special tool.

🌼 **SERVICE TIP** According to the major supplier of drive chains, it is impossible to accurately measure chain stretch without expensive gauges. Therefore, they suggest that drive chains be replaced during an overhaul of a transaxle or transfer case.

The shafts and gears of the assembly have numerous seals and thrust washers (**Figure 8-25**). The seals must be replaced whenever the unit is disassembled.

Figure 8-25 The shafts and gears of a chain drive assembly have numerous seals that must be replaced whenever the unit is disassembled.

TRANSFER GEARS

Classroom Manual
Chapter 8, page 271

Some transaxles use gears, instead of a drive chain, to move or transfer the output of the transmission to the final drive unit (**Figure 8-26**). The shafts and gears in these transaxles must be carefully inspected and replaced if they are damaged.

To remove and install the transfer shaft gear, a holding tool must be used to stop the transfer gear from turning while loosening or tightening the retaining nut (**Figure 8-27**). The nut is typically tightened to 200 lbs.-ft. To remove the transfer shaft from the transaxle case, the retaining snap ring must be removed (**Figure 8-28**). Then the shaft can be pulled out with its bearing (**Figure 8-29**). The shaft's bearing must be pressed on and off the shaft.

A puller must be used to remove the transfer gear from the output shaft (**Figure 8-30**) after its retaining bolt has been removed.

Behind the transfer shaft gear is a selective shim used to provide correct meshing of the teeth of the transfer shaft gear and the output shaft transfer gear (**Figure 8-31**) and to control transfer shaft end play. End play is measured with a dial indicator. To ensure good contact with the indicator's plunger and the end of the transfer shaft, Chrysler

Figure 8-26 The transfer shafts and gears must be carefully inspected and replaced if they are damaged.

Figure 8-27 To remove and install the transfer shaft gear, a holding tool must be used to stop the transfer gear from turning while loosening or tightening the retaining nut.

Figure 8-28 To remove the transfer shaft from the transaxle case, the retaining snap ring must first be removed.

Figure 8-29 The transfer shaft can be pulled out with its bearing with a suitable puller tool.

Figure 8-30 A puller must be used to remove the transfer gear from the output shaft, after its retaining bolt has been removed.

Figure 8-31 Behind the transfer shaft gear is a selective shim used to provide correct meshing of the teeth of the transfer shaft gear and the output shaft transfer gear and to control transfer shaft end play.

recommends that a steel check ball be placed between the plunger tip and the shaft (**Figure 8-32**). To help hold the ball in place during the check, coat the ball in heavy grease. If the end play is not within specifications, select a thrust washer with the correct thickness and install it behind the transfer gear.

The output shaft should also be checked for end play. If the end play is incorrect, a different size shim should be installed behind the output shaft transfer gear (**Figure 8-33**). Since the gear is splined to the shaft, a thrust bearing is not needed here. After end play has been corrected and after the output shaft gear has been reinstalled, the turning torque of the shaft should be measured. If the turning torque is too high, a slightly thicker shim should be installed. If the turning torque is too low, a slightly thinner shim should be installed. The bearings for the transfer and output shafts are pressed on and off the shafts.

It is very important that the transfer gears be tightened to specifications. The torque setting not only ensures that the gears will remain on the shafts, but it also maintains the correct bearing adjustments.

Special Tools

Bushing pullers and drivers

Hydraulic press

Dial indicator

Figure 8-32 Chrysler recommends that a steel check ball coated in heavy grease be placed between the plunger tip and the shaft to ensure good contact with the indicator's plunger and the end of the transfer shaft.

Figure 8-33 A selective shim is installed behind the output shaft transfer gear to control output shaft end play.

FINAL DRIVE UNITS

Classroom Manual
Chapter 8, page 270

Transaxle final drive units should be carefully inspected. Examine each gear, thrust washer, and shaft for signs of damage. If the gears are chipped or broken, they should be replaced. Also inspect the gears for signs of overheating or scoring on the bearing surface of the gears.

Carefully check the differential pinion shaft and gear for wear. Also check to make sure the gear is not seized to the shaft. Seized pinion gears are a common cause of shaft failure and can result in the total destruction of the final drive unit and transaxle case.

Helical Gear Drives

Special Tools

Inch-pound
torque wrench

Helical-type final drive units should be checked for worn or chipped teeth, overloaded tapered roller bearings, and excessive differential side gear wear. Excessive play in the differential is a cause of engagement clunk (**Figure 8-34**). Be sure to measure the clearance between the side gears and the differential case and to check the fit of the gears on the gear shaft. Proper clearances can be found in the appropriate shop information.

Carefully check the condition of the bearings. Replace them if there are signs of wear or damage. Normally the bearings are pressed on and off the case using special tools (**Figure 8-35**). The side bearings of some final drive units are preloaded with shims (**Figure 8-36**). Select the correct size shim to bring the unit into specifications. With a torque wrench, measure the amount of rotating torque. Compare your readings against specifications (**Figure 8-37**).

If the bearing preload and end play is fine, as is the condition of the bearings, the parts can be reused. However, always install new seals during assembly. New bearings require that preload be set to the specifications for a new bearing. Used bearings should be set to the amount found during teardown or about one-half the preload of a new bearing.

Planetary Gear Assemblies

Planetary-type final drives (**Figure 8-38**) are also checked for the same differential case problems that the helical type would encounter. All planetary gear teeth should be inspected for chips and damage. Pay particular attention to the planetary carriers. Look obvious problems like blackened gears or pinion shafts. These conditions indicate severe overloading and require that the carrier be replaced. When looking at the gears, a bluish condition can be a normal condition as this could be from the heat-treating process used during

Figure 8-34 The backlash of the side gears in a final drive unit should be checked prior to assembling the transaxle.

Figure 8-35 The side bearings are normally pressed on and off the unit.

Figure 8-36 The side bearings of some final drive units are preloaded with selective shims.

Figure 8-37 Using an inch-pound torque wrench, the turning torque of the transaxle assembly should be checked after assembly.

manufacture. Check the planetary pinion gears for loose bearings. Check each gear by rolling it on its shaft to feel for roughness or binding of the needle bearings. Wiggle the gear to be sure it is not loose on the shaft, which can cause the gear to whine when it is loaded. Also, inspect the gears' teeth for chips or imperfections, as these will also cause whine.

Check the gear teeth around the inside of the planetary ring gear. Check the carrier assembly for cracks or other defects. Check the end play of the pinion gears by placing a feeler gauge between the planetary carrier and each pinion gear (**Figure 8-39**). Compare the end play to specifications. If the end play is too low, the pinion assembly must be removed and the correct thrust washer installed. If the end play is too much, the differential assembly must be replaced.

Any problems normally result in the replacement of the carrier as a unit as most pinion bearings and shafts are not sold as separate parts. **Photo Sequence 13** covers a typical procedure for servicing a planetary gear-type final drive unit.

Figure 8-38 Planetary-type final drives should be checked in the same basic way as helical-type final drives.

Figure 8-39 The end play of the planetary pinion gears should be checked with a feeler gauge.

PHOTO SEQUENCE 13
Servicing Planetary Gear-Type Final Drive Units

P13-1 Place the final drive unit into the transaxle's oil pan. Doing this will lessen the chances of losing bearings while disassembling the unit. Make sure the pan is clean.

P13-2 With a pin punch and hammer, remove the differential pinion shaft retaining ring.

P13-3 Remove the retaining ring from the end of the output shaft. This ring must not be reused!

P13-4 Pull the output shaft from the differential carrier.

P13-5 Remove the final drive sun gear.

P13-6 Remove the differential pinion shaft retaining pin, the differential pinion shaft, pinion gears, and thrust washers and place them aside. Keep these in the order and orientation in which they were assembled.

P13-7 Remove the differential side gears and thrust washers.

P13-8 Using a screwdriver, carefully remove the final drive carrier retaining ring.

P13-9 Remove the planet pinion pins, then the planet pinion gears, thrust washers, needle bearings, and the bearing spacers.

PHOTO SEQUENCE 13 (CONTINUED)

P13-10 Inspect the needle bearings, thrust washers, pinion gears, and planet pinion pins.

P13-11 Remove the final drive sun gear to carrier thrust bearing.

P13-12 To begin reassembly, install the pinion gear needle bearing spacer, thrust washer, and pinion gear onto the planet pinion gear pin.

P13-13 Install the needle bearings, one at a time, into the top and bottom of the planet pinion gear. Use transmission assembly lube to help hold the bearings in position.

P13-14 Coat the sun gear to final drive carrier thrust washer with transmission assembly lube. Then install the thrust bearing onto the final drive carrier.

P13-15 Separate the pinion pin from the gear. Assemble the planet pinion gear thrust washers and gears into the final drive carrier. Make sure the gears face in the same direction as they faced when they were removed. Also be careful not to let the bearings fall out.

P13-16 Install the planet pinion gear pins into the carrier. Then, install the final drive carrier retaining ring.

P13-17 With a feeler gauge, check the end play of the pinion gears. If the end play is too low, the planet pinion assembly must be removed and the correct thickness thrust washer installed. If the end play is too much, the differential assembly must be replaced.

P13-18 Install the differential side gears and thrust washers into the carrier.

P13-19 Apply transmission assembly lube onto the thrust washers for the pinion gears, and then install them into the carrier.

P13-20 Align the bores of the pinion gears with the bore in the carrier for the pinion pin, and then install the pin. Make sure the hole in the pivot pin, for its retaining pin, is aligned with the retaining pin bore.

P13-21 Install the pinion pin's retaining pin.

P13-22 Insert the sun gear into the carrier, making sure the gear is placed in the correct direction.

P13-23 Install the output shaft into the final drive assembly.

P13-24 Install a new retaining snap ring onto the end of the output shaft.

Adjusting Side Gear End play

Special Tools

Adapter kit for transaxle

Dial indicator

0- to 1-inch micrometer

Most of the procedures for servicing front-wheel-drive (FWD) final drive units are the same as for rear-wheel-drive (RWD) units. Because FWD units typically use the end of the output shaft as the drive pinion gear, all normal pinion shaft adjustments are not necessary. However, ring gear and side bearing adjustments are still necessary. These are typically made with the differential case assembled and out of the transaxle case. Always refer to your service information before proceeding to make these adjustments on a transaxle. What follows is a typical procedure for adjusting side gear end play:

1. Install the correct adapter into the differential bearings (**Figure 8-40**).
2. Mount the dial indicator to the ring gear with the plunger resting against the adapter.
3. With your fingers or a screwdriver, move the ring gear up and down (**Figure 8-41**).
4. Record the measured end play.

Dial indicator

Special
tool

Differential
assembly

Move
side gear
up and down

Figure 8-40 Tool setup for measuring side gear end play.

Figure 8-41 Move the ring gear up and down and observe the dial indicator.

Thrust
washer

Side
gear

Figure 8-42 Location of thrust washers.

5. Measure the old thrust washer with the micrometer.
6. Install the correct size of thrust washer (**Figure 8-42**).
7. Repeat the procedure for the other side.

Adjusting Bearing Preload

The following procedure is typical for the measurement and adjustment of the differential bearing preload in a transaxle. Always refer to the applicable service information before proceeding to make these adjustments on a particular transaxle.

1. Remove the bearing cup and existing shim from the differential bearing retainer.
2. Select a gauging shim that will allow for 0.001- to 0.010-inch end play.
3. Install the gauging shim into the differential bearing retainer (**Figure 8-43**).
4. Press in the bearing cup.
5. Lubricate the bearings and install them into the case.
6. Install the bearing retainer.
7. Tighten the retaining bolts.

 Special Tools

Set of gauging shims
Dial indicator
Fresh lubricant
Inch-pound torque
wrench

Figure 8-43 Typical location of preload shim.

Figure 8-44 Setup for measuring differential bearing preload.

8. Mount the dial indicator with its plunger touching the differential case (**Figure 8-44**).
9. Apply medium pressure in a downward direction while rolling the differential assembly back and forth several times.
10. Zero the dial indicator.
11. Apply medium pressure in an upward direction while rotating the differential assembly back and forth several times.
12. The required shim-to-set preload is the thickness of the gauging shim plus the recorded end play.
13. Remove the bearing retainer, cup, and gauging shim.
14. Install the required shim.
15. Press the bearing cup into the bearing retainer.
16. Install the bearing retainer and tighten the bolts.
17. Check the rotating torque of the transaxle. If this is less than specifications, install a thicker shim. If the torque is too great, install a slightly thinner shim.
18. Repeat the procedure until the desired torque is reached.

PART REPLACEMENT

Planetary-type final drives, like helical final drives, are available in more than one possible ratio for a given type of transaxle, so care should be taken to assure that the same gear ratios are used during assembly. This is not normally a problem when overhauling a single unit; however, in a shop where many transmissions are being repaired, it is possible to mix up parts causing problems during the rebuild.

Side Bearings

The side bearings of most final drive units must be pulled off and pressed onto the differential case. Always be sure to use the correct tools for removing and installing the bearings.

Figure 8-45 Typical setup for removing side bearings from a differential case.

Final Drive Bearing Replacement

The final drive unit is positioned either in the transaxle case or in a separate housing mounted to the transaxle case. Regardless of its location, it is supported by bearings. Normally, tapered roller bearings are used. These bearings are removed with pullers (**Figure 8-45**) and installed with a press and driver. When replacing the bearings, make sure the bearing seats are free from nicks and burrs. These defects will not allow the bearing to seat properly and will give false end play measurements. If the bearings were replaced, end play and gear clearances must be checked and corrected before installing the final drive unit into the housing.

Ring Gear Replacement

The ring gear of many transaxles is riveted to the differential case. The rivets must be drilled and then driven out with a hammer and drift to separate the ring gear from the case (**Figure 8-46**). To install a new ring gear to the case, new special nuts and bolts are used (**Figure 8-47**). These nuts and bolts must be of the specified hardness and should be tightened in steps and to the specified torque.

Figure 8-46 Ring gear rivets must be drilled and then driven out to separate the ring gear from the differential case.

Figure 8-47 The ring gear should be fastened to the differential case with special nuts and bolts.

HOUSING AND CASE SERVICE

Critical to all functions of a transmission is the condition of the housing. The housing may actually comprise several units, sealed together by gaskets, sealant, or seals (**Figure 8-48**). Each case is part of the transmission housing and should be carefully inspected.

While removing the parts from inside the case, make sure you note the location of the check balls that may be in the hydraulic network. Also, many housings have a small filtering screen placed into a hydraulic pathway. These are typically easily removed and should be removed before cleaning the case. Low air pressure (approximately 30 psi) applied to the back of the screen will typically remove them. Keep in mind that some cleaning solvents will destroy the plastic screens. Once removed, the screens should be inspected for debris and cleaned with hot water and soap.

Signs of leakage through parts of the housing can be caused by the porosity of the metal or by a crack. If the case is cracked, it should be replaced. If the problem is porosity, the problem may be corrected by applying an epoxy sealer over the spot of leakage after the case has been thoroughly cleaned.

After all parts contained in the housing are removed, the housing should be thoroughly cleaned and all passages (**Figure 8-49**) blown out with compressed air. These passages can be checked for restrictions by applying compressed air to each one. If air flows out the other end, there is no restriction. Compressed air can also be used to check for internal leaks. Plug off one end of the passage and apply air to the other. If pressure builds in that passage, there are probably no leaks in it.

Check the passages in the case for cross-tracking from one circuit to another. Fill the circuit with solvent and watch to see if the solvent disappears or leaks away. If the level of the solvent changes, you should check each part of the circuit to find where the leak is. If there is evidence of cross-tracking, the case should be replaced.

Figure 8-48 The transmission/transaxle housing may be comprised of several different cases, each sealed to one another.

Figure 8-49 The oil passages in the housing should be cleaned and checked for blockage and cross-tracking. Make sure to remove all check balls before cleaning.

Vents are located in the oil pump body or transmission case to equalize pressure in the transmission. These vents can be checked by blowing low-pressure air through them, squirting solvent or brake-cleaning spray through them, or pushing a small-diameter wire through the vent passage.

All sealing surfaces must be flat, smooth, and without damage that would lead to an oil leak. To keep all of the rotating parts in alignment, the surfaces must also be true. The flatness of the surfaces can be checked by laying a straightedge across them in several places.

Check the valve body mounting area for warpage with a straightedge and feeler gauge. This should be done in several locations. If there is a slight burr or high spot, it can be removed by flat-filing the surface. A straightedge should be laid across the lower flange of the case to check for distortion. Any warpage found here may result in circuit leakage, causing any number of hydraulically related problems.

All bores where a seal is installed should be inspected for surface roughness, nicks, or scratches. Imperfections in steel or cast iron parts can usually be polished out with crocus cloth.

All other internal and external bores must also be inspected. Check the fit of the servo pistons in their bore without their seals to be sure they have free travel. There should be no tight spots or binding over the whole range of travel. Any deep scratches or gouges that cause binding of the piston will require case replacement. Accumulator bores are checked in the same way as servo bores.

Case-mounted hydraulic clutch bores are prone to the same problems as servo bores. Look for any scratches or gouges in the sealing area that would affect the rubber seals. It is possible to damage these areas during disassembly, so be careful with tools used during overhaul.

Check all bellhousing bolt holes and dowels. Cracks around the bolt holes indicate that the case bolts were tightened with the case out of alignment with the engine block.

All internal threads should be carefully inspected. If the threads are distorted or stripped, they should be replaced with a threaded insert, or the case should be replaced.

Sometimes a slightly damaged thread can be restored by running a thread chaser through the bore (**Figure 8-50**). Always make sure to check the pitch and size of the threads (**Figure 8-51**) and use a matching chaser or tap. After the threads have been repaired, make sure you thoroughly clean the case.

All nuts, bolts, and studs should also be inspected for damage and replaced if necessary. Check all of the bolts that were removed during disassembly for aluminum on the threads. If aluminum is present, you can assume the thread bore is damaged and should be repaired.

Bushings in a transmission case are normally found in the rear of the case and require the same inspection and replacement techniques as other bushings in the transmission. Always be sure that the oil passage to a pressure-fed bushing or bearing is open and is free of dirt and foreign material.

Figure 8-50 A thread chaser should be used to clean out internal threads.

Figure 8-51 A thread pitch gauge ought to be used to determine the size and pitch of a bolt or screw before the threads are tapped or inserted.

CASE STUDY

A customer had his 2008 Ford Taurus towed into the shop. He stated that nothing happened when the transmission was in any gear. The technician looked things over and attempted to verify the complaint. Sure enough, the car wouldn't move in forward or in reverse. He did notice a slight change in noise when he moved the lever into drive and reverse, so he knew something was happening.

He proceeded to do the normal checks with the scan tool and found nothing unusual except for the codes that said there was no signal from the output sensor. That didn't alarm him because he knew there wouldn't be a signal since the car wouldn't move. He then checked the fluid. The level was correct and smelled fine.

He then raised the car and checked the front axle shafts to see if they were stuck. He could turn the wheels easily; in fact, he noticed how easy it was to rotate them. But he wasn't sure what to make of that.

He then ran a pressure test and found the pressure to be normal at idle speed and observed pressure changes as he selected drive and reverse. From this he knew it was not a hydraulic problem.

Since the problem didn't appear to be electronic or hydraulic, he knew it had to be mechanical. Thinking about what he had observed, he came to the conclusion that the problem must be in the input or the output of the transaxle. He strongly suspected the output because of the change of noise he noticed. Giving it more thought, he was convinced the problem was the output because of the little effort it took to rotate the wheels.

He proceeded to pull the transaxle and began his disassembly and inspection. He carefully checked the drive chain and sprocket assembly and found no evidence of a problem. He then pulled the output shaft from the transaxle. As soon as he had it out, he saw the problem; the splines were stripped and rounded off.

He proceeded to overhaul the transmission and clean every part thoroughly. He replaced the output shaft and transfer gear, then put the transaxle back together. After the transaxle was in, he put it in Drive and it moved. During the road test he found no problems. He told the customer it was now working fine and the customer was pleased.

ASE-STYLE REVIEW QUESTIONS

1. While inspecting a planetary gearset:
   Technician A says blackened pinion shafts indicate severe overloading.
   Technician B says bluish gears indicate overheating.
   Who is correct?

   A. A only
   B. B only
   C. Both A and B
   D. Neither A nor B

2. *Technician A* says loose planetary pinion gear bearings will cause the gear to whine when it is loaded.
   Technician B says damaged teeth on a planetary pinion gear will cause the gear to whine.
   Who is correct?

   A. A only
   B. B only
   C. Both A and B
   D. Neither A nor B

3. *Technician A* removes bushings by carefully cutting one side of the bushing and collapsing it. Once collapsed, the bushing can be easily removed with a pair of pliers.
   Technician B removes small-bore bushings by tapping the inside bore of the bushing with threads that match a selected bolt, which fits into the bushing. After tapping the bushing, the bolt is inserted and a slide hammer is used to pull the bolt and bushing out of its bore.
   Who is correct?

   A. A only
   B. B only
   C. Both A and B
   D. Neither A nor B

Chapter 8

4. *Technician A* says bushings can be heated with a torch to remove them easily.

   Technician B says bushings can be removed with a slide hammer and the correct attachment.

   Who is correct?

   A. A only
   B. B only
   C. Both A and B
   D. Neither A nor B

5. While checking a planetary gearset:

   Technician A says the end clearance of the pinion gears should be checked with a feeler gauge.

   Technician B says the end clearance of the long pinions in a Ravigneaux gearset should be checked at both ends.

   A. A only
   B. B only
   C. Both A and B
   D. Neither A nor B

6. *Technician A* says most FWD final drive units require that ring and pinion backlash and pinion depth be set to specifications.

   Technician B says ring gear and side gear bearing adjustments are required on all FWD final drive units.

   Who is correct?

   A. A only
   B. B only
   C. Both A and B
   D. Neither A nor B

7. *Technician A* says a blocked oil delivery passage will cause a shaft to score.

   Technician B says that if a shaft is fitted with a check ball and the check ball does not seat properly, low oil pressure will result.

   Who is correct?

   A. A only
   B. B only
   C. Both A and B
   D. Neither A nor B

8. *Technician A* says thrust washers should be inspected for scoring, flaking, and wear through to the base material.

   Technician B says plastic thrust washers will not wear unless they are damaged.

   Who is correct?

   A. A only
   B. B only
   C. Both A and B
   D. Neither A nor B

9. While inspecting a parking pawl assembly:

   Technician A says the pivot pin must fit loosely in its bracket and tightly in the pawl.

   Technician B says the engagement lug on the pawl must be square in order to fully engage into the parking gear.

   Who is correct?

   A. A only
   B. B only
   C. Both A and B
   D. Neither A nor B

10. While checking the end play of a transfer gear assembly:

    Technician A says that if the end play of the output transfer shaft is incorrect, a different size thrust washer should be installed behind the output shaft transfer gear.

    Technician B says that after the end play of the output shaft has been corrected, the turning torque of the shaft should be measured. If the turning torque is too high, a slightly thinner shim should be installed. If the turning torque is too low, a slightly thicker shim should be installed.

    Who is correct?

    A. A only
    B. B only
    C. Both A and B
    D. Neither A nor B

ASE CHALLENGE QUESTIONS

1. While discussing housing and case service:

   Technician A says while removing the parts from inside the case, note the location of the check balls.

   Technician B says many housings have a small filtering screen placed into a hydraulic pathway that should be inspected and cleaned

   A. A only
   B. B only
   C. Both A and B
   D. Neither A nor B

2. *Technician A* says a drive chain that is too loose should be shortened by removing a pair of links in the chain.

   Technician B says the drive sprockets should be replaced if the gear teeth are polished or show any other signs of wear.

   Who is correct?

   A. A only
   B. B only
   C. Both A and B
   D. Neither A nor B

3. *Technician A* checks bushing wear by observing the lateral movement of the shaft that fits into the bushing. Any noticeable lateral movement indicates wear and therefore the bushing should be replaced.

   Technician B checks for bushing wear by measuring the inside diameter of the bushing and the outside diameter of the shaft with a dial caliper or micrometer.

   Who is correct?

   A. A only   C. Both A and B

   B. B only   D. Neither A nor B

4. *Technician A* says sprag and roller clutches should be inspected in the same way as bearings.

   Technician B says to check their operation by attempting to rotate them in both directions. If working properly, they will allow rotation in both directions.

   Who is correct?

   A. A only   C. Both A and B

   B. B only   D. Neither A nor B

5. *Technician A* says all hubs, drums, and shells should be carefully examined for wear and damage.

   Technician B says minor scoring or burrs on band application surfaces of a drum can be removed by lightly polishing the surface with a 200-grit crocus cloth.

   Who is correct?

   A. A only   C. Both A and B

   B. B only   D. Neither A nor B

Name _____   Date _____

CHECKING THRUST WASHERS, BUSHINGS, AND BEARINGS

Upon completion of this job sheet, you should be able to inspect, measure, and replace thrust washers and bearings and inspect the bushings in a transmission/transaxle.

ASE Education Foundation Correlation

This job sheet addresses the following **MAST** tasks:

C.12   Inspect, measure, and replace thrust washers and bearings. (**P-2**)

C.14   Inspect bushings; determine needed action. (**P-2**)

We Support

ASE | Education Foundation

Tools and Materials

- Wire-type feeler gauge set
- Bushing driver set
- Bushing puller tool

Describe the Vehicle Being Worked On:

Year _____ Make _____ Model _____

VIN _____ Engine type and size _____

Model and type of transmission _____

Procedure

Task Completed

1. The best time to inspect thrust washers, bearings, and bushings is during disassembly. The bushings should be inspected for pitting and scoring. Describe their condition.

   _____

   _____

2. Check the depth that bushings are installed to and the direction of the oil groove, if so equipped, before you remove them. Many bushings that are used in the planetary gearing and output shaft areas have oiling holes in them. Be sure to line these up correctly during installation or you may block off oil delivery and destroy the geartrain. Describe their condition.

   _____

   _____

3. Observe the lateral movement of the shaft that fits into the bushing. Any noticeable lateral movement indicates wear and the bushing should be replaced. Describe your findings.

   _____

   _____

4. The amount of clearance between the shaft and the bushing can be checked with a wire-type feeler gauge. Insert the wire between the shaft and the bushing; if the gap is greater than the maximum allowable, the bushing should be replaced. What are the specifications for this gap and how do they compare to your measurement?

   _____

   _____

5. Measure the inside diameter of the bushing and the outside diameter of the shaft with a vernier-type caliper or micrometer. Compare the two and state your conclusions.

_____

_____

6. Most bushings are press-fit into a bore. Remove them by driving them out of the bore with a properly sized bushing tool. Some bushings can be removed with a slide hammer fitted with an expanding or threaded fixture that grips the inside of the bushing. Another way to remove bushings is to carefully cut one side of the bushing and collapse it. Once collapsed, the bushing can be easily removed with a pair of pliers. What did you use to remove the bushing?

_____

_____

7. Small-bore bushings located in areas where it is difficult to use a bushing tool can be removed by tapping the inside bore of the bushing with threads that match a bolt that fits into the bushing. After the bushing has been tapped, insert the bolt and use a slide hammer to pull the bolt and bushing out of its bore. □

8. All new bushings should be pre-lubed during transmission assembly and installed with the proper bushing driver. Make sure they are not damaged and are fully seated in their bores. □

9. The purpose of a thrust washer is to support a thrust load and keep parts from rubbing together. Selective thrust washers come in various thicknesses to take up clearances and adjust shaft end play. Flat thrust washers and bearings should be inspected for scoring, flaking, and wear through to the base material. Describe their condition.

_____

_____

10. Flat thrust washers should also be checked for broken or weak tabs. These tabs are critical for holding the washer in place. On metal flat thrust washers, the tabs may appear cracked at the bend of the tab; however, this is a normal appearance. Describe their condition.

_____

_____

11. Only damaged plastic thrust washers will show wear. The only way to check their wear is to measure the thickness and compare it to a new part. Describe their condition.

_____

_____

12. Proper thrust washer thicknesses are important to the operation of an automatic transmission. Always follow the recommended procedure for selecting the proper thrust plate. Use transmission assembly lube to hold thrust washers in place during assembly. □

13. All bearings should be checked for roughness before and after cleaning. □

Task Completed

14. Carefully examine the inner and outer races, and the rollers, needles, or balls for cracks, pitting, etching, or signs of overheating. Describe their condition.

_____

_____

15. Give a summary of your inspection.

_____

_____

_____

Problems Encountered

_____

_____

_____

Instructor's Comments

_____

_____

_____

Name _____   Date _____

SERVICING OIL DELIVERY SEALS

Upon completion of this job sheet, you should be able to inspect oil delivery seal rings, ring grooves, and sealing surface areas.

ASE Education Foundation Correlation ————————————————————

This job sheet addresses the following **MAST** task:

C.13   Inspect oil delivery circuits, including seal rings, ring grooves, and sealing surface areas, feed pipes, orifices, and check valves/balls. **(P-2)**

We Support
Education Foundation

Tools and Materials

- TransJel
- Seal driver tools
- Crocus cloth
- Feeler gauge set

Procedure Guidelines

- Three types of seals are used in automatic transmissions: O-ring and square-cut (lathe-cut), lip, and sealing rings. These seals are designed to stop fluid from leaking out of the transmission and to stop fluid from moving into another circuit of the hydraulic circuit.

- O-ring and square-cut seals are used to seal nonrotating parts. When installing a new O-ring or square-cut seal, coat the entire surface of the seal with assembly lube or petroleum jelly. Make sure you don't stretch or distort the seal while you are working it into its holding groove. After a square-cut seal is installed, double-check it to make sure it is not twisted. The flat surface of the seal should be parallel with the bore. If it is not, fluid will easily leak past the seal.

- Lip seals that are used to seal a shaft typically have a metal flange around their outside diameter. The shaft rides on the lip seal at the inside diameter of the seal assembly. The rigid outer diameter provides a mounting point for the lip seal and is pressed into a bore. Once pressed into the bore, the outer diameter of the seal prevents fluid from leaking into the bore, while the inner lip seal prevents leakage past the shaft.

- Piston lip seals are set into a machined groove on the piston. This type of lip seal is not housed in a rigid metal flange. They are designed to be flexible and provide a seal while the piston moves up and down. While the piston moves, the lip also flexes up and down. The most important thing to keep in mind while installing a lip seal is to make sure the lip is facing the correct direction. The lip should always be aimed toward the source of pressurized fluid. If installed backward, fluid under pressure will easily leak past the seal. Also remember to make sure the surfaces to be sealed are clean and not damaged.

- Teflon or metal sealing rings are commonly used to seal servo pistons, oil pump covers, and shafts. These rings may be designed to provide for a seal, but they may also be designed to allow a controlled amount of fluid leakage. Sealing rings are either solid rings or cut. Cut sealing rings are one of three designs: open-end, butt-end, or locking-end.

- Solid sealing rings are made of a Teflon-based material and are never reused. To remove them, carefully cut the seal after it has been pried out of its groove.

- Installing a new solid sealing ring requires special tools. These tools allow you to stretch the seal while pushing it into position. Never attempt to install a solid seal without the proper tools. Because these seals are soft, they are easily distorted and damaged.

- Open-end sealing rings fit loosely into a machined groove. The ends of the rings do not touch when they are installed. This type of ring is typically removed and installed with a pair of snap-ring pliers. The ring should be expanded just enough to move it off or onto the shaft.

▪ Butt-end sealing rings are designed so that their ends butt up or touch each other once the seal is in place. This type of seal can be removed with a small screwdriver. The blade of the screwdriver is used to work the ring out of its groove. To install this type of ring, use a pair of snap-ring pliers and expand the ring to move it into position.

▪ Locking-end rings may have either hooked ends that connect or ends that are cut at an angle to hold the ends together. These seals are removed and installed in the same way as butt-end rings. After these rings are installed, make sure the ends are properly positioned and touching.

▪ All seals should be checked in their own bores prior to installation. They should be slightly smaller or larger (63 percent) than their groove or bore. If a seal is not the proper size, find one that is. Do not assume that because a particular seal came with the overhaul kit it is the correct one.

▪ Never install a seal when it is dry. The seal should slide into position and allow the part it seals to slide into it. A dry seal is easily damaged during installation.

▪ Install only genuine seals recommended by the manufacturer of the transmission.

Describe the Vehicle Being Worked On:

Year _____ Make _____ Model _____

VIN _____ Engine type and size _____

Model and type of transmission _____

Procedure   **Task Completed**

1. Before installing seals, clean the shaft and/or bore area.   ☐

2. Carefully inspect these areas for damage. File or stone away any burrs or bad nicks and polish the surfaces with a fine crocus cloth, then clean the area to remove the metal particles. Describe your findings.

   _____

   _____

3. Lubricate the seal, especially any lip seals, to ease installation.   ☐

4. All metal sealing rings should also be checked for proper fit. Since these rings seal on their outer diameter, the seal should be inserted in its bore and should feel tight there. If the seal has some form of locking ends, these should be interlocked prior to trying the seal in its bore. Describe your findings.

   _____

   _____

5. Check the fit of the sealing rings in their shaft groove. Describe your findings.

   _____

6. Check the side clearance of the ring by placing the ring into its groove and measuring the clearance between the ring and the groove with a feeler gauge. Describe your findings.

   _____

   _____

7. While checking the clearance, look for nicks in the grooves and for evidence of groove taper or stepping. Describe your findings.

   _____

   _____

8. Use the correct driver when installing a seal and be careful not to damage the seal during installation.

_____

Problems Encountered

_____

_____

_____

Instructor's Comments

_____

_____

_____

Name _____   Date _____

SERVICING PLANETARY GEAR ASSEMBLIES

Upon completion of this job sheet, you should be able to inspect and measure planetary gear assemblies.

ASE Education Foundation Correlation ⎯⎯⎯⎯⎯⎯⎯⎯⎯⎯⎯⎯⎯⎯⎯⎯

This job sheet addresses the following **MAST** task:

C.15   Inspect and measure planetary gear assembly components; determine needed action. **(P-2)**

We Support
Education Foundation

Tools and Materials

• Snap-ring pliers
• Feeler gauge set

Describe the Vehicle Being Worked On:

Year _____ Make _____ Model _____

VIN _____ Engine type and size _____

Model and type of transmission: _____

Procedure   **Task Completed**

1. The planetary gears used in automatic transmissions are the helical type and all gear teeth should be inspected for chips or stripped teeth. Describe their general condition before disassembling the gearset.

   _____

   _____

2. Any gear that is mounted to a splined shaft needs the splines checked for mutilation or shifted splines. Record your findings.

   _____

   _____

3. Note any discoloration of the parts and explain the cause for it.

   _____

   _____

4. Check the planetary pinion gears for loose bearings. Record your findings.

   _____

   _____

5. Check each gear individually by rolling it on its shaft to feel for roughness or binding of the needle bearings. Wiggle the gear to be sure it is not loose on the shaft. Looseness will cause the gear to whine when it is loaded. Record your findings.

   _____

   _____

6. Inspect the gears' teeth for chips or imperfections, as these will also cause whine. Record your findings.

_____

_____

7. Check the gear teeth around the inside of the front planetary ring gear. Record your findings.

_____

_____

8. Check the fit between the front planetary carrier to the output shaft splines. Record your findings.

_____

_____

9. Remove the snap ring and thrust washer from the front planetary ring gear. Record your findings.

_____

_____

10. Examine the thrust washer and the outer splines of the front drum for burrs and distortion. Record your findings.

_____

_____

11. With the snap ring removed, the front planetary carrier can be removed from the ring gear. Check the planetary carrier gears for end play by placing a feeler gauge between the planetary carrier and the planetary pinion gear. Compare the end play to specifications. Record your findings.

_____

_____

12. Check the splines of the sun gear. Record your findings.

_____

_____

13. Sun gears should have their inner bushings inspected for looseness on their respective shafts. Record your findings.

_____

_____

14. Check the fit of the sun shell to the sun gear and inspect the shell for cracks, especially at the point where the gears mate with the shell. Record your findings.

_____

_____

15. Check the sun shell for a bell-mouthed condition where it is tabbed to the clutch drum. Any variation from a true round should be considered junk and should not be used. Record your findings.

   _____

   _____

16. Look at the tabs and check for the best fit into the clutch drum slots. This involves trial fitting the shell and drum at all the possible combinations and marking the point where they fit the tightest. Record your findings.

   _____

   _____

17. Check the gear carrier for cracks and other defects. Record your findings.

   _____

   _____

18. Check the thrust bearings for excessive wear and, if required, correct the input shaft thrust clearance by using a washer with the correct thickness. Record your findings.

   _____

   _____

19. To determine the correct thickness, measure the thickness of the existing thrust washer and compare it to the measured end play. All the pinions should have about the same end play. Record your findings.

   _____

   _____

20. Replace all defective parts and reassemble the gearset.   ☐

Problems Encountered

_____

_____

_____

Instructor's Comments

_____

_____

_____

Name _____   Date _____

SERVICING INTERNAL TRANSAXLE DRIVES

Upon completion of this job sheet, you should be able to inspect the transaxle drive, link chains, sprockets, gears, bearings, and bushings.

ASE Education Foundation Correlation

This job sheet addresses the following **MAST** task:

C.17   Diagnose and inspect transaxle drive, link chains, sprockets, gears, bearings, and bushings; perform needed action. (**P-2**)

We Support
ASE | **Education Foundation**

Tools and Materials

- Marking tool
- Machinist's rule

Describe the Vehicle Being Worked On:

Year _____ Make _____ Model _____

VIN _____ Engine type and size _____

Model and type of transmission: _____

Procedure   **Task Completed**

Chain Drives

1. This inspection is done with the transaxle on a bench and partially disassembled   ☐
   and should be repeated to double-check during reassembly. Begin by checking chain
   deflection between the centers of the two sprockets. Deflect the chain inward on one
   side until it is tight.

2. Mark the housing at the point of maximum deflection.   ☐

3. Deflect the chain outward on the same side until it is tight.   ☐

4. Again mark the housing in line with the outside edge of the chain at the point of   ☐
   maximum deflection.

5. Measure the distance between the two marks. If this distance exceeds specifications,
   replace the drive chain. Describe your findings.

   _____

   _____

6. Be sure to check for an identification mark on the chain during disassembly. The
   mark may be a painted or dark-colored link, and may indicate either the top or the
   bottom of the chain, so be sure you remember which side was up. How was your chain
   marked?

   _____

   _____

7. The sprockets should be inspected for tooth wear and for wear at the point where they ride. If the chain was found to be too slack, it may have worn sprockets in the same manner that engine timing gears wear when the timing chain stretches. A slightly polished appearance on the face of the gears is normal. Describe your findings.

_____

_____

8. Check the bearings and bushings used on the sprockets for damage. Describe your findings.

_____

_____

9. The radial needle thrust bearings must be checked for any deterioration of the needles and cage. Describe your findings.

_____

_____

10. The running surface in the sprocket must also be checked, as the needles may pound into the gear's surface during abusive operation. Describe your findings.

_____

_____

11. The bushings should be checked for any signs of scoring, flaking, or wear. Describe your findings.

_____

_____

12. Based on the above, what parts need to be replaced?

_____

_____

_____

_____

Final Drive Units

1. Final drive units may be helical or planetary gear units. A careful inspection of the assembly is done with the transaxle disassembled. The helical type should be checked for worn or chipped teeth, overloaded tapered roller bearings, and excessive differential side gear and spider gear wear. Describe your findings.

_____

_____

2. Measure the clearance between the side gears and the differential case. Compare your measurement to specifications. Describe your findings.

_____

_____

3. Check the fit of the spider gears on the spider gear shaft. Describe your findings.

_____

_____

4. Check the assembly's end play. How do your measurements compare to specifications?

_____

_____

5. What is used to preload the side bearings on this transaxle?

_____

_____

6. With a torque wrench, measure the amount of rotating torque. Compare your readings against specifications. Describe your findings.

_____

_____

7. If the bearing preload, end play, and the condition of the bearings are fine, the parts can be reused. However, always install new seals during assembly. ☐

8. Planetary-type final drives are checked for the same problems as helical-type ones. Check for worn or chipped teeth, overloaded tapered roller bearings, and excessive differential side gear and spider gear wear. Describe your findings.

_____

_____

9. The planetary pinion gears need to be checked for looseness or roughness on their shafts and for end play. Describe your findings.

_____

_____

10. Check the end play of the assembly. How did you do this and how do your measurements compare to specifications?

_____

_____

11. What is used to preload the side bearings on this transaxle?

_____

_____

12. With a torque wrench, measure the amount of rotating torque. Compare your readings against specifications. Describe your findings.

_____

_____

13. What are your conclusions about the final drive unit?

_____

_____

_____

_____

_____

Problems Encountered

_____

_____

_____

Instructor's Comments

_____

_____

_____

Name _____   Date _____

SERVICING FINAL DRIVE COMPONENTS

Upon completion of this job sheet, you will be able to inspect, measure, and adjust transaxle final drive units.

ASE Education Foundation Correlation ——————————————————

This job sheet addresses the following **MAST** task:

C.18   Inspect, measure, repair, adjust or replace transaxle final drive components. **(P-2)**

We Support
| Education Foundation

Tools and Materials

- Basic hand tools

Describe the Vehicle Being Worked On:

Year _____ Make _____ Model _____

VIN _____ Engine type and size _____

Model and type of transmission _____

Procedure

Task Completed

1. Final drive units may be helical gear or planetary gear units. A careful inspection of the assembly is done with the transaxle disassembled. The helical type should be checked for worn or chipped teeth, overloaded tapered roller bearings, and excessive differential side gear and spider gear wear. Describe your findings.

   _____

   _____

2. Measure the clearance between the side gears and the differential case. Compare your measurement to specifications. Describe your findings.

   _____

   _____

3. Check the fit of the spider gears on the spider gear shaft. Describe your findings.

   _____

   _____

4. Check the assembly's end play. How do your measurements compare to specifications?

   _____

   _____

5. What is used to preload the side bearings on this transaxle?

   _____

   _____

6. With a torque wrench, measure the amount of rotating torque. Compare your readings against specifications. Describe your findings.

   _____

   _____

7. If the bearing preload and end play are fine, as is the condition of the bearings, the ☐
parts can be reused. However, always install new seals during assembly.

8. Planetary-type final drives are checked for the same problems as helical types.
Check for worn or chipped teeth, overloaded tapered roller bearings, and excessive
differential side gear and spider gear wear. Describe your findings.

_____

_____

9. The planetary pinion gears need to be checked for looseness or roughness on their
shafts and for end play. Describe your findings.

_____

_____

10. Check the end play of the assembly. How did you do this and how do your
measurements compare to specifications?

_____

_____

11. What is used to preload the side bearings on this transaxle?

_____

_____

12. With a torque wrench, measure the amount of rotating torque. Compare your
readings against specifications. Describe your findings.

_____

_____

13. What are your conclusions about the final drive unit?

_____

_____

_____

_____

Problems Encountered

_____

_____

_____

Instructor's Comments

_____

_____

_____

Name _____   Date _____

TRANSMISSION CASE SERVICE

Upon completion of this job sheet, you will be able to inspect case bores, passages, vents, bushings, and mating surfaces.

ASE Education Foundation Correlation

This job sheet addresses the following **MAST** task:

C.16   Inspect case bores, passages, bushings, vents, and mating surfaces; determine needed action. (**P-2**)

We Support
ASE | Education Foundation

Tools and Materials

- Air nozzle
- Crocus cloth
- Straightedge
- Feeler gauge set

Describe the Vehicle Being Worked On:

Year _____ Make _____ Model _____

VIN _____ Engine type and size _____

Model and type of transmission _____

Procedure

Task Completed

1. The transmission case should be thoroughly cleaned and all passages blown out.   ☐

2. The passages can be checked for restrictions by applying compressed air to each one. If air flows from the other end, there is no restriction. Describe your findings.

   _____

   _____

3. To check for leaks, plug off one end of the passage and apply air to the other. If pressure builds up in that passage, there are probably no leaks in it. Describe your findings.

   _____

   _____

4. Check the fit of the servo piston in the bore without the seal to be sure it has free travel. There should be no tight spots or binding over the whole range of travel. Any deep scratches or gouges that cause binding of the piston will require case replacement. Describe your findings.

   _____

   _____

5. Accumulator bores are checked the same as servo bores. Describe your findings.

   _____

   _____

6. Check the oil pump bore at the front of the case. Describe your findings.

   _____

   _____

7. Case-mounted hydraulic clutch bores are prone to the same problems as servo bores. Look for any scratches or gouges in the sealing area that would affect the rubber seals. It is possible to damage these areas during disassembly, so be careful with tools used during overhaul. Describe your findings.

_____

_____

8. Sealing surfaces of the case should be inspected for surface roughness, nicks, or scratches where the seals ride. Imperfections in steel or cast iron parts can usually be polished out with crocus cloth. Describe your findings.

_____

_____

9. Check the passages in the case for cross-tracking of one circuit to another. Fill the circuit with solvent and watch to see if the solvent disappears or leaks away. If the solvent goes down, you should check each part of the circuit to find where the leak is. Describe your findings.

_____

_____

10. Make sure all necessary check balls were in position during disassembly.   ☐

11. Check the valve body mounting area for warpage with a straightedge and feeler gauge. This should be done in several locations. If there is a slight burr or high spot, it can be removed by flat-filing the surface. Describe your findings.

_____

_____

12. A long straightedge should be laid across the lower flange of the case to check for distortion. Any warpage found here may result in circuit leakage, causing any number of hydraulically related problems. Describe your findings.

_____

_____

13. Check all bellhousing bolt holes and dowel pins. Cracks around the bolt holes indicate that the case bolts were tightened with the case out of alignment with the engine block. Describe your findings.

_____

_____

14. Check all of the bolts that were removed during disassembly for aluminum on the threads. If so, the thread bore is damaged and should be repaired. Thread repair entails the installation of a thread insert or by retapping the bore. After the threads have been repaired, make sure you thoroughly clean the case. Describe your findings.

_____

_____

15. The small screens found during teardown should be inspected for foreign material. Describe your findings.

_____

_____

16. Most screens can be removed easily. Care should be taken when cleaning because some cleaning solvents will destroy the plastic screens. Low-pressure air (approximately 30 psi) can be used to blow the screens out in a reverse direction. ☐

17. Bushings in a transmission case are normally found in the rear of the case and require the same inspection and replacement techniques as other bushings in the transmission. Always be sure that the oil passage to a pressure-fed bushing or bearing is open and is free of dirt and foreign material. Describe your findings.

_____

_____

18. Vents are located in the pump body or transmission case and provide for equalization of pressures in the transmission. These vents can be checked by blowing low-pressure air through them, squirting solvent or brake cleaning spray through them, or by pushing a small-diameter wire through the vent passage. Describe your findings.

_____

_____

_____

Problems Encountered

_____

_____

_____

Instructor's Comments

_____

_____

_____

CHAPTER 9
FRICTION AND REACTION UNIT SERVICE

Upon completion and review of this chapter, you should be able to:

- Describe the purpose of friction and reaction units.
- Inspect servo and accumulator bores, pistons, seals, pins, springs, and retainers; determine necessary action.
- Measure and adjust clutch pack clearance.

- Air test the operation of the clutch pack and servo assemblies.
- Inspect and service roller and sprag clutch, races, rollers, sprags, springs, cages, and retainers.

Terms To Know

Clutch Volume Index (CVI)
Selective servo apply pins

FRICTION AND REACTION UNITS

To provide for the various forward speed gears and a reverse, transmissions use compound planetary gears controlled by friction and reaction devices. Although diagnosis can lead you to the device that is not holding or driving properly, recognition of those suspected parts becomes difficult when the transmission is apart. Clutch and brake application charts are a must during diagnosis and transmission overhaul.

During inspection and disassembly of friction units and their hub or drum, pay close attention to which planetary gearset member each is attached. Keep in mind that some planetary members are connected to each other and as one member rotates, it rotates the other at the same speed. Sometimes the drum or housing for a clutch is also the drum for a brake band (**Figure 9-1**). Often this means the same planetary member can be driven or held to achieve different gears.

Servos and accumulators control the application of bands and clutches to achieve a desirable shift feel. There are many different types and designs of each. Each has an important function to the operation of the various designs of transmissions and transaxles. Fortunately, service for all servos and accumulators is very similar.

Many of the late-model transmissions use a clutch-to-clutch shifting design that does not use accumulators or bands; rather, they utilize variable-force shift solenoids to apply and release clutch packs with enough control that accumulators, bands, and servos are, in most cases, not necessary. There are millions of vehicles on the road currently that do use bands, servos, and accumulators, so this information is still very relevant.

Classroom Manual
Chapter 9, page 282

Direct clutch housing

Band

Figure 9-1 The drum for this clutch also serves as the clamping surface for the band. The tool shown in the clutch assembly is used to pull the assembly out of the transmission housing.

BAND SERVICE

Servicing bands and their components include inspection of the bands as well as the drums that the bands are wrapped around. Before the introduction of overdrive automatic transmissions, most bands operated in a free condition during most driving conditions. This means the band was not applied in the cruising gear range. However, many overdrive automatic transmissions use a band in the overdrive cruise range, which puts an additional load on the band and subsequently causes additional wear on the band. For this reason, a thorough inspection of the bands is very important (**Figure 9-2**).

The bands in a transmission are either single or double wrap, depending on the application. Both types can be the heavy-duty cast iron type or the normal strap type. The frictional material used on clutches and bands is quite absorbent. This characteristic can be used to tell if there is much life left in the lining. Simply squeeze the lining with your fingers to see if any fluid appears. If fluid appears, this tells you the lining can still hold fluid and has some life left in it. It is hard to tell exactly how long the band will last, but at least you have an indication that it is still useable. Strap- or flex-type bands should never be twisted or flattened out. This may crack the lining and lead to flaking of the lining.

Inspection

Band failures found during overhaul are easy to spot. Look for chipping, cracks, burn marks, glazing, and non-uniform wear patterns and flaking. If any of these defects are apparent, the band should be replaced.

Also inspect brake band friction material for wear. If the linings show wear, carefully check the band struts (**Figure 9-3**), levers, and anchors for wear. Replace any worn or damaged parts. Look at the linings of heavy-duty bands to see if the lining is worn evenly. A twisted band will show tapered wear on the lining. If the friction material is blackened,

Classroom Manual
Chapter 9, page 284

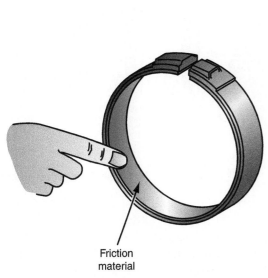

Figure 9-2 The friction material of the bands must be carefully inspected.

Figure 9-3 While removing the band, carefully inspect the band strut.

this is caused by an excessive buildup of heat. High heat may weaken the bonding of the lining and allow the lining to come loose from the metal portion of the band. On double-wrap bands, check both segments of the lining for cracks and other damage. Make sure you carefully check the band lugs for damage as well.

The drum surface should be checked for discoloration, scoring, glazing, and distortion. The drums will be either iron castings or steel stampings. Cast iron drums that are not scored can generally be restored to service by sanding the running surface with 180-grit emery paper in the drum's normal direction of rotation. A polished surface is not desirable on cast iron drums.

The surface of the drum must be flat (**Figure 9-4**). This is not usually a problem with a cast iron drum, but it can affect a stamped steel drum. It is possible for the outer surface of the drum to dish outward during its normal service life. This is a common problem on the GM 4L60 and should be inspected on any transmission that has a stamped steel band surface. Check the drum for flatness across the outer surface where the band runs. Any

Figure 9-4 Check the flatness of the drum with a straightedge.

Figure 9-5 The band apply pin is measured during overhaul with a special tool.

Figure 9-6 Band apply pin lengths are selective.

⚠ **Caution**

It is important to follow the manufacturers' recommendations on every aspect of transmission repair. Serious damage can result due to improper service techniques.

dishing here will cause the band to distort as it attempts to get a full grip on the drum. Distortion of the band weakens the bond of the friction material to the band and will cause early failure due to flaking of the frictional lining. A dished stamped steel drum should be replaced. Check the service information for maximum allowable tolerances.

Band Adjustments

The vast majority of transmissions in service have no provisions for externally adjusting the band-running clearance. This clearance is established during the rebuild process (**Figure 9-5**). In most cases the band apply pin is selective and is fitted when the transmission is rebuilt (**Figure 9-6**). Modern transmissions have no provisions for band adjustment other than selectively sized servo apply pins and struts.

Classroom Manual
Chapter 9, page 285

SERVO AND ACCUMULATOR SERVICE

On some transmissions, the servo and accumulator assemblies are serviceable with the transmission in the vehicle (**Figure 9-7**). Others require the disassembly of the transmission and/or removal of the valve body. Internal leaks at the servo or clutch seal will cause excessive pressure drops during gear changes.

Before disassembling a servo or an accumulator, carefully inspect the area to determine the exact location of any leak. Do this before cleaning the area around the seal. Look at the path of the fluid and identify possible sources. These sources could be worn gaskets, loose bolts, cracked housings, or loose line connections.

Band servos and accumulators are basically pistons with seals in a bore held in position by springs and retaining bolts or snap rings (**Figure 9-8**). Pistons may have seals made of cast iron, rubber or elastomer seals that should always be replaced. Even though some seals might look OK, it is cheap insurance to replace all the seals when rebuilding a transmission.

> ⚙ **SERVICE TIP** Depending on the transmission, a special tool may be required to keep the spring compressed so the cover can be removed.

Removal

On many vehicles it is possible to remove a servo or an accumulator while the transmission is in the vehicle. To do this, the ATF needs to be drained and the oil pan removed. Some transmissions require that the valve body be removed to gain access to the accumulators

Figure 9-7 The servos in some transmissions are serviceable while the transmission is in the vehicle and are contained in its own bore.

Figure 9-8 Servos and accumulators may be retained by a snap ring.

or servos. Removal of the valve body, on some models, may require the removal of the transmission first.

To disassemble an accumulator, depress the piston to remove the accumulator cover snap ring. If the piston assembly is retained by a cover, unbolt the cover (**Figure 9-9**). After removing the accumulator cover, remove the springs and pistons.

Many accumulator pistons can be installed upside down. This results in free travel of the piston, or excessive compression of the accumulator spring. Note the direction of the piston during the teardown as you will not always find a good reference to follow during reassembly. It is quite common for manufacturers to mate servo piston assemblies with accumulators. This takes up less space in the transmission case and, because they have the same basic shape, can reduce some of the machining during manufacture.

A servo (**Figure 9-10**) is disassembled in a similar fashion. Be careful when disassembling a servo, some have two or more pistons and four or five fluid passages and fluid control orifices or check balls inside them. Make sure you do not mix the parts. Also note the exact location of the check balls.

In some servos, a strong return spring is under the piston; therefore, it is advisable to use the appropriate compressing tool (**Figure 9-11**). The tool will push down on the piston to allow removal of the retaining ring. Often, the servo assembly will contain more than one spring and two or more seals.

Inspection

Inspect the outside area of the seal. If it is wet, determine if the oil is leaking out or if it is merely a lubricating film of oil. When removing the servo, continue to look for causes of the leak. Check both the inner and outer parts of the seal for wet oil, which means leakage. When removing the seal, inspect the sealing surface, or lips before washing. Look for unusual wear, warping, cuts, and gouges, or particles embedded in the seal.

The accumulator or servo' piston, spring, piston rod, and guide should be cleaned and dried. Check the servo piston for cracks, burrs, scores, and wear. Aluminum servo pistons should be carefully checked for cracks and their fit on the guide pins. Cracked pistons will allow for a pressure loss and being loose on the guide pin may allow the piston to bind in

⚠ **Caution**

If the appropriate spring compressor is not available, be very careful while pushing down the piston with something else. The spring pressure can force the servo piston assembly out of case and at you. Also, the bore in the transmission case can be damaged by trying to pry on the retaining ring to remove it.

Figure 9-9 Note the location of the various seals in this servo/accumulator assembly.

Cover

Seals

Snap ring

Piston

Cushion springs

Piston pin

Return spring

Figure 9-10 A servo assembly.

Retaining ring

Compressor tool

Figure 9-11 On some units, a spring compressor must be used to depress the piston enough to remove the retaining ring.

its bore. The seal groove should be free of nicks or any imperfection that might pinch or bind the seal. Clean up any problems with a small file. The piston ring should rotate freely in its groove. If it does not, clean and inspect the grooves. Replace the piston ring during reassembly.

Also, check the condition of the servo/accumulator pins. Look for signs of wear and damage. Also, check the fit of the pins and pistons in the case. The bores in the case should not allow the pins and pistons to wobble. If they do, they are worn or the bores in the case are worn.

Inspect the servo or accumulator spring for cracks. Also, check the area where the spring rests against the case or piston. The spring may wear a groove, make sure the piston or case material has not worn too thin. Some servos have a steel chafing plate in order to eliminate this problem.

Inspect the servo cylinder for scores or other damage. Move the piston rod through the piston rod guide and check for freedom of movement. Check band servo components for wear and scoring. Replace all other components as necessary, and then reassemble the servo assembly.

Classroom Manual
Chapter 9, page 303

> ⚙️ **SERVICE TIP** Some specialty shops have equipment that bores the worn out servo pin bore and installs a bushing with the correct inside diameter. This corrects any leakage past the pin and makes case replacement unnecessary.

Seals

A faulty servo piston or cover seal can cause a loss in servo apply pressure which results in band slippage. A faulty accumulator piston or cover seal can cause abnormal shift quality. A cover seal is either a gasket or an O-ring, while the piston seal is either a metal ring or rubber rings (**Figure 9-12**). Molded rubber seals are usually replaced during overhaul as rubber is subject to deterioration.

Most original equipment servo seals are of Teflon type. These seals will exhibit no feeling of drag in the bore due to the slipperiness of Teflon. A majority of replacement transmission gasket sets will supply cast iron hook-end seal rings, in place of the Teflon seals. When installed, the cast iron seals will have a noticeable drag as the servo piston is moved through the bore. This is not a problem, and in some cases may even improve operation.

Check the cast iron seal rings to make sure they are able to turn freely in the piston groove. These seal rings are not typically replaced unless they are damaged.

Installation

The springs used in accumulator and servo assemblies are specific to that application. Also, to see if one of the springs is weaker or stronger (both of which will change shift quality), compare the springs in the transmission you are working on to the specifications given in the service information (**Figure 9-13**).

Lubricate and install
lip seals as shown

Figure 9-12 Proper installation of servo piston seals.

SPRING SPECIFICATIONS

Spring	Standard (New)–Unit: mm (in.)			
	Wire Dia.	O.D.	Free Length	No. of Coils
1st accumulator spring	2.1 (0.083)	16.0 (0.630)	89.1 (3.508)	16.2
4th accumulator spring A	2.6 (0.102)	17.0 (0.669)	88.4 (3.480)	14.2
4th accumulator spring B	2.3 (0.091)	10.2 (0.402)	51.6 (2.031)	13.8
3rd accumulator spring	2.9 (0.114)	17.5 (0.689)	89.2 (3.512)	13.4
2nd accumulator spring A	2.4 (0.094)	14.5 (0.571)	68.0 (2.677)	12.5
3rd subaccumulator spring	2.7 (0.106)	17.0 (0.669)	39.0 (1.535)	6.3

Figure 9-13 Specifications for some of the springs used in servos and accumulators.

Servos and accumulators must be assembled in the correct sequence. Incorrect assembly can result in dragging bands and harsh shifting. If the servo pin has Teflon seals, these will need to be cut off with a knife and a new Teflon seal installed with the correct sleeve and sizing tools.

If rubber seal rings are installed on the piston, replace them whenever you are servicing an accumulator or a servo. When reassembling the servo or accumulator, lubricate the seal with ATF and install it on the piston rod. On spring-loaded lip seals, make sure the spring is seated around the lip, and that the lip is not damaged during installation. Lubricate and install the piston rod guide with its snap ring into the piston. Lubricate the servo or accumulator cylinder walls. Then, install the piston assembly and return spring into the cylinder. Use a spring compressor tool to depress the piston enough to install the piston retaining ring. If the assembly had a cover plate, tighten it to specifications.

Selective Servo Apply Pins

Classroom Manual
Chapter 9, page 286

Transmissions without a band adjusting screw use **selective servo apply pins**, which maintain the correct clearance between the band and the drum. These servo apply pins must be checked for correct length to assure proper stroking action of the servo piston, as well as shift timing. The travel needed to apply a band relates to the timing of band application. Although valving and orificing determine true shift timing, the adjustment of the servo pin completes the job by providing the correct clearance of the band to the drum. To select the correct apply pin, most transmissions require the use of special tools that simulate a fully applied band.

Selective servo apply pins are called "selective" when they are available in a variety of lengths. The correct length must be used to prevent transmission damage and ensure proper band operation.

To select the correct apply pin for some Ford transmissions, follow the below procedure: (This procedure does not compensate for any band wear and should only be used with new bands.)

1. Lubricate the piston seal.
2. Install the servo return spring and piston. Do not install the piston cover.
3. Install the special tool over the top of the piston (**Figure 9-14**).
4. Tighten the center bolt of the tool to the specified torque (normally 50 lbs.-in. or 5.6 Nm).
5. Attach the dial indicator holding tool to the transmission.

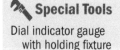

Special Tools

Dial indicator gauge with holding fixture
Servo piston remover/installer

Figure 9-14 Special tool for checking servo piston movement.

Figure 9-15 Measure the total movement of the piston after the piston stop on the special tool is loosened.

6. Position the stem of the dial indicator on the flat portion of the servo piston and zero the indicator.
7. Loosen the apply bolt of the piston tool until the piston stops against the tool.
8. Measure the amount of piston travel on the dial indicator and compare that to specifications (**Figure 9-15**).
9. If piston travel is not within specification, select and install the correct servo piston assembly to bring the servo piston travel within specification. The lengths are identified by the number of grooves near the piston (**Figure 9-16**).
10. Remove the dial indicator and servo selection tool.
11. Install the servo assembly. Using the special tool, install the servo retaining ring.

To select the servo apply pin in a 4L80-E transmission, two special tools and a torque wrench are needed. A gauge pin (**Figure 9-17**) is placed into the bore for the servo. Then the checking tool (**Figure 9-18**) is positioned over the bore with its hex nut facing the linkage for the parking pawl. The checking tool is fastened with two servo cover bolts that are torqued to a specified amount. The hex nut on the checking tool is then torqued

1 groove	2 groove	3 groove
2.936 in.	2.989 in.	3.043 in.
(74.56 mm)	(75.92 mm)	(77.29 mm)

Figure 9-16 The length of some servo pistons is identifiable by the grooves cut into the stem.

Figure 9-17 A special gauge pin that is placed into the bore for the servo to determine which selective servo pin should be used.

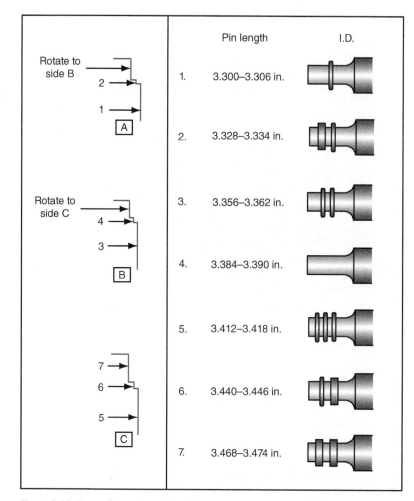

Figure 9-19 The exposed part of the gauge pin should be compared to the chart given in the service information to determine the correct pin length.

	Pin length	I.D.
1.	3.300–3.306 in.	
2.	3.328–3.334 in.	
3.	3.356–3.362 in.	
4.	3.384–3.390 in.	
5.	3.412–3.418 in.	
6.	3.440–3.446 in.	
7.	3.468–3.474 in.	

Figure 9-18 The checking tool is positioned over the bore and gauge pin with its hex nut facing the linkage for the parking pawl.

to 25 lbs.-ft. and the exposed part of the pin gauge is studied to determine the proper apply pin length. A chart in the service information (**Figure 9-19**) relates the appearance of the exposed part of the pin gauge to the correct pin length.

If a replacement band, drum, or even case has been used, an apply pin check must be made. Since these pins are selective, you will have to start by checking the length of the pin already in the transmission. If it is too long, or too short, a new length pin must be installed.

Classroom Manual
Chapter 9, page 286

⚙ **SERVICE TIP** New friction discs should not be used with used steel plates unless the steel plates are deglazed.

MULTIPLE-FRICTION DISC ASSEMBLIES

Two types of multiple-friction disc assemblies are found in transmissions: rotating drum (**Figure 9-20**) and case grounded (**Figure 9-21**). Both are serviced in the same way but require slightly different inspection procedures.

All friction disc and steel plate packs are held in place by snap rings. These snap rings may be selective in thickness and must be kept with the clutch pack during disassembly.

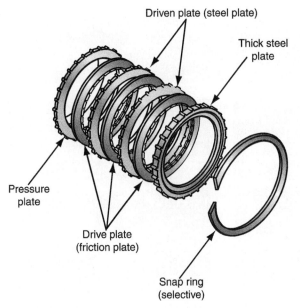

Figure 9-20 Clutch assembly from a rotating drum-type clutch.

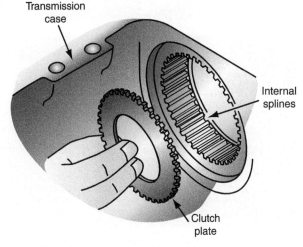

Figure 9-21 A case ground clutch assembly.

It is common to use the same diameter snap ring on more than one clutch in a single transmission. However, the rings can be of differing thicknesses. This thickness variation can be used to set the clearance in the clutch pack. The snap rings may also have a distinct shape that will be effective only when used in the correct groove.

An example of complete disassembly, inspection, assembly, and clearance checks is shown in **Photo Sequence 14**. Always refer to the manufacturer's recommendations for a particular transmission.

Disassembly

Using a screwdriver, remove the large clutch-retaining-plate snap ring and remove the thick steel clutch pressure plate (**Figure 9-22**). Now remove the clutch pack. Some clutch packs have a wave snap ring that must be looped through the clutch pack until the steel plates can be removed.

Once the snap ring is removed, the backing plate will be the first steel plate you will take out. This plate, like the snap ring, may be a selective part. Since the backing plate is

Many transmission rebuilders replace all friction discs and steel plates as a part of the overhaul process. This can actually save time because it eliminates the inspection time and there is no doubt about whether any discs or plates were of questionable condition.

Figure 9-22 Removing the large snap ring to disassemble clutch pack.

PHOTO SEQUENCE 14
Proper Procedure for Disassembling, Inspecting, Assembling, and Clearance Checking a Direct Clutch

P14-1 Set the direct clutch on the bench.

P14-2 Pry out the snap ring.

P14-3 Remove the pressure plate and the clutch pack assembly.

P14-4 Install the piston compressor and compress the piston.

P14-5 Remove the snap ring.

P14-6 Remove the piston compressor, the spring retainer assembly, and the piston.

P14-7 Install new seals on the piston.

P14-8 Check the movement of the check ball.

P14-9 Lubricate the outer piston seal with TransJel.

PHOTO SEQUENCE 14 (CONTINUED)

P14-10 Reinstall the piston.

P14-11 Place the piston spring retainer into the drum.

P14-12 Install the compressor and snap ring.

P14-13 Alternately install friction and steel discs.

P14-14 Install the pressure plate and snap ring.

P14-15 Check the clearance with a feeler gauge.

P14-16 Air test the assembly.

Figure 9-23 When removing the clutch discs from the drum, keep them in the original order.

 Special Tools

Snap-ring pliers
Feeler gauge
OSHA-approved air
  nozzle

If the return spring
assembly is installed
upside down, the unit
will pass an air check
but will not apply the
clutches.

thicker than the other steel plates in the clutch pack, it is easy to spot. Now remove the remainder of the friction discs and steel plates, keeping them in the order in which they were in the drum (**Figure 9-23**).

Using a clutch spring compressor tool, compress the clutch return springs. Then, using snap-ring pliers, remove the clutch hub retainer snap ring, retaining plate, and springs. Most retainers have small tabs that prevent you from removing the snap ring without using a compressor (**Figure 9-24**).

There are many types of clutch spring compressor tools available. The many different locations and depths of the clutch springs in their drums dictate having more than one simple compressor (**Figures 9-25–9-27**).

Figure 9-24 Retainer ring for captive return springs.

Figure 9-25 A spring compressor tool for a multiple-friction disc assembly in a GM 4T80E transaxle.

Figure 9-26 A spring compressor tool for a multiple-friction disc assembly in Honda transaxles.

Figure 9-27 A spring compressor tool for a multiple-friction disc assembly that sets into the transmission housing.

With the compressor installed on the spring retainer (**Figure 9-28**), compress the spring and retainer just enough to allow the snap ring to be removed. Pushing the retainer down too much may bend or distort it. Make sure the snap ring is not partially caught in its groove before releasing the spring compressor. Also, be very careful when releasing the spring compressor tool; some springs have very high tensions and can injure you if they get a chance to fly out.

⚠️ **WARNING** **Be careful when compressing the springs. Careless procedures can allow the springs and retainer to fly into your face.**

With the retainer removed, a single large coil spring or multiple small coil springs will be exposed. Note the number and placement of the multiple springs for assembly.

If no coil springs were encountered when the clutch pack was removed, then a Belleville- or disc-type return spring is used (**Figure 9-29**). This type of clutch will have a heavy steel plate with one rounded side in the bottom of the clutch pack. This heavy plate is a pressure plate. The released position of these springs has very little or no outward force on the retaining snap ring; therefore, the snap ring can be removed without the use of a compressor. The snap ring for this application may be selective, so remember which snap ring goes where.

Classroom Manual
Chapter 9, page 293

Belleville-type return springs are usually found in the forward clutch.

Figure 9-28 Proper position of the spring compressor will prevent damage to the piston and seals.

Figure 9-29 A clutch assembly fitted with a Belleville spring.

Figure 9-30 Use an air nozzle with a rubber tip and apply air pressure at the feed hole in the drum to pop out the piston from the drum.

Some clutches can have two or more snap rings, so pay attention to their placement in the clutch. Additionally, a wavy snap ring is sometimes used to retain the Belleville spring. This is used to give some cushion to the application of the clutch.

For easy removal of the piston from its drum, mount the clutch on the oil pump. Then lift the piston out of the bore. If it does not come out, the drum can be inverted and slammed squarely against a hardwood bench top. This will dislodge a tight piston, but it will not damage the drum.

Another way to remove the piston is to charge the apply circuit of the clutch with compressed air. Use an air nozzle with a rubber tip and apply air pressure to pop out the piston from the drum. Air is blown in either at the feed hole in the drum or by placing the drum on the clutch support and using its normal feed passage (**Figure 9-30**). Wrap a rag around the clutch drum to catch the piston and the fluid overspray. Make sure your fingers are out of the way of the moving piston.

⚡ **WARNING** **Air pressure should be reduced to 25–30 psi. to avoid expelling the piston at a high rate of speed, which could cause injury.**

Pistons located in the transmission case can be removed by blowing compressed air into the apply passage in the case.

With the piston removed, take note of the types of seals used and their position. This is important with lip seals, as the lip will face the direction the fluid comes from. Lip seals installed in the wrong direction will not hold pressure. Seals are generally replaced during overhaul. The reuse of old seals is not recommended.

After the clutch pack has been disassembled, check the clutch drum, clutch housing, or transmission housing for defects. The splined area in the transmission housing should be carefully inspected for broken or chipped splines and other damage. If these splines are damaged, the case should be replaced. Clutch drums and housings should be checked for worn or damaged bushings. These should be replaced if they show signs of wear or damage. The splines inside the drum should be inspected for broken or chipped splines and other damage. If these splines are damaged, the drum or housing should be replaced. The outside surface of the housing should be inspected for signs of overheating and other defects. The flatness of the surface should also be checked by placing a straightedge across the surface (**Figure 9-31**). Look straight at the housing and the straightedge. If light shines between these two surfaces, the surface is distorted and the housing should be replaced.

Inspection and Cleaning

Once a clutch assembly has been taken apart, you may wish to inspect the clutch components or continue to disassemble the remainder of the clutch units in the transmission. If you choose the latter, make sure you keep the parts of each clutch separate from the others.

Figure 9-31 Checking the flatness of a housing with a straightedge.

> ✿✿ **SERVICE TIP** Used steels can be deglazed by sanding their faces with 320-grit emery paper, which can produce a dull surface on the plates.

Clean the components of the clutch assembly. Make sure all clutch parts are free of any residue of varnish, burned disc facing material, or steel filings. Take special care to wash out any foreign material from the inside of drums and the hub disc splines. If left in, the material can be washed out by the fresh transmission fluid and sent through the transmission. This can ruin the rebuild.

The clutch splines must be in good shape with no excessively rounded corners or shifted splines. Test their fit by trial fitting each new clutch disc on the splines. Move the discs up and down the splines to check for binding. If they bind, this can cause dragging of the discs during a time when they should be free floating. Replace the hubs if the discs drag during this check. Check the spring retainer. It should be flat and not distorted at its inner circumference. Check all springs for height, cracks, and straightness (**Figure 9-32**). Any springs that are not of correct height or that are distorted should be replaced. Many retainers have the springs attached to them by crimping. This speeds up production at the assembly line. Turning this type of retainer upside down is a quick check of spring length. Closely examine the Belleville spring for signs of overheating or cracking, and replace it if it is damaged.

Figure 9-32 Check the height of each one of the return springs.

The steel plates are commonly called **steels** by the trade.

Classroom Manual
Chapter 9, page 295

The steel plates should be checked to be sure they are flat and not worn too thin. Check all **steels** against the thickest one in the pack or a new one. Most steels will have an identifying notch or mark on the outer tabs. If the plates pass inspection, remove the polished surface finish and the steels are ready for reuse. The steel plates should also be checked for flatness by placing one plate on top of the other and checking the space on the inside and outside diameters. Clutch plates must not be warped or cone-shaped. Also, check the steel plates for burning and scoring and for damaged driving lugs. Check the grooves inside the clutch drum and check the fit of the steel plates, which should travel freely in the grooves.

Close inspection of the friction discs is simple. The discs will show the same types of wear as bands will. Disc facings should be free of chunking, flaking, and burned or blackened surfaces. Discs that are stripped of their facing have been overheated and subjected to abuse. In some cases, the friction discs and steel plates can be welded together. This occurs when the facing comes off the disc due to a loosening of the facing's bonding because of extreme heat. As the facing comes off, metal-to-metal contact is made between the discs and steel plates. The friction involved as the clutch tries to engage causes them to fuse together. This may lock the clutch in an engaged condition. Depending on which clutch is affected, driveability problems can occur, including driving in neutral, binding up in reverse (forward clutch seized), starting in direct drive, binding up in second (high/reverse clutch seized), and other problems that are less common.

If the discs do not show any signs of deterioration, squeeze each disc to see if fluid is still trapped in the facing material. If fluid comes to the surface, the disc is not glazed. Glazing seals the surface of the disc and prevents it from holding fluid. Holding fluid is basic to proper disc operation. It allows the disc to survive engagement heat, which would burn the facing and cause glazing. Fluid stored in the friction material cools and lubricates the facings as it transfers heat to the steel plates and also carries heat away as some oil is spun out of the clutch pack by centrifugal force. This helps avoid the scorching and burning of the disc.

Clutch discs must not be charred, glazed, or heavily pitted. If a disc shows signs of flaking friction material or if the friction material can be scraped off easily, replace the disc.

A black line around the center of the friction surface also indicates that the disc should be replaced. Examine the teeth on the inside diameter of each friction disc for wear and other damage (**Figure 9-33**).

Figure 9-33 Carefully examine the clutch discs.

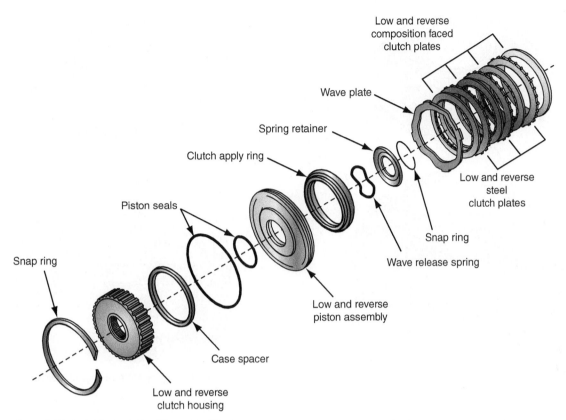

Figure 9-34 A clutch assembly with a wave plate.

Wave plates are used in some clutch assemblies to cushion the application of the clutch (**Figure 9-34**). These should be inspected for cracks and other damage. Never mix wave plates from one clutch assembly to another. As an aid in assembly, most wave plates have identifying marks.

The clutch pistons (**Figure 9-35**) are checked for cracks, warpage, and fit in their bores. Also check the check ball and the check ball bore for damage. Carefully examine the seal ring grooves and inside diameter of the piston for cracks, nicks, and burrs. Groove wear can be accelerated by excessive pump pressures. The excess pressure forces the seal rings against the sides of the grooves so hard that fluid cannot get between the ring and groove to lubricate the ring. If the bores of the clutch are severely grooved, a stuck pressure-regulator valve could be the problem.

The reverse side of the pump cover is the clutch support that incorporates seal rings for the fluid circuits leading to the clutch drums. The seal rings fit loosely into the grooves in the clutch support and rely on pump oil pressure to push them against the side of the groove to make the seal. The seal rings should be checked for side play in the grooves and for proper fit into the drum. Check the grooves for burrs, step wear, or pinched groove conditions. It should be noted that these seals rotate with the drum. Any condition that hinders rotation will cause the ring seals to bind, resulting in drum wear. This can destroy the drum if it is not corrected.

Carefully inspect aluminum pistons, which may have hairline cracks that will cause pressure leakage during use. This could cause clutch slipping as a result of lost hydraulic force pushing the piston against the discs. If burned discs were found in the clutch pack, be sure to check for cracks in the piston.

Stamped steel pistons have replaced most aluminum pistons because they are cheaper to produce. Aluminum pistons require a casting process followed by machining, whereas stamped pistons can be made by the thousands and may only require a simple spot-weld

Classroom Manual
Chapter 9, page 295

Piston
(note 2
anti-spin tabs)

Figure 9-35 A clutch piston with anti-spin tabs.

Check ball

Figure 9-36 A clutch piston with an annular check ball.

operation to prepare them for use. Stamped pistons show cracking more easily than aluminum. Also, look for any separation where spot welding is used on these pistons.

It is common to find annular check balls in clutch pistons (**Figure 9-36**). These balls allow for an air release from the bore while it is being filled with ATF and for a quick release of pressure when the clutch is released. Inspect each check ball to be sure it is free in its bore. Even after a thorough cleaning in solvent, the check balls may not be free. A fine wire can be pushed into the bore, followed by a spray of carburetor cleaner, to remove any stubborn deposits. The bore should then be blown out with compressed air.

A check ball may also be located in the drum and should be checked. However, it is often very difficult to use the same cleaning techniques as used on pistons. A quick way to determine if the check ball is free is to shake the clutch drum to hear the relief check ball rattle. If the check ball does not rattle, replace the drum.

The ability of the check ball to seal is also important. To check how well the ball seats in its bore, pour clean solvent into the bore. Observe the other end of the bore. If fluid leaks out, the ball is not seating and the piston or drum should be replaced.

Examine the outside surface of the drum for glazing. Glazing can be removed with emery cloth. Also check the drum's cylinder walls for deep scratches and nicks.

Inspect the front clutch bushing for wear and scores. If the bushing is worn, replace it. Also inspect any bushings found in the clutch drums for excessive wear, scoring, or looseness in the drum bore. Replace as needed.

Clutch Pack Reassembly

⚠ **Caution**

Use caution when installing the piston to prevent damage to the seals. Be careful not to stretch the seals during installation.

Begin assembly of a clutch unit by gathering the new seals and other new parts that may be necessary. Prior to installation, all clutch discs and bands are to be soaked in the type of transmission fluid that will be used in the transmission. The minimum soak time is 15 minutes. Be sure that all discs are submerged in the fluid and that both sides are coated.

Before fitting the new rubber seals to the piston, check them against the old seals. This will ensure correct sizing and shape of the new seal. Most overhaul kits include more seals than are required to complete the job. This is because changes in transmission design may

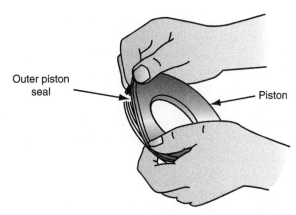

Outer piston
seal

Piston

Figure 9-37 Install new seals around the outside of the piston before reassembling the clutch.

dictate the use of a seal or gasket of unique design or size. Therefore, both the old design seal and the new design seal are included in the kit. The rebuilder must check to be sure the correct seal is being used. This also holds true for cast iron and Teflon ring seals.

Once the correct seals are chosen from the kit, they can be installed on the piston (**Figure 9-37**). Remember to position lip seals so they face the same direction that the fluid pressure comes into the drum.

Seals should be lubricated with ATF or TransJel. Never use chassis lube, "white lube," or motor oil. These will not melt into the transmission fluid as it heats up, but will clog filters and orifices or cause valves and check balls to stick.

Manufacturers switched to Teflon seals for these positions because they helped reduce the wear at the bore in the drum. During an overhaul, they may be replaced with hook-end-type steel rings.

Care must be taken to be sure that the ends of scarf-cut Teflon seals are installed correctly. These rings must also be checked for fit into the drum. They should have a snug, but not too tight, fit into the bore.

Some manufacturers use an endless type of Teflon seal. These seals must be installed with special sleeves and pushing tools to avoid overstretching the seal. First the seals are pushed over the installing sleeve to their location in the groove, then a sizing tool is slipped over the sleeve and seal ring to fit the seal to the groove.

 Caution

Never assemble a clutch assembly when it is dry. Always lubricate its components thoroughly with clean ATF.

Some rebuilders use an old credit card in place of the feeler gauge. Its wide surface and plastic edge will not cut into the rubber seals.

> **SERVICE TIP** Some rebuilders use a wax stick to coat lathe-cut seals for installation. This is available under the trade name Door-Ease. Its original use was to stop squeaks on rubber door bumpers and latches. If Door-Ease is used, do not coat the drum bore with ATF. The Door-Ease works fine by itself.

Assemble the piston, being careful not to allow the seal to kink or become damaged during installation. The piston can now be installed. Several methods may be used to aid piston installation. Lathe-cut seals can be fit into their bores by using a thin feeler gauge mounted on a handle. These are available from most automotive tool suppliers. Pistons with lathe-cut seals are installed by positioning the piston in the bore of the clutch drum. Then slowly work the piston down in the bore until resistance is felt. Using the feeler-type seal installer, work your way around the outer circumference of the seal, using a downward action followed by a clockwise pulling motion as you push the seal back into the groove in the piston.

Occasionally, a piston will not allow access to the outer seal area. A large chamfered edge is at the top of the seal bore to allow the seal to be worked into the bore without the help of any special tools. The piston can be installed by rotating the piston as you push down. Use even pressure to avoid binding the piston or cocking the piston in its bore. Uneven pressure can also cause the ring seal to be pushed out or to tear.

Pistons with lip seals require a more delicate installation. The lips can be bent back or torn unless proper caution is taken during installation. The basic shape of a lip seal makes it necessary to use an installation tool (**Figure 9-38**). The lip must be pushed back toward the piston body in order to allow the seal to enter the bore. Lip seals will often stick in snap-ring grooves as you try to slip the piston and seals into the drum. Piano wire installers can be used to roll lip seals back away from the snap-ring grooves or the bore of the drum (**Figure 9-39**). The round cross section of the wire prevents cutting or tearing the seal lip during installation.

While holding the piston as squarely in the drum as you can, work the tool around the lip to allow the seal to enter the bore or around the center of the drum. Do not apply too much downward force as the seal lip is worked into the bore. The piston will fall into place after the lip is fully inserted.

Pistons with multiple seals require special care to avoid damaging the other seals as you work on one seal. Multiple-seal piston installation is made simple by using plastic ring seal installers. These rings compress the seals back into the piston grooves so they will not hang up. Two installers are often used at the same time.

Regardless of the type of tool used to install the piston seals, always take your time to avoid tearing or rolling the new seals during installation.

When working with stamped steel pistons, install the ring spacer on the top side of the piston, making sure it has the correct thickness for the application.

Figure 9-38 Using a seal protector to install a piston into a drum without damaging the seals.

Figure 9-39 Piano-wire-type piston installation tool.

Once the piston seals enter the bore, push the piston all the way down until it stops. Then lift the piston up slightly. Push it back down to the bottom of the bore to be sure it is all the way down. After the piston is installed, rotate the piston by hand to ensure that there is no binding.

Reassemble the springs and retainer after the piston has been installed. Place the single or multiple spring set onto the piston using the pockets provided to locate them. Be sure the loose spring sets are spaced as they were during teardown. Set the retainer plate over the springs. Make sure the retainer is facing in the correct direction. On some transmissions, if the retaining ring and springs are faced in the wrong direction, the clutch assembly will work well during an air check but it will not apply the clutches during use. Captive spring retainers are simply set on the piston and require no other setup.

Position the spring compressor on the retainer plate and compress the springs. Be careful not to allow the retainer to catch in the snap ring groove while compressing the spring. This will bend or distort the plate, making it unsuitable for use. Remember to compress the springs only enough to get the snap ring into its groove. Use snap-ring pliers to expand or contract the snap ring. Once the snap ring is installed and fully seated, release the compressor.

Clutches with a Belleville return spring may not require a compressor for assembly. The Belleville spring is merely laid on the piston and centered in the bore. A wire ring is sometimes inserted on the piston where the Belleville spring touches the piston. This is used to prevent the steel spring from chafing the aluminum piston. If left out, there will be too much endplay and piston damage will occur.

The large snap ring that retains the Belleville spring is now inserted in its groove. This snap ring may be either flat or wavy construction. There may also be a plastic spacer ring used under the snap ring to center the spring. Be certain the snap ring seats firmly against the drum. Then install the pressure plate on top of the Belleville spring.

Continue clutch assembly by stacking the clutch pack. Begin by installing the dish plate, with the dish facing outward. If a wavy plate or cushion spring is used, it will normally be installed next to the piston. Alternately stack the steels and friction discs until the correct number of plates have been installed. Place the backing plate (the thickest steel plate) on top and install the retainer ring in its groove. Assemble the remaining clutch packs.

As mentioned before, all models of a transmission do not always have the same number of discs and steels in their clutches. Most overhaul kits supply enough discs and steels to rebuild all models. They are likely to have more discs and steels than are required. This is why a technician should always note how many discs and steels are in each pack while disassembling the transmission.

Trying to fit all of the supplied discs and steels may result in clutches with no free play or no room for the snap ring. If these problems come up while stacking the pack, refer to a service information. It may list the number of discs and steels for the model you are building. After the plates are installed, position the retainer plate and install the retaining snap ring. Proceed by measuring the clearance between the plate and snap ring.

Loose springs are used mostly with aluminum pistons.

⚠ **Caution**

Steel plates and friction discs from the same transmission may look the same but have different thicknesses. Compare the new plates and discs to the ones removed from each clutch during teardown. Incorrect disc and steel thicknesses can cause headaches. The new disc facings should match the ones that were removed in both thickness and grooving, if correct clutch engagement is to be maintained.

⚙ **SERVICE TIP** There are three different direct clutch counts for the A500 and A518. The only reliable way to determine the correct number of clutches for the transmission you are working on is to measure the snap-ring groove height in the drum. If the distance from the groove to the top of the drum is 0.485 inch, there should be five clutches and four steels. If the measurement is 0.35 inch, there should be six clutches and five steels. If the measurement is 0.1 inch, eight clutches and seven steels should be installed.

Clearance Checks

The clearance check of a multiple-disc pack is critical for correct transmission operation. Excessive clearance causes delayed gear engagements, while too little clearance causes the clutch to drag. Adjusting the clearance of multiple-disc packs can be done with the large outer snap ring in place.

With the clutch pack and pressure plate installed, use a feeler gauge to check the distance between the pressure plate and the outer snap ring (**Figure 9-40**). Clearance can also be measured between the backing plate and the uppermost friction disc. If the clutch pack has a waved snap ring, place the feeler gauge between the flat pressure plate and the wave of the snap ring farthest away from the pressure plate. Compare the distance to specifications. Attempt to set pack clearance to the smallest dimension shown in the chart.

Clearances can also be checked with a dial indicator and hook tool (**Figure 9-41**). The hook tool is used to raise one disc from its downward position and the amount that it is able to move is recorded on the dial indicator. This represents the clearance.

Another way to measure clearance is to mount the clutch drum on the clutch support and use 25–35 psi of compressed air through the oil pump body channels to charge the clutch. Clearance can be measured by mounting a dial indicator so that it reads backing plate movement as the air forces the piston to apply the clutch (**Figure 9-42**).

If the clearance is greater than specified, install a thicker snap ring to take up the clearance. If the clutch clearance is insufficient, install a thinner snap ring.

Another way to adjust clutch clearance is to vary the thickness of the clutch pressure plate. By using a pressure plate of desired thickness, you can obtain adequate clutch clearance (**Figure 9-43**).

Proper clearance can also be achieved through the use of special steel clutch plates. These are available from some manufacturers of transmission overhaul kits.

If specifications are not available, a general clutch pack clearance can be used. Allow 0.01 inch clearance for each friction disc in the clutch pack. For example, if the pack has four friction discs, set the minimum clearance to 0.04 inch. Since this is the minimum clearance, some transmission specialists then add 0.03 inch to the clearance. Therefore, in the example given, the maximum clearance would be 0.07 inch.

Figure 9-40 Measuring clutch clearance with a feeler gauge.

Figure 9-41 Using a hook tool to measure clutch clearance.

End plate

Top disc

Figure 9-42 Dial indicator setup for measuring clutch clearances.

Service Limit		
1st	0.65–0.85 mm	(0.026–0.033 in.)
2nd	0.65–0.85 mm	(0.026–0.033 in.)
3rd	0.40–0.60 mm	(0.016–0.024 in.)
4th	0.40–0.60 mm	(0.016–0.024 in.)
Low-hold	0.80–1.00 mm	(0.031–0.039 in.)

P/N	Plate No.	Thickness mm (in.)
22551–PX4–003	1	2.1 (0.082)
22552–PX4–003	2	2.2 (0.086)
22553–PX4–003	3	2.3 (0.090)
22554–PX4–003	4	2.4 (0.094)
22555–PX4–003	5	2.5 (0.098)
22556–PX4–003	6	2.6 (0.102)
22557–PX4–003	7	2.7 (0.106)
22558–PX4–003	8	2.8 (0.110)
22559–PX4–003	9	2.9 (0.114)

Number

End plate

Figure 9-43 A typical chart of the various pressure plate thicknesses available to correct clutch clearances.

⚙ **SERVICE TIP** FWD transaxles that utilize a chain to connect the torque converter output to the transaxle input shaft do not have a common shaft for the clutches and the oil pump. Therefore, it is impossible to test the drums from these transaxles while they are mounted on the oil pump. Instead, the drums should be tested while they are mounted on the front sprocket support. These are tested in the same way as any other drum.

Air Testing

After the clearance of the clutch pack is set, perform an air test on each clutch. This test will verify that all the seals and check balls in the hydraulic component are able to hold and release pressure.

Air checks can also be made with the transmissions assembled. This is the absolute best way to check the condition of the circuit, because there are very few components missing from the circuit. The manufacturers of different transmissions have designed test plates that are available to test different hydraulic circuits. Testing with the transmission assembled also allows for testing of the servos.

The 4T60-E and AX4S transmissions are common transaxles that have a chain link and an off-center input shaft.

Figure 9-44 To test a clutch, apply air to the hole in the test plate designated for that clutch.

To test a clutch assembly, install the oil pump assembly with its reaction shaft support over the input shaft and slide it into place on the front clutch drum. When the clutch drums are mounted on the oil pump, all components in the circuit can be checked. If the clutch cannot be checked in this manner, blocking off apply ports with your finger and applying air pressure through the other clutch apply port will work.

To conduct an air test with the transmission assembled, invert the entire assembly and place it in an open vise or transmission support tool. Pour clean ATF into the circuits that will be tested. Try to get as much fluid into the circuit as you can. Then, install the air test plate. On many transmissions, it is necessary to remove the valve body in order to mount the test plate. Then, air test the circuit using the test hole designated for that clutch (**Figure 9-44**). Be sure to use low-pressure compressed air (25–35 psi) to avoid damage to the seals. Higher pressures may blow the rubber seals out of the bore or roll them on the piston.

While applying air pressure, you may notice some escaping air at the metal or Teflon seal areas. This is normal, but not necessarily desirable, as these seals have a controlled amount of leakage designed into them. Basically, the less a seal leaks, the better it is. There should be no air escaping from the piston seals. The clutch should apply with a dull but positive thud. It should also release quickly without any delay or binding. Examine the check ball seat for evidence of air leakage.

> Only use compressed air that is free of dirt and moisture.

CLUTCH VOLUME INDEX

Many electronically controlled transmissions regulate shift feel by regulating the volume of fluid used to apply or release a clutch or brake. The computer monitors gear ratio changes by monitoring the input and output speed of the transmission. By comparing the signal from the turbine or input speed sensor to the signal from the output speed sensor, the computer can determine the operating gear ratio. The computer also monitors the **Clutch Volume Index (CVI)**.

CVIs represent the volume of fluid needed to compress a clutch pack or apply a brake band. Based on its constant calculation of gear ratio, the computer can monitor how long it takes to change a gear (**Figure 9-45**). If the time is too long, more fluid is sent to the apply device. The volume of the fluid required to apply the friction elements is continuously updated as part of the transmission's adaptive learning. As friction material wears, the volume increases.

> CVIs can be monitored with a scan tool and the readings compared to charts given in the service information. Keep in mind that the **Clutch Volume Index (CVI)** lets you know how much fluid is needed to apply or release a friction component. When it is out of spec, a problem is indicated.

The TCM times how long it takes to compress the clutch pack to change the gear ratio.

Input speed sensor

Output speed sensor

Input shaft

Output shaft

Piston and seal

Separator plate

Friction discs

Figure 9-45 Things considered by the system's computer (TCM) when it calculates CVI.

Clutch volume updates, as shown on a scan tool, are valuable in diagnosis of slippage problem and are typically updated as follows.

- The volume of the low/reverse clutch volume is updated during a 2–1 or 3–1 manual downshift to low gear with a throttle angle below 5 degrees and the transmission temperature above 110°F. The clutch volume should be between 82 and 134.
- The volume of the second clutch is updated during a 3–2 kickdown shift with a throttle angle between 20 degrees and 54 degrees and the transmission temperature above 110°F. The clutch volume should be between 25 and 64.
- The volume of the overdrive clutch is updated during a 2–3 upshift with a throttle angle between 5 degrees and 54 degrees and the transmission temperature above 110°F. The clutch volume should be between 30 and 64.
- The volume of the fourth clutch is updated during a 3–4 upshift with a throttle angle between 5 degrees and 54 degrees and a transmission temperature above 110°F. The clutch volume should be between 30 and 64.
- The volume of the underdrive clutch is updated during a 4–3 kickdown shift with the throttle angle between 20 degrees and 54 degrees and the transmission temperature above 110°F. The clutch volume should be between 44 and 92.

For the correct CVI specifications for the vehicle you are working on, refer to the appropriate service information.

GM's Transmission Adaptive Pressure

Similar to the CVI, GM's Transmission Adaptive Pressure (TAP) control system uses a line pressure control system during upshifts to compensate for the normal wear of the transmission. As the transmission's parts wear, shift times change. To compensate for these changes, the transmission control module (TCM) adjusts its commands to the various pressure control solenoids, to maintain ideal shift timing. The TCM monitors the signals from the input shaft speed sensor (ISS) and the output shaft speed sensor (OSS) during shifts. This data is used to determine if the shifts occurred too quickly or slowly. If the shifts did not take place according to the original calibration of the TCM, the TCM will change its commands to the solenoids.

ONE-WAY CLUTCH ASSEMBLIES

Classroom Manual
Chapter 9, page 289

⚠ **Caution**

Be careful not to damage the inner race of the roller clutch while removing it.

One-way clutches are holding devices. They allow a gearset member to rotate in one direction only and are used to ground or effectively hold a member by preventing it from rotating in the direction it would otherwise tend to rotate. One-way clutches are also used to transmit torque from one member to another. In these cases, the inner race is splined to one member of the gearset. The outer race is splined to another member.

An example of this is shown in **Figure 9-46**. The inner race of this overdrive clutch is splined to the coast clutch cylinder, which is splined to the overdrive sun gear. The one-way clutch's outer race is splined to the overdrive ring gear. When the overdrive ring gear has a counterclockwise torque from the movement of the vehicle, the gearset attempts to rotate the sun gear clockwise. The attempted counterclockwise rotation of the ring gear locks the one-way clutch, which in turn locks the ring gear with the sun gear. This action provides for direct drive through the gearset.

When overdrive gear is selected, the coast clutch cylinder is held by the inner race of the one-way clutch, which in turn holds the overdrive sun gear. This causes the planet gear assembly to walk around the sun gear and drive the ring gear clockwise with an overdrive. During coast, the clutch releases and allows the ring gear to rotate faster than the sun gear.

Because they are purely mechanical in nature, one-way clutches are relatively simple to inspect and test. They are not always easy to remove and the proper procedure must be followed. The following is given as an example and is the recommended procedure for removing and installing a reverse input and second roller clutch in a GM 4T45E transaxle:

1. Remove the reverse clutch housing.
2. Remove the roller clutch and the roller clutch retainer by prying up on the assembly with two screwdrivers (**Figure 9-47**). Note that the retainer is pressed onto the housing and should not be reused.
3. Rotate the roller clutch clockwise until it lines up with the roller clutch cam ramps (**Figure 9-48**).
4. Remove the roller clutch from the inside of the roller clutch cam. Inspect the roller clutch for excessive wear and damage.
5. Clean and dry each of the components.
6. To reassemble, reassemble the roller clutch assembly.
7. Press the roller clutch assembly onto the housing.

The durability of these clutches relies on constant fluid flow during operation. If a one-way clutch has failed, a thorough inspection of the hydraulic fluid feed circuit to the clutches must be made to determine if the failure was due to fluid starvation. The rollers

Figure 9-46 The inner race of this overdrive one-way clutch is splined to the coast clutch cylinder, which is splined to the overdrive sun gear. The outer race is splined to the overdrive ring gear.

Figure 9-47 Remove the roller clutch and the roller clutch retainer by prying up on the assembly with two screwdrivers.

Retainer

Roller cam ramps

Roller clutch

Clutch housing

Figure 9-48 To separate the cam ramp assembly from the rollers, rotate the roller clutch clockwise until it lines up with the roller clutch cam ramps.

and sprags ride on a wave of fluid, when they overrun. Since most of these clutches spend most of their time in the overrunning state, any loss of fluid can cause rapid failure of the components.

Sprags, by design, produce the fluid wave effect as they slide across the inner and outer races making them somewhat less prone to damage. Rollers, due to their spinning action, tend to throw off fluid, which allows more chance for damage during fluid starvation. During the check of the hydraulic circuit, take a look at the feedholes in the races of the clutch. Use a small diameter wire and spray brake cleaner to be certain the feedholes are clear. Push the wire through the feedholes and spray the cleaner into them. Blowing through them with compressed air after cleaning is recommended.

One-way clutches should be inspected for wear or damage, spline damage, surface finish damage, and damaged retainers and grooves. All smooth surfaces should be checked for any imperfections. If there is damage, the clutch should be replaced. Some clutch units have a plastic race that need to be carefully inspected for chafing or other damage. Many one-way clutches are retained by a snap ring that must be removed to remove the clutch.

Figure 9-49 One way to make sure you have installed the clutch in the correct direction is to determine the direction of lockup before installing the clutch. The operation of a one-way clutch is checked in the same way.

Classroom Manual
Chapter 9, page 289

Roller Clutches

Roller clutches should be disassembled for inspection. The surface of the rollers should have a smooth finish with no evidence of any flatness. Likewise, the race should be smooth and show no sign of brinelling, as this indicates severe impact loading. This condition may also cause the roller clutch to "buzz" as it overruns. The roller springs should be carefully checked for signs of overheating, wear, or damage. A bad spring may allow a roller to fall out of its race.

All rollers and races that show any type of damage or surface irregularities should be replaced. Check the folded springs for cracks, broken ends, or flattening out. All of the springs should have approximately the same shape. Replace all distorted or otherwise damaged springs. The cam race of a roller clutch may show the same brinelled wear due to impact overloading. The cam surface, like the smooth race, must be free of all irregularities.

When reassembling a roller clutch, make sure the rollers and springs are facing the correct direction (**Figure 9-49**). If they are reversed, the clutch will not lock in either direction.

Sprag Clutches

Sprag clutches (**Figure 9-50**) cannot easily be disassembled; therefore, a complete and thorough inspection of the assembly is necessary. Pay particular attention to faces of the sprags. If the faces are damaged, the clutch unit should be replaced. Sprags and races with scored or torn faces are an indication of dry running and require the replacement of the complete unit. Also check the sprags for looseness in their race. When assembling a sprag-type clutch, it is easy to place the sprags in the wrong direction. Make sure the unit is assembled correctly.

Mechanical Diodes

Some transmissions use a mechanical diode, rather than a sprag or roller clutch. These units use rectangular struts that engage a race when the unit is rotating in the direction of the torque. The struts are positioned between a plate with pockets for the retracted

Figure 9-50 A typical sprag one-way clutch assembly.

struts and a second plate with notches for the engagement of the struts. These units are often called "ratcheting" units because they make a ratcheting or clicking sound when they are freewheeling. This is how these are best checked. In one direction, the sound should be heard and the unit should not rotate in the opposite direction.

Installation

Once the one-way clutches are ready for installation, verify that they overrun in the proper direction. In some cases, it is possible to install these clutches backward; this would allow them to lock in wrong direction. This would result in some definite driveability problems. One way to make sure you have installed the clutch in the correct direction is to determine the direction of lockup before installing the clutch, then study the transmission's power flow and match the direction of the clutch with it. Most one-way clutches have some marking that indicates the direction in which the clutch should be set (**Figure 9-51**). During installation, use a new retaining snap ring.

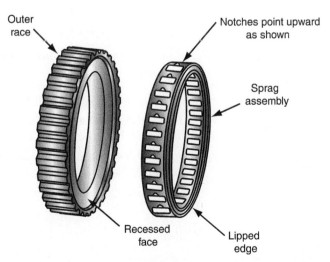

Figure 9-51 Most one-way clutches have some marking that indicates in which direction the clutch should be set.

CASE STUDY

A car dealer called to see if the automotive department at the college could help with a diagnostic problem. His customer had a new Ford pickup. The owner complained of a buzzing noise while driving. The technician at the dealership disassembled the transmission and installed an overhaul kit. The next day, the customer picked up his truck. He returned the next day, slightly annoyed, with the same complaint of buzzing while driving. The same technician once again disassembled the transmission and installed another overhaul kit. This time the technician road tested the truck after installing the kit. The noise was still there. By this time, both the owner of the truck and the technician were not very happy. After two rebuilds, the transmission still had the same noise.

The transmission was brought to the college and was checked out on the transmission "dyno." In the forward ranges, the buzzing was only apparent in second and third gear ranges. There was no buzzing in low or reverse gear ranges. The students, using a power flow chart, determined that the only geartrain member not in motion in low and reverse was the rear carrier. Because the rear carrier is held by the one-way clutch, we decided to disassemble the transmission to check the clutch. The inner race of the roller clutch had a flat spot from an apparent machining mishap during manufacture. Every time a roller went past the flat spot, it clicked into the flat, producing the buzz at road speeds. We installed a new roller inner race, which was supplied by the dealer, and, like magic, the buzz was gone.

Why were students able to determine the problem while an experienced technician could not find it? Very simply, the technician did not understand the basic theories of transmission operation. The road test very clearly showed that the buzzing occurred only in second and third gears. A basic knowledge of the Simpson geartrain and knowing how to read and use a flow chart would have helped the technician diagnose the problem. Unfortunately, this technician, like many others, only knew how to "put a kit in it" and hoped that the noise would go away. Just as a point of interest, according to the service manager, the technician loaded up her tools and quit when she learned that students, armed with a power flow chart and a shop manual, had diagnosed and repaired the problem that had her stumped.

ASE-STYLE REVIEW QUESTIONS

1. While visually inspecting a roller-type one-way clutch:

   Technician A replaces the clutch because the rollers have a smooth finish.

   Technician B replaces the clutch because the race shows signs of brinelling.

   Who is correct?

   A. A only   C. Both A and B

   B. B only   D. Neither A nor B

2. *Technician A* says most original equipment servo seals are of Teflon type.

   Technician B says a majority of replacement transmission gasket sets will supply cast iron hook-end seal rings in place of Teflon seals, and these are acceptable replacements.

   Who is correct?

   A. A only   C. Both A and B

   B. B only   D. Neither A nor B

3. Which of the following is *not* a good reason to replace a band?

   A. The band is chipped.

   B. The friction material is glazed.

   C. The friction material is soaked with fluid.

   D. The friction material has random wear patterns.

4. While conducting an air test on a transaxle:

   Technician A notices escaping air at the metal or Teflon seals. He proceeds to replace them.

   Technician B says the clutch should apply with a slight delay, then a dull thud.

   Who is correct?

   A. A only   C. Both A and B

   B. B only   D. Neither A nor B

5. While inspecting a clutch assembly that has an aluminum piston:

   Technician A carefully inspects the piston for hairline cracks that will cause pressure leakage, resulting in clutch slippage.

   Technician B says that if burned discs were found in the clutch pack, the piston should be checked for defects and cracks.

   Who is correct?

   A. A only   C. Both A and B

   B. B only   D. Neither A nor B

6. *Technician A* says a band apply pin may be selective.

   Technician B says most bands are not adjustable as routine maintenance.

   Who is correct?

   A. A only   C. Both A and B

   B. B only   D. Neither A nor B

7. *Technician A* says transmissions use compound planetary gears controlled by friction and reaction devices.

   Technician B says clutch and brake application charts are a must during diagnosis and transmission overhaul.

   Who is correct?

   A. A only   C. Both A and B

   B. B only   D. Neither A nor B

8. While assembling a multiple-friction disc pack:

   Technician A alternately stacks the steels and friction discs until the correct number of plates has been installed.

   Technician B installs all of the discs contained in the overhaul kit and corrects the clearance with a selective pressure plate or snap ring.

   Who is correct?

   A. A only   C. Both A and B

   B. B only   D. Neither A nor B

9. *Technician A* says an air test can be used to check servo action.

   Technician B says an air test can be used to check for internal fluid leaks.

   Who is correct?

   A. A only   C. Both A and B

   B. B only   D. Neither A nor B

10. While checking clutch discs:

    Technician A says the steel plates should be replaced if they are worn flat.

    Technician B says the friction discs should be squeezed to see if they can hold fluid. If they hold fluid and look OK, they are serviceable.

    Who is correct?

    A. A only   C. Both A and B

    B. B only   D. Neither A nor B

ASE CHALLENGE QUESTIONS

1. *Technician A* says friction discs that have blackened surfaces have been subject to overheating.

   Technician B says that if the friction facing has been burned or worn off, the discs can become welded together and can cause the vehicle to creep when the transmission is in neutral.

   Who is correct?

   A. A only   C. Both A and B

   B. B only   D. Neither A nor B

2. *Technician A* says excessive groove wear in a clutch piston can be caused by excessive pump pressures.

   Technician B says excessive wear, scoring, and grooving in the bore of a clutch can be caused by a stuck pressure-regulator valve.

   Who is correct?

   A. A only   C. Both A and B

   B. B only   D. Neither A nor B

3. *Technician A* says incorrect assembly of servos and accumulators can result in dragging bands and harsh shifting.

   Technician B says internal leaks at the servo or clutch seal will cause excessive pressure drops during gear changes.

   Who is correct?

   A. A only

   B. B only

   C. Both A and B

   D. Neither A nor B

4. *Technician A* says that if a one-way clutch has failed, a thorough inspection of the hydraulic fluid feed circuit to the clutch must be made to determine if the failure was due to fluid starvation.

   Technician B says that if the race of a roller-type one-way clutch has random hard spots, it is likely that the roller clutch will "buzz" as it overruns.

   Who is correct?

   A. A only

   B. B only

   C. Both A and B

   D. Neither A nor B

5. While inspecting a servo:

   Technician A says the servo needs new seals because both the inside and outside of the cover seal are wet.

   Technician B says that whenever the servo is disassembled, all rubber, cast iron, and Teflon seals should be replaced.

   Who is correct?

   A. A only

   B. B only

   C. Both A and B

   D. Neither A nor B

Name _____   Date _____

INSPECTING APPLY DEVICES

Upon completion of this job sheet, you should be able to inspect various apply devices of a transmission.

ASE Education Foundation Correlation ————————————————————————

This job sheet addresses the following **MAST** tasks:

C.19   Inspect clutch drum, piston, check-balls, springs, retainers, seals, and friction and pressure plates, bands and drums; determine needed action. (**P-2**)

C.22   Inspect roller and sprag clutch, races, rollers, sprags, springs, cages, retainers; determine needed action. (**P-2**)

We Support

ASE | **Education Foundation**

Tools and Materials

- Compressed air and air nozzle
- Lint-free shop towels
- Supply of clean solvent
- Service information

Describe the Vehicle Being Worked On

Year _____ Make _____ Model _____

VIN _____ Engine type and size _____

Model and type of transmission _____

Procedure

Task Completed

1. Disassemble the transmission into major units. Set each unit aside until this job sheet refers to it. Describe any problems you encountered while disassembling the transmission. Be sure to follow the procedures given in the appropriate service information while taking the transmission apart.

   _____

   _____

   _____

2. Clean each overrunning clutch assembly in fresh solvent. Allow the assemblies to air dry.   ☐

3. Check the rollers and sprags for signs of wear or damage. Describe your findings.

   _____

   _____

4. Check the springs for distortion, distress, and damage. Describe your findings.

   _____

   _____

5. Check the inner race and the cam surfaces for scoring and other damage. Describe your findings.

   _____

   _____

Task Completed

6. Check the condition of the snap rings. Describe your findings.

_____

_____

7. Summarize the condition of the overrunning clutch units.

_____

_____

_____

_____

8. Wipe the transmission's bands clean with a dry, lint-free cloth.   ☐

9. Check the bands for damage, wear, distortion, and lining faults. Describe your findings.

_____

_____

10. Inspect the band apply struts for distortion and other damage. Describe your findings.

_____

_____

11. Summarize the condition of the bands.

_____

_____

_____

_____

12. Clean servo and accumulator parts in fresh solvent and allow them to air dry.   ☐

13. Inspect their bores for scoring and other damage. Describe their condition.

_____

_____

14. Check the piston and piston rod for wear, nicks, burrs, and scoring. Describe your findings.

_____

_____

15. Inspect all springs for damage and distortion. Describe their condition.

_____

_____

16. Check the mating surfaces between the piston and the walls of their bores for scoring, wear, nicks, and other damage. Describe your findings.

_____

_____

17. Check the movement of each piston in its bore. Describe that movement.

_____

_____

18. Check the fluid passages for restrictions and clean out any dirt present in the passages. Describe your findings.

_____

_____

19. Summarize the condition of the servos and accumulators in this transmission.

_____

_____

_____

Problems Encountered

_____

_____

_____

Instructor's Comments

_____

_____

_____

Name _____   Date _____

CHECKING AND OVERHAULING A MULTIPLE-FRICTION DISC ASSEMBLY

Upon completion of this job sheet, you should be able to air test clutch packs and servos and disassemble, inspect, and reassemble a multiple-friction disc assembly.

ASE Education Foundation Correlation ——————————————————

This job sheet addresses the following **MAST** tasks:

C.20   Measure clutch pack clearance; determine needed action. (**P-1**)

C.21   Air test operation of clutch and servo assemblies. (**P-1**)

We Support
ASE | Education Foundation

Tools and Materials

- Feeler gauge set
- Special tools for the assigned transmission
- Emery cloth
- Crocus cloth
- Clean solvent
- Lint-free shop towels
- OSHA-approved air nozzle
- Air test plate
- Service information

Describe the Transmission Being Worked On

Model and type of transmission _____

Vehicle the Transmission Is From

Year _____ Make _____ Model _____

VIN _____ Engine type and size _____

Procedure

Task Completed

1. With the transmission disassembled, set aside each clutch pack for inspection.   ☐

2. Disassemble each clutch pack. Keep each assembly separate from the others. Record the number of steel and friction plates in each assembly.

   _____

   _____

   _____

3. Check the steel plates for discoloration, scoring, distortion, and other damage. Describe their condition.

   _____

   _____

   _____

4. If the steel plates are in good condition, rough up the shiny surface with emery (80-grit) cloth.   ☐

5. Inspect the friction plates for wear, damage, and distortion. Describe their condition.

   _____

   _____

   _____

6. Soak all new and reusable friction plates in ATF for at least 15 minutes before reassembling the clutch pack.   ☐

7. Check the pressure plate for discoloration, scoring, and distortion. Describe its condition.

   _____

   _____

   _____

8. Inspect the coil springs for distortion and damage. Describe their condition.

   _____

   _____

   _____

9. Inspect the Belleville spring for wear and distortion. Make sure to check the inner fingers. Describe your findings.

   _____

   _____

10. Check the clutch drums for damaged lugs, grooves, and splines. Also check them for score marks. Describe their condition.

    _____

    _____

    _____

11. Check the movement of the check balls in the drums. Describe your findings.

    _____

    _____

    _____

12. Summarize the condition of the clutch packs.

    _____

    _____

    _____

13. Reassemble the clutch packs with the required new parts and with new seals. Follow the recommended procedure for doing this.

    _____

    _____

    _____

14. Insert a feeler gauge between the pressure plate and the snap ring. What is the measured clearance? _____

15. What is the specified clearance? _____

16. If the measured clearance is not the same as the specified clearance, what should you do?

_____

_____

_____

17. Do what is necessary to correct the clearance.   ☐

18. Tip the clutch assembly on its side and insert a feeler gauge between the pressure plate and the adjacent friction plate. What is the measured clearance? _____

19. What is the specified clearance? _____

20. If the measured clearance is not the same as the specified value, what should you do?

_____

_____

21. Do what is necessary to correct the clearance.   ☐

22. After the clearance is set, check the service information for the proper procedure for air testing the clutch pack. Summarize the procedure.

_____

_____

_____

23. Place the pack or the partially assembled transmission in a vise or on a stand.   ☐

24. Apply low-pressure air to the designated test port. Pay attention to the sound of the air leaking and the activation of the disc pack. Describe what you heard.

_____

_____

25. Release the air and listen for the release of the pack. Describe what happened.

_____

_____

_____

26. Based on the above, what are your conclusions?

_____

_____

27. When the transmission is assembled, install the air test plate to the designated area on the transmission. What does the plate attach to?

_____

_____

28. What components can be checked with the test plate?

_____

_____

_____

29. Apply low-pressure air to each of the test ports and record the results.

_____

_____

_____

30. What is indicated by the results of this test?

_____

_____

_____

Problems Encountered

_____

_____

_____

Instructor's Comments

_____

_____

_____

Name _____   Date _____

CHECKING END CLEARANCE OF A MULTIPLE-FRICTION DISC PACK

Upon completion of this job sheet, you should be able to use a feeler gauge and a service information to check end clearance of a multiple-friction disc pack.

ASE Education Foundation Correlation

This job sheet addresses the following **MAST** task:

C.20   Measure clutch pack clearance; determine needed action. **(P-1)**

We Support
ASE | **Education Foundation**

Tools and Materials

- Feeler gauge set
- A multiple-friction disc pack
- Service information
- Goggles or safety glasses with side shields

Describe the Transmission the Pack Is From

Model and type of transmission _____

Vehicle the Transmission Is From

Year _____ Make _____ Model _____

VIN _____ Engine type and size _____

Procedure

1. Insert a feeler gauge between the pressure plate of the pack and the snap ring. What is the measured clearance? _____

2. What is the specified clearance? _____

3. If the measured clearance is not the same as the specified clearance, what should be done?

   _____

   _____

4. Tip the clutch assembly on its side and insert a feeler gauge between the pressure plate and the adjacent friction plate. What is the measured clearance? _____

5. What is the specified clearance? _____

6. If the measured clearance is not the same as the specified value, what should be done?

   _____

   _____

7. What other measuring instrument can be used to measure the end clearance of this pack?

   _____

   _____

Problems Encountered

_____

_____

_____

Instructor's Comments

_____

_____

_____

CHAPTER 10

REBUILDING COMMON TRANSMISSIONS

Upon completion and review of this chapter, you should be able to:

- Disassemble, service, and reassemble a Chrysler 41TE transaxle.
- Disassemble, service, and reassemble a Ford 4R44E transmission.
- Disassemble, service, and reassemble a Ford AX4S transaxle.
- Disassemble, service, and reassemble a GM 4L60 transmission.
- Use the typical procedure for overhauling an Allison series 1000 transmission.
- Disassemble, service, and reassemble a GM 4T60 transaxle.

- Use the typical procedure for overhauling a 6T70 transaxle.
- Disassemble, service, and reassemble a Nissan L4N71B/E4N71B transmission.
- Disassemble, service, and reassemble a Toyota A541E transaxle.
- Disassemble, service, and reassemble a Toyota A340 transmission.
- Disassemble, service, and reassemble a Toyota U660/U760 transaxle.

This chapter contains many photo sequences that show the step-by-step procedures for overhauling common transmissions and transaxles. Although a particular model of transmission was chosen for each set of photographs, variations of the same transmission can be used. The procedures in this chapter are for educational purposes and do not include the many changes that a different model would require.

In addition, please note that these sequences do not include detailed procedures for rebuilding subassemblies such as the valve body and clutch packs. Detailed information on these, as well as other subassemblies, is given in other chapters of this manual.

Basic Tools

Mechanic's basic tool set

Appropriate service information

Transmission holding fixture

Torque wrench

Rubber-tipped air nozzle

Pan of clean ATF

Feeler gauge

CHRYSLER TRANSMISSIONS

Many different transaxle models have been used by Chrysler over the years. The 41TE, also called the A604, is commonly found in minivans up until around 2010. This is a four-speed unit with electronic controls. The compound planetary gearset in the A604 does not have a common member; rather, the two planetary units are connected in tandem, with the front carrier connected to the rear ring gear and the rear carrier connected to the front ring gear.

To control the forward gear ranges, the A604 relies on an input clutch assembly, a 2nd/4th clutch, and a low/reverse clutch. The input clutch assembly houses the three different input clutches: the underdrive, overdrive, and reverse clutches.

The underdrive clutch is applied in first, second, and third gears. When applied, the underdrive hub drives the rear sun gear. The overdrive clutch is applied in third and

Classroom Manual
Chapter 10, page 314

Special Tools

Dial indicator and holding fixture

Clutch compressor tool

Oil pump puller

Seal remover/ installer

Output shaft

Holding tool

overdrive gears. When the overdrive clutch is applied, it drives the front planet carrier. The underdrive clutch and the overdrive clutch are applied in third gear, giving two inputs to the gearset, which provides for direct drive. The reverse clutch is only applied in reverse and drives the front sun gear assembly.

The 2nd/4th gear and low/reverse units are holding devices and are placed at the rear of the transmission case. The 2nd/4th clutch is engaged in second and fourth gears and grounds the front sun gear to the case. The low/reverse clutch is applied in Park, Reverse, Neutral, and first gear. When this clutch is engaged, the front planet carrier and rear ring gear assembly is grounded to the case. The overhaul procedure for this transaxle is shown in **Photo Sequence 15**.

PHOTO SEQUENCE 15
Typical Procedure for Overhauling a 36RH Transmission

P15-1 Check and record the endplay of the input shaft, then unbolt and remove the oil pan and gasket.

P15-2 Unbolt and remove the oil filter. Then unbolt and remove the valve body by first disconnecting the parking lock rod from the manual lever and unbolting the TCC solenoid.

P15-3 Remove the accumulator piston spring and lift piston from the case.

P15-4 Remove parking lock rod. Then unbolt and remove the extension housing. There is a cover plate in the bottom of the extension housing that once removed will allow access to the output shaft's snap ring. This ring must be expanded while pulling the extension housing off the case.

P15-5 Tighten the front band adjustment screw to prevent damage to the discs during disassembly.

P15-6 Remove the oil pump bolts and pull the oil pump out of the case with special pullers.

PHOTO SEQUENCE 15 (CONTINUED)

P15-7 Loosen the front band adjustment screw and remove the band strut and anchor. Then remove the band.

P15-8 Remove the retaining or snap ring, then the front clutch assembly, with the input shaft, from the case.

P15-9 Remove the front planetary gearset.

P15-10 Loosen the rear band adjustment screw, and then remove the band strut and band. Then remove the output shaft with the governor attached to it.

P15-11 Compress the return spring for the front servo and remove the retaining snap ring.

P15-12 Remove the servo assembly from the case.

P15-13 Compress the return spring for the rear servo and remove the retaining snap ring.

P15-14 Remove the servo assembly from the case.

P15-15 Inspect all servo piston seals.

PHOTO SEQUENCE 15 (CONTINUED)

P15-16 Reinstall the servo assemblies after careful inspection.

P15-17 Inspect the planetary gearset and drive shells. Replace components as needed.

P15-18 Inspect the bands and their struts and anchors. Replace as needed.

P15-19 Disassemble and inspect the clutch units. Replace parts as needed. Remember to allow the discs to soak in ATF before assembling the clutch pack.

P15-20 Replace the pistons' oil seal rings in the clutch assemblies.

P15-21 Reassemble the clutch packs, and then check the clearances.

P15-22 Unbolt the stator support from the oil pump housing. Separate the support from the oil pump.

P15-23 Check the wear of the gears and the pump itself. Replace parts as needed.

P15-24 Replace the pump seals and reassemble the pump. Torque the bolts to specifications.

PHOTO SEQUENCE 15 (CONTINUED)

P15-25 Disassemble, clean, and inspect the valve body.

P15-26 Bolt valve body together and tighten bolts to specifications.

P15-27 Install the output shaft with the governor into the case. Then, install the rear band and anchor assembly and tighten the adjustment screw.

P15-28 Install the rear planetary gearset and the drum. Make sure all thrust washers are being installed in their correct locations.

P15-29 Install the front clutch assembly and the front band assembly, and then tighten the adjustment screw.

P15-30 Assemble servos with new seals. Then install the oil pump and torque the bolts in the proper sequence and to the proper torque.

P15-31 Adjust both bands according to specifications.

P15-32 Air test the entire transmission before installing the oil filter and the oil pan.

Special Tools

Dial indicator and
  holding fixture

Clutch compressor tool

Oil pump puller

Seal remover/installer

Servo cover
  compressor tool

Depth micrometer
  and fixture

Seal protector

Classroom Manual
Chapter 10, page 323

FORD MOTOR COMPANY TRANSMISSIONS

Some of the most popular transmissions in cars and trucks are the 4R70W, 4R100, 4R44E, and 5R55. Another commonly used transmission is a heavy-duty version of the 4R44E, referred to as the 4R55E. The 4R44E is a Simpson gearset–based four-speed transmission. It has full electronic controls and relies on three multiple-friction disc packs, three brake bands, and two one-way clutches to provide the various gear ratios. The five-speed 5R44/5R55 series is based on the 4R55E. The overhaul procedures for this transmission are shown in **Photo Sequence 16**

PHOTO SEQUENCE 16
Typical Procedure for Overhauling a 4R44E Transmission

P16-1 Remove the torque converter.

P16-2 Remove the input shaft. Note that the two splined ends of the shaft are different.

P16-3 Remove the bellhousing retaining bolts, and then the bellhousing and oil pump as an assembly. Also remove the thrust washer and the gasket.

P16-4 Separate the oil pump from the bell-housing.

P16-5 Remove the steel plate from the housing.

P16-6 Remove the oil pan bolts, and then the oil pan.

P16-7 Remove the retaining bolt for the filter screen. Remove filter screen, then remove the detent spring under the screen.

P16-8 Disconnect the wires at the solenoids.

P16-9 Remove the valve body retaining bolts. While moving the valve body away, unlock and remove the selector lever connecting link. Then remove the valve body and the gasket. Note and mark the valve body bolts; they are of different lengths.

P16-10 Remove the Allen head retaining bolt holding the center support from within the case's worm tracks.

P16-11 Remove the solenoid harness and the connector from the case.

P16-12 Remove the bolts retaining the extension housing to the case, then remove the extension housing, parking pawl, and pawl spring.

P16-13 Loosen the overdrive band locknut and back off the adjustment screw. Discard the locknut.

P16-14 Remove the anchor and the apply struts.

P16-15 Lift out the clutch assembly and the band. Mark the band as "Overdrive" and label the anchor end of the band.

PHOTO SEQUENCE 16 (CONTINUED)

P16-16 Remove the overdrive one-way clutch.

P16-17 Pull the overdrive planetary assembly out of the case.

P16-18 Remove the center support retaining snap ring.

P16-19 Remove the overdrive apply lever and shaft. Then remove the overdrive control bracket from the valve body side of the case. The overdrive apply lever shaft is longer than the intermediate apply lever shaft.

P16-20 Remove and mark the thrust washer from the top of the center support.

P16-21 Carefully remove the center support bearing by prying on it evenly and upward.

P16-22 Remove and mark the thrust washer from below the center support.

P16-23 Loosen the intermediate band locknut and back off the adjusting screw. Discard the locknut. Remove the anchor and apply struts. Then remove the reverse/high and forward clutch assembly.

P16-24 Remove the intermediate band. Mark the band as "Intermediate" and label the anchor end of the band.

PHOTO SEQUENCE 16 (CONTINUED)

P16-25 Remove the forward planetary gear assembly with its thrust washer. Mark the thrust washer.

P16-26 Remove the sun gear shell.

P16-27 Remove the planet and ring gear assembly.

P16-28 Remove the large snap ring from the case to release the reverse planetary gear assembly.

P16-29 Remove the reverse planetary gear assembly with its thrust washer. Mark the thrust washer.

P16-30 Remove the small snap ring on the output shaft and remove the output shaft ring gear.

P16-31 Remove the low/reverse drum and one-way clutch assembly.

P16-32 Remove the low/reverse servo from the valve body side of the case.

P16-33 Remove the low/reverse band. Then remove and mark the thrust washer.

P16-34 Remove the intermediate band apply lever and the shaft.

P16-35 Turn the transmission so that the output shaft points upward. Then lift out the output shaft.

P16-36 Remove the park gear/collector body assembly from the rear of the case. Then remove and mark the thrust washer.

P16-37 Remove the intermediate and over-drive servos' cover snap rings, covers, pistons, and springs. Mark the overdrive servo cover.

P16-38 Remove the neutral/safety switch.

P16-39 Remove the linkage centering pin, manual lever nut, manual lever, internal park pawl rod, and detent plate assembly. Then remove the lever shaft oil seal.

P16-40 Remove TCC and 3–4 shift solenoid connector. To remove the connector, a tab on the outside of the case must be depressed while the connector is pulled.

P16-41 Before installing the center support into the case, install new high clutch seals onto the support hub. You must size these seals. Failure to do so will cause the seals to be cut or rolled over during installation. Use the overdrive brake drum for sizing. Carefully rotate the center support while inserting it into the drum housing. Apply a liberal amount of transmission assembly lube to the center support hub and seals.

P16-42 Make sure the center support is seated fully into the overdrive drum. Set the assembly aside to allow the assembly to stand for several minutes; this will allow the seals to seat in their grooves.

PHOTO SEQUENCE 16 (CONTINUED)

P16-43 Position the thrust washer in the rear of the case. Install the collector body and the output shaft. Tighten the bolts to specifications.

P16-44 Position the thrust washer into the case from the front. Then install the low/reverse brake drum, output shaft ring gear, and snap ring. Position the reverse planet assembly and thrust washers.

P16-45 Install the snap ring to retain the center support. Ends of the snap ring should be positioned in the wide shallow cavity at the five o'clock position. Then install the Allen head bolt that retains the center support to the case. Note: Two types of snap rings are used. One has no notches and the other has a notch on its outer and inner diameters. The outer diameter notch should be positioned on the left.

P16-46 Replace the piston's O-ring and install the intermediate servo piston assembly, piston cover, and snap ring.

P16-47 Install the intermediate servo apply lever and shaft into the case. Then install the complete forward clutch and reverse/high clutch assemblies. Rotate the transmission so that the output shaft points downward.

P16-48 Install the intermediate band, apply strut, and anchor strut. Temporarily install the input shaft.

PHOTO SEQUENCE 16 (CONTINUED)

P16-49 Insert the input shaft with its short splines facing down, through the center support, and into the splines of the forward clutch cylinder. Carefully place the center support into the case. Do not seat it into the intermediate brake drum but make sure it is in line with the retaining bolt hole in the worm tracks.

P16-50 Install the sun gear and support into the overdrive planet assembly and one-way clutch. Make sure the needle bearing race is centered inside the planetary assembly.

P16-51 Install the overdrive planet assembly and one-way clutch into the case. Install the overdrive drum assembly, overdrive bracket, apply lever, and shaft. Then install the overdrive band, apply strut, and anchor strut.

P16-52 Make sure the needle bearing race in the overdrive planetary is centered and the overdrive clutch is fully seated. Then place the selective washer on top of the overdrive clutch drum and temporarily install the pump assembly (without its gasket) into the case.

P16-53 Check the endplay using a dial indicator. If the endplay exceeds limits, replace the selective washer with one that will allow for proper endplay. After the endplay is correct, remove the oil pump and the selective thrust washer.

P16-54 Install a new oil pump seal and position the separator plate onto the converter housing. Position the pump assembly onto the separator plate and converter housing. Install the retaining bolts finger-tight.

P16-55 With the recommended tool, align the pump in the converter housing to prevent seal leakage, pump gear breakage, or bushing failure.

P16-56 Install and tighten the retaining bolts. Then remove the alignment tool.

P16-57 Install the input shaft into the pump and install the converter into the pump gears. Rotate the converter to check for free movement.

PHOTO SEQUENCE 16 (CONTINUED)

P16-58 Install a new lock nut on the overdrive band adjusting screw. Tighten the adjusting screw to 10 lbs. ft. and back it off exactly two turns. Then tighten the lock nut to 35–45 lbs. ft. Repeat this procedure for the intermediate band. Perform air pressure tests to ensure proper transmission operation.

P16-59 Install the shift lever oil seal. Then install the internal shift linkage, external manual control lever, and centering pin. Then tighten the nut to specifications. Install the O-ring and retaining nut. Install the neutral/safety switch.

P16-60 Using tapered punches, align the valve body to the separator plate and gasket. Use the transmission assembly lube to hold the gasket in place, and then tighten the retaining bolts to specifications. Attach and lock the link to the manual valve. Carefully ease the valve body into the case and install and tighten the retaining bolts. Make sure the correct length bolts are installed in their proper locations.

P16-61 Connect the wires to the various solenoids.

P16-62 Then install the reverse servo piston assembly and spring. Make sure the piston rod

is correctly seated into the reverse band apply end. Then install the correct servo rod and cover, using a new servo cover gasket. Then tighten the attaching bolts. Install new O-rings on the filter screen. Then install the filter screen and the oil pan with a new gasket. Tighten the pan bolts to specifications in two steps.

P16-63 Install the parking pawl and return spring into the extension housing. Then install the extension housing with a new gasket. Make sure the parking pawl rod is fully seated. Then install and tighten the extension housing bolts.

P16-64 Complete the assembly by installing the input (pump drive) shaft and the torque converter. Make sure the torque converter is properly seated.

Ford Transaxles

The AX4S four-speed transaxle is used in many FWD vehicles and is commonly found in Taurus, Windstar, and Sable models. It relies on two simple planetary units that operate in tandem. The planet carriers of each planetary unit are locked to the other planetary unit's ring gear. Each planetary set has its own sun gear and set of planet pinion gears. The AX4S uses four multiple-friction disc assemblies, two band assemblies, and two one-way clutches to control the operation of the planetary gearset. All the multiple-disc packs are applied hydraulically and are released by several small coil springs when hydraulic pressure is diverted from the clutch's piston. The overhaul procedure for this transaxle is shown in **Photo Sequence 17**.

Classroom Manual
Chapter 10, page 322

PHOTO SEQUENCE 17
Typical Procedure for Overhauling an AX4S Transaxle

P17-1 Remove the torque converter and mount the transaxle in a vertical position on a holding fixture. Then drain the fluid.

P17-2 Move the transaxle to a horizontal position. Remove the two governor cover bolts, and then remove the cover and seal. Discard the seal. Lift the governor, speedometer drive gear assembly, and bearing from the case.

P17-3 Remove the bolts from the overdrive servo cover. Then mark the alignment of the cover and remove it, the piston assembly, and the spring. Discard the O-ring from the cover. Then remove the low-intermediate servo cover bolts. Mark the alignment of the cover and remove it, the piston assembly, and the spring. Remove and discard the gasket.

P17-4 Remove the neutral safety switch bolts and then remove the switch.

P17-5 Remove the dipstick tube attaching bolt and pull the tube from the case. Then remove the chain cover bolts from inside the torque converter housing.

P17-6 Remove the valve body cover bolts, valve body cover, and gasket. Discard the gasket.

P17-7 Unplug the electrical connectors from the pressure switches and solenoid. Use both hands to do this and do not pull on the wires; pull on the connector. Compress the tabs on both sides of the five-pin bulkhead connector from inside the chain cover and remove the connector and wiring.

P17-8 Using a 9-mm wrench on the flats of the manual shaft, rotate the shaft clockwise so that the manual valve is all the way in. Note the location, according to length, of the oil pump and valve body bolts. Then remove the bolts. Do not remove the two bolts that hold the oil pump and the valve body together. The oil pump cover bolts should also not be removed at this time.

P17-9 Push in the TV plunger. Pull the pump and the valve body assembly outward. Rotate the valve body clockwise and remove the manual valve link from the manual valve and disconnect it from the detent lever. Now remove the oil pump and the valve body assembly.

P17-10 Remove the throttle valve bracket bolts and then remove the bracket. Pull the oil pump drive shaft out of the case and remove and discard the Teflon seals from the shaft.

P17-11 Place the transaxle in a vertical position. Then remove and discard the circlip for the left output shaft. Then remove the shaft's seal with a seal remover.

P17-12 Mark the location and length of the chain cover bolts. Then remove them and the chain cover. Discard the gasket.

P17-13 Mark and remove the accumulator springs from the chain cover.

P17-14 Simultaneously lift out both sprockets with the chain assembly. Mark and remove the thrust washers from the drive and driven sprocket supports. Inspect the drive sprocket support bearing to determine if it needs replacing.

P17-15 Remove the lock pin and roll pins from the manual shaft. Be careful not to damage the machined surfaces. Slide the shaft out of the case, and then pry the seal from the case.

PHOTO SEQUENCE 17 (CONTINUED)

P17-16 Use a straightedge and note whether the machined bolt hole surface of the driven sprocket support is above or below the case's machined surface. Remember this for reassembly. Then remove the driven sprocket support assembly. It may be necessary to back out the reverse clutch anchor bolt.

P17-17 Remove the Teflon seals from the support assembly. Mark and remove the thrust washers and needle bearing. Then remove the plastic overdrive band retainer and the overdrive band.

P17-18 Lift the front sun gear and the shell assembly out of the case.

P17-19 Remove the oil pan cover bolts and the cover. Discard the gasket. Remove the reverse apply tube/oil filter bolt, bracket, and oil filter screen.

P17-20 Remove the tube retaining bracket bolts and brackets. Note: For complete transaxle disassembly, the reverse apply tube MUST be removed prior to removing the reverse clutch. The rear lube tube must also be removed and the seal replaced whenever the differential is removed.

P17-21 Remove the park rod abutment bolts. Remove the roll pin for the park pawl shaft. Then, using a magnet, remove the park pawl shaft, park pawl, and return spring.

P17-22 Place the transaxle in a horizontal position. Grasp the outer diameter of the reverse clutch cylinder with your fingertips and slide the assembly out of case.

P17-23 Place the transaxle in a vertical position. Grasp the front planetary shaft and lift out both the front and rear planetary assemblies. Lift out the low-intermediate drum and sun gear assembly. Then remove the low-intermediate band.

P17-24 Remove the snap ring for the final drive assembly from the case using a screwdriver inserted through the side of the case. Lift out the final drive assembly using the output shaft.

PHOTO SEQUENCE 17 (CONTINUED)

P17-25 Remove the final drive ring gear, thrust washer, and needle bearing. Remove and discard the rear lube tube seal by tapping it toward the inside of the case. Then install the converter oil seal and right output shaft seal using a seal installer.

P17-26 Place the transaxle in a vertical position. Install the needle bearing over the case boss with its flat side facing up and outer lip facing down.

P17-27 Install the final drive ring gear with its external splines up. Lightly tap the ring gear to fully seat it in the case.

P17-28 Reassemble the governor drive gear, differential assembly, final drive sun gear, parking gear, needle bearings, rear planetary support, and thrust washer.

P17-29 Lower the final drive assembly into the case. Install the snap ring and align it with the low-intermediate band anchor pin.

P17-30 Mount a dial indicator with its plunger on the end of the output shaft. Check end clearance. If the clearance is not within specifications, replace the thrust washer with one that will bring the clearance to specifications. Available thicknesses are: 0.04520.049 inch (1.15–1.25 mm) Orange, 0.05520.059 inch (1.40–1.50 mm) Purple, and 0.06420.069 inch (1.65–1.75 mm) Yellow

P17-31 Install the park pawl, return spring, park pawl shaft, and locator pin. Make sure the park pawl engages park gear and returns freely. Install the park rod actuating lever and park rod into the case. Install the park rod abutment and start the abutment bolts. Push in the park pawl and locate the rod between the pawl and the abutment.

P17-32 Using a 3/8 inch drift, gently install lube tube seal flush against the rear case support. Install the low-intermediate band and align the anchor pin pocket with the anchor pin. Install the low-intermediate drum and the sun gear assembly.

P17-33 Reassemble the ring gear and shell assembly, rear planetary, needle bearing, front planetary, and snap ring. Carefully slide the planetary assembly over the output shaft.

PHOTO SEQUENCE 17 (CONTINUED)

P17-34 Lower the reverse clutch into the case and start clutch plate engagement. Align the clutch cylinder anchor pin pocket with the anchor pin case hole. Use the intermediate clutch hub to complete clutch plate engagement and fully seat the reverse clutch. Rotate the planet carrier with the hub to engage splines.

P17-35 Start the reverse anchor pin bolt, but do not tighten. Reassemble the forward, direct, and intermediate clutch assembly. Lower the assembly into the case. Align the shell and sun gear splines with those in the forward planetary.

P17-36 Install the overdrive band into the case. Install the plastic retainer with the cross-hairs facing up.

P17-37 Check the drive sprocket end clearance to determine required thrust washer thicknesses. To do this, you must first determine if the machined bolt hole surfaces on the driven sprocket support are above or below the case machined surface. If they are above the surface, place a depth micrometer on the machined bolt hole surface and measure the distance to the case's machined surface. If they are below the case's surface, place the depth micrometer on the case's surface and measure the distance to the machined bolt hole surface. Install the correct thrust washer. Repeat this procedure for the drive sprocket.

P17-38 Tap the seal for the manual shaft into the case. Start the manual shaft through the seal and slide the manual detent lever onto the shaft. Then slide the shaft through the park rod actuating lever and tap it into the bore in the case. Install a new lock pin through the case hole. Make sure it is aligned with the groove in the shaft. Install new roll pins.

P17-39 Align the tabs of the thrust washers and install them onto the drive and driven sprocket supports. Lubricate and install the cast iron sealing ring onto the input shaft. Install the chain over the sprockets.

PHOTO SEQUENCE 17 (CONTINUED)

P17-40 Lower the chain assembly onto sprocket supports. Rotate the supports while lowering the chain assembly to ensure that the sprockets are fully seated on the supports.

P17-41 Install and align the thrust washers into the chain cover. Install a new chain cover gasket onto the cover. Then install the correct accumulator springs in their proper location. Carefully apply downward pressure to overcome accumulator spring pressure, and start two of the chain cover bolts. Position these bolts 180 degrees apart.

P17-42 Start the remaining chain cover bolts and tighten them in sequence and to specifications. The input shaft should now have little endplay and should rotate freely. If it does not rotate freely, remove the chain cover and check for a damaged cast iron seal.

P17-43 Tighten the park rod abutment bolts, reverse anchor pin bolt, and locknut to specifications. Lightly tap the lube tubes until they are fully seated in their bores. Install the tube retaining brackets.

P17-44 Install the O-rings onto the oil filter and press the filter into the case. Install the reverse apply tube/oil filter bracket and bolt. Install the oil pan with a new gasket. Tighten the bolts to specifications.

P17-45 Install new Teflon seals onto the pump drive shaft and install the shaft.

P17-46 Install the TV bracket and tighten the bracket bolts to specifications. Connect the manual valve link to the detent lever.

P17-47 Start the oil pump and the valve body assembly over the pump shaft and connect the manual valve link to the manual valve.

P17-48 Check the alignment of the valve body to the pump and make sure the gasket is the correct one for the application. Then install the oil pump and the valve body assembly.

P17-49 Install the bulkhead connector and other electrical connectors. Install the neutral start switch. With the manual shaft in neutral, align the switch using a 0.089-inch drill bit. Then tighten the switch to specification.

P17-50 Install the valve body cover with a new gasket. Then tighten the bolts to specifications. Install the dipstick tube grommet and dipstick tube in the case.

🔧 Special Tools

Dial indicator and
holding fixture

Clutch compressor tool

Oil pump puller

Seal remover/
installer

Servo cover
compressor tool

Output shaft support
tool

GENERAL MOTORS TRANSMISSIONS

Most General Motors transmissions and transaxles are based on the Simpson geartrain. Many different versions of the same basic design can be found. For many designs, the differences lie in the electronic controls and their intended use. The 4L60-E remains very popular with those in high performance and truck enthusiasts. It is a four-speed unit. The available gear ratios are provided by two planetary gearsets. The ring gear of one gearset is splined to the carrier of the other. One gearset is referred to as the input gearset, and the other is called the reaction gearset. To control gearset activity, five multiple-friction disc assemblies, one sprag one-way clutch, one roller one-way clutch, and a band are used. Earlier 4L60-Es had a one-piece case casting that incorporated the bellhousing; later 4L60-Es had separate bellhousing. The extension housing is a separate casting bolted to the rear of the case. The overhaul procedures for this transmission are outlined in **Photo Sequence 18**.

Classroom Manual
Chapter 10, page 335

PHOTO SEQUENCE 18
Typical Procedure for Overhauling a 4L60 Transmission

P18-1 Remove the torque converter.

P18-2 Mount the transmission in a holding fixture and position it so that the oil pan is facing up. Position a servo cover compressor on two oil pan bolts. Then compress the servo cover and remove the retaining ring, servo cover, and O-ring.

P18-3 Remove the 2-4 servo assembly. Servo pin length should be checked prior to disassembly. This helps to identify any wear on the 2-4 band and/or reverse input drum.

PHOTO SEQUENCE 18 (CONTINUED)

P18-4 Remove the governor cover and O-ring, and then remove the governor assembly.

P18-5 Remove the speed sensor assembly or the speedometer driven gear. This type of transmission will be equipped with one or the other.

P18-6 Remove the extension housing bolts, then pull the housing away from the case.

P18-7 Remove the oil pan bolts, pan, and gasket.

P18-8 Remove the oil filter.

P18-9 Remove the solenoid retaining bolts. Then remove the solenoid and O-ring.

P18-10 Remove the wiring harness. Note the location and position for reference during reassembly.

P18-11 Loosen and remove the valve body and auxiliary valve body attaching bolts. Mark the location of each bolt, as they are of different lengths. Also mark the location of any check balls you encounter.

P18-12 Carefully remove the 1–2 accumulator cover retaining bolts, cover and pin assembly, piston, seal, and spring.

PHOTO SEQUENCE 18 (CONTINUED)

P18-13 Remove spacer plate, and note the location of the check balls and filter.

P18-14 Remove the 3–4 accumulator spring, piston, and pin. Note the location of the spacer plate, gasket, check balls, and filters.

P18-15 Remove the oil pump retaining bolts. Then using the correct puller, remove the oil pump.

P18-16 Remove the oil pump seal and gasket. Then remove the reverse input clutch-to-pump thrust washer from the pump.

P18-17 Remove the reverse and input clutch assembly by lifting it out with the input shaft.

P18-18 Remove the 2–4 band anchor pin. Then remove the band from the case.

P18-19 Remove the input sun gear. Install an Output Shaft Support tool onto the output shaft, then remove the input carrier to output shaft retaining ring.

P18-20 Remove the input carrier and the output shaft. Then remove the input carrier thrust washer from the reaction carrier shaft.

P18-21 Remove the input ring gear and the reaction carrier shaft.

PHOTO SEQUENCE 18 (CONTINUED)

P18-22 Remove the reaction sun shell and thrust washer. Then remove the sun shell-to-clutch race thrust washer and the support-to-case retaining ring.

P18-23 Remove the spring retainer from the low/reverse support, then remove the reaction sun gear, low/reverse clutch race, clutch rollers, support assembly, and reaction carrier assembly.

P18-24 Remove the low/reverse clutch assembly.

P18-25 Remove the reaction ring gear and bearing assembly. Then remove the reaction ring gear support-to-case bearing assembly.

P18-26 Remove the parking lock bracket retaining bolts and the lock bracket.

P18-27 Remove the parking pawl shaft, parking pawl, and return spring by using a screw extractor to remove the shaft plug from the case.

P18-28 Using the correct spring compressor, compress the low/reverse clutch spring retainer and remove the retaining ring and the spring assembly.

P18-29 Remove the low/reverse piston assembly by applying air to the apply passage.

P18-30 Remove the manual shaft nut, shaft, and retainer. Then remove the parking lock actuator assembly and the inner detent lever.

PHOTO SEQUENCE 18 (CONTINUED)

P18-31 Pry the shaft seal from the case.

P18-32 Install a new shaft seal into the case. Then install the parking lock actuator assembly and detent lever. Install the manual shaft, nut, and retainer.

P18-33 Install new seals onto the low/reverse piston and coat them with transmission assembly lube.

P18-34 With the transmission in a vertical position, install the piston into the transmission case. Make sure the piston is properly aligned and fully seated.

P18-35 Install the clutch springs. Using the spring compressor, compress the springs past the ring groove in the case. Then install the retaining ring.

P18-36 Coat the bearing assembly with transmission assembly lube and install it into the case. Then install the reaction ring gear and support.

P18-37 Install the bearing onto the support, then install the oil deflector and reaction carrier assembly into the case.

P18-38 Install the clutch pack with the correct number of plates.

P18-39 Install the low/reverse support into the case, then install the inner race by pushing down while rotating it until it is fully engaged.

PHOTO SEQUENCE 18 (CONTINUED)

P18-40 Install the spring retainer, then install the low/reverse retainer ring.

P18-41 Install the snap ring onto the reaction sun gear, then install the sun gear into the reaction carrier.

P18-42 Install the thrust washer onto the low/reverse clutch race and install the reaction sun gear shell onto the sun gear.

P18-43 Install the thrust washer onto the reaction sun gear shell, make sure the thrust washer tangs are positioned in the slots in the shell.

P18-44 Install the input ring gear and reaction carrier shaft into the sun gear shell. The carrier shaft splines must engage with the reaction carrier.

P18-45 Install the thrust washer onto the reaction carrier shaft. Then install the output shaft into the transmission case.

P18-46 Install the output shaft support tool so that the shaft is positioned as high up as possible. Then install the input carrier assembly onto the output shaft.

P18-47 Install a new retaining ring onto the output shaft, then remove the output shaft support tool.

P18-48 Install the input sun gear.

PHOTO SEQUENCE 18 (CONTINUED)

P18-49 Install the selective thrust washer and bearing assembly onto the input housing.

P18-50 Install new input shaft seals, and then size them according to the procedures given in the service information.

P18-51 Position the reverse input assembly onto the input clutch assembly. Then install the reverse and input clutch assemblies into the case as a single unit. Align the 3–4 clutch plates of the input assembly with the input ring gear.

P18-52 Install the 2–4 band into the case. Align the anchor pin end with the case pin hole and install the anchor pin into the case. Make sure the anchor pin lines up with the end of the band. Then install the 2–4 servo assembly into the case and index the apply pin onto the end of the band. Install the servo cover with a new O-ring.

P18-53 Check the endplay. Install the thrust washer of correct thickness at the rear of the oil pump, and then install the oil pump after installing a new pump O-ring and gasket. Make sure all holes, especially the filter and pressure regulator holes, are lined up.

P18-54 Install the retaining bolts and torque them to specifications.

P18-55 Install the 3–4 accumulator piston pin into the case. Then install the piston seal and piston onto the pin. Install the accumulator spring, check balls, filters, retainer, and the spacer plate and gasket.

P18-56 Install the 1–2 accumulator spring, oil seal ring, piston, cover, and bolts. Tighten the bolts to specifications.

P18-57 Install the valve body and the auxiliary valve body. Then install the retaining bolts and tighten them to specifications.

PHOTO SEQUENCE 18 (CONTINUED)

P18-58 Install the speedometer gear or the speed sensor rotor onto the output shaft. Then install a new O-ring into the output shaft sleeve. Install the extension housing and torque the bolts to specifications. Using a seal installer, install a new oil seal in the extension housing.

P18-59 Install the speedometer driven gear or the speed sensor into the extension housing. Then install all external electrical connectors. Install the manual shift lever and the torque converter.

Allison Series 1000

The Allison series 1000 transmission is found in many GM trucks, such as the Silverado, GMC Sierra, and Hummer H1. They are built to handle heavy loads and high-torque engines. They also are very heavy weighing more than 350 lbs. (160 kg).

The 1000 series is very similar to the 2000 series. These transmissions are either five- or six-speed units and are electronically controlled. Shifting is controlled by a transmission control module (TCM) and up to six different solenoids. They are manufactured by the Allison Division of GM.

Some 1000 series units had a sixth gear, which was an additional overdrive. The unit in **Photo Sequence 19** is a five-speed unit.

⚠ **Caution**

This transmission is very heavy. Always use caution when installing or removing it. Use lifting equipment and an assistant while moving the transmission. Also, make sure the lifting equipment can support the weight of the transmission.

P19-1 Mount the unit on a suitable stand. If the unit has PTO covers, remove them.

P19-2 Remove the various speed sensors and the park/neutral position switch.

P19-3 Remove the vent and the torque converter assembly. Gently pry the vent out of the housing. Then rotate the transmission so that the converter is facing up. Using the correct tools, remove the torque converter and place it on a workbench. The converter is heavy, so make sure you are using the proper lifting equipment.

P19-4 Lift and remove the 1–2–3–4, 4–5 clutch piston assembly. Then remove the thrust bearing assembly from the 1–2–3–4, 4–5 clutch piston assembly or the input carrier assembly.

P19-5 Install the special main shaft holding tool or an equivalent tool to keep the main shaft in the housing. Then rotate so the transmission's pan is facing up. Remove the oil pan bolts. Then remove the pan with its gasket. Now remove the internal oil filter and seal.

P19-6 Depress the tabs on the internal wiring harness connector and gently push the connector into and through the transmission case. Disconnect the wiring harness from the various solenoids. Then remove the wiring harness from the valve body.

PHOTO SEQUENCE 19 (CONTINUED)

P19-7 Remove the manual shift detent spring bolts and its spring. Unbolt the valve body but do not attempt to remove it. Lift the valve body until its locating pins are disengaged from the transmission. Then slide the valve body to the side and disconnect the manual valve pin from the manual shift shaft detent lever. Remove the control valve body. Then remove the shift valve body and manual valve as a unit.

P19-8 Rotate the transmission so that the rear is facing up. Remove the flange from the output shaft. Then remove the oil seal assembly.

P19-9 Using the correct tool, remove the retaining ring from the low and reverse clutch housing. Then remove the bearings from the end of the output shaft, along with the selective spacer.

P19-10 Unbolt the low and reverse clutch housing and remove the gasket.

P19-11 Remove the park pawl and its return spring from its support pin. Then pull the output shaft and the output carrier assembly out of the transmission. Separate the shaft from the carrier. Then remove the low and reverse clutch piston return spring assembly.

P19-12 Keeping them as a single unit, remove the main shaft, intermediate sun gear, and sun gear spacer. Then separate the spacer and sun gear from the main shaft.

P19-13 Remove the intermediate carrier assembly. Then remove the thrust bearing from the intermediate carrier assembly.

P19-14 Remove the low and reverse clutch selective steel plate. Then remove the friction and steel clutch plates. Now remove the clutch backing plate. Set these aside and carefully inspect them for wear and damage.

P19-15 Remove the input carrier assembly. Then separate the thrust bearing from the carrier.

P19-16 Using the correct tool, compress the second clutch return spring and remove the low and reverse clutch retaining ring. Then remove the second clutch retaining ring. Remove the tool and remove the second clutch backing plate.

P19-17 Remove the friction and steel clutch plates. Then, as a unit, remove the second clutch spring plate and the piston return spring assemblies. Remove the second clutch piston assembly. Now remove the input ring gear.

P19-18 Rotate the transmission case so that the front is facing up. Install the correct compressing tool and attachments and compress the third, fifth, and reverse clutch piston return spring assemblies. Then remove the retaining ring. Remove the tools and then remove the third, fifth, and reverse clutch backing plate assemblies.

P19-19 Remove the friction and steel clutch plates. Then, as a unit, remove the third, fifth, and reverse clutch spring plate and the piston return spring assemblies. Remove the third, fifth, and reverse clutch piston assembly.

P19-20 Remove the oil seal assembly from the input shaft. Unbolt and remove the converter housing from the oil pump cover. Remove the channel plate gasket and the seals at the rear hub of the oil pump cover and the stator shaft. Remove the thrust washer and unbolt and remove the oil pump assembly.

P19-21 On a workbench, position the 1-2-3-4, 4-5 clutch assembly so that the rear of the assembly is up. Remove the retaining ring from the clutch assembly. Then remove the input sun gear with the input drive flange attached. Remove the bronze thrust washer from the 4-5 drive hub and remove the hub. Remove the thrust bearing assembly from the rear of the 1-2-3-4 clutch drive hub and remove the hub.

PHOTO SEQUENCE 19 (CONTINUED)

P19-22 Remove the friction and steel clutch from the housing. Then remove the retaining ring from the 1–2–3–4 clutch backing plate. Now remove the retaining ring from the turbine shaft and the 1–2–3–4 clutch backing plate from the turbine shaft.

P19-23 Remove the friction and steel clutch plates from the turbine shaft. Remove the retaining ring from the turbine shaft. Then remove the clutch housing and seals from the turbine shaft. Set the turbine shaft aside.

P19-24 Install the correct spring compressor onto the 1–2–3–4 clutch piston return spring assembly. Using the correct attachments, compress the spring assembly so that the retaining ring can be removed.

P19-25 Remove the piston return spring assembly from the 1–2–3–4 clutch piston. Then remove the 1–2–3–4 clutch piston assembly. Using the correct tools, compress the return ring assembly enough to remove the retaining ring from the 1–2–3–4, 4–5 clutch housing. Then separate the 1–2–3–4 clutch piston housing from the 1–2–3–4, 4–5 clutch housing.

P19-26 Remove the 4–5 clutch piston return spring assembly from the 4–5 clutch piston. Then remove the retaining ring from the 1–2–3–4, 4–5 clutch housing, which will allow the removal of the 4–5 clutch piston assembly.

P19-27 To begin reassembling the 1–2–3–4, 4–5 clutch housing, install the piston seal rings onto the 4–5 clutch piston. Lubricate all seals and the piston. Install the O-ring into the bottom groove on the inner hub of the housing. Install the 4–5 clutch assembly into the housing. Make sure the piston is properly seated on the housing. Push the balance piston below the retaining ring groove on the housing and install the retaining ring.

PHOTO SEQUENCE 19 (CONTINUED)

P19-28 Align the 4–5 clutch piston return spring assembly so that its notches are facing and are aligned with the tangs of the piston. Then install the return spring assembly into the piston. Install the 1–2–3–4 clutch piston housing into the 4–5 clutch piston.

P19-29 Position the 1–2–3–4, 4–5 clutch housing on the base of the special compressing tool. Compress the 4–5 clutch piston return spring assembly until the retaining ring can be installed. Remove the tools. Lubricate and install the clutch piston seals onto the clutch piston.

P19-30 Align the 1–2–3–4 clutch piston so that the spring indents are facing up. Install the clutch piston into the housing. Align the piston return spring assembly so that the springs are facing and are aligned with the spring pockets of the piston and install the spring assembly.

P19-31 Turn the housing over so that its rear is facing up. Then set the retaining ring into the inner hub. Now set up the spring compressor tool and compress the 1–2–3–4 clutch spring assembly until the retaining ring is able to be seated. Remove the tools. Then lubricate the seals and piston and install them in the housing. Insert the retaining ring. It may be necessary to press on the assembly to do this.

P19-32 Install the overlap seal rings onto the turbine shaft. Position the turbine shaft so that the rear of the shaft is facing up. Align the 1–2–3–4, 4–5 clutch housing so that the rear of the housing is facing up. Install the housing onto the rear of turbine shaft and then install the retaining ring.

P19-33 Alternately install the friction and steel clutch discs onto the turbine shaft. Align the pressure plate with the turbine shaft so that the countersunk holes of the pressure plate are facing up. Then install it onto the turbine shaft. Align the ridge plate with the turbine shaft so that its inner ridge is facing down and then slide the inner ridge into the turbine shaft's retaining ring groove. Align the holes in the ridge plate with the holes of the pressure plate and install the retaining bolts. Tighten these to specifications.

PHOTO SEQUENCE 19 (CONTINUED)

P19-34 Select the correct clutch backing plate according to the manufacturer's guidelines. Align the backing plate so that its beveled edge is facing up. Then install it onto the turbine shaft. Install the retaining ring onto the turbine shaft. Lift the backing plate and install the spiral retaining ring into the backing plate.

P19-35 Position the thrust bearing assembly according to specifications and install it onto the turbine shaft. Center the clutch plates and install the drive hub over the clutch plates. Install the 4–5 clutch drive hub so that the smaller end is facing up. Now, alternately install the friction and steel clutch discs into the housing.

P19-36 Align the tangs of the bronze thrust washer with the holes in the 4–5 clutch drive hub and install it. Install the input sun gear into the input sun gear flange with its teeth facing the concave face of the flange. Hold the sun gear and align the double wide splines of the flange with the double wide spaces in the housing. Then install the sun gear and flange into the housing. Install the sun gear external retaining ring into the housing.

P19-37 Position the oil pump assembly so that the rear of the pump is facing up. Align the locating pin holes in the channel plate with the locating pins in the oil pump assembly. Install the channel plate onto the oil pump assembly.

P19-38 Position the oil pump cover assembly so that the stator shaft is facing up. Align the oil pump assembly with the channel plate and install the oil pump assembly and channel plate onto the oil pump cover. Tighten the retaining bolts to specifications.

P19-39 Align the tangs of the thrust washer with the slots at the base of the oil pump cover's rear hub. Then install it onto the oil pump cover. Install new overlap seal rings onto the rear hub and the stator shaft.

P19-40 Install two guide bolts into the oil pump cover module and align the converter housing bolt holes with the bolt holes in the oil pump cover. Install the converter housing onto the guide bolts and onto the cover. Remove the guide bolts and install new seals onto the pump retaining bolts. Install and tighten the bolts in a crisscross pattern to specifications. Using the correct tool, install the seal over the oil pump's hub.

P19-41 Rotate the transmission so that the front of the case is facing up. Apply lubricant to the inside and outside diameter of the third, fifth, and reverse clutch piston assembly. Align the piston with the transmission case so that the seal is facing down and the lubrication/cooling orifice is at the top of the case. Gently tap the piston assembly into position.

P19-42 Install the piston return spring assemblies onto the third, fifth, and reverse spring plate. Then install the spring plate into the transmission case. Alternately install the friction and steel clutch discs into the case. Install the third, fifth, and reverse clutch backing plate assembly with gear thrust plates down. With the compressor tool, compress the clutch pack until the retaining ring is able to be installed.

P19-43 Rotate the transmission so that the rear of the case is facing up. Apply lubricant to the inside and outside diameters of the second clutch piston assembly. Align the piston with the transmission case so that the seal is facing down and the lubrication/cooling orifice is at the top of the case. Gently tap the piston assembly into position. Install the input ring gear so that the outside splines are facing down. Then rotate the ring gear until the friction clutch plates are engaged.

P19-44 Install the input carrier module. Rotate it while doing this until the carrier drops into place. Install the thrust bearing onto the input carrier. Install the piston return spring assemblies onto the second spring plate. Then install the spring plate into the transmission case. Alternately install the friction and steel clutch discs into the case. Install the third, fifth, and reverse clutch backing plate assembly with gear thrust plates down. With the compressor tool, compress the clutch pack until the retaining ring is able to be installed. Then install the low and reverse clutch retaining ring with its gap 180 degrees from the second clutch retaining ring gap. Remove the compressor tools.

P19-45 Locate low and reverse clutch housing, alternately install the friction and steel clutch discs into the housing. Then install the correct low and reverse clutch steel plate.

PHOTO SEQUENCE 19 (CONTINUED)

P19-46 Install the intermediate carrier module. Rotate it until it drops into place. Install the thrust washer onto the carrier. Then install the main shaft holding tool.

P19-47 Install the intermediate sun gear and its spacer onto the main shaft. Then install this assembly into the intermediate carrier. Install the output sun gear onto the main shaft with its flat side facing the spacer. Install the thrust bearing assembly into the sun gear.

P19-48 Install the low and reverse clutch piston return spring assembly, making sure the tang aligns with the notch in the transmission case. Install the park pawl cam guide and park pawl support pin into the transmission case. Install the park pawl return spring onto the park pawl and install the assembly onto the park pawl support pin. Then connect the park pawl return spring to the case.

P19-49 Install the output shaft into the output carrier assembly. While holding the output shaft, place the carrier assembly into the top of the output ring gear. Rotate the output shaft until the carrier drops into the ring gear. Gently place the output shaft onto the main shaft and rotate the carrier until it seats.

P19-50 Prepare to install the low and reverse clutch housing. Install a new gasket onto the transmission case. Install the low and reverse clutch housing making sure it is aligned with the locating pin in the case and the park pawl pin. Tighten four of the retaining bolts by hand to seat the housing in the case. Then install the rest of the bolts and tighten all of them to specifications.

P19-51 Rotate the transmission so that the valve body area of the case is facing up. Align the manual valve pin with the manual selector valve. Install the manual valve pin into the manual selector valve and then install the valve into the shift valve body. Align the control valve assembly with the case while installing the manual valve pin into the manual shift shaft detent lever. Then install the control valve body assembly onto the transmission case, using the alignment pins as a guide. Install the retaining bolts and tighten them to specifications. Install the manual shift shaft detent spring onto the valve body. Then install and tighten the manual shift shaft detent spring bolts.

P19-52 With your hand, install the internal wiring harness connector into the transmission case. Then, using the special tool, make sure the connector is securely in place and the four tabs on the connector are snapped into place. Then fasten the harness to the valve body and connect the individual connectors to the appropriate solenoid. Then install the oil filter and pan. Make sure to tighten the bolts to specifications.

P19-53 To install the 1-2-3-4, 4-5 clutch housing assembly, place the transmission so that the front of the case is facing up. Then remove the main shaft holding tool. Apply lubricant to the thrust bearing assembly and install it into the input carrier assembly. Install the 1-2-3-4, 4-5 clutch assembly. Rotate the assembly until it seats.

P19-54 The torque converter housing is very heavy. Make sure you have assistance while installing it. Install two guide bolts into the transmission case. Place a new converter housing to case gasket onto the transmission case. Apply lubrication to the seal rings on the rear of the oil pump cover assembly. Lower the converter housing/ oil pump cover assembly over the turbine shaft and onto the case. Tap the housing until it seats on the case. Remove the guide bolts from the transmission case and install the retaining bolts. Make sure the correct length is placed in its appropriate bore. Then tighten them to specifications.

P19-55 Rotate the transmission so that its rear is facing up. Using the proper tool, install the rear bearing. Install the retaining ring with its beveled edge facing up. Now rotate the transmission so that the front is facing up. Set a dial indicator onto the end of the turbine shaft. Zero the dial indicator. Then lift the rear of the 1-2-3-4, 4-5 clutch housing with a screwdriver through the PTO hole in the case. This is the amount of endplay and your measurement should be compared to specifications. If the endplay is not within the specification, a different selective spacer must be installed.

P19-56 Raise the torque converter over the torque converter housing and lower it over the turbine and stator shafts and into the oil pump assembly. Make sure the torque converter is seated in the oil pump assembly.

P19-57 Set the special switch alignment tool onto the park/neutral position switch. Align the PNP switch with the case so that the tool is facing outward. While maintaining the correct switch-to-manual-shift-shaft alignment, slide the switch over the manual shift shaft and install the retaining bolts. Remove the tool and install the switch's splash shield over the manual shift shaft. Install the shift lever, and then install the shift lever nut by hand, on the end of manual shift shaft. Tighten the shift lever nut to specification. Reconnect the shift selector linkage/cable to the shift lever and the external wiring harness to the PNP switch.

P19-58 Using the correct driver, install a new drive shaft slip yoke oil seal. Make sure the seal is installed to the correct depth below the rear face of the low and reverse clutch housing. Then install the vent by gently tapping on it.

P19-59 Install the ISS, TSS, and OSS in their designated bores. Make sure to install them with new seals that are lubricated. Tighten the retaining bolts to specifications.

P19-60 Attach a sling to the transmission. Fasten the sling to a hoist. Use the hoist to remove the weight of the transmission off the transmission stand. Then unbolt the transmission from the stand. Place the transmission on a work table with the oil pan facing down. Install the PTO covers with new gaskets. Tighten the cover bolts to specifications.

GM Transaxles

Through the years, GM has used many different transaxles. Early units were mostly based on the Simpson geartrain. Later, the transaxles relied on two simple planetary gears connected in series. The commonly used series of this type of transaxle is the 4Txx series. The 4T60 was introduced in 1984 and was the replacement for the 440-T4. In 1991, new electronic controls were added and the unit was referred to as the 4T60E. The latter was used in many GM vehicles. It was replaced with an updated version, the 4T65E, which is still found in some vehicles.

The geartrain of the 4T60 is based on two simple planetary gearsets operating in tandem. The combination of the two planetary units functions much like a compound unit. The two tandem units do not share a common member; rather, certain members are locked together or are integral with each other. The front planet carrier is locked to the rear ring gear and the front ring gear is locked to the rear planet carrier. The transaxle houses a third planetary unit, which is used only as the final drive unit and not for overdrive.

The 4T60 uses a variable-displacement vane-type pump, and four multiple-friction disc assemblies, two bands, and two one-way clutches to provide the various gear ranges. One of the one-way clutches is a roller clutch, and the other is a sprag type. The multiple-disc packs are released by several small coil springs when hydraulic pressure is diverted from the clutch's piston. **Photo Sequence 20** covers the overhaul procedures for this transaxle.

6T70/75

The 6T70/75 is a fully automatic, six-speed, electronic-controlled transaxle. The planetary gearsets provide the six forward gear ratios and reverse. Changing gear ratios is fully automatic and is accomplished through the use of a TCM inside the transmission. The TCM receives and monitors various electronic sensor inputs.

Classroom Manual
Chapter 10, page 333

Special Tools

Dial indicator and holding fixture

Clutch compressor tool

Oil pump puller

Seal remover/installer

Servo cover compressor tool

Output shaft support tool

PHOTO SEQUENCE 20
Typical Procedure for Overhauling a 4T60 Transaxle

P20-1 Remove the torque converter, then place the transaxle in a holding fixture. Remove the speedometer sensor and the governor assembly.

P20-2 Remove the bottom oil pan, oil filter, modulator, and modulator valve.

P20-3 Remove the accumulator cover with governor feed and return pipes and accumulator pistons, gaskets, retainers, and pipes from their bores.

P20-4 Mark the cover for the reverse servo so that you know from where it came. Then remove the cover by applying pressure to it and removing the retaining ring. Then remove the servo assembly from the case.

P20-5 Mark the cover for the 1–2 servo so that you know from where it came. Then apply pressure to the cover and remove the servo cover retaining ring. Then remove the cover and the servo assembly.

P20-6 Remove the side cover bolts, nuts, and washers. Then remove the side covers and gaskets. Disconnect and remove the wiring harness to the pressure switches, solenoid, and case connector.

P20-7 Remove the TV lever, linkage, and bracket from the valve body assembly. Remove the pump assembly cover bolts, then the pump cover. Remove the servo pipe retainer bolt, retainer plate, mounting bolts, and valve body.

P20-8 Remove the oil reservoir weir. Then mark the location of and remove the check balls between the spacer plate and the valve body and between the channel plate and the spacer plate.

P20-9 Disconnect the manual valve link from the manual valve. Place the detent lever in the Park position and remove the retaining clip. Then remove the channel plate with its gaskets.

PHOTO SEQUENCE 20 (CONTINUED)

P20-10 Remove the oil pump drive shaft. Remove the input clutch accumulator and converter clutch piston assemblies. Remove the fourth clutch plates and the apply plate, then remove the fourth clutch's thrust bearing, hub, and shaft.

P20-11 Rotate the final drive unit until both ends of the output shaft retaining ring are showing.

P20-12 With a C-ring removal tool, loosen the ring from the output shaft. Rotate the shaft 180 degrees and remove the ring from the shaft. Then remove the output shaft.

P20-13 Remove and discard the O-ring from the input shaft, then remove the drive sprocket, driven sprocket, and chain as an assembly with the selective fit thrust washers. Place the chain assembly on a bench so that it is in the same direction and position it was in the transmission.

P20-14 Remove the driven sprocket support and thrust washer from between the sprockets and the channel plate. Remove the scavenging scoop and driven sprocket support with the second clutch thrust washer.

P20-15 Using the correct tool, remove the second clutch and input shaft clutch housings as an assembly. Remove the reverse band.

P20-16 Remove the thrust washers. Then measure the endplay of the input shaft clutch housing. Select the correct thickness of thrust washer and set aside for reassembly.

P20-17 Remove input clutch sprag assembly, third clutch assembly, and the input sun gear. Then remove the reverse reaction drum, input carrier assembly, and thrust washer.

P20-18 Remove the reaction carrier, reaction sun gear/drum assembly, and forward band.

PHOTO SEQUENCE 20 (CONTINUED)

P20-19 Remove the reaction sun gear thrust bearing and the final drive sun gear shaft.

P20-20 Check the final drive endplay and select the correct size thrust washer for the unit.

P20-21 Using the proper tool, remove the final drive assembly and selective thrust washers and bearings.

P20-22 Clean and inspect the transaxle case and all of the transaxle components. Then position the correct size thrust washers and bearings onto the final drive assembly and install the unit. Transmission assembly lube can be used to hold the washers and bearings in place while positioning the unit.

P20-23 Install the final drive sun gear shaft through the final drive ring gear. The splines must engage with the parking gear and the final drive sun gear.

P20-24 Install the forward band into the case, making sure the band is aligned with the anchor pin.

P20-25 Install reaction sun gear-to-final drive ring gear thrust bearing. Then assemble the reaction sun gear and drum assembly onto the final drive ring gear.

P20-26 Check the endplay of the carriers in the reaction planetary gearset. Then install the thrust washer and carrier assembly into the case. Rotate the carrier until the pinions engage with the reaction sun gear.

P20-27 Install the input carrier with its thrust bearing into the case. Then install the reverse reaction drum, making sure its splines engage with the input carrier.

PHOTO SEQUENCE 20 (CONTINUED)

P20-28 Inspect the roller and sprag clutches, then put the spacer onto the sun gear followed by the input sprag retainer, sprag assembly, and roller clutch. Make sure the input sprag and third roller clutch hold and freewheel in opposite directions while holding the input sun gear.

P20-29 Lubricate the inner seal on the input clutch's piston, then install the seal and assemble the input piston into the input housing. Install a new O-ring onto the input shaft.

P20-30 Install spring retainer and guide into the piston, then install the third clutch piston housing into the input housing. Using the proper compressor, install the retaining snap ring.

P20-31 Install the inner seal for the third clutch, then install the third clutch piston into the housing. Compress the spring retainer and install the retaining snap ring.

P20-32 Install the wave plate, and then install the correct number of clutch plates in the correct sequence. Install the input clutch backing plate and the retaining snap ring. Air check the operation of the clutch.

P20-33 Assemble the second clutch piston in the housing. Install the apply ring, return spring, snap ring, and wave plate. Then assemble the correct number of clutch plates in the correct sequence. Install the backing plate and snap ring, then air check the operation of the clutch.

P20-34 Install the thrust washers. Using the correct tool, install the second clutch and input shaft clutch housings as an assembly. Install the reverse band.

P20-35 Install the scavenging scoop and driven sprocket support with the second clutch thrust washer. Install the driven sprocket support and thrust washer between the sprockets and the channel plate.

P20-36 Install the drive sprocket, driven sprocket, and chain as an assembly.

P20-37 Install the output shaft. Start the C-ring on the output shaft, then rotate the shaft 180 degrees and fully seat the ring onto the shaft.

P20-38 Install the fourth clutch's thrust bearing, hub, and shaft. Then install the fourth clutch plates and the apply plate. Install the input clutch accumulator and converter clutch piston assemblies. Install the oil pump drive shaft.

P20-39 Install the channel plate with new gaskets; connect the manual valve link to the manual valve. Place the detent lever in the Park position and install the retaining clip.

P20-40 Install the oil reservoir weir. Install the check balls in their proper location between the spacer plate and the valve body and between the channel plate and the spacer plate.

P20-41 Install the servo pipe retainer bolt, retainer plate, mounting bolts, and valve body. Install the pump assembly cover bolts, then the pump cover. Tighten the bolts to specifications. Install the TV lever, linkage, and bracket onto the valve body assembly.

P20-42 Connect and install the wiring harness to the pressure switches, solenoid, and case connector. Then install the side covers and gaskets. Install the side cover bolts, nuts, and washers, and then tighten them to specifications.

P20-43 Install the 1–2 servo cover, the servo assembly, and the retaining ring.

P20-44 Install the reverse servo cover, the servo assembly, and the retaining ring.

P20-45 Install the accumulator cover with governor feed and return pipes and accumulator pistons, gaskets, retainers, and pipes into their bores.

PHOTO SEQUENCE 20 (CONTINUED)

P20-46 Install the bottom oil pan, oil filter, modulator, and modulator valve.

P20-47 Install the torque converter, speedometer sensor, and governor assembly.

The TCM commands shift solenoids and variable bleed pressure control solenoids to control shift timing and quality. The TCM also controls the torque converter clutch. All of the solenoids and the TCM are packaged into a single unit, called the control solenoid valve assembly.

The hydraulic system primarily consists of a vane-type pump and two control valve body assemblies. These units rely on five multiple-disc clutches and a one-way clutch. The gearsets then transfer torque through the transfer drive gear, transfer driven gear and differential assembly.

This series of transaxle was jointly designed with Ford which fits a similar unit in many of its FWD vehicles. The basic construction of the GM and Ford units are similar. The most noticeable differences are in the application of electronic controls. **Photo Sequence 21** covers the basic steps for overhauling this series of transaxle.

 Special Tools

Dial indicator and
   holding fixture
Clutch compressor
   tool
Oil pump puller
Seal remover/installer

CUSTOMER CARE Because this model uses a sliding reverse gear, damage to the transaxle can occur if the car is towed or moved while the transaxle's gear selector is in Neutral but the transaxle is still mechanically in reverse gear. A servo normally is used to delay forward gear engagement after the transaxle had been in Reverse. Simply shifting into Neutral may not disengage reverse gear. Any time the vehicle is to be moved, start the engine and place the transaxle into Drive, then shift it into Neutral. Turn off the engine. Now the vehicle is ready to be moved.

PHOTO SEQUENCE 21
Typical Procedure for Overhauling a 6T70 Transaxle

P21-1 Remove the torque converter before mounting the transaxle on its holding fixture. Rotate it so that the valve cover body is facing up. Then unbolt the cover and remove it with its gasket. While lifting off the cover, make sure not to pull on the wires leading to the solenoids and TCM.

P21-2 Unbolt and remove the control solenoid valve assembly. Then remove the filter plate.

P21-3 Unbolt and remove the control valve body assembly. Keep track of the location of the shorter bolts. Then unbolt the retaining bolt for the OSS and remove the sensor.

P21-4 Rotate the unit so that the torque converter side is facing down. Unbolt and remove the case cover with its gasket. Set this assembly aside. Then remove the input shaft thrust bearing. Now, the 3-5—reverse and 4-5-6 clutch housing can be lifted out of the case.

P21-5 Now remove the reaction carrier hub assembly and its thrust washer. Then remove the 2-6 clutch spring, plate, friction and steel discs, and its backing plate.

P21-6 Remove the reaction carrier assembly. Then remove the thrust bearings for the input carrier and the input sun gear. Remove the input carrier assembly.

P21-7 Remove the input sun gear. Then remove the output carrier with its thrust washers.

P21-8 Remove the output sun gear assembly. Then remove the output carrier transfer drive gear hub assembly with its bearings.

P21-9 Lift the spring, apply plate, friction and steel discs, and backing plate for the low and reverse clutch out of the case.

PHOTO SEQUENCE 21 (CONTINUED)

P21-10 Remove the low and reverse clutch retaining ring. Then remove the clutch assembly. Now the friction and steel discs for the 1–2–3–4 clutch can be removed. Then remove the clutch plate and the waved plate. Remove the clutch spring retainer ring and the spring. Remove the clutch piston by pulling it with pliers.

P21-11 Rotate the transaxle so that the torque converter side is facing up. Unbolt and remove the torque converter and oil pump housing with its outer seal.

P21-12 Remove the lubrication tube for the differential's drive pinion. Then remove the differential's drive pinion gear assembly and set it aside.

P21-13 Remove the differential's carrier assembly and set it aside.

P21-14 Using the correct pullers, remove the differential carrier and drive pinion bearing cups from the case. Then remove the thrust washers and the drive pinion's lubricant dam.

P21-15 Using the correct puller and holding tool, remove the bearings on the carrier assembly.

P21-16 Then install new bearings with the correct driver.

P21-17 Remove the transfer drive gear support assembly. Inspect it for damage or wear to the splines, bushings, machined surfaces, and threaded holes. Also, check the gear for damage or wear. Make sure the drive gear bearing rotates smoothly.

P21-18 Unbolt the fluid passage tube from the support. Check the tubes for cracks or other damage. Remove the old gasket and place a new one onto the support. Reinstall the passage tube and tighten the retaining bolts to specifications.

P21-19 Remove the old drive gear support seal. Using the correct driving tools, install a new seal. Turn the transfer gear support assembly over and remove and install the seal on that side. Then replace the torque converter fluid seal and install the support assembly into the housing.

P21-20 Place the new bearing cups over the lower bearings on the differential carrier and drive pinion assemblies. Carefully lower the gear assemblies into the case. Then place the bearing cups onto the upper bearings of the assemblies.

P21-21 Install the special selection sleeves over the top of the bearing cups on the gears. Place a housing gasket over the case. Then install the long guide pins and spacers that are part of the gauging set. Lower the torque converter housing onto the case and over the guide pins. Make sure the spacers and gauges are lined up.

P21-22 Remove the guide pins and install spacers in their place. Make sure all bolt holes in the housing and case have a spacer. Then install the bolts that are part of the gauging set and tighten them in the prescribed sequence and to the specified torque. Rotate the differential assembly at least 10 times to allow the bearings to seat into the cups.

P21-23 Hold up the top sleeve of the gauging tool. Measure the gap between the gauging tool sleeves. To do this, use a feeler gauge to identify the largest leaf that will fit into the gap. Take two measurements, each 180 degrees apart, at each gear. The two measurements should be averaged, and the correct thrust washer to install behind the bearing cups is the one that has a thickness that is closest to that average. Do this at both gears.

P21-24 Unbolt the cover from the housing and remove the special tools and spacers. Remove the bearing cups from the gearsets. Remove the housing gasket. Place the correct selective washers into the case's bores for the bearing cups, and with the correct tool install the cups. Then install the cups in their bores in the housing.

PHOTO SEQUENCE 21 (CONTINUED)

P21-25 Place the gear assemblies into the case. Place a new housing gasket on the case. Install the retaining bolts and tighten them to specifications.

P21-26 Rotate the transaxle so that the torque converter side is facing down. Gather the 1-2-3-4 clutch assembly and the proper tools for assembling it. Install a new seal on the clutch's piston and lubricate it. Place the seal protector into the piston's bore and push the piston in place.

P21-27 Install the clutch's spring and place its retainer ring into the bore. Using the correct spring compressor, compress the spring enough to allow the retaining ring to be installed.

P21-28 Install the 1-2-3-4 clutch's wave plate, pressure plate, and friction and steel discs into the bore. Install the low and reverse clutch assembly and retaining ring. Make sure the gap of the ring is facing the valve body side of the case.

P21-29 Install the low and reverse clutch backing plate with the clutch plate facing the bottom of the case. Then install the friction and steel clutch discs, the apply plate, and the cushion spring.

P21-30 Install the input shaft thrust bearing, transfer gear hub assembly, and hub bearing.

P21-31 Install the output sun gear and carrier assemblies. Then install the thrust bearings.

P21-32 Install the input carrier assembly, followed by the input sun gear and thrust bearing. Then install the input carrier thrust bearing and the reaction carrier assembly.

P21-33 Now install the 2-6 clutch backing plate, friction and steel discs, apply plate, and cushion spring. Then install the reaction sun gear and 2-6 clutch hub thrust bearing.

PHOTO SEQUENCE 21 (CONTINUED)

P21-34 Gather the 4-5-6 clutch housing and the tools required to service it. With the compressor tool, press down on the reluctor just enough to remove the retaining ring. Then remove the reluctor wheel. Remove the 2-5 reverse clutch piston and check the seals on the piston. Replace them if necessary.

P21-35 Turn the input shaft over and remove the backing plate retaining ring. Then remove the reaction hub carrier assembly and the 4-5-6 clutch backing plate, apply plate, friction and steel discs, and the thrust bearing.

P21-36 Remove the 4-5-6 clutch dam retaining ring. Then remove the dam and the 4-5-6 clutch spring assembly, piston, and seals. Replace the piston seals. Then reinstall the piston, spring assembly, and dam. Now install the retaining ring.

P21-37 Install the 4-5-6 clutch hub thrust bearing, reaction carrier hub assembly, 4-5-6 apply plate, 4-5-6 friction and steel plates, 4-5-6 backing plate, and retaining ring.

P21-38 Now install the 3-5 reverse clutch wave plate, apply plate, friction and steel discs, backing plate, and retaining ring.

P21-39 Turn the input shaft over and install the dam seal, 3-5 reverse clutch spring, and the 3-5 reverse piston and seals. Then install the reluctor wheel. Using the correct tool, compress the assembly enough to install the reluctor wheel retaining ring.

P21-40 Put the case in an area suitable for work and remove the retainer for the ISS, then remove the sensor. Now remove the seals for the 3-5 reverse and 4-5-6 clutches from the center of the case.

P21-41 Remove the low/reverse clutch spring retaining ring, spring, and piston. Then, with the correct tool, compress the 2-6 spring assembly to remove the retaining ring, spring assembly, and piston.

P21-42 Now, using the seal protector, install the low and reverse clutch piston and spring assemblies. Make sure the air bleed on the piston is aligned with the mark on the cover.

PHOTO SEQUENCE 21 (CONTINUED)

P21-43 Then compress the springs to install the retaining ring.

P21-44 Insert the wiring from the ISS through its bore in the cover and fasten the ISS to the cover.

P21-45 Install the input shaft thrust washer. Then place a new case cover gasket on the housing. Route the ISS wire harness through its bore in the case. Using guide pins, lower the case cover onto the housing. Then remove the guide pins and install the cover bolts. Tighten them to specifications.

P21-46 Move the control valve body to a suitable working space. Separate the upper and lower assemblies by unbolting them. Make sure to keep track of the check ball locations. After doing this, inspect the upper channel plate for damage. Discard the old lower spacer plate assembly.

P21-47 Carefully clean and inspect the upper and lower valve body assemblies. If either of these is damaged, the complete assembly must be replaced.

P21-48 Assemble the upper and lower valve bodies with a new spacer plate. Make sure the check balls are in their proper locations. Tighten the bolts to specifications. Then install the channel plate.

P21-49 Install the OSS and tighten the retaining nut to specifications. Then align the manual valve to the detent lever assembly. Make sure the detent assembly is properly aligned before tightening the bolt.

P21-50 Place the valve body onto the case and insert the retaining bolts. Make sure the shorter bolts are in their proper bores. Then tighten the bolts in the correct sequence and to the specified torque.

P21-51 Place a new filter plate over the top of the valve body and make sure it is secured by the retaining clips.

PHOTO SEQUENCE 21 (CONTINUED)

P21-52 Place the control solenoid valve assembly in position over the filter plate. Then place the control solenoid valve spring in position and install the retaining bolts for the assembly. Make sure the different-length bolts are in their proper locations.

P21-53 Connect the wiring harness connectors to the appropriate places. Make sure the harness is correctly routed and is secure.

P21-54 Install a new wiring connector seal into the valve body cover. Then place a new cover gasket onto the case. Make sure the connector is seated in its seal. Install the retaining nuts and bolts and tighten them to specifications.

NISSAN MOTOR COMPANY TRANSMISSIONS

Classroom Manual
Chapter 10, page 350

Widely used Nissan rear-wheel-drive (RWD) transmissions are the 3N71B, which is a three-speed unit, and the L4N71B/E4N71B series, which provides four forward gears through the use of a Simpson gearset and an additional planetary unit mounted in front of the Simpson gearset. The primary difference between the "L" and the "E" is that the E model provides electronic control for shifting and torque converter clutch operation. Nissan transaxles are similar to other transaxles covered in this chapter.

These transmissions use four multiple-friction disc assemblies, two servos and bands, and a one-way clutch to provide for the different ranges of gears. All the clutches except the low/reverse unit are released by several small coil springs. The low/reverse unit utilizes a Belleville-type spring for greater clamping pressures. **Photo Sequence 22** covers the typical overhaul procedures for this model transmission. Always follow the recommended procedure for the transmission or transaxle you are working on.

PHOTO SEQUENCE 22
Typical Procedure for Overhauling A Nissan L4N71B Transmission

P22-1 Place the transmission on a suitable workbench. Remove the torque converter and check the input shaft endplay and record your readings.

P22-2 Remove the bellhousing.

P22-3 Unscrew and remove the electrical kickdown solenoid and O-ring.

P22-4 Unscrew and remove the throttle modulator valves and its diaphragm rod and O-ring.

P22-5 Remove the speedometer drive assembly, with gear and O-ring.

P22-6 Remove the oil pan and inspect its contents. An analysis of foreign material will provide clues about problems to look for during the overhaul procedure. Check the service information for specific details.

P22-7 Remove the valve body from the case.

P22-8 Remove the manual valve from the valve body to prevent the valve from dropping out.

P22-9 Back off the servo piston stem locknut and tighten the piston stem snug to prevent the front clutch drum from dropping out when removing the front pump.

PHOTO SEQUENCE 22 (CONTINUED)

P22-10 Use the correct puller and remove the front pump from the case.

P22-11 Remove the front clutch thrust washer and the bearing race.

P22-12 Remove the overdrive servo cover.

P22-13 Loosen the overdrive servo locknut, and back off the overdrive band's servo piston stem to release the band.

P22-14 Remove the front and rear clutch assemblies. Note the positions of the front pump thrust washers and rear clutch thrust washer.

P22-15 Remove the brake band strut and overdrive brake band.

P22-16 Remove the overdrive housing.

P22-17 Remove the extension housing. Take care not to lose the parking pawl, spring, pin, and parking actuator.

P22-18 Remove the governor valve assembly.

PHOTO SEQUENCE 22 (CONTINUED)

P22-19 Remove the front drum support and input shaft.

P22-20 Remove the second brake band strut and brake band.

P22-21 Remove the front clutch and planetary gear pack.

P22-22 Remove the output shaft snap ring.

P22-23 Remove the output shaft.

P22-24 Remove the snap ring, connecting drum, and one-way clutch assembly.

P22-25 Use a screwdriver to remove the large retaining snap ring. Then remove the low and reverse brake assembly.

P22-26 Install the low and reverse brake assembly starting with the steel-dished plate and then alternating steel and friction plates.

P22-27 Install the retaining plate and snap ring. Check for proper clearance between the snap ring and the retaining plate. Select the proper thickness of retaining plate that will give the correct ring to plate clearance if the measurement does not meet the specified limits. Check the operation of the low and reverse brake by using compressed air.

P22-28 Install the one-way clutch assembly, connecting drum, and snap ring.

P22-29 Install the governor needle bearing, thrust washer, output shaft, and oil distributor in the case.

P22-30 Install the planetary gear pack and front clutch assembly.

P22-31 Install the snap ring on the output shaft.

P22-32 Install the second brake band, band strut, and band servo. Lubricate the O-ring seals with ATF.

P22-33 Lubricate the drum support gasket with ATF and install the drum support into the housing.

P22-34 Install the governor valve assembly and torque the retaining bolts to specifications.

P22-35 Install the parking actuator and parking pawl assemblies in the extension housing. Then install the extension housing. Torque its mounting bolts to specifications.

P22-36 Install the front drum support gasket and new O-ring onto the overdrive housing. Gently tap it into place using a rubber mallet. Make sure the bolt holes are aligned.

PHOTO SEQUENCE 22 (CONTINUED)

P22-37 Install the needle bearing race and direct clutch thrust washer. Then install the overdrive brake band, strut, and servo assembly. Lubricate the servo O-ring before installing it.

P22-38 Install the overdrive pack on the drum support.

P22-39 Adjust the overdrive band. Tighten the piston stem to specifications. Then back off two full turns. Tighten the servo locknut to specifications. Test the overdrive servo with compressed air.

P22-40 Install the overdrive servo cover and tighten to specifications.

P22-41 Adjust the second brake band. Tighten the piston stem to specifications and then back off three full turns.

P22-42 Secure the piston stem while you tighten the servo locknut to specifications.

P22-43 Install the oil pump bearing and thrust washer.

P22-44 Install the oil pump assembly using the special tool. Be sure to lubricate the gasket and O-ring with ATF. Align the mounting bolt holes in the pump housing with the bolt holes in the overdrive housing.

P22-45 Lubricate and install the manual valve into the control valve body.

P22-46 Install the control valve body and tighten the valve body attaching bolts to the specified torque.

P22-47 Install the oil pump gasket, oil pan, and retaining bolts. Tighten the bolts to specifications, while using a star pattern to tighten them.

P22-48 Install the speedometer drive assembly with its drive gear and lubricated O-ring.

P22-49 Install the vacuum diaphragm, diaphragm rod, and O-ring.

P22-50 Install the kickdown (or downshift) solenoid and O-ring.

P22-51 Install the torque converter housing. Torque the mounting bolts to specifications in a star pattern.

Classroom Manual
Chapter 10, page 353

Special Tools

Dial indicator and holding fixture

Clutch spring compressor tool

Air nozzle

Seal remover/installer

Gear pullers and drivers

Small chisel

Punch and drift set

TOYOTA TRANSAXLES

Toyota led the way with electronically controlled transmissions, and since the first model was released they have continued to refine their basic model. That base model, the A-140E transaxle, was a three-speed transmission with an overdrive assembly. The basis for operation is a Simpson gearset in line with a single overdrive planetary gearset. The transaxle uses four multiple-disc clutches, two band and servo assemblies, and three one-way clutches to provide the various gear ranges.

The A541E transaxle is a revised copy of the A-140E. The A541E has adaptive capability. This transaxle has been used in many different Toyota and Lexus models, including the Toyota Camry, Toyota Avalon, and the Lexus ES-300. This transaxle uses six multiple-disc clutches, three one-way clutches, and one brake band. The operation of the transaxle and the lockup converter is totally controlled by the PCM. However, a throttle pressure cable is used to mechanically modulate line pressure. **Photo Sequence 23** covers the typical overhaul procedures for this transmission model.

P23-1 With the transaxle mounted in a stand and its external electrical components, oil pan, and linkages removed, remove the upper cover of the transaxle case.

P23-2 Remove the oil pipe bracket.

P23-3 Remove the oil strainer.

P23-4 Remove the retaining bolts for the detent spring, and then remove the spring.

P23-5 Unbolt and remove the retaining bolts for the manual valve body and then remove the valve body. Remove the retaining bolt for the oil pipes, and then remove the pipes.

P23-6 Disconnect the electrical connectors to the solenoids, and then remove the connector clamp.

P23-7 Remove the oil pipes' retaining bolts, and then pry out the apply pipes.

P23-8 Unbolt the valve body retaining bolts, and then disconnect the throttle cable from the valve body while removing the valve body.

P23-9 Remove the throttle cable and solenoid wiring.

PHOTO SEQUENCE 23 (CONTINUED)

P23-10 Remove the accumulator pistons and springs.

P23-11 Measure the piston stroke for the second coast brake. Use a divider to transfer the distance to a vernier caliper or micrometer for measurement.

P23-12 Remove the second coast brake piston's retaining ring and use air pressure to remove the second coast brake piston.

P23-13 Position the transaxle so that it is upright, then remove the oil pump with a puller.

P23-14 Remove the direct and forward clutch assemblies.

P23-15 Separate the direct clutch from the forward clutch.

P23-16 Remove the guide for the second/coast brake band.

P23-17 Remove the second/coast brake band.

P23-18 Remove the ring gear for the front planetary gearset.

PHOTO SEQUENCE 23 (CONTINUED)

P23-19 Remove the front planetary gearset with its bearings.

P23-20 Remove the sun gear and the sun gear input drum.

P23-21 Remove the second brake drum, and then remove the second brake piston return spring, and the one-way clutch #1.

P23-22 Remove the friction discs and plates from the housing.

P23-23 Remove the snap ring and the one-way clutch #2.

P23-24 Then remove the thrust washer and the rear planetary ring gear.

P23-25 Remove the flange, discs, and plates of the first/reverse brake.

P23-26 Rotate the transaxle and remove the retaining bolts for the overdrive unit.

P23-27 Remove the overdrive cover.

PHOTO SEQUENCE 23 (CONTINUED)

P23-28 Then pull the overdrive planetary gearset from the transaxle.

P23-29 Remove the output gear and nut.

P23-30 Remove the retaining ring, output shaft, and bearing race.

P23-31 Install the output shaft with a new race and retaining ring.

P23-32 Install the output gear with a new nut and tighten the nut to specifications.

P23-33 Stake the output gear's retaining nut to the output shaft.

P23-34 Remove the overdrive clutch pack by compressing the first/reverse brake piston return spring and remove the snap ring. Then remove the piston.

P23-35 Remove the parking lock pawl bracket and guide.

P23-36 Remove the manual shaft and seal.

PHOTO SEQUENCE 23 (CONTINUED)

P23-37 Loosen the final drive and axle seal covers.

P23-38 Remove the covers, then remove and service the differential unit.

P23-39 Begin reassembly by installing the differential and drive axle unit. Make sure the unit is properly set up.

P23-40 Install new seals and gaskets, then position the final drive housing and tighten the bolts to specifications.

P23-41 Install the parking pawl lock, pawl bracket and guide, and the manual shaft. Use a new seal and collar when doing this. Then check the operation of the parking pawl lock.

P23-42 Install the first/reverse brake piston into the case and install the piston return spring and retaining snap ring.

P23-43 Install the overdrive brake band and overdrive planetary gearset into the case.

P23-44 Install the overdrive cover with a new gasket. Tighten the retaining bolts to specifications.

P23-45 Check the endplay of the intermediate shaft. If it is within specifications, coat the thrust washer with transmission assembly lube and install the rear planetary gearset.

PHOTO SEQUENCE 23 (CONTINUED)

P23-46 Install the discs, plates, and flange for the first/reverse brake. Then install the retaining snap ring.

P23-47 Check the operation of the brake with compressed air.

P23-48 Check the clearance of the disc pack with a feeler gauge.

P23-49 Install the one-way clutch #2 into the case with the shiny side of the flange facing out.

P23-50 Install the second/coast brake band guide.

P23-51 Then coat the thrust washer for the one-way clutch #1 and install it with the clutch.

P23-52 Install the discs and flange for the second brake, and then install the snap ring.

P23-53 Check the operation of the second brake with compressed air.

P23-54 Coat the thrust washer with transmission assembly lube, and then install the sun gear with its drum.

P23-55 Turn the drum clockwise until it seats into the one-way clutch #1.

P23-56 Install a new O-ring onto the intermediate shaft, and then install the front planetary gearset.

P23-57 Coat the thrust washers with transmission assembly lube and assemble the forward and direct clutch assemblies.

P23-58 Install the second/coast brake band and the forward and direct clutch assemblies.

P23-59 Position the transaxle with the oil pump opening up. Coat the new pump O-ring with ATF and install it on the pump.

P23-60 Install the pump into the transmission housing.

P23-61 Check the endplay of the input shaft.

P23-62 Install the piston for the second/coast brake. Use a new snap ring to retain it.

P23-63 Identify the correct accumulator and piston spring for each of the accumulator bores.

P23-64 Install the accumulator pistons and springs. Use a new gasket and torque the bolts to specifications.

P23-65 Connect the wires to the solenoids and connect the throttle cable.

P23-66 Put the oil pipes into their proper position and then mount the valve body to the case. Install and tighten the retaining bolts to specifications.

P23-67 Connect the solenoids to the electrical harness. Then install the manual valve body and detent spring. Be sure to tighten the retaining bolts to specifications.

P23-68 Install and properly tighten the oil pipe bracket and retaining bolts.

P23-69 Install the oil strainer.

P23-70 Complete the assembly by installing the upper cover of the transaxle case and its external electrical components, oil pan, and linkages.

A340

The A340 is a four-speed transmission has been popular in Toyota trucks and was used in the Tacoma until 2015. It was also used in the 4Runner, Tundra, and T-100. The A340 E transmission uses three sets of planetary gears, six multiple-disc clutches, one band, and three one-way clutches. The operation of the transaxle and the lockup converter is totally controlled by the PCM. The A340 is disassembled and reassembled in **Photo Sequence 24**.

Special Tools

Dial indicator and holding fixture

Front-pump puller

Clutch spring compressor tool

Air nozzle

Seal remover/installer

Gear pullers and drivers

Small chisel

Punch and drift set

PHOTO SEQUENCE 24
Typical Procedure for Overhauling A Toyota A340 Transaxle

P24-1 The A340 transmission.

P24-2 Remove the shift lever and neutral safety switch.

P24-4 Remove the six converter housing bolts and then the converter housing.

P24-3 Remove the vehicle speed sensor and the input shaft speed sensors.

P24-5 Remove the extension housing bolts and then the extension housing.

PHOTO SEQUENCE 24 (CONTINUED)

P24-6 Remove the vehicle speed sensor reluctor wheel.

P24-7 Remove the oil pan.

P24-8 Once the oil pan has been removed, inspect the pan for signs of foreign material. Some particles are to be expected, but nothing too heavy.

P24-9 Remove the oil strainer.

P24-10 Remove the solenoid wires.

P24-11 Remove the valve body.

P24-12 Remove the accumulator pistons and the check valve body and spring.

P24-13 Remove the oil pump.

P24-14 Remove the overdrive planetary gear unit with overdrive direct clutch and one-way clutch. Remove the bearing and race from under the assembly.

P24-15 Remove the overdrive planetary ring gear.

P24-16 Remove the overdrive brake clutch plates.

P24-17 Remove the second coast brake cover, piston assembly, and spring.

PHOTO SEQUENCE 24 (CONTINUED)

P24-18 Remove the two bolts that hold the overdrive support in the case.

P24-19 Remove the overdrive support snap ring.

P24-20 Remove the overdrive support. (Note: A special tool to remove the overdrive support is available.)

P24-21 Remove the direct clutch.

P24-22 Remove the band retainer pin.

P24-23 Remove the band.

P24-24 Remove the front planetary and one-way clutch from the case.

P24-25 Remove the snap ring from the output shaft, then the planetary gearset.

P24-26 Remove the second brake clutches from the case.

PHOTO SEQUENCE 24 (CONTINUED)

P24-27 Remove the rear planetary gear unit with the second brake drum, first, and reverse brake pack and output shaft.

P24-28 The first and reverse clutch piston can be removed at this time with a special tool to compress the piston to allow removal of the snap ring.

P24-29 A special service tool to compress the first and reverse clutch.

P24-30 Use the special tool to remove the reaction sleeve.

P24-31 Remove the #1 brake piston using the special tool.

P24-32 Disassembling the direct clutch (during a normal rebuild, all clutches would be rebuilt or inspected).

PHOTO SEQUENCE 24 (CONTINUED)

P24-33 The clutch housing is placed onto a special fixture to compress the piston return springs, which will allow removal of the snap ring.

P24-34 The return springs can now be removed.

P24-35 Now the snap ring holding the clutch plates is removed. This snap ring does not have to be compressed.

P24-36 Now the clutch plates can be inspected. Note the thicker backing plate on top.

P24-37 Now the clutch piston can be taken out and the seals replaced. After replacement of seals, the clutch can be reassembled and installed in the transmission.

P24-38 Install the rear planetary gear unit with second brake drum, first and reverse brake pack, and output shaft. If the clutches are aligned on installation, they can be installed with the other components. The rear planetary is installed on the output shaft with a snap ring to retain. The output shaft can be lifted slightly to facilitate installation of the snap ring. Next the snap ring to retain the clutches can be installed into the case.

P24-39 Install the #1 one-way clutch.

P24-40 Install the flange, discs, and plates of second clutch. After the clutch plates are installed, install the retaining snap ring.

P24-41 Install the planetary sun gear and the front planetary carrier into the transmission's one-way clutch.

P24-42 Install the second coast brake band.

P24-43 Install the pin into the band and the retaining e-clip.

P24-44 Install the forward clutch and the front planetary ring gear as a unit.

P24-45 Install the direct clutch.

P24-46 Install the second coast brake servo assembly and cover.

P24-47 Install the overdrive support assembly, install the bolts into the support from the valve body area, and install the snap ring that retains the support.

P24-48 Install the overdrive planetary ring gear and the overdrive clutches.

P24-49 Install the overdrive planetary gear with overdrive direct clutch and one-way clutch.

P24-50 Install the front pump with a new gasket. After installation and torqueing, make sure the input shaft turns easily.

PHOTO SEQUENCE 24 (CONTINUED)

P24-51 With the pump installed, the transmission can be air checked with a rubber-tipped blowgun to ensure that the clutches and seals are working. Pictured is forward clutch being checked. Check with service information for more clutch checking locations.

P24-52 Install the accumulators and springs.

P24-53 Install the check ball body and spring.

P24-54 Install the valve body.

P24-55 Install the valve body wiring.

P24-56 Install the oil strainer.

P24-57 Install the oil pan.

P24-58 Install the vehicle speed sensor rotor and extension housing.

P24-59 Install the bellhousing, neutral safety switch and lever, and the vehicle speed sensor.

Special Tools

Dial indicator and holding fixture

Clutch spring compressor tool

Air nozzle

Seal remover/installer

Gear pullers and drivers

Small chisel

Punch and drift set

U660/U760

The U660/U760 is a six-speed transaxle used in the Camry, Venza, and present RAV4. There are only minor differences between the U660/U760 transmissions. The U760 is generally used with four-cylinder engines, while the U660 is used behind six-cylinder engines. These transmissions use five multiple-disc clutches: two driving clutches (C1 and C2) and three brake clutches (B1, B2, and B3). There is a one-way clutch called F1 that is applied in first gear. Two planetary gearsets are used—a Ravigneaux and a simple planetary gearset called the underdrive clutch—which makes the U660/U760 a Lepellitier-style geartrain. The U660/U760 basic teardown is shown in **Photo Sequence 25**.

PHOTO SEQUENCE 25
Typical Procedure for Overhauling a Toyota U660/760 Transaxle

P25-1 Remove the oil fill plug from the transmission pan; then remove the pan itself. Remove the magnets from inside the pan, and inspect them for debris.

P25-2 Remove the two bolts from the oil strainer; remove the strainer.

P25-3 Remove the valve body and gaskets.

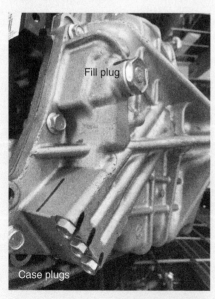

P25-4 Remove the transmission fill and case plugs; then remove the rear transaxle cover.

P25-5 Remove the thrust needle roller bearing and race, the oil seals, and the two T-30 Torx bolts from the rear transaxle cover. A special service tool is available to remove the bearing.

P25-6 Remove multiple-disc clutch assembly. This contains the C1 and C2 clutches.

PHOTO SEQUENCE 25 (CONTINUED)

Remove snap ring

P25-7 Remove the snap ring from the apply sleeve that holds the C2 clutch.

Retainer housing

Return springs

Molded clutch piston

Clutch housing

P25-8 Disassemble and inspect the C2 clutch.

P25-9 Remove the snap ring for the C2 clutches and remove them. Remove the rear planetary sun gear from the multiple disc direct clutch assembly by removing the snap ring.

P25-10 There are two snap rings on the C1 clutches. Remove both snap rings and then the C1 clutches.

Snap ring

Balance piston

Return spring

Apply piston

Forward C1 / Direct C2 Drum

P25-11 Disassembly and inspection of the C1 clutch and balance piston.

P25-12 Rear planetary sun gear.

P25-13 Remove the 20 bolts from the transmission case.

P25-14 Remove the oil pump subassembly by removing the bolts marked in red. Remove the O-ring from the front of the pump.

Remove this snap ring

P25-15 Remove the snap rings and #2 and #3 clutches from the rear of the pump subassembly.

P25-16 Remove the underdrive planetary gear assembly from the front oil pump assembly.

P25-17 Remove the #3 brake hub from the underdrive planetary gear assembly.

Snap-ring

Brake ring stopper

P25-18 Remove the two snap rings and brake snap ring stopper from the front oil pump assembly.

P25-19 Remove the #1 brake piston. Remove from the oil pump subassembly. It might be helpful to carefully apply compressed air to the oil feed passage.

P25-20 Remove the input shaft from the front pump assembly, shown with the underdrive planetary sun gear attached.

P25-21 Remove the planetary sun gear from the underdrive planetary gear assembly by removing the snap ring.

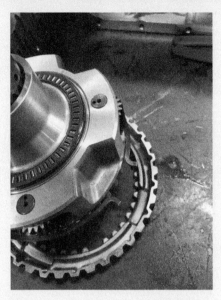

P25-22 Remove the underdrive planetary ring gear by spreading the snap ring as shown.

P25-24 Remove the differential case from the transaxle case sub assembly.

P25-23 After removal of the #1 brake clutch, the seven Torx bolts can be removed and the pump inspected. Make sure the pump gears are oriented with the dimples facing up on reassembly.

P25-25 Place the parking pawl in P.

P25-26 Remove the nut from the counter gear with the special tool.

P25-27 From the case of the transmission, remove the snap ring from the low one-way clutch.

PHOTO SEQUENCE 25 (CONTINUED)

P25-28 Be careful not to lose the anti-rattle spring sitting next to the one-way clutch.

P25-29 Remove the ring gear, low one-way clutch, and B2 brake hub as a unit.

P25-30 Remove the B2 brake clutch discs from the case.

Ravigneaux gearset

B2 Brake apply piston

P25-31 Remove the snap ring that retains the B2 brake apply piston and return springs.

SST

P25-32 Remove the output planetary gearset (Ravigneaux) from the case.

P25-33 Remove the #3 brake piston snap ring and the piston from the case.

P25-34 Case after removal of the #3 brake piston.

P25-35 It is not necessary to remove the counter drive gear unless the counter drive gear bearing or the transmission case is being replaced.

P25-36 Inspect the bearings and gear faces of the counter driven gear for wear or damage.

P25-37 Inspect the differential side bearings and gears for wear or damage.

P25-38 Replace the seals and reinstall #3 brake piston.

Reassembly of the transaxle begins from this figure 25–38.

Ravigneaux gearset

B2 Brake apply piston

P25-39 Reinstall the Ravigneaux gearset into the case, install new seals on the B2 piston assembly and install into the case.

P25-40 Replace the B2 clutches.

P25-41 Install the one-way clutch and planetary ring gear. Do not forget to install the anti-rattle spring.

PHOTO SEQUENCE 25 (CONTINUED)

P25-42 Reinstall the snap ring for the one-way clutch.

P25-43 Place the drive linkage into Park to hold the counter gear for installation of the nut.

P25-44 Install the new counter gear drive nut onto the threaded end of the Ravigneaux gearset. Move the parking linkage back to the Neutral position.

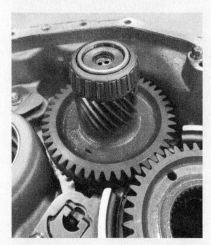

P25-45 Install the driven counter gear into the case.

P25-47 Reinstall the pump into the pump housing. Orient the pump gears with the dimple up.

P25-46 Install the differential into the case.

P25-48 Install the #1 brake piston into the oil pump subassembly.

P25-49 Install the second brake piston return springs.

PHOTO SEQUENCE 25 (CONTINUED)

P25-51 Install the brake snap ring stopper.

P25-50 Install the snap ring with the ends aligned with the cutout shown at #1.

P25-52 Reinstall the input shaft and bearings into the front pump subassembly, shown here with underdrive planetary sun gear installed.

P25-53 Install planetary sun gear sub assembly onto the planetary carrier if removed.

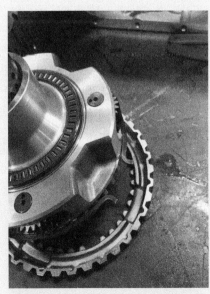

P25-54 Install the planetary ring gear onto the planetary carrier by spreading the snap ring.

PHOTO SEQUENCE 25 (CONTINUED)

P25-55 Install the third brake hub on to the planetary gearset.

P25-56 Install the underdrive planetary into the oil pump subassembly.

Clutches for #2 and #3 brakes

P25-57 Install the #2 and #3 brake discs into the pump subassembly.

P25-58 Install the oil pump subassembly back into the transmission case. Replace the O-ring on the pump housing.

P25-59 Bolt the transmission housing and the oil pump subassembly together.

PHOTO SEQUENCE 25 (CONTINUED)

P25-60 Reassemble the forward C1 and direct C2 clutches and balance piston assembly.

P25-61 Install the C2 clutches and snap ring; then install the rear planetary sun gear.

Retainer housing

Return springs

Molded clutch piston

Reinstall the snap ring

P25-63 Install the snap ring from the apply sleeve that holds the C2 clutch.

P25-64 Reinstall the assembled multiple clutch disc assembly into the case.

Clutch housing

P25-62 Reassemble the C2 clutch.

P25-65 Reinstall the rear cover.

P25-66 Install the valve body, valve body wiring, and filter.

P25-67 Reinstall the oil pan.

U241

The Toyota U241 transmission is a four-speed automatic that was used with four-cylinder engines in a variety of models including the RAV4, Camry, and Scion tC and xB. The U241 utilizes 6 multiple-disc clutches, of which three are driving clutches, three are braking clutches used to hold components, and two are one-way clutches. This unit has three planetary gearsets, the front, rear, and underdrive. The U250 is the five-speed version of the U240. The U241 transmission teardown and reassembly is shown in **Photo Sequence 26**.

Special Tools

Dial indicator and holding fixture
Clutch spring compressor tool
Air nozzle
Seal remover/installer
Gear pullers and drivers
Small chisel
Punch and drift set

PHOTO SEQUENCE 26
U241 transmission teardown and reassembly

P26-1 Remove the nut retaining the control shaft lever. Remove the control shaft lever.

P26-2 Use a screwdriver to bend the retaining tabs on the nut and remove the nut, and bolts on the park-neutral position switch. Remove the remaining two bolts and remove the switch from the case.

P26-3 Remove the oil cooler unions from the case.

P26-4 Remove the 4 case plugs from the case. The "O" rings will be replaced later.

P26-5 Remove the two speed sensors from the top of the case.

PHOTO SEQUENCE 26 (CONTINUED)

P26-6 Remove the transmission oil pan.

P26-7 Once the pan has been removed, carefully inspect residue left in the pan. The magnet will hold metal particles. Brass particles can be bushings. Black material is generally clutch plates. All transmissions will have a small amount of material which does not necessarily indicate a serious problem.

P26-8 Remove the transmission filter bolts and the filter. The filter can be cut apart and inspected for debris if necessary, as it will be replaced.

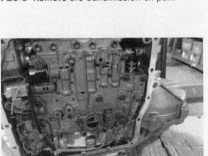

P26-9 Disconnect the electrical components.

P26-10 Remove the transmission electrical connector from the case.

P26-11 The wiring and connector are removed together.

P26-12 Remove the transmission valve body.

P26-13 The valve body removed from the transmission with the solenoids attached.

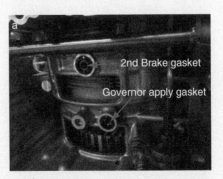

P26-14 Remove the governor and 2nd brake gasket from the transmission case. Note that the 2nd brake gasket is made from two parts.

PHOTO SEQUENCE 26 (CONTINUED)

P26-15 Remove the check ball and spring from the transmission case.

P26-16 Remove all three accumulators, make sure to keep the appropriate springs and pistons together for reinstallation later.

P26-17 Remove the two detent spring retaining bolts and the detent spring assembly.

P26-18 Remove the park lock pawl bracket bolts and then the bracket.

P26-19 Position the transaxle with the pump facing upwards. Install a dial indicator and measure the endplay of the assembly. Make note of the measurement and check it against specifications.

P26-20 Remove the 16 bolts retaining the transaxle case. It may be necessary to tap the case lightly to break the sealer holding the case together after the bolts are removed.

P26-21 Remove the bolts retaining the oil pump to the transaxle case and remove the pump.

P26-22 Remove thrust needle roller bearing and race from underdrive planetary gear.

P26-23 Remove the differential assembly from the case.

P26-24 Remove the overdrive brake gasket.

P26-25 Remove forward clutch assembly.

P26-26 Remove multiple-disc clutch hub.

P26-27 Remove the bolt and parking pawl shaft clamp.

P26-28 Disconnect spring under parking pawl and move to the side.

P26-29 Remove underdrive planetary gear assembly. After the planetary gear assembly is removed finish taking the parking pawl assembly out of the case.

P26-30 Remove the underdrive clutch and the associated thrust bearings and races.

P26-31 Remove the snap ring holding the underdrive one-way clutch snap ring.

P26-32 Remove the underdrive one-way clutch.

PHOTO SEQUENCE 26 (CONTINUED)

P26-33 Remove underdrive clutch disc snap ring.

P26-34 Remove the number 2 underdrive clutch plates from the case.

P26-35 Special spring compressor used to remove the clutch return springs and clutch seals from the case.

P26-36 Remove the bolts from the rear cover. It might be necessary to break the seal by tapping lightly on the case after the bolts have been removed.

P26-37 The outer clutch rear oil seal and rear cover thrust bearings are on the inside of the rear case cover.

P26-38 A special service tool is required to replace the needle roller bearings in the rear case if necessary.

P26-39 Remove brake apply tube.

P26-40 The governor apply gaskets will be replaced during the overhaul. Here the seal on the right has already been removed.

P26-41 Remove the direct clutch and the associated thrust bearing.

P26-42 Remove the rear planetary sun gear and the associated thrust bearings.

P26-43 Remove the one-way clutch from the transaxle case.

P26-44 Remove the 2nd brake clutch snap ring and then the clutch discs.

P26-45 Remove the 2nd brake clutches.

P26-46 Remove the 2nd brake clutch piston snap ring and the 2nd clutch piston.

P26-47 Remove the snap ring from the brake hub.

P26-48 Remove the rear planetary gear assembly and the associated thrust washers and bearings.

P26-49 The next component to remove is the front planetary gear from the counter drive gear. To start the process, the retainer tabs that

surround the retainer nut must be bent down with a chisel and hammer to allow the SST to engage the nut completely.

PHOTO SEQUENCE 26 (CONTINUED)

P26-50 Using the SST 09387–00030 and SST 09387–00080, remove the retainer for the front planetary gear.

P26-51 Using Toyota SST 09950–60010 (09951–00450) press the front planetary from the counter gear.

P26-52 Remove the front planetary gearset.

P26-53 The front planetary ring gear as removed from the front planetary.

P26-54 Remove the first and reverse clutches.

P26-55 Using Toyota SST 09387–00070 compress the first and reverse brake piston. Do not compress the piston more than about 1 mm to prevent damage to the return spring assembly.

P26-56 Remove the first and reverse brake return spring assembly.

PHOTO SEQUENCE 26 (CONTINUED)

P26-57 If the piston cannot be easily removed, carefully apply a small amount of air pressure to the passage shown to push the piston out of the case.

P26-58 Replace the seals on the first and reverse brake piston assembly, and reinstall it in the case of the transmission.

P26-59 Here the first and reverse clutch piston and springs have been reassembled into the transaxle case. The springs and snap ring are reinstalled using the Toyota SSTs in Figure 56 previously shown.

P26-60 Install the front planetary ring gear into the transaxle case.

P26-61 Reinstalling the front planetary gearset into the front planetary ring gear.

2nd brake hub w/ front planetary

Gear assy

SST

Transaxle case

Counter drive gear

P26-62 Using the Toyota SST to press the front planetary into the counter gear.

SST

P26-63 Tighten the nut that holds the front planetary to the counter gear with the SST shown in Figure 63 until the turning torque is
Turning torque at 60 rpm:
New Bearing
0.5 – 1.0 N·m (5.1 – 10.0 kgf·cm, 4.4 – 8.7 in.·lbf)
Used Bearing
0.3 – 0.5 N·m (3.1 – 5.1 kgf·cm, 2.7 – 4.4 in.·lbf)

PHOTO SEQUENCE 26 (CONTINUED)

P26-64 After the correct turning torque is achieved, then the nut retainer cage should be staked back on the nut.

P26-65 Install the planetary input sun gear and thrust washers.

P26-66 Install rear planetary gearset and thrust washers.

P26-67 Install the rear planetary gear to the rear planetary ring gear and install the retaining snap ring.

P26-68 Install the 2nd brake clutch piston (NOTE: shown without planetary gearsets for clarity).

P26-69 Install first and reverse brake disc clutches and backing plate. The backing plate (on top) is selective fit.

P26-70 Install the one way clutch outer sleeve and one way clutch.

P26-71 Install the rear planetary sun gear and thrust washers.

P26-72 Install 2nd brake clutch discs and snap ring along with thrust washers.

P26-73 Install direct clutch assembly.

P26-74 Install the front clutch and brake apply tubes and bracket.

P26-75 Install the necessary thrust washers and gaskets on the rear cover.

P26-76 The rear cover will be sealed with THREE BOND 1281.

P26-77 Apply Loctite 242 to the bolts marked "A".

P26-78 Reinstall the underdrive clutch piston into the case after replacing the seals. Install the return springs and snap ring with the help of the Toyota SST as shown.

P26-79 Underdrive clutch piston installed into the case.

P26-80 Install the three underdrive clutches and the snap ring into the case.

P26-81 Install the underdrive one-way clutch and snap ring into the case, along with the appropriate trust washers. Make sure the marks on the one-way clutch are facing upward.

P26-82 Install the underdrive clutch assembly into the case.

P26-83 Temporarily install the parking pawl assembly as shown.

P26-84 Install the underdrive planetary gear set.

P26-85 Install the parking pawl shaft and retainer.

P26-86 Install the thrust needle roller bearing in the underdrive planetary.

P26-87 Install the front differential assembly.

P26-88 Install the oil pump assembly.

P26-89 Install the transaxle housing assembly.

PHOTO SEQUENCE 26 (CONTINUED)

P26-90 Check transaxle end play and insure it is within specifications.
End play: 0.26 – 1.25 mm (0.0103 – 0.0492 in.)

P26-91 Working from the bottom of the transmission, install the Park-Lock bracket.

P26-92 Install the manual detent spring assembly.

B3

C1

Replace all seals on the accumulators and reinstall along with correct springs

C3

P26-93 Reinstall all the accumulators as shown.

P26-94 Reinstall the check ball into the case.

P26-95 Replace the governor apply and 2nd brake gaskets on the case.

P26-96 Reinstall the transaxle valve body.

P26-97 Reinstall the transaxle wiring to the valve body.

P26-98 Reinstall the oil strainer.

P26-99 Reinstall the transaxle oil pan.

P26-100 Install the speed and rotational sensors into the case.

P26-101 Replace the sealing gaskets and reinstall all the transaxle case plugs.

P26-102 Replace sealing gaskets and reinstall oil cooler unions.

P26-103 Reinstall the park neutral switch to the case.

Special Tools

Dial indicator and
  holding fixture
Clutch compressor tool
Front-pump puller
Air nozzle
Seal remover/installer
Gear pullers and drivers
Small chisel
Punch and drift set

AB60

The AB60 six-speed transmission is used in the Toyota Sequoia and Tundra, and comes in two variations, the AB60E two-wheel-drive version, while the AB60F is the four-wheel-drive version. The basic difference between the two is the extension housing and output shaft, since the AB60F is designed to match-up with a transfer case. The AB60 has three planetary gearsets, four multiple-disc clutches, and four one-way clutches. To help improve drivability when towing a trailer, the AB60 has a tow/haul mode that alters shifts to provide better performance while towing. The torque converter clutch can be engaged in both fifth and sixth gears. Both fifth and sixth gears are also overdrive ratios. The AB60 transmission teardown and reassembly is shown in **Photo Sequence 27**.

PHOTO SEQUENCE 27
AB60 Transmission teardown and reassembly

P27-1 Remove the park neutral switch.

P27-2 Remove the speed sensors from the case.

P27-3 Remove the transmission breather tube from the case.

P27-4 Check and record the transmission endplay reading.

P27-5 Remove the transmission bell housing.

P27-6 Remove the transmission oil pan.

P27-7 Remove the transmission filter.

P27-8 Remove the transmission wiring from the valve body, including the two temperature sensors.

P27-9 The wiring harness can now be completely removed from the transmission case by removing the two external connectors.

P27-10 Remove the detent spring from the valve body.

P27-11 Remove the bolts from the valve body and carefully remove the valve body assembly.

P27-12 Remove the valve body seals, they will be replaced during before reinstallation of the valve body.

P27-13 Remove the check ball and housing, taking care that the spring does not fall into the case.

P27-14 Remove the C2 and B3 accumulator pistons by carefully applying a small amount of air to the passage shown. Keep the springs and their associated pistons together.

P27-15 Remove C3 piston with a small amount of air as shown. Remove B3 piston as well. Again keep the pistons and springs together for reinstallation later.

P27-16 Remove the bolts holding the front pump assembly in place.

P27-17 Remove the front pump by using the special Toyota service tool or equivalent.

PHOTO SEQUENCE 27 (CONTINUED)

P27-18 Front pump removed.

P27-19 Remove the clutch drum and input shaft assembly from the case.

P27-20 Remove the number 2 one-way clutch from the clutch drum along with the thrust washers on both sides

P27-21 Remove snap ring from the clutch drum. Remove the reaction sleeve and reverse clutch hub. Remove the clutches from the hub.

P27-22 Remove the coast clutch hub and the number 4 one-way clutch.

PHOTO SEQUENCE 27 (CONTINUED)

P27-23 Remove forward and coast clutches from the input shaft hub, and remove the input shaft hub from the clutch drum.

P27-24 Remove the number 1 clutch balancer and forward clutch return springs from the clutch drum.

P27-25 Remove snap ring from clutch hub, remove direct clutches, clutch balancer, reverse clutch piston and return springs, then remove direct clutch piston.

P27-26 Remove the snap ring from the number 3 disc set in the case.

P27-27 Number 3 brake clutches removed from the case.

P27-28 Remove the snap ring from the case that retains the number 3 brake return piston and springs.

PHOTO SEQUENCE 27 (CONTINUED)

P27-29 Remove number 3 clutch piston and return springs.

P27-30 Remove one-way clutch from the case.

P27-31 Remove one-way clutch inner race and planetary thrust washers.

P27-32 Remove the front planetary gearset, along with the thrust washers.

P27-33 Remove the planetary sun gear from the case.

P27-34 Remove the front planetary ring gear from the case and remove the center planetary gear from the front planetary ring gear.

P27-35 Remove the number one brake disc set.

P27-36 To remove the snap-ring shown with the return springs, a Toyota special tool is required.

SST

P27-37 Special tool used to remove number 1 brake cylinder piston return springs. After the return springs and piston are removed, then the clutch plates can be removed.

PHOTO SEQUENCE 27 (CONTINUED)

P27-38 Remove the snap ring at the case.

SST

P27-41 Remove the first and reverse brake piston return springs using the Toyota special tool, then remove the first and reverse clutch piston.

P27-43 If the rear planetary gearset is inspected and OK, install the appropriate thrust washers and install the planetary gearset into the case.

P27-39 Remove the intermediate shaft with the number 3 one-way clutch, rear planetary flange and ring gear. Remove the number 4 clutch discs after the input shaft assembly has been removed.

P27-42 Replace the seals on the first and reverse clutch piston and install it into the case. Install the return springs and compress them with the special tool shown in Figure 41. Install the new number 4 brake clutch discs after the piston and return springs are replaced.

P27-40 Remove the rear planetary gear from the case.

P27-44 After an inspection of the components involved, install the intermediate shaft with the number 3 one-way clutch, rear planetary flange and ring gear. Install the new number 4 clutch discs after the input shaft assembly has been installed.

PHOTO SEQUENCE 27 (CONTINUED)

P27-45 Intermediate shaft and related parts reinstalled into the case.

P27-46 Install the number 2 brake clutch piston after replacing the seals.

P27-47 Install the number 2 brake multiple clutch disc set, then the return spring assembly.

P27-48 Use the special tool to compress the return springs and reinstall the snap ring.

P27-49 Reinstall the number 1 brake set.

P27-50 Reinstall the front planetary ring gear after reinstalling the center planetary gear on the front planetary ring gear.

P27-51 Install the planetary sun gear into the case.

P27-52 Install the front planetary gearset, along with the thrust washers.

P27-53 Install the one-way clutch inner race and planetary thrust washers.

P27-54 Reinstall the one-way clutch on the race- the one-way clutch can also be installed on the race and then placed into the case.

P27-55 Install the number 3 clutch piston and return springs.

P27-56 Remove the snap ring from the case that retains the number 3 brake return piston and springs.

PHOTO SEQUENCE 27 (CONTINUED)

P27-57 Number 3 brake clutches can be installed into the case.

P27-58 Install the snap ring onto the number 3 disc set in the case.

P27-59 Remove snap ring from clutch hub, remove direct clutches, clutch balancer, reverse clutch piston, and return springs, then remove direct clutch piston.

P27-60 Install the number 1 clutch balancer and forward clutch return springs from the clutch drum.

P27-61 Install the forward and coast clutches into the input shaft hub, and install the input shaft hub into the clutch drum.

P27-62 Install the coast clutch hub and the number 4 one-way clutch.

P27-63 Install the clutches into the hub. Install the reaction sleeve and reverse clutch hub. Install the snap ring to the hub.

P27-64 Install the one-way clutch onto the clutch drum.

P27-65 Reinstall the clutch drum and input shaft assembly into the case.

P27-66 Replace the appropriate seals and install the front pump assembly in place.

P27-67 Install new accumulator seals and reinstall the accumulators in the appropriate location.

P27-68 Install the check ball and housing, taking care that the spring does not fall into the case.

P27-69 Install the valve body seals with new seals.

P27-70 Install the valve body.

P27-71 Install the detent spring from the valve body.

PHOTO SEQUENCE 27 (CONTINUED)

P27-72 Install the solenoid wiring and then the filter. Reinstall the electrical connectors, with new seals, into the case.

P27-73 Install the transmission oil pan with a new gasket.

P27-74 Install the transmission bellhousing.

P27-75 Check the transmission endplay reading and make sure it is within specification.

P27-76 Install the transmission breather tube from the case.

P27-77 Install the speed sensors from the case.

P27-78 Install the park neutral switch.

CASE STUDY

A very irate customer had his late-model Honda towed back to the shop. Just three weeks had passed since the transmission had been rebuilt and now it had failed again.

The technician who had done the overhaul was assigned the car. He conducted a visual inspection and found nothing obviously wrong with the electrical system or the transaxle itself. However, the fluid had a burnt smell. He conducted an oil pressure test and found extremely low pressures. He suspected that the pump or pressure regulator had failed.

He pulled the transaxle and checked the pressure regulator and oil pump. He found the pump's gears had seized together. This was the first Honda transaxle he had ever rebuilt and he was disappointed that it had failed. Wondering how this could happen, he reviewed the service information and Technical Service Bulletins. In bold print in the service information he found the answer. He had failed to correctly align the oil pump shaft during assembly. Based on his experiences with other transmissions he had rebuilt, he hadn't known this was critical and had assumed that the shaft would only fit one way.

He aligned the shaft, replaced the oil pump gears, and reassembled the transmission. He road tested the car and conducted a pressure test. Everything was fine. He then gave the car back to the customer, with an apology, knowing that the transaxle was now right. The customer appreciated his honesty and left with confidence that the transmission was okay. The technician learned a lesson: always refer to and follow the directions given in the service information.

ASE-STYLE REVIEW QUESTIONS

1. *Technician A* says some seals must be cut to size before they are installed.

   Technician B says square-cut seals are designed to roll over when a part is fit over them.

   Who is correct?

   A. A only

   B. B only

   C. Both A and B

   D. Neither A nor B

2. While discussing Chrysler transmissions:

   Technician A says the 41TE is also called the A604.

   Technician B says the A604 was used until around 2010 in Chrysler minivans.

   Who is correct?

   A. A only

   B. B only

   C. Both A and B

   D. Neither A nor B

3. While discussing differential side bearing preload on a 6T70 transmission:

   Technician A says preload is often corrected by selective snap rings.

   Technician B says preload is often corrected by selective thrust washers.

   Who is correct?

   A. A only

   B. B only

   C. Both A and B

   D. Neither A nor B

4. *Technician A* says the Allison 1000 transmission is designed and built by Ford Motor Company.

   Technician B says 1000-series transmissions were built as five- and six-speed units.

   Who is correct?

   A. A only

   B. B only

   C. Both A and B

   D. Neither A nor B

5. *Technician A* says that a special main shaft holding tool must be installed while overhauling an Allison 1000 transmission.

   Technician B says a common set of snap ring pliers is used to remove the low and reverse clutch housing snap ring on an Allison 1000 transmission.

   Who is correct?

   A. A only
   B. B only
   C. Both A and B
   D. Neither A nor B

6. While discussing the 4R44E transmission:

   Technician A says the bellhousing and oil pump are removed from the transmission as a unit.

   Technician B says the solenoid harness connector is removed after the valve body.

   Who is correct?

   A. A only
   B. B only
   C. Both A and B
   D. Neither A nor B

7. *Technician A* says the 4L60-E uses five multiple-friction disc assemblies, one sprag-type one-way clutch, one roller-type one-way clutch, and a band.

   Technician B says late-model 4L60-E transmissions have a one-piece case.

   Who is correct?

   A. A only
   B. B only
   C. Both A and B
   D. Neither A nor B

8. *Technician A* says Toyota led the way with electronically controlled transmissions.

   Technician B says the A140 was the basis for the A541 transmission.

   Who is correct?

   A. A only
   B. B only
   C. Both A and B
   D. Neither A nor B

9. *Technician A* says A340 E transmission uses three sets of planetary gears, six multiple-disc clutches, one band, and three one-way clutches.

   Technician B says the operation of the transaxle and the lockup converter is totally mechanically controlled.

   Who is correct?

   A. A only
   B. B only
   C. Both A and B
   D. Neither A nor B

10. While discussing the 6T70 transmission:

    Technician A says the planetary gearsets provide the five forward gear ratios and reverse.

    Technician B says the 6T70 shifting is accomplished through the use of a TCM inside the transmission.

    Who is correct?

    A. A only
    B. B only
    C. Both A and B
    D. Neither A nor B

ASE CHALLENGE QUESTIONS

1. *Technician A* says the 4T60 uses a variable-displacement vane-type pump, four multiple-friction disc assemblies, and two bands.

   Technician B says it also has two one-way clutches to provide the various gear ranges.

   Who is correct?

   A. A only
   B. B only
   C. Both A and B
   D. Neither A nor B

2. While discussing the 6T70 transmission:

   Technician A says the TCM commands shift solenoids and variable bleed pressure control solenoids to control shift timing and quality.

   Technician B says all of the solenoids and the TCM are packaged into a single unit, called the control solenoid valve assembly.

   Who is correct?

   A. A only
   B. B only
   C. Both A and B
   D. Neither A nor B

3. *Technician A* says the U660/U760 has three brake clutches: B1, B2, and B3.

   Technician B says the U660/U760 has a one-way clutch—the F1 that is applied in third gear.

   Who is correct?

   A. A only

   B. B only

   C. Both A and B

   D. Neither A nor B

4. *Technician A* says the L4N71Bs/E4N71Bs are Nissan transmissions.

   Technician B says that the E model has electronic controls.

   Who is correct?

   A. A only

   B. B only

   C. Both A and B

   D. Neither A nor B

5. While discussing the Ford AX4S:

   Technician A says it uses three simple planetary gearsets in tandem.

   Technician B says each planetary gearset has two sun gears.

   Who is correct?

   A. A only

   B. B only

   C. Both A and B

   D. Neither A nor B

Name _____   Date _____

DISASSEMBLE AND INSPECT A TRANSMISSION

Upon completion of this job sheet, you should be able to use service information to determine what special tools and procedures are required to correctly service a particular transmission or transaxle.

ASE Education Foundation Correlation ──────────────────────

This job sheet addresses the following **MAST** tasks:

C.6 Disassemble, clean, and inspect transmission/transaxle. (**P-2**)

C.9 Assemble transmission/transaxle. (**P-1**)

We Support

ASE | Education Foundation

Tools and Materials

• Appropriate service information

Procedure

Your instructor will assign you to a particular transmission or transaxle. You will use service information to identify and describe the special tools recommended for the complete overhaul of the transmission. You will also identify any tools your shop has that are acceptable substitutes for the factory-recommended special tools. After you have assembled the list of special tools, go through the steps for complete overhaul and identify those procedures that are more than remove and install types (such as checking input shaft endplay).

1. Describe the transmission that was assigned to you.

   Model and type of transmission _____

   Describe the vehicle it is from: Year _____

   Make _____ Model _____ VIN _____

2. What service information or electronic database are you using for this assignment? Be specific.

   _____

   _____

   _____

   _____

   _____

   _____

3. List the required special tools that are available in your shop.

   _____

   _____

   _____

   _____

   _____

   _____

4. Review the overhaul procedures and describe the special procedures that must be known in order to properly service the transmission assigned to you. Include in your description the parts or subsystems that require this procedure. Do not simply say, "check endplay." Describe what needs to be done to check the endplay.

_____

_____

_____

_____

_____

Problems Encountered

_____

_____

_____

Instructor's Comments

_____

_____

_____

Name _____   Date _____

CONTINUOUSLY VARIABLE TRANSMISSIONS

Upon completion of this job sheet, you will be able to explain how a normal CVT works.

ASE Education Foundation Correlation ————————————————

This job sheet addresses the following **MLR** task:

C.1   Describe the operational characteristics of a continuously variable transmission (CVT). (**P-1**)

This job sheet addresses the following **AST/MAST** task:

C.4   Describe the operational characteristics of a continuously variable transmission (CVT). (**P-3**)

We Support
ASE | Education Foundation

Tools and Materials

• Service information

Describe the Vehicle Being Worked On

Model and type of transmission _____

Year _____ Make _____ Model _____

VIN _____ Engine type and size _____

Procedure

1. Why is this type of transmission called a continuously variable transmission?

   _____

   _____

2. What is the basic goal of a CVT?

   _____

   _____

3. Describe the major components of this transmission.

   _____

   _____

   _____

   _____

4. Explain how this CVT works.

   _____

   _____

   _____

   _____

5. Does the driver feel a change in ratio as the vehicle is driven? Why or why not?

_____

_____

_____

_____

6. Using the service information as a guide, what inputs does the transmission rely on to change ratios?

_____

_____

_____

_____

Instructor's Response

_____

_____

_____

Name _____   Date _____

HYBRID TRANSMISSIONS

Upon completion of this job sheet, you will be able to explain how a transmission in a hybrid vehicle works.

ASE Education Foundation Correlation ————————————————————

This job sheet addresses the following **MLR** task:

C.2   Describe the operational characteristics of a hybrid vehicle drive train. (**P-3**)

This job sheet addresses the following **AST/MAST** task:

C.5   Describe the operational characteristics of a hybrid vehicle drive train. (**P-3**)

We Support
| Education Foundation

Tools and Materials

• Service information

Describe the Vehicle Being Worked On

Model and type of transmission _____

Year _____ Make _____ Model _____

VIN _____ Engine type and size _____

Procedure

1. Using the service information, briefly describe the hybrid system used on this vehicle.

   _____

   _____

   _____

   _____

2. Describe the key characteristics that make this a hybrid vehicle.

   _____

   _____

   _____

   _____

3. Where is the electric motor(s) located?

   _____

   _____

   _____

   _____

4. What are the principal purposes of the electric motor?

   _____

   _____

_____

_____

5. What is the voltage rating of the hybrid battery?

_____

_____

6. Does this vehicle use a conventional transmission?

_____

_____

7. If the vehicle has a unique transmission, describe why it is unique.

_____

_____

_____

8. Can this vehicle run on electric power only? If so, when?

_____

_____

_____

9. Briefly explain the action of the electric motor(s), when are they turned on and when do they function as a generator.

_____

_____

10. Briefly describe the precautions that must be followed when servicing these transmissions and vehicles.

_____

_____

_____

Problems Encountered

_____

_____

_____

Instructor's Comments

_____

_____

_____

Final Exam Automatic Transmission/Transaxle A2

1. Which of the following is the *least* likely cause for a buzzing noise from a transmission in Park?

   A. Improper fluid level or condition

   B. Defective oil pump

   C. Defective flexplate

   D. Damaged planetary gearset

2. A vehicle experiences engine flare in low gear only.
   Technician A says the torque converter lockup clutch is slipping.
   Technician B says the transmission oil pump is not providing the required pressure.
   Who is correct?

   A. A only   C. Both A and B

   B. B only   D. Neither A nor B

3. The results of a hydraulic pressure test are being discussed:
   Technician A says low idle pressure may be caused by a defective exhaust gas recirculation system.
   Technician B says low neutral and park pressures may indicate a fluid leakage past the clutch and servo seals.
   Who is correct?

   A. A only   C. Both A and B

   B. B only   D. Neither A nor B

4. The vehicle creeps in NEUTRAL.
   Technician A says a too high engine idle speed could be the cause.
   Technician B says a too tight clutch pack may be the problem.
   Who is correct?

   A. A only   C. Both A and B

   B. B only   D. Neither A nor B

5. *Technician A* says overtorqued valve body fasteners may cause erratic shifting.
   Technician B says debris in the valve body can cause erratic shifting.
   Who is correct?

   A. A only   C. Both A and B

   B. B only   D. Neither A nor B

6. The transmission's output shaft and its sealing components are being discussed:
   Technician A says the shaft and all of its sealing components must be replaced if nicks and scratches are found in the shaft's sealing area.
   Technician B says all of the shaft's seals and rings must be replaced during a rebuild.
   Who is correct?

   A. A only   C. Both A and B

   B. B only   D. Neither A nor B

7. *Technician A* says harsh shifts could be caused by a transmission that has not been properly "relearned" after TCM replacement.
   Technician B says a harsh shift could be caused by using the incorrect fluid type.
   Who is correct?

   A. A only   C. Both A and B

   B. B only   D. Neither A nor B

8. The vehicle does not move in any gear.
   Technician A says a misadjusted shift cable could be the cause.
   Technician B says leakage at the oil pump and/or valve body could cause this condition.
   Who is correct?

   A. A only   C. Both A and B

   B. B only   D. Neither A nor B

9. Sensors are being discussed:
Technician A says most speed sensors are AC generators.
Technician B says most speed sensors use a stationary magnet, rotor, and a voltage sensor.
Who is correct?

A. A only
B. B only
C. Both A and B
D. Neither A nor B

10. While discussing hybrid automatic transmissions:
Technician A says some hybrid transmissions have an auxiliary electric oil pump.
Technician B says some hybrid transmissions do not use a conventional torque converter.
Who is correct?

A. A only
B. B only
C. Both A and B
D. Neither A nor B

11. Shift solenoids are being discussed:
Technician A says some solenoids cannot be serviced separately.
Technician B says delayed shifts may be caused by open solenoid windings.
Who is correct?

A. A only
B. B only
C. Both A and B
D. Neither A nor B

12. When diagnosing the cause of no engine braking during Manual Low operation:
Technician A checks for a defective oil pump.
Technician B suspects a damaged drive link.
Who is correct?

A. A only
B. B only
C. Both A and B
D. Neither A nor B

13. The customer complains of transmission noises in all gears except PARK and NEUTRAL.
Technician A says the oil pump may be the cause.
Technician B says the drive chain or sprocket may be the source of the noise.
Who is correct?

A. A only
B. B only
C. Both A and B
D. Neither A nor B

14. Converter installation is being discussed:
Technician A says converter depth must be measured before it is removed from the transmission.
Technician B says the converter must be engaged to the turbine (input) shaft, stator shaft, and pump drive to be properly installed.
Who is correct?

A. A only
B. B only
C. Both A and B
D. Neither A nor B

15. A vehicle with an electronic automatic transmission exhibits erratic shifting. The speed sensor is suspected.
Technician A says the permanent magnet style vehicle speed sensor (VSS) can be checked on the vehicle with an ohmmeter.
Technician B says there should be a specified resistance between the leads when checked with an ohmmeter.
Who is correct?

A. A only
B. B only
C. Both A and B
D. Neither A nor B

16. The vehicle cranks in REVERSE or DRIVE, but not in any other gear.
Technician A says the park/neutral switch is open.
Technician B says the transmission range switch may be misadjusted.
Who is correct?

A. A only
B. B only
C. Both A and B
D. Neither A nor B

17. While diagnosing the cause of sluggish acceleration:
Technician A suspects a possible engine performance problem.
Technician B checks the level and condition of the transmission fluid.
Who is correct?

A. A only
B. B only
C. Both A and B
D. Neither A nor B

18. Which of the following is the *most* likely cause of slipping in all forward gear ranges?
    A. Faulty transmission range switch
    B. Clogged transmission oil filter
    C. Faulty band or clutch
    D. Sticking valve in the valve body

19. A vehicle with a newly rebuilt electronic automatic transmission shifts directly to Third from Low. The most probable cause for this concern is:
    A. 1–2 servo is defective
    B. 2–3 shift valve is sticking
    C. Manual linkages are out of adjustment
    D. Throttle linkage is out of adjustment

20. While diagnosing the cause of no forced down-shifts during full throttle operation:
    Technician A suspects a misadjusted manual shift linkage.
    Technician B suspects a dirty valve body and valves.
    Who is correct?
    A. A only   C. Both A and B
    B. B only   D. Neither A nor B

21. Which of the following is the *least* likely cause for a transmission not upshifting and only operating in first gear?
    A. Improper fluid level or condition
    B. Faulty valve body
    C. Defective oil pump
    D. Faulty Throttle Position (TP) sensor

22. The vehicle exhibits a scraping or grating noise in every gear position anytime engine torque is changed up or down.
    Technician A says this could be caused by a damaged planetary gearset.
    Technician B says cracks around the flexplate mounting holes would cause this noise.
    Who is correct?
    A. A only   C. Both A and B
    B. B only   D. Neither A nor B

23. A MAP sensor test reveals below specified voltage signals at all engine speeds.
    Technician A says a transmission may have late upshifts based on the signals.
    Technician B says a possible vacuum leak could cause this condition.
    Who is correct?
    A. A only   C. Both A and B
    B. B only   D. Neither A nor B

24. The valve body bores are being discussed:
    Technician A says the bores should be measured with small bore (hole) gauges.
    Technician B says the bores only need to be cleaned if the damage to the valves can be corrected by polishing.
    Who is correct?
    A. A only   C. Both A and B
    B. B only   D. Neither A nor B

25. During disassembly, some extension-to-case bolts were found to have aluminum strips in the threads.
    Technician A says the extension housing should be replaced.
    Technician B says to clean and repair the internal threads with a thread insert.
    Who is correct?
    A. A only   C. Both A and B
    B. B only   D. Neither A nor B

26. *Technician A* says to use two flare nut wrenches to disconnect transmission cooler lines.
    Technician B says to use vice grip pliers to hold a cooler line that is twisting as the flare nut is turned and then replace the line and fitting.
    Who is correct?
    A. A only   C. Both A and B
    B. B only   D. Neither A nor B

27. Air checking the transmission is being discussed:
    Technician A says using a special adapter plate is necessary on some transmissions.
    Technician B says a special style blow gun should be used.
    Who is correct?
    A. A only   C. Both A and B
    B. B only   D. Neither A nor B

28. During inspection, the valve body mounting surface of the case is found to be slightly warped.
   Technician A says the problem may be corrected by using a fine file on the valve body.
   Technician B says the case must be replaced if the mounting surface for the valve body is warped.
   Who is correct?

   A. A only   C. Both A and B
   B. B only   D. Neither A nor B

29. A bushing was nicked and scratched when the shaft was being removed. There was no other apparent wear or damage.
   Technician A says fine emery cloth could be used to repair the damage.
   Technician B says sanding or polishing the bushing may cause the shaft to wobble in the bushing bore.
   Who is correct?

   A. A only   C. Both A and B
   B. B only   D. Neither A nor B

30. While diagnosing the cause of gear slipping in Second gear only:
   Technician A suspects a worn or damaged clutch assembly.
   Technician B suspects a damaged drive link.
   Who is correct?

   A. A only   C. Both A and B
   B. B only   D. Neither A nor B

31. While conducting a pressure test:
   Technician A says the cause of low pressure in all operating ranges could be a clogged filter.
   Technician B says the cause of high pressure in all operating ranges could be a defective electronic pressure regulator solenoid.
   Who is correct?

   A. A only   C. Both A and B
   B. B only   D. Neither A nor B

32. *Technician A* says pump side clearance is determined using straightedge and feeler gauges.
   Technician B says the pump's gear side clearance may be measured with a dial indicator.
   Who is correct?

   A. A only   C. Both A and B
   B. B only   D. Neither A nor B

33. *Technician A* says the transmission fluid level may have to be checked with a special procedure if there is no dipstick.
   Technician B says if the ATF has a dark color and a burnt smell it indicates that a transmission rebuild is needed.
   Who is correct?

   A. A only   C. Both A and B
   B. B only   D. Neither A nor B

34. There are traces of transmission fluid in the engine coolant.
   Technician A says the leak may be from a poor line connection to the radiator.
   Technician B says a cracked cooler line may be the source of the leak.
   Who is correct?

   A. A only   C. Both A and B
   B. B only   D. Neither A nor B

35. *Technician A* says when removing the drive chain, input gear, and output gear from certain transaxles, a special tool must be used to spread the gears.
   Technician B says a special tool must be used to draw the gears together in order to remove the drive chain from this type of transaxle.
   Who is correct?

   A. A only   C. Both A and B
   B. B only   D. Neither A nor B

36. While discussing a vibration:
   Technician A says if the vibration changes with a change in engine speed, the cause of the problem is most probably the torque converter.
   Technician B says if the vibration changes with vehicle speed, the cause of the problem is probably the driveshaft.
   Who is correct?

   A. A only   C. Both A and B
   B. B only   D. Neither A nor B

37. A shudder is noticeable at speeds around 40–55 mph in third or fourth gear.
   Technician A says the TCC solenoid may be partially blocked.
   Technician B says TCC oil pressure may be too high.
   Who is correct?

   A. A only   C. Both A and B
   B. B only   D. Neither A nor B

38. While diagnosing the cause of no torque converter clutch engagement on a transmission that seems to shift fine:
Technician A suspects a damaged clutch pressure plate into all gears.
Technician B suspects a severely worn input clutch.
Who is correct?
A. A only
B. B only
C. Both A and B
D. Neither A nor B

39. Inspection of the torque converter is being discussed:
Technician A says to replace the converter if the transmission's front pump is damaged.
Technician B says to reweld or repair loose converter drive lugs.
Who is correct?
A. A only
B. B only
C. Both A and B
D. Neither A nor B

40. Technician A says some thrust bearings are called Torrington bearings.
Technician B says some thrust bearings are selective, and used to control endplay.
Who is correct?
A. A only
B. B only
C. Both A and B
D. Neither A nor B

41. An ohmmeter is being used to check several shift solenoids:
Technician A says an open solenoid will prevent the transmission from shifting properly.
Technician B says an infinite reading means the solenoid's windings have more resistance than the meter can measure.
Who is correct?
A. A only
B. B only
C. Both A and B
D. Neither A nor B

42. Technician A says some transmissions have to re-learn the shift adapts after the battery has been disconnected.
Technician B says this adaptive learning mode must be manually activated using a scan tool.
Who is correct?
A. A only
B. B only
C. Both A and B
D. Neither A nor B

43. Transmission/transaxle removal is being discussed:
Technician A says the engine must be supported on some front-wheel-drive (FWD) vehicles when the transaxle is being removed.
Technician B says the converter access plate must be removed before removing a transmission or transaxle.
Who is correct?
A. A only
B. B only
C. Both A and B
D. Neither A nor B

44. The first thing a technician must do to repair a problem is to
A. do a visual inspection
B. verify the complaint
C. do a pressure test
D. see if the vehicle has been maintained

45. Transmission/transaxle installation is being discussed:
Technician A says to always check for damage to the pilot bushing when the converter is removed.
Technician B says the converter drive hub should be lubricated with ATF to help prevent damage to the pump's front seal.
Who is correct?
A. A only
B. B only
C. Both A and B
D. Neither A nor B

46. While diagnosing the cause of transmission overheating:
Technician A checks for contaminated fluid.
Technician B suspects a plugged cooler.
Who is correct?
A. A only
B. B only
C. Both A and B
D. Neither A nor B

47. Technician A says slippage in only First gear can only be caused by a faulty vehicle speed sensor.
Technician B says a defective one-way clutch can cause slippage in all gears.
Who is correct?
A. A only
B. B only
C. Both A and B
D. Neither A nor B

48. *Technician A* cleans the valve body in carburetor cleaner, rinses it with water, and blow dries it with compressed air.
    Technician B says valves that are dragging in their bores should have the bores cleaned by a brake hone in an electric drill.
    Who is correct?

    A. A only   C. Both A and B
    B. B only   D. Neither A nor B

49. Transmission oil pumps are being discussed:
    Technician A says high line pressure would result if the relief valve spring is broken or missing.
    Technician B says the relief valve is designed to restrict the fluid flow until operating pressures are achieved.
    Who is correct?

    A. A only   C. Both A and B
    B. B only   D. Neither A nor B

50. *Technician A* says that transmission input shaft endplay measurements are needed at the completion of a transmission rebuild.
    Technician B says that transmission endplay measurements are needed before a rebuild is started.
    Who is correct?

    A. A only   C. Both A and B
    B. B only   D. Neither A nor B

APD Transmission Parts
Atlanta, GA

ATEC Trans-Tool
San Antonio, TX

Automatic Transmission Rebuilders Association (ATRA)
Ventura, CA

Automotive Specialty Tools
Green Valley Lake, CA

Baum Tools Unlimited, Inc.
Longboat Key, FL

Carquest Corp.
Tarrytown, NY

Drivetrain Specialists
Las Vegas, NV

Hastings Manufacturing
Hastings, MI

Kent-Moore, Div. SPX Corp.
Warren, MI

Lisle Corp.
Clarinda, IA

Mac Tools
Washington Courthouse, OH

MATCO Tool Co.
Stow, OH

NAPA Hand/Service Tools
Lancaster, PA

OTC, Div. SPX Corp.
Owatonna, MN

Parts Plus
Memphis, TN

Snap-on Tools Corp.
Kenosha, WI

Transtar Industries, Inc.
Cleveland, OH

Whatever It Takes Transmission Parts
Louisville, KY

To convert these	to these,	multiply by:
TEMPERATURE		
Centigrade Degrees	Fahrenheit Degrees	1.8, then add 32
Fahrenheit Degrees	Centigrade Degrees	0.556, then minus 32
LENGTH		
Millimeters	Inches	0.03937
Inches	Millimeters	25.4
Meters	Feet	3.28084
Feet	Meters	0.3048
Kilometers	Miles	0.62137
Miles	Kilometers	1.60935
AREA		
Square Centimeters	Square Inches	0.155
Square Inches	Square Centimeters	6.45159
VOLUME		
Cubic Centimeters	Cubic Inches	0.06103
Cubic Inches	Cubic Centimeters	16.38703
Cubic Centimeters	Liters	0.001
Liters	Cubic Centimeters	1000
Liters	Cubic Inches	61.025
Cubic Inches	Liters	0.01639
Liters	Quarts	1.05672

To convert these	to these,	multiply by:
Quarts	Liters	0.94633
Liters	Pints	2.11344
Pints	Liters	0.47317
Liters	Ounces	33.81497
Ounces	Liters	0.02957
WEIGHT		
Grams	Ounces	0.03527
Ounces	Grams	28.34953
Kilograms	Pounds	2.20462
Pounds	Kilograms	0.45359
WORK		
Centimeter-Kilograms	Inch-Pounds	0.8676
Inch-Pounds	Centimeter-Kilograms	1.15262
Meter-Kilograms	Foot-Pounds	7.23301
Foot-Pounds	Newton-Meters	1.3558
PRESSURE		
Kilograms/ Square Centimeter	Pounds/Square Inch	14.22334
Pounds/Square Inch	Kilograms/ Square Centimeter	0.07031
Bar	Pounds/Square Inch	14.504
Pounds/Square Inch	Bar	0.06895

Note: **Terms are highlighted in color,** followed by **Spanish translation in bold.**

Abrasion Wearing or rubbing away of a part.

Abrasión El desgaste o consumo por rozamiento de una parte.

Acceleration An increase in velocity or speed.

Aceleración Un incremento en la velocidad.

Accumulator A device used in automatic transmissions to cushion the shock of shifting between gears, providing a smoother feel inside the vehicle.

Acumulador Un dispositivo que se usa en las transmisiones automáticas para suavizar el choque de cambios entre las velocidades, así proporcionando una sensación más uniforme en el interior del vehículo.

Adhesives Chemicals used to hold gaskets in place during the assembly of an engine. They also aid the gasket in maintaining a tight seal by filling in the small irregularities on the surfaces and by preventing the gasket from shifting due to engine vibration.

Adhesivo Los productos químicos que se usan para sujetar a los empaques en una posición correcta mientras que se efectua la asamblea de un motor. También ayuden para que los empaques mantengan un sello impermeable, rellenando a las irregularidades pequeñas en las superficies y previniendo que se mueva el empaque debido a las vibraciones del motor.

Aeration The process of mixing air into a liquid.

Aireación El proceso de mezclar el aire en un líquido.

Alignment An adjustment to a line or to bring into a line.

Alineación Un ajuste que se efectúa en una linea o alinear.

Ammeter A test instrument used to measure electrical current.

Amperímetro Un aparato que mide corriente eléctrica.

Amplitude The height of a waveform is called its amplitude.

Amplitud La altura de onda eléctrica se llama su amplitud.

Antiseize Thread compound designed to keep threaded connections from damage due to rust or corrosion.

Antiagarrotamiento Un compuesto para filetes diseñado para proteger a las conecciones fileteados de los daños de la oxidación o la corrosión.

Apply devices Devices that hold or drive members of a planetary gearset. They may be hydraulically or mechanically applied.

Dispositivos de aplicación Los dispositivos que sujeten o manejan los miembros de un engranaje planetario. Se pueden aplicar mecánicamente o hidráulicamente.

Arbor press A small, hand-operated shop press used when only a light force is required against a bearing, shaft, or other part.

Prensa para calar Una prensa de mano pequeña del taller que se puede usar en casos que requieren una fuerza ligera contra un cojinete, una flecha u otra parte.

ASE The National Institute for Automotive Service Excellence (ASE) has a voluntary certification program for automotive, heavy-duty truck, auto body repair, and engine machine shop technicians.

ASE El Instituto Nacional para la Excelencia en el Servicio Automotriz (ASE) tiene un programa voluntario de certificación para técnicos automotrices, de camiones de servicio pesado, de reparación de carrocerías y de talleres mecánicos.

ATF Automatic transmission fluid.

ATF Fluido de transmisión automática.

Automatic transmission A transmission in which gear or ratio changes are self-activated, eliminating the necessity of hand-shifting gears.

Transmisión automática Una transmisión en la cual un cambio deengranajes o los cambios en relación son por mando automático, así eliminando la necesidad de cambios de velocidades manual.

Auxiliary transmission fluid pump (ATFP) The auxillary transmission fluid pump supplies pressurized fluid to the transmission on hybrid vehicles when the gasoline engine is not running.

Bomba de aceite de transmisión auxiliar (ATFP) La bomba de aceite de transmisión auxiliar suministra aceite presurizado a la transmisión en vehículos híbridos cuando el motor de gasolina no está encendido.

Axial Parallel to a shaft or bearing bore.

Axial Paralelo a una flecha o al taladro del cojinete.

Axis The centerline of a rotating part, a symmetrical part, or a circular bore.

Eje La linea de quilla de una parte giratoria, una parte simétrica, o un taladro circular.

Axle The shaft or shafts of a machine on which the wheels are mounted.

Semieje El eje o los ejes de una máquina sobre los cuales se montan las ruedas.

Axle ratio The ratio between the rotational speed (rpm) of the driveshaft and that of the driven wheel; gear reduction through the differential, determined by dividing the number of teeth on the ring gear by the number of teeth on the drive pinion.

Relación del eje La relación entre la velocidad giratorio (rpm) del árbol propulsor y la de la rueda arrastrada; reducción de los engranajes por medio del diferencial, que se determina por dividir el número de dientes de la corona por el número de los dientes en el piñón de ataque.

Axle shaft A shaft on which the road wheels are mounted.

Flecha del semieje Una flecha en la cual se monta las ruedas.

Backlash The amount of clearance or play between two meshed gears.

Juego La cantidad de holgura o juego entre dos engranajes endentados.

Balance Having equal weight distribution. The term is usually used to describe the weight distribution around the circumference and between the front and back sides of a wheel.

Equilibrio Lo que tiene una distribución igual de peso. El término suele usarse para describir la distribución del peso alrededor de la circunferencia y entre los lados delanteros y traseros de una rueda.

Balance valve A regulating valve that controls a pressure of just the right value to balance other forces acting on the valve.

Válvula niveladora Una válvula de reglaje que controla a la presión del valor correcto para mantener el equilibrio contra las otras fuerzas que afectan a la válvula.

Ball bearing An antifriction bearing consisting of a hardened inner and outer race with hardened steel balls that roll between the two races, and supports the load of the shaft.

Rodamiento de bolas Un cojinete de antifricción que consiste de una pista endurecida interior e exterior y contiene bolas de acero endurecidos que ruedan entre las dos pistas, y sostiene la carga de la flecha.

Ball joint A suspension component that attaches the control arm to the steering knuckle and serves as the lower pivot point for the steering knuckle. The ball joint gets its name from its ball-and-socket design. It allows both up-and-down motion as well as rotation. In a MacPherson strut FWD suspension system, the two lower ball joints are nonload carrying.

Articulación esférica Un componente de la suspensión que une el brazo de mando a la articulación de la dirección y sirve como un punto pivote inferior de la articulación de la dirección. La articulación esférica derive su nombre de su diseño de bola y casquillo. Permite no sólo el movimiento de arriba y abajo sino también el de rotación. En un sistema de suspensión tipo FWD con poste de MacPherson, las articulaciones esféricas inferiores no soportan el peso.

Ballooning A condition in which the torque converter has been blown up like a balloon caused by excessive pressure in the converter.

Inflación Una condición en el qual el convertidor del esfuerzo de torsión se ha inflado como un baloon, que es una condición causada por la presión excesiva dentro del convertidor.

Band A steel band with an inner lining of frictional material. A device used to hold a clutch drum at certain times during transmission operation.

Banda Una banda de acero que tiene un forro interior de una materia de fricción. Un dispositivo que retiene al tambor del embrague en algunos momentos durante la operación de la transmisión.

Bearing The supporting part that reduces friction between a stationary and rotating part or between two moving parts.

Cojinete La parte portadora que reduce la fricción entre una parte fija y una parte giratoria o entre dos partes que muevan.

Bearing cage A spacer that keeps the balls or rollers in a bearing in proper position between the inner and outer races.

Jaula del cojinete Un espaciador que mantiene a las bolas o a los rodillos del cojinete en la posición correcta entre las pistas interiores e exteriores.

Bearing caps In the differential, caps held in place by bolts or nuts which, in turn, hold bearings in place.

Tapones del cojinete En un diferencial, las tapas que se sujeten en su lugar por pernos o tuercas, los cuales en su turno, retienen y posicionan a los cojinetes.

Bearing cone The inner race, rollers, and cage assembly of a tapered roller bearing. Cones and cups must always be replaced in matched sets.

Cono del cojinete La asamblea de la pista interior, los rodillos, y el jaula de un cojinete de rodillos cónico. Se debe siempre reemplazar a ambos partes de un par de conos del cojinete y los anillos exteriores a la vez.

Bearing cup The outer race of a tapered roller bearing or ball bearing.

Anillo exterior La pista exterior de un cojinete cónico de rodillas o de bolas.

Bearing race The surface on which the rollers or balls of a bearing rotate. The outer race is the same thing as the cup, and the inner race is the one closest to the axle shaft.

Pista del cojinete La superficie sobre la cual rueden los rodillos o las bolas de un cojinete. La pista exterior es lo mismo que un anillo exterior, y la pista interior es la más cercana a la flecha del eje.

Belleville spring A tempered spring steel cone-shaped plate used to aid the mechanical force in a pressure plate assembly.

Resorte de tensión Belleville Un plato de resorte del acero revenido en forma cónica que aumenta a la fuerza mecánica de una asamblea del plato opresor.

Bell housing A housing that fits over the clutch components and connects the engine and the transmission.

Concha del embrague Un cárter que encaja a los componentes del embrague y conecta al motor con la transmisión.

Bloodborne pathogens Pathogenic microorganisms that are present in human blood and can cause disease. These pathogens include, but are not limited to, hepatitis B virus (HBV) and human immunodeficiency virus (HIV).

Patógenos transportados en la sangre Microorganismos patógenos que se encuentran en la sangre humana y que pueden causar enfermedades. Estos son, entre otros, el virus de la hepatitis B y el virus de inmunodeficiencia humana (VIH).

Blowgun A blowgun is used to supply regulated compressed air for cleaning and air checking of transmission parts.

Pistola de aire La pistola de aire se usa para suministrar aire comprimido regulado para limpiar y verificar el aire de las piezas de la transmisión.

Bolt head The part of a bolt that the socket or wrench fits over to torque or tighten the bolt.

Cabeza del perno La pieza de un perno que el socket o la llave ajusta sobre la cabeza del perno para apretarlo.

Bolt shank The smooth area on a bolt from the bottom surface of the head to the start of the threads.

Asta del perno El área lisa entre el fondo de la cabeza y el principio de las roscas se llama la asta del perno.

Bolt torque The turning effort required to offset resistance as the bolt is being tightened.

Torsión del perno El esfuerzo de torsión que se requiere para compensar la resistencia del perno mientras que esté siendo apretado.

Brinelling Rough lines worn across a bearing race or shaft due to impact loading, vibration, or inadequate lubrication.

Efecto brinel Lineas ásperas que aparecen en las pistas de un cojinete o en las flechas debido al choque de carga, la vibración, o falta de lubricación.

Burr A feather edge of metal left on a part being cut with a file or other cutting tool.

Rebaba Una lima espada de metal que permanece en una parte que ha sido cortado con una lima u otro herramienta de cortar.

Bus A common connector used as an information source for the vehicle's various control units.

Bus Un conector común que se usa como un fuente de información para los varios aparatos de control del vehículo.

Bushing A cylindrical lining used as a bearing assembly; can be made of steel, brass, bronze, nylon, or plastic.

Buje Un forro cilíndrico que se usa como una asamblea de cojinete que puede ser hecho del acero, del latón, del bronze, del nylon, o del plástico.

Butt-end locking ring A locking ring whose ends are cut to butt up against or contact each other once in place. There is no gap between the ends of a butt-end ring.

Anillo de enclavamiento a tope Un anillo de enclavamiento cuyos extremidades son cortadas para toparse o ajustarse una contra la otra en lugar. No hay holgura entre las extremidades de un anillo de retén a tope.

C-clip A C-shaped clip used to retain the drive axles in some rear axle assemblies.

Grapa de C Una grapa en forma de C que retiene a las flechas motrices en algunas asambleas de ejes traseras.

Cage A spacer used to keep the balls or rollers in proper relation to one another. In a constant-velocity joint, the cage is an open metal framework that surrounds the balls to hold them in position.

Jaula Una espaciador que mantiene una relación correcta entre los rodillos o las bolas. En una junta de velocidad constante, la jaula es un armazón abierto de metal que rodea a las bolas para mantenerlas en posición.

Cap An object that fits over an opening to stop flow.

Tapón Un objecto que tapa a una apertura para detener el flujo.

Carbon monoxide (CO) Carbon monoxide is the product of incomplete combustion that occurs in vehicle exhaust. If breathed in sufficient quantities it can cause death.

Monóxido de carbono (CO) El monóxido de carbono es el producto de la combustión incompleta que tiene lugar en el escape del vehículo. Si se inhala más de cierta cantidad, puede ocasionar la muerte.

Case porosity Leaks caused by tiny holes that are formed by trapped air bubbles during the casting process.

Porosidad del cárter Las fugas que se causan por los hoyitos pequeños formados por burbújas de aire entrapados durante el proceso del moldeo.

Category III multimeters Category III multimeters are used to provide safe measurements on hybrid vehicles by being specially designed to handle high voltages that might be encountered.

Multímetros de categoría III Los multímetros de categoría III se usan para mediciones seguras en vehículos híbridos ya que están diseñados especialmente para aceptar los altos voltajes que se podrían encontrar.

Caustic Something that causes corrosion.

Cáustico Algo que provoca corrosión.

Chamfer A bevel or taper at the edge of a hole or a gear tooth.

Chaflán Un bisél o cono en el borde de un hoyo o un diente del engranaje.

Chamfer face A beveled surface on a shaft or part that allows for easier assembly. The ends of FWD driveshafts are often chamfered to make installation of the CV joints easier.

Cara achaflanada Una superficie biselada en una flecha o una parte que facilita la asamblea. Los extremos de los árboles de mando de FWD suelen ser achaflandos para facilitar la instalación de las juntas CV.

Chase To straighten up or repair damaged threads.

Embutir Enderezar o reparar a los filetes dañados.

Chasing To clean threads with a tap.

Embutido Limpiar a los filetes con un macho.

Chassis The vehicle frame, suspension, and running gear. On FWD cars, it includes the control arms, struts, springs, trailing arms, sway bars, shocks, steering knuckles, and frame. The driveshafts, constant-velocity joints, and transaxle are not part of the chassis or suspension.

Chasis El armazón de un vehículo, la suspensión, y el engranaje de marcha. En los coches de FWD, incluye los brazos de mando, los postes, los resortes (chapas), los brazos traseros, las estabilizadoras, las articulaciones de la dirección y el armazón. Los árboles de mando, las juntas de velocidad constante, y la flecha impulsora no son partes del chasis ni de la suspensión.

Circlip A split steel snap ring that fits into a groove to hold various parts in place. Circlips are often used on the ends of FWD driveshafts to retain the constant-velocity joints.

Grapa circular Un seguro partido circular de acero que se coloca en una ranura para posicionar a varias partes. Las grapas circulares se suelen usar en las extremidades de los árboles de mando en FWD para retener las juntas de velocidad constante.

Circuit tester A circuit tester is a simple lamp that can indicate the presence of voltage in a circuit. Also called a test lamp

Probador de circuitos Un probador de circuitos es una simple lámpara que indica la presencia de voltaje en un circuito. También se llama lámpara de prueba.

Class A fire A fire in which wood, paper, and other ordinary materials are burning.

Fuego de la Clase A Un fuego en el cual la madera, el papel, y otros materiales ordinarios se están quemando se llama un fuego de la Clase A.

Class B fire A fire involving flammable liquids, such as gasoline, diesel fuel, paint, grease, oil, and other similar liquids.

Fuego de la Clase B Un fuego que quema líquidos inflamables, tales como gasolina, el combustible diesel, la pintura, la grasa, el aceite, y otros líquidos similares se llama un fuego de la clase B.

Class C fire An electrical fire.

Fuego de la Clase C Un fuego eléctrico se llama un fuego de la clase C.

Class D fire A unique type of fire because the material burning is a metal.

Fuego de la Clase D Un fuego de la clase D es un fuego inusual en el cual un metal se está quemando.

Clearance The space allowed between two parts, such as between a journal and a bearing.

Holgura El espacio permitido entre dos partes, tal como entre un muñón y un cojinete.

Clutch A device for connecting and disconnecting the engine from the transmission or for a similar purpose in other units.

Embrague Un dispositivo para conectar y desconectar el motor de la transmisión o para tal propósito en otros conjuntos.

Clutch packs   A series of clutch discs and plates installed alternately in a housing to act as a driving or driven unit.

Conjuntos de embrague   Una seria de discos y platos de embrague que se han instalado alternativamente en un cárter para funcionar como una unedad de propulsión o arrastre.

Clutch slippage   A situation in which engine speed increases but increased torque is not transferred through to the driving wheels.

Resbalado del embrague   Una situatión en el qual la velocidad del motor aumenta pero la torsión aumentada del motor no se transfere a las ruedas de marcha.

Clutch volume index (CVI)   This represents the volume of fluid needed to compress a clutch pack or apply a brake band.

Índice del Volumen del Embrague   Esto representa el volumen de líquido necesitado para comprimir un paquete del embrague o para aplicar una banda del freno.

Coefficient of friction   The ratio of the force resisting motion between two surfaces in contact to the force holding the two surfaces in contact.

Coeficiente de la fricción   La relación entre la fuerza que resiste al movimiento entre dos superficies que tocan y la fuerza que mantiene en contacto a éstas dos superficies.

Coil preload springs   Coil springs are made of tempered steel rods formed into a spiral that resist compression; located in the pressure plate assembly.

Muelles de embrague   Los muelles espirales son fabricadas de varillas de acero revenido y resisten la compresión; se ubican en el conjunto del plato opresor.

Coil spring   A heavy wire-like steel coil used to support the vehicle weight while allowing for suspension motions. On FWD cars, the front coil springs are mounted around the MacPherson struts. On the rear suspension, they may be mounted to the rear axle, to trailing arms, or around rear struts.

Muelles de embrague   Un resorte espiral hecho de acero en forma de alambre grueso que soporte el peso del vehículo mientras que permite a los movimientos de la suspensión. En los coches de FWD, los muelles de embrague delanteros se montan alrededor de los postes Macpherson. En la suspensión trasera, pueden montarse en el eje trasero, en los brasos traseros, o alrededor de los postes traseros.

Compound   A mixture of two or more ingredients.

Compuesto   Una combinación de dos ingredientes o más.

Concentric   Two or more circles having a common center.

Concéntrico   Dos círculos o más que comparten un centro común.

Constant-velocity joint   A flexible coupling between two shafts that permits each shaft to maintain the same driving or driven speed regardless of operating angle, allowing for a smooth transfer of power. The constant-velocity joint (also called CV joint) consists of an inner and outer housing with balls in between, or a tripod and yoke assembly.

Junta de velocidad constante   Un acoplador flexible entre dos flechas que permite que cada flecha mantenga la velocidad de propulsión o arrastre sin importar el ángulo de operación, efectuando una transferencia lisa del poder. La junta de velociadad constante (también llamado junta CV) consiste en un cárter interior e exterior entre los cuales se encuentran bolas, o de un conjunto de trípode y yugo.

Control arm   A suspension component that links the vehicle frame to the steering knuckle or axle housing and acts as a hinge to allow up-and-down wheel motions. The front control arms are attached to the frame with bushings and bolts and are connected to the steering knuckles with ball joints. The rear control arms attach to the frame with bushings and bolts and are welded or bolted to the rear axle or wheel hubs.

Brazo de mando   Un componente de la suspención que une el armazón del vehículo al articulación de dirección o al cárter del eje y que se porta como una bisagra para permitir a los movimientos verticales de las ruedas. Los brazos de mando delanteros se conectan al armazón por medio de pernos y bujes y se conectan al articulación de dirección por medio de los articulaciones esféricos. Los brazos de mando traseros se conectan al armazón por medio de pernos y bujes y son soldados o empernados al eje trasero o a los cubos de la rueda.

Corrosion   Chemical action, usually by an acid, that eats away (decomposes) a metal.

Corrosión   Un acción químico, por lo regular un ácido, que corroe (descompone) un metal.

Corrosivity   The characteristic of a material that enables it to dissolve metals and other materials or burn the skin.

Corrosividad   Característica de un material que le permite disolver metales y otros materiales, o que quemar la piel.

Cotter pin   A type of fastener, made from soft steel in the form of a split pin, that can be inserted in a drilled hole. The split ends are spread to lock the pin in position.

Pasador de chaveta   Un tipo de fijación, hecho de acero blando en forma de una chaveta que se puede insertar en un hueco tallado. Las extremidades partidas se despliegen para asegurar la posición de la chaveta.

Counterclockwise rotation   Rotating in the opposite direction of the hands on a clock.

Rotación en sentido inverso   Girando en el sentido opuesto de las agujas de un reloj.

Coupling   A connecting means for transferring movement from one part to another; may be mechanical, hydraulic, or electrical.

Acoplador   Un método de conección que transfere el movimiento de una parte a otra; puede ser mecánico, hidráulico, o eléctrico.

Coupling phase   Point in torque converter operation in which the turbine speed is 90 percent of impeller speed and there is no longer any torque multiplication.

Fase del acoplador   El punto de la operación del convertidor de la torsión en el cual la velocidad de la turbina es el 90% de la velocidad del impulsor y no queda ningún multiplicación de la torsión.

Cover plate   A stamped steel cover bolted over the service access to the manual transmission.

Cubrejuntas   Un cubierto de acero estampado que se emperna en la apertura de servicio de la transmisión manual.

Crocus cloth   A very fine polishing paper designed to remove very little metal; therefore, it is safe to use on critical surfaces.

Tela de óxido férrico   Un papel muy fino para pulir. Fue diseñado para raspar muy poco del metal; por lo tanto, se suele emplear en las superficies críticas.

Cycle   One set of changes in a signal that repeats itself several times.

Ciclo   Un conjunto de cambios en una señal que se repite varias veces.

Default mode   A mode of operation that allows for limited use of the transmission in the case of electronic failure.

Modo del uso limitado Un modo de operación que permite el uso limitado de la transmisión en el caso del incidente electrónico.

Deflection Bending or movement away from normal due to loading.

Desviación Curvación o movimiento fuera de lo normal debido a la carga.

Degree A unit of measurement equal to 1/360th of a circle.

Grado Una uneda de medida que iguala al 1/360 parte de un círculo.

Density Compactness; relative mass of matter in a given volume.

Densidad La firmeza; una cantidad relativa de la materia que ocupa a un volumen dado.

Detent A small depression in a shaft, rail, or rod into which a pawl or ball drops when the shaft, rail, or rod is moved. This provides a locking effect.

Detención Un pequeño hueco en una flecha, una barra o una varilla en el cual cae una bola o un linguete al moverse la flecha, la barra o la varilla. Esto provee un efecto de enclavamiento.

Detent mechanism A shifting control designed to hold the manual transmission in the gear range selected.

Aparato de detención Un control de desplazamiento diseñado a sujetar a la transmisión manual en la velocidad selecionada.

Diagnosis A systematic study of a machine or machine parts to determine the cause of improper performance or failure.

Diagnóstico Un estudio sistemático de una máquina o las partes de una máquina con el fín de determinar la causa de una falla o de un operación irregular.

Dial caliper A measurement tool that can measure outside diameter, inside diameter, and depth in a single unit.

Calibrador de dial, herramienta de medición que mide el diámetro externo, el diámetro interno y la profundidad en una única unidad.

Dial indicator A measuring instrument with the readings indicated on a dial rather than on a thimble as on a micrometer.

Indicador de carátula Un instrumento de medida cuyo indicador es en forma de muestra en contraste al casquillo de un micrómetro.

Differential A mechanism between drive axles that permits one wheel to run at a different speed than the other while turning.

Diferencial Un mecanismo entre dos semiejes que permite que una rueda gira a una velocidad distincta que la otra en una curva.

Differential action An operational situation in which one driving wheel rotates at a slower speed than the opposite driving wheel.

Acción del diferencial Una situación durante la operación en la cual una rueda propulsora gira con una velocidad más lenta que la rueda propulsora opuesta.

Differential case The metal unit that encases the differential side gears and pinion gears, and to which the ring gear is attached.

Caja de satélites La unedad metálica que encaja a los engranajes planetarios (laterales) y a los satélites del diferencial, y a la cual se conecta la corona.

Differential drive gear A large circular helical gear that is driven by the transaxle pinion gear and shaft and drives the differential assembly.

Corona Un engranaje helicoidal grande circular que es arrastrado por el piñon de la flecha de transmisión y la flecha y propela al conjunto del diferencial.

Differential housing Cast iron assembly that houses the differential unit and the drive axles. Also called the rear axle housing.

Cárter del diferencial Una asamblea de acero vaciado que encaja a la unedad del diferencial y los semiejes. También se llama el cárter del eje trasero.

Differential pinion gears Small beveled gears located on the differential pinion shaft.

Satélites Engranajes pequeños biselados que se ubican en la flecha del piñon del diferencial.

Differential pinion shaft A short shaft locked to the differential case. This shaft supports the differential pinion gears.

Flecha del piñon del diferencial Una flecha corta clavada en la caja de satélites. Esta flecha sostiene a los satélites.

Differential ring gear A large circular hypoid-type gear enmeshed with the hypoid drive pinion gear.

Corona Un engranaje helicoidal grande circular endentado con el piñon de ataque hipoide.

Differential side gears The gears inside the differential case that are internally splined to the axle shafts and are driven by the differential pinion gears.

Planetarios (laterales) Los engranajes adentro de la caja de satélites que son acanalados a los semiejes desde el interior, y que se arrastran por los satélites.

Digital multimeter (DMM) A device that measures voltage, resistance, and current with a digital readout. Some can also measure frequency and capacitance.

Multímetro digital (DMM) Dispositivo que mide voltaje, resistencia y corriente con una lectura digital. Algunos también miden la frecuencia y la capacitancia.

Digital volt ohmmeter (DVOM) A device that measures volts and ohms and usually amperage with a digital display.

Ohmímetro de voltaje digital (DVOM) dispositivo que mide el voltaje y los ohmios y, generalmente, el amperaje con una pantalla digital.

Dipstick A metal rod used to measure the fluid in an engine or transmission.

Varilla de medida Una varilla de metal que se usa para medir el nivel de flúido en un motor o en una transmisión.

Direct drive System in which one turn of the input driving member corresponds to one complete turn of the driven member, such as when there is direct engagement between the engine and driveshaft in which the engine crankshaft and the driveshaft turn at the same rpm.

Mando directo Una vuelta del miembro de ataque o propulsión que se compara a una vuelta completa del miembro de arrastre, tal como cuando hay un enganchamiento directo entre el motor y el árbol de transmisión en el qual el cigueñal y el árbol de transmisión giran al mismo rpm.

Disengage When the operator moves the clutch pedal toward the floor to disconnect the driven clutch disc from the driving flywheel and pressure plate assembly.

Desembragar Cuando el operador mueva el pedal de embrague hacia el piso para desconectar el disco de embrague del volante impulsor y del conjunto del plato opresor.

Distortion A warpage or change in form from the original shape.

Distorción Abarquillamiento o un cambio en la forma original.

DLC The data link connector. This is the connector used to connect into a vehicle's computer system for the purpose of diagnostics. Prior to J1930, this was commonly referred to as the ALDL.

DLC El conector de enlaces de datos. Este es el conector que se usa en conectarse al sistema computerizado del vehículo con el propósito de efectuar los diagnósticos. Antes del J1930, esto solía referirse como el ALDL.

Dowel A metal pin attached to one object that, when inserted into a hole in another object, ensures proper alignment.

Espiga Una clavija de metal que se fija a un objeto, que al insertarla en el hoyo de otro objeto, asegura una alineación correcta.

Dowel pin A pin inserted in matching holes in two parts to maintain those parts in fixed relation one to another.

Clavija de espiga Una clavija que se inserte en los hoyos alineados en dos partes para mantener ésos dos partes en una relación fijada el uno al otro.

Downshift To shift a transmission into a lower gear.

Cambio descendente Cambiar la velocidad de una transmisión a una velocidad más baja.

Driveline torque Relates to rear-wheel driveline; the transfer of torque between the transmission and the driving axle assembly.

Potencia de la flecha motríz Se relaciona a la flecha motríz de las ruedas traseras y transfere la potencia de la torsión entre la transmisión y el conjunto del eje trasero.

Driven gear The gear meshed directly with the driving gear to provide torque multiplication, reduction, or a change of direction.

Engranaje de arrastre El engranaje endentado directamente al engranaje de ataque para proporcionar la multiplicación, la reducción, o los cambios de dirección de la potencia.

Drive pinion gear One of the two main driving gears located within the transaxle or rear driving axle housing. Together, the two gears multiply engine torque.

Engranaje de piñon de ataque Uno de dos engranajes de ataque principales que se ubican adentro de la flecha de transmisión o en el cárter del eje de propulsión. Los dos engranajes trabajan juntos para multiplicar la potencia.

Driveshaft An assembly of one or two universal joints connected to a shaft or tube; used to transmit power from the transmission to the differential. Also called the propeller shaft.

Árbol de mando Una asamblea de uno o dos uniones universales que se conectan a un árbol o un tubo; se usa para transferir la potencia desde la transmisión al diferencial. También se le refiere como el árbol de propulsión.

Dry friction The friction between two dry solids.

Fricción seca Fricción entre dos s-lidos secos.

DSO A common acronym for a digital storage oscilloscope.

DSO Siglas comunes para un osciloscopio con memoria digital.

DTC The acronym for diagnostic trouble code.

DTC La sigla en inglés por un código diagnóstico de averías.

Duty cycle A term for the percentage of time in one cycle that a device is turned on.

Ciclo de trabajo es la denominación para el porcentaje de tiempo de un ciclo en el que un dispositivo está encendido.

Dynamic In motion.

Dinámico En movimiento.

EEPROM An electrically erasable programmable read only memory chip.

EEPROM Una placa de memoria, leer solamente, erasible y programable.

Elastomer Any rubber-like plastic or synthetic material used to make bellows, bushings, and seals.

Elastómero Cualquiera materia plástic parecida al hule o una materia sintética que se utiliza para fabricar a los fuelles, los bujes y las juntas.

End clearance Distance between a set of gears and their cover, commonly measured on oil pumps.

Holgura del extremo La distancia entre un conjunto de engranajes y su placa de recubrimiento, suele medirse en las bombas de aceite.

Endplay The amount of axial or end-to-end movement in a shaft due to clearance in the bearings.

Juego de las extremidades La cantidad del movimiento axial o del movimiento de extremidad a extremidad en una flecha debido a la holgura que se deja en los cojinetes.

Engage When the vehicle operator moves the clutch pedal up from the floor, this engages the driving flywheel and pressure plate to rotate and drive the driven disc.

Accionar Cuando el operador del vehículo deja subir el pedal del embrague del piso, ésto acciona la volante de ataque y el plato opresor para impulsar al disco de arrastre.

Engagement chatter A shaking, shuddering action that takes place as the driven disc makes contact with the driving members. Chatter is caused by a rapid grip and slip action.

Chasquido de enganchamiento Un movimiento de sacudo o temblor que resulta cuando el disco de ataque viene en contacto con los miembros de propulsión. El chasquido se causa por una acción rápida de agarrar y deslizar.

Engine torque A turning or twisting action developed by the engine, measured in foot-pounds or kilogram-meters.

Torsión del motor Una acción de girar o torcer que crea el motor, ésta se mide en librasópie o kilosómetros.

EP toxicity The characteristic of a material that enables it to leach one or more of eight heavy metals in concentrations greater than 100 times standard drinking water concentrations.

Toxicidad por el procedimiento de extracción Característica de un material que permite que se filtren uno o más de ocho metales pesados en concentraciones mayores a 100 veces las concentraciones estándares en el agua potable.

Essential tool kit A set of special tools designed for a particular model of car or truck.

Estuche de herramientas principales Un conjunto de herramientas especiales diseñadas para un modelo específico de coche o camión.

Etching A discoloration or removal of some material caused by corrosion or some other chemical reaction.

Grabado por ácido Una descoloración o remueva de una materia que se efectua por medio de la corrosión u otra reacción química.

Extension housing An aluminum or iron casting of various lengths that encloses the transmission output shaft and supporting bearings.

Cubierta de extensión Una pieza moldeada de aluminio o acero que puede ser de varias longitudes que encierre a la flecha de salida de la transmisión y a los cojinetes de soporte.

External gear A gear with teeth across the outside surface.

Engranaje exterior Un engranaje cuyos dientes estan en la superficie exterior.

Externally tabbed clutch plates Clutch plates that are designed with tabs around the outside periphery to fit into grooves in a housing or drum.

Placas de embrague de orejas externas Las placas de embrague que se diseñan de un modo para que las orejas periféricas de la superficie se acomoden en una ranura alrededor de un cárter o un tambor.

Extreme-pressure lubricant A special lubricant for use in hypoid-gear differentials; needed because of the heavy wiping loads imposed on the gear teeth.

Lubricante de presión extrema Un lubricante especial que se usa en las diferenciales de tipo engranaje hipóide; se requiere por la carga de transmisión de materia pesada que se imponen en los dientes del engranaje.

Face The front surface of an object.

Cara La superficie delantera de un objeto.

Fatigue The buildup of natural stress forces in a metal part that eventually causes it to break. Stress results from bending and loading the material.

Fatiga El incremento de tensiones y esfuerzos normales en una parte de metal que eventualmente causen una quebradura. Los esfuerzos resultan de la carga impuesta y el doblamiento de la materia.

Feeler gauge A metal strip or blade finished accurately with regard to thickness used for measuring the clearance between two parts; such gauges ordinarily come in a set of different blades graduated in thickness by increments of 0.001 inch.

Calibrador de laminillas Una lámina o hoja de metal que ha sido acabado precisamente con respecto a su espesor que se usa para medir la holgura entre dos partes; estas galgas típicamente vienen en un conjunto de varias espesores graduados desde el 0.001 de una pulgada.

Fillet The small, smooth curve on a bolt, where the shank flows into the bolt head.

Filete La curva pequeña, lisa en un perno, adonde la asta fluye en la cabeza del perno se llama un filete.

Final drive ratio The ratio between the drive pinion and ring gear.

Relación del mando final La relación entre el piñon de ataque y la corona.

Fit The contact between two machined surfaces.

Ajuste El contacto entre dos superficies maquinadas.

Fixed-type constant-velocity joint A joint that cannot telescope or plunge to compensate for suspension travel. Fixed joints are always found on the outer ends of the driveshafts of FWD cars. A fixed joint may be of either Rzeppa or tripod type.

Junta tipo fijo de velocidad constante Una junta que no tiene la capacidad de los movimientos telescópicos o repentinos que sirven para compensar en los viajes de suspensión. Las juntas fijas siempre se ubican en las extremidades exteriores de los árboles de mando en los coches de FWD. Una junta tipo fijo puede ser de un tipo Rzeppa o de trípode.

Flammability A statement of how well a substance supports combustion.

Inflamabilidad Declaración acerca de cuánto resiste a la combustión una sustancia.

Flange A projecting rim or collar on an object that keeps it in place.

Reborde Una orilla o un collar sobresaliente de un objeto cuyo función es de mantenerlo en lugar.

Flash programming Flash programming is the programming or reprogramming of a control module.

Programación flash La programación flash es la programación o reprogramación de un módulo de control.

Flat rate Flat rate is the established amount of time a technician gets paid for a particular job.

Tarifa fija la tarifa fija es el tiempo establecido que se le paga a un técnico por un trabajo determinado.

Flexplate A lightweight flywheel used only on engines equipped with an automatic transmission. A flexplate is equipped with a starter ring gear around its outside diameter and also serves as the attachment point for the torque converter.

Placa articulada Un volante ligera que se usa solamente en los motores que se equipan con una transmisión automática. El diámetro exterior de la placa articulada viene equipado con un anillo de engranajes para arrancar y también sirve como punto de conección del convertidor de la torsión.

Fluid coupling A device in the powertrain consisting of two rotating members; transmits power from the engine, through a fluid, to the transmission.

Acoplamiento de fluido Un dispositivo en el tren de potencia que consiste de dos miembros rotativos; transmite la potencia del motor, por medio de un fluido, a la transmisión.

Fluid drive A drive in which there is no mechanical connection between the input and output shafts, and power is transmitted by moving oil.

Dirección fluido Una dirección en la cual no hay conecciones mecánicas entre las flechas de entrada o salida, y la potencia se transmite por medio del aceite en movimiento.

Flywheel A heavy metal wheel that is attached to the crankshaft and rotates with it; helps smooth out the power surges from the engine power strokes; also serves as part of the clutch and engine-cranking system.

Volante Una rueda pesada de metal que se fija al cigueñal y gira con ésta; nivela a los sacudos que provienen de la carrera de fuerza del motor; también sirve como parte del embrague y del sistema de arranque.

Flywheel ring gear A gear fitted around the flywheel that is engaged by teeth on the starting motor drive to crank the engine.

Engranaje anular del volante Un engranaje, colocado alrededor del volante que se acciona por los dientes en el propulsor del motor de arranque y arranca al motor.

Foot-pound (ft.-lb.) A measure of the amount of energy or work required to lift 1 pound a distance of 1 foot.

Pie libra Una medida de la cantidad de energía o fuerza que requiere mover una libra a una distancia de un pie.

Force Any push or pull exerted on an object; measured in pounds and ounces, or in newtons (N) in the metric system.

Fuerza Cualquier acción empujado o jalado que se efectua en un objeto; se mide en pies y onzas, o en newtones (N) en el sistema métrico.

Four-wheel-drive On a vehicle, driving axles at both front and rear, so that all four wheels can be driven.

Tracción a cuatro ruedas En un vehículo, se trata de los ejes de dirección fronteras y traseras, para que cada uno de las ruedas puede impulsar.

Freeze frame Freeze frame is a record of activity that was occurring when a DTC was set that can be used to aid diagnosis.

Imagen congelada La imagen congelada o fija es un registro de la actividad que estaba teniendo lugar cuando se estableció un código de diagnóstico de problema (DTC), y puede usarse para ayudar a diagnosticar el problema.

Frequency Frequency is the number of times per second that a signal completes a cycle in a second.

Frecuencia La frecuencia es el número de veces por segundo que una señal completa un ciclo en un segundo.

Frame The main understructure of the vehicle to which everything else is attached. Most FWD cars have only a subframe for the front suspension and drivetrain. The body serves as the frame for the rear suspension.

Armazón La estructura principal del vehículo al cual todo se conecta. La mayoría de los coches FWD sólo tiene un bastidor auxiliar para la suspensión delantera y el tren de propulsión. El carrocería del coche sirve de chassis par la suspensión trasera.

Freewheel To turn freely and not transmit power.

Volante libre Da vueltas libremente sin transferir la potencia.

Freewheeling clutch A mechanical device that will engage the driving member to impart motion to a driven member in one direction but not the other. Also known as an "overrunning clutch."

Embrague de volante libre Un dispositivo mecánico que acciona el miembro de tracción y da movimiento al miembro de tracción en una dirección pero no en la otra. También se conoce bajo el nombre de un "embrague de sobremarcha."

Friction The resistance to motion between two bodies in contact with each other.

Fricción La resistencia al movimiento entre dos cuerpos que estan en contacto.

Friction bearing A bearing in which there is sliding contact between the moving surfaces. Sleeve bearings, such as those used in connecting rods, are friction bearings.

Rodamientos de fricción Un cojinete en el cual hay un contacto deslizante entre las superficies en movimiento. Los rodamientos de manguitos, como los que se usan en las bielas, son rodamientos de fricción.

Friction disc In the clutch, a flat disc, faced on both sides with frictional material and splined to the clutch shaft. It is positioned between the clutch pressure plate and the engine flywheel. Also called the clutch disc or driven disc.

Disco de fricción En el embrague, un disco plano al cual se ha cubierto ambos lados con una materia de fricción y que ha sido estriado a la flecha del embrague. Se posiciona entre el plato opresor del embrague y el volante del motor. También se llama el disco del embrague o el disco de arrastre.

Friction facings A hard-molded or woven asbestos or paper material that is riveted or bonded to the clutch driven disc.

Superficie de fricción Un recubrimiento remachado o aglomerado al disco de arrastre del embrague que puede ser hecho del amiento moldeado o tejido o de una materia de papel.

Front pump Pump located at the front of the transmission. It is driven by the engine through two dogs on the torque converter housing. It supplies fluid whenever the engine is running.

Bomba delantera Una bomba ubicado en la parte delantera de la transmisión. Se arrastre por el motor al través de dos álabes en el cárter del convertidor de la torsión. Provee el fluido mientras que funciona el motor.

Front-wheel-drive (FWD) Describes a vehicle with all drivetrain components located at the front.

Tracción de las ruedas delanteras (FWD) El vehículo tiene todos los componentes del tren de propulsión en la parte delantera.

Galling Wear caused by metal-to-metal contact in the absence of adequate lubrication. Metal is transferred from one surface to the other, leaving behind a pitted or scaled appearance.

Desgaste por fricción El desgaste causado por el contacto de metal a metal en la ausencia de lubricación adecuada. El metal se transfere de una superficie a la otra, causando una aparencia agujerado o con depósitos.

Gasket A layer of material, usually made of cork, paper, plastic, composition, or metal, or a combination of these, placed between two parts to make a tight seal.

Empaque Una capa de una materia, normalmente hecho del corcho, del papel, del plástico, de la materia compuesta o del metal, o de cualquier combinación de éstos, que se coloca entre dos partes para formar un sello impermeable.

Gasket cement A liquid adhesive material or sealer used to install gaskets.

Mastique para empaques Una substancia líquida adhesiva, o una substancia impermeable, que se usa para instalar a los empaques.

Gear A wheel with external or internal teeth that serves to transmit or change motion.

Engranaje Una rueda que tiene dientes interiores o exteriores que sirve para transferir o cambiar el movimiento.

Gear lubricant A type of grease or oil blended especially to lubricate gears.

Lubricante para engranaje Un tipo de grasa o aceite que ha sido mezclado específicamente para la lubricación de los engranajes.

Gear ratio The number of revolutions of a driving gear required to turn a driven gear through one complete revolution. For a pair of gears, the ratio is found by dividing the number of teeth on the driven gear by the number of teeth on the driving gear.

Relación de los engranajes El número de las revoluciones requeridas del engranaje de propulsión para dar una vuelta completa al engranaje arrastrado. En una pareja de engranajes, la relación se calcula al dividir el número de los dientes en el engranaje de arrastre por el número de los dientes en el engranaje de propulsión.

Gear reduction When a small gear drives a large gear, there is an output speed reduction and a torque increase that results in a gear reduction.

Velocidad descendente Cuando un engranaje pequeño impulsa a un engranaje grande, hay una reducción en la velocidad de salida y un incremento en la torsión que resulta en una cambio descendente de los velocidades.

Gearshift A linkage-type mechanism by which the gears in an automobile transmission are engaged and disengaged.

Varillaje de cambios Un mecanismo tipo eslabón que acciona y desembraga a los engranajes de la transmisión.

Gear whine A high-pitched sound developed by some types of meshing gears.

Ruido del engranaje Un sonido agudo que proviene de algunos tipos de engranajes endentados.

Glitches Abnormal, slight movements of a waveform on a lab scope. These can be caused by circuit problems or noise in the circuit.

Irregularidades espontáneas ("glitch") Los movimientos pequeños e abnormales en una onda en una pantalla de laboratorio. Estos pueden ser causados por los problemas de los circuitos o el ruido dentro del circuito.

Grade marks Marks on fasteners that indicate strength.

Marcas del grado Las marcas en los sujetadores que indican fuerza se llaman las marcas del grado.

Graphing multimeter (GMM) A graphing multimeter digitally stores electrical signals for display in a waveform view on a digital screen.

Multímetro gráfico (GMM) El multímetro gráfico almacena digitalmente señales eléctricas y las muestra como ondas en una pantalla digital.

Hazardous waste Any used material or any byproduct that can be classified as potentially hazardous to one's health and/or the environment.

Desechos Peligrosos Cualquier material usado, o cualquier subproducto, que se pueda clasificar como potencialmente dañoso al medio ambiente o parjudicial para salud.

Heat exchanger Heat exchanger is a device used to disperse heat energy.

Intercambiador de calor El intercambiador de calor es un dispositivo que se usa para dispersar la energía del calor.

Hub The center part of a wheel, to which the wheel is attached.

Cubo La parte central de una rueda, a la cual se monta la rueda.

Hydraulic press A piece of shop equipment that develops a heavy force by use of a hydraulic piston-and-jack assembly.

Prensa hidráulica Una herramienta del taller que provee una fuerza grande por medio de una asamblea de gato con un pistón hidráulico.

Hydraulic pressure Pressure exerted through the medium of a liquid.

Presión hidráulica La presión esforzada por medio de un líquido.

Hydrocarbons Particles of gasoline, present in the exhaust and in crankcase vapors, that have not been fully burned.

Hidrocarburos Partículas de la gasolina, presentes en los vapores del cárter del cigüeñal y el tubo de escape, que no han sido completamente quemadas.

ID Inside diameter.

DI Diámetro interior.

Idle Engine speed when the accelerator pedal is fully released and there is no load on the engine.

Marcha lenta La velocidad del motor cuando el pedal accelerador esta completamente desembragada y no hay carga en el motor.

Ignitability The characteristic of a solid that enables it to spontaneously ignite. Any liquid with a flash point below 140°F is also said to possess ignitability.

Incendiabilidad Característica de un sólido que le permite encenderse espontáneamente. Cualquier líquido con un punto de inflamabilidad por debajo de los 140°F también se dice que posee incendiabilidad.

Impedance The operating resistance of an electrical device.

Impedancia La resistencia operativa de un dispositivo eléctrico.

Impeller The pump or driving member in a torque converter.

Impulsor La bomba o el miembro impulsor en un convertidor de torsión.

Increments Series of regular additions from small to large.

Incrementos Una serie de incrementos regulares que va de pequeño a grande.

Index To orient two parts by marking them. During reassembly, the parts are arranged so the index marks are next to each other. Used to preserve the orientation between balanced parts.

Índice Orientar a dos partes marcándolas. Al montarlas, las partes se colocan para que las marcas de índice estén alinieadas. Se usan los índices para preservar la orientación de las partes balanceadas.

Input shaft The shaft carrying the driving gear by which the power is applied, as to the transmission.

Flecha de entrada La flecha que porta el engranaje propulsor por el cual se aplica la potencia, como a la transmisión.

Inspection cover A removable cover that permits entrance for inspection and service work.

Cubierta de inspección Una cubierta desmontable que permite a la entrada para inspeccionar y mantenimiento.

Integral Built into, as part of the whole.

Integral Incorporado, una parte de la totalidad.

Internal gear A gear with teeth pointing inward, toward the hollow center of the gear.

Engranaje interno Un engranaje cuyos dientes apuntan hacia el interior, al hueco central del engranaje.

International System (SI) The International System of weights and measures. The system is normally called the metric system.

Sistema Internacional (SI) El sistema internacional de pesos y de medidas. Este sistema normalmente se llama la sistema métrico.

Jack (safety) stands Used to hold a vehicle up while using a floor jack.

Gato de seguridad Se usa para sostener un vehículo mientras se está usando un gato patín.

Jam nut A second nut tightened against a primary nut to prevent it from working loose. Used on inner and outer tie-rod adjustment nuts and on many pinion-bearing adjustment nuts.

Contra tuerca Una tuerca secundaria que se aprieta contra una tuerca primaria para prevenir que ésta se afloja. Se emplean en las tuercas de ajustes interiores e exteriores para las barras de acoplamiento y también en muchas de las tuercas de ajuste de portapiñones.

Journal A bearing with a hole in it for a shaft.

Manga de flecha Un cojinete que tiene un hoyo para una flecha.

Key A small block inserted between the shaft and hub to prevent circumferential movement.

Chaveta Un tope pequeño que se meta entre la flecha y el cubo para prevenir un movimiento circunferencial.

Keyway A groove or slot cut to permit the insertion of a key.

Ranura de chaveta Un corte de ranura o mortaja que permite insertar una chaveta.

Knock A heavy metallic sound usually caused by a loose or worn bearing.

Golpe Un sonido metálico fuerte que suele ser causado por un cojinete suelto o gastado.

Lab scope A device that displays voltage and or current as a waveform. Also called an oscilliscope.

Osciloscopio de diagnóstico Dispositivo que muestra voltaje o corriente como ondas. También llamado osciloscopio.

Lapping The process of fitting one surface to another by rubbing them together with an abrasive material between the two surfaces.

Pulido El proceso de ajustar a una superficie con otra por frotarlas juntas con una materia abrasiva entre las dos superficies.

Lash The amount of free motion in a geartrain, between gears, or in a mechanical assembly, such as the lash in a valve train.

Juego La cantidad del movimiento libre en un tren de engranajes, entre los engranajes o en una asamblea mecánica, tal como el juego en un tren de vávulas.

LED The common acronym for a light-emitting diode.

LED La sigla común en inglés por un diódo emisor de luz.

Linkage Any series of rods, yokes, levers, and so on used to transmit motion from one unit to another.

Biela Cualquiera serie de barras, yugos, palancas, y todo lo demás, que se usa para transferir los movimientos de una unedad a otra.

Locking ring A type of sealing ring that has ends that meet or lock together during installation. There is no gap between the ends when the ring is installed.

Anillo de enclavamiento Un tipo de anillo obturador que tiene las extremidades que se tocan o se enclavan durante la instalación. No hay holgura entre las extremidades cuando se ha instalado el anillo.

Locknut A second nut turned down on a holding nut to prevent loosening.

Contra tuerca Una tuerca segundaria apretada contra una tuerca de sostén para prevenir que ésta se afloja.

Lock pin Used in some ball sockets (inner tie-rod end) to keep the connecting nuts from working loose. Also used on some lower ball joints to hold the tapered stud in the steering knuckle.

Clavija de cerrojo Se usan en algunas rótulas (las extremidades interiores de la barra de acoplamiento) para prevenir que se aflojan las tuercas de conexión. También se emplean en algunas juntas esféricas inferiores para retener al perno cónico en la articulación de dirección.

Lockplates Metal tabs bent around nuts or bolt heads.

Placa de cerrojo Chavetas de metal que se doblan alrededor de las tuercas o las cabezas de los pernos.

Lockwasher A type of washer which, when placed under the head of a bolt or nut, prevents the bolt or nut from working loose.

Arandela de freno Un tipo de arandela que, al colocarse bajo la cabeza de un perno, previene que el perno o la tuerca se aflojan.

Low speed The gearing that produces the highest torque and lowest speed of the wheels.

Velocidad baja La velocidad que produce la torsión más alta y la velocidad más baja a las ruedas.

Lubricant Any material, usually a petroleum product such as grease or oil, that is placed between two moving parts to reduce friction.

Lubricante Cualquier substancia, normalmente un producto de petróleo como la grasa o el aciete, que se coloca entre dos partes en movimiento para reducir la fricción.

Machinist's rule A ruler used to measure short distances.

Regla de mecánico Regla que se usa para medir distancias cortas.

Mainline pressure The hydraulic pressure that operates apply devices and is the source of all other pressures in an automatic transmission. It is developed by pump pressure and regulated by the pressure regulator.

Línea de presión La presión hidráulica que opera a los dispositivos de applicación y es el orígen de todas las presiones en la transmisión automática. Proviene de la bomba de presión y es regulada por el regulador de presión.

Main oil pressure regulator valve Regulates the line pressure in a transmission.

Válvula reguladora de la linea de presión Regula la presión en la linea de una transmisión.

Manifold absolute pressure (MAP) sensor The sensor that measures changes in the intake manifold pressure that result from changes in engine load and speed.

Sensor de presión absoluta del colector de escape Un sensor que mide los cambios en la presión del colector de escape que resultan de cambios en carga y velocidad del motor.

Manual control valve A valve used to manually select the operating mode of the transmission. It is moved by the gearshift linkage.

Válvula de control manual Una válvula que se usa para escojer a una velocidad de la transmisión por mano. Se mueva por la biela de velocidades.

Manual shaft Manual shaft part of the linkage that moves the manual valve.

Eje manual Parte de eje manual del varillaje que mueve la válvula manual.

Meshing The mating, or engaging, of the teeth of two gears.

Engrane Embragar o endentar a los dientes de dos engranajes.

Meter 1/10,000,000 of the distance from the North Pole to the Equator, or 39.37 inches.

Metro Un 1/10,000,000 de la distancia del polo del norte al ecuador.

Micrometer A precision measuring device used to measure small bores, diameters, and thicknesses. Also called a mike.

Micrómetro Un dispositivo de medida precisa que se emplea a medir a los taladros pequeños y a los espesores. También se llama un mike (mayk).

MIL The malfunction indicator lamp for a computer control system. Prior to J1930, the MIL was commonly called a Check Engine or Service Engine Soon lamp.

MIL La lámpara de indicación de averías para un sistema de control de computadora. Antes del J1930, la MIL se llamaba la lámpara de Revise Motor o Servicia el Motor Pronto.

Millimeter (mm) The third and most commonly used division of a meter.

Milímetro (mm) Un milímetro es el tercero y la división lo más comúnmente posible usada de un metro, una medida lineal en la sistema métrico, conteniendo la milésima pieza de un metro; igual al 0.03934 de una pulgada.

Mineral spirits An alcohol-based liquid that leaves no surface residue after it dries.

Alcoholes mineral Un líquido a base de alcohol que no sale de ningún residuo superficial después de que se seque.

Misalignment When bearings are not on the same centerline.

Desalineamineto Cuando los cojinetes no comparten la misma linea central.

Mounts Made of rubber to insulate vibrations and noise while they support a powertrain part, such as engine or transmission mounts.

Monturas Hecho de hule para insular a las vibraciones y a los ruidos mientras que sujetan una parte del tren de propulsión, tal como las monturas del motor o las monturas de la transmisión.

Multimeter A tool that combines the voltmeter, ohmmeter, and ammeter together in a diagnostic instrument.

Multímetro Una herramienta que combina el voltímetro, el ohmímetro, y el amperímetro junta en uno instrumento de diagnóstico.

Multiple disc A clutch with a number of driving and driven discs as compared to a single plate clutch.

Discos múltiples Un embrague que tiene varios discos de propulsión o de arraste al contraste con un embrague de un sólo plato.

Needle bearing An antifriction bearing using a great number of long, small-diameter rollers. Also known as a quill bearing.

Rodamiento de agujas Un rodamiento (cojinete) antifricativo que emplea un gran cantidad de rodillos largos y de diámetro muy pequeños.

Needle deflection Distance of travel from zero of the needle on a dial gauge.

Desviación de la aguja La distancia del cero que viaja una aguja de un indicador.

Neoprene A synthetic rubber that is not affected by the various chemicals that are harmful to natural rubber.

Neoprene Un hule sintético que no se afecta por los varios productos químicos que pueden dañar al hule natural.

Neutral In a transmission, the setting in which all gears are disengaged and the output shaft is disconnected from the drive wheels.

Neutral En una transmisión, la velocidad en la cual todos los engranajes estan desembragados y el árbol de salida esta desconectada de las ruedas de propulsión.

Neutral-start switch A switch wired into the ignition switch to prevent engine cranking unless the transmission shift lever is in neutral or the clutch pedal is depressed.

Interruptor de arranque en neutral Un interruptor eléctrico instalado en el interruptor de encendido que previene el arranque del motor al menos de que la palanca de cambio de velocidad esté en una posición neutral o que se pisa en el embrague.

Newton-meter (Nm) Metric measurement of torque or twisting force.

Metro newton (Nm) Una medida métrica de la fuerza de torsión.

Nominal shim A shim with a designated thickness.

Laminilla fina Una cuña de un espesor especificado.

Nonhardening A gasket sealer that never hardens.

Sinfragua Un cemento de empaque que no endurece.

Nut A removable fastener used with a bolt to lock pieces together; made by threading a hole through the center of a piece of metal that has been shaped to a standard size.

Tuerca Un retén removable que se usa con un perno o tuerca para unir a dos piezas; se fabrica al filetear un hoyo taladrado en un pedazo de metal que se ha formado a un tamaño especificado.

Occupational Safety and Health Administration (OSHA) The government agency charged with ensuring safe work environments for all workers.

Administraion de la seguridad y de la salud ocupacionales (OSHA) La agencia de estatal cargada con asegurar los ambientes seguros del trabajo para todos los trabajadores.

OD Outside diameter.

DE Diámetro exterior.

Ohmmeter A test meter used to measure electrical resistance and/or continuity.

Contador de ohmios Un aparato que mide resistencia y/o continuidad eléctrica.

Oil seal A seal placed around a rotating shaft or other moving part to prevent leakage of oil.

Empaque de aciete Un empaque que se coloca alrededor de una flecha giratoria para prevenir el goteo de aceite.

On-board diagnostics On board diagnostics are the self diagnostic programming of the control modules used to aid diagnostics.

Diagnóstico a bordo el diagnóstico a bordo es la programación de autodiagnóstico de los módulos de control que se usa para tratar diagnósticos.

One-way clutch See Sprag clutch.

Embrague de una via Vea Sprag clutch.

Open An open is a break in the circuit that stops current flow.

Abierto Una rotura en el circuito que para el flujo de la corriente eléctrica.

O-ring A type of sealing ring, usually made of rubber or a rubber-like material. In use, the O-ring is compressed into a groove to provide the sealing action.

Anillo en O Un tipo de sello anular, suele ser hecho de hule o de una materia parecida al hule. Al usarse, el anillo en O se comprime en una ranura para proveer un sello.

Oscillate To swing back and forth like a pendulum.

Oscilar Moverse alternativamente en dos sentidos contrarios como un péndulo.

Oscilloscope A device used to display a voltage or current waveform on a screen. Also known as a lab scope.

Osciloscopio Dispositivo que sirve para mostrar ondas de voltaje o corriente en una pantalla.

Outer bearing race The outer part of a bearing assembly on which the balls or rollers rotate.

Pista exterior de un cojinete La parte exterior de una asamblea de cojinetes en la cual ruedan las bolas o los rodillos.

Out-of-round Wear of a round hole or shaft which, when viewed from an end, will appear egg-shaped.

Defecto de circularidad Desgaste de un taladro o de una flecha circular, que al verse de una extremidad, tendrá una forma asimétrica, como la de un huevo.

Output shaft The shaft or gear that delivers the power from a device, such as a transmission.

Flecha de salida La flecha o la velocidad que transmite la potencia de un dispositivo, tal como una transmisión.

Overall ratio The product of the transmission gear ratio multiplied by the final drive or rear axle ratio.

Relación global El producto de multiplicar la relación de los engranajes de la transmisión por la relación del impulso final o por la relación del eje trasero.

Overdrive Any arrangement of gearing that produces more revolutions of the driven shaft than of the driving shaft.

Sobremultiplicación Un arreglo de los engranajes que produce más revoluciones de la flecha de arrastre que los de la flecha de propulsión.

Overdrive ratio Identified by the decimal point indicating less than one driving input revolution compared to one output revolution of a shaft.

Relación del sobremultiplicación Se identifica por el punto decimal que indica menos de una revolución del motor comparado a una revolución de una flecha de salida.

Overrun coupling A freewheeling device to permit rotation in one direction but not in the other.

Acoplamiento de sobremarcha Un dispositivo de marcha de rueda libre que permite las giraciones en una dirección, pero no en la otra dirección.

Overrunning clutch A device consisting of a shaft or housing linked together by rollers or sprags operating between movable and fixed races. As the shaft rotates, the rollers or sprags jam between the movable and fixed races. This jamming action locks together the shaft and housing. If the fixed race should be driven at a speed greater than the movable race, the rollers or sprags will disconnect the shaft.

Embrague de sobremarcha Un dispositivo que consiste de una flecha o un cárter eslabonados por medio de rodillos o palancas de detención que operan entre pistas fijas y movibles. Al girar la flecha, los rodillos o palancas de detención se aprietan entre las pistas fijas y movibles. Este acción de apretarse enclava el cárter con la flecha. Si la pista fija se arrastra en una velocidad más alta que la pista movible, los rodillos o palancas de detención desconectarán a la flecha.

Oxidation Burning or combustion; the combining of a material with oxygen. Rusting is slow oxidation, and combustion is rapid oxidation.

Oxidación Quemando o la combustión; la combinación de una materia con el oxígeno. El orín es una oxidación lenta, la combustión es la oxidación rápida.

Parallel The quality of two items being the same distance from each other at all points; usually applied to lines and, in automotive work, to machined surfaces.

Paralelo La calidad de dos artículos que mantienen la misma distancia el uno del otro en cada punto; suele aplicarse a las líneas y, en el trabajo automotívo, a las superficies acabadas a máquina.

Parameter identification (PID) A parameter identification or PID is a piece of data displayed on a scan tool about a specific sensor, reading, or actuator on a scan tool. A scan tool may be able to read hundreds of PIDS on a single module.

Identificación de parámetros (PID) la identificación de parámetros o PID son datos que se muestran en una herramienta de exploración sobre un sensor específico, lectura o actuador en una herramienta de exploración. Una herramienta de exploración puede leer cientos de PIDS en un único módulo.

Pascal's law The law of fluid motion.

Ley de pascal La ley del movimiento del fluido.

Pawl A lever that pivots on a shaft. When lifted, it swings freely and when lowered, it locates in a detent or notch to hold a mechanism stationary.

Trinquete Una palanca que gira en una flecha. Levantado, mueve sín restricción, bajado, se coloca en una endentación o una muesca para mantener sín movimiento a un mecanismo.

PCM The powertrain control module of a computer control system. Prior to J1930, the PCM was commonly called a ECA, ECM, or one of many acronyms used by the various manufacturers.

PCM El módulo de control del sistema de transmisión de fuerza de un sistema de control de una computadora. Antes d3l J1930, el PCM se llamaba un ECA, un ECM, o una de varias siglas usadas por los varios fabricantes.

Pitch The number of threads per inch on any threaded part.

Paso El número de filetes por pulgada de cualquier parte fileteada.

Pivot A pin or shaft on which another part rests or turns.

Pivote Una chaveta o una flecha que soporta a otra parte o sirve como un punto para girar.

Planetary gearset A system of gearing that is modeled after the solar system. A pinion is surrounded by an internal ring gear and planet gears are in mesh between the ring gear and pinion around which all revolve.

Conjunto de engranajes planetarios Un sistema de engranaje cuyo patrón es el sistema solar. Un engranaje propulsor (la corona interior) rodea al piñon de ataque y los engranajes satélites y planetas se endentan entre la corona y el piñon alrededor del cual todo gira.

Planet carrier The carrier or bracket in a planetary gear system that contains the shafts on which the pinions or planet gears turn.

Perno de arrastre planetario El soporte o la abrazadera que contiene las flechas en las cuales giran los engranajes planetarios o los piñones.

Planet gears The gears in a planetary gearset that connect the sun gear to the ring gear.

Engranajes planetarios Los engranajes en un conjunto de engranajes planetario que connectan al engranaje propulsor interior (el engranaje sol) con la corona.

Planet pinions In a planetary gear system, the gears that mesh with, and revolve about, the sun gear; they also mesh with the ring gear.

Piñones planetarios En un sistema de engranajes planetarios, los engranajes que se endentan con, y giran alrededor, el engranaje propulsor (sol); también se endentan con la corona.

Plug Anything that will fit into an opening to stop fluid or airflow.

Tapón Cualquier cosa que se ajuste en una apertura para prevenir el goteo o el escape de un corriente del aire.

Pneumatic tools Power tools that rely on compressed air for power.

Herramientas neumáticas Las herramientas de motor cuyo energía proviene del aire comprimido.

Porosity A statement of how porous or permeable to liquids a material is.

Porosidad Una expresión de lo poroso o permeable a los líquidos es una materia.

Powertrain The mechanisms that carry the power from the engine crankshaft to the drive wheels; these include the clutch, transmission, driveline, differential, and axles.

Tren impulsor Los mecanismos que transferen la potencia desde el cigueñal del motor a las ruedas de propulsión; éstos incluyen el embrague, la transmisión, la flecha motríz, el diferencial y los semiejes.

Preload A load applied to a part during assembly so as to maintain critical tolerances when the operating load is applied later.

Carga previa Una carga aplicada a una parte durante la asamblea para asegurar sus tolerancias críticas antes de que se le aplica la carga de la operación.

Press-fit Forcing a part into an opening that is slightly smaller than the part itself to make a solid fit.

Ajustamiento a presión Forzar a una parte en una apertura que es de un tamaño más pequeño de la parte para asegurar un ajustamiento sólido.

Pressure Force per unit area, or force divided by area. Usually measured in pounds per square inch (psi) or in kilopascals (kPa) in the metric system.

Presión La fuerza por unedad de una area, o la fuerza divida por la area. Suele medirse en libras por pulgada cuadrada (lb/pulg2) o en kilopascales (kPa) en el sistema métrico.

Pressure plate That part of the clutch that exerts force against the friction disc; it is mounted on and rotates with the flywheel.

Plato opresor Una parte del embraque que aplica la fuerza en el disco de fricción; se monta sobre el volante, y gira con éste.

Propeller shaft See Driveshaft.

Flecha de propulsion Vea Flecha motríz.

Prussian blue A blue pigment; in solution, useful in determining the area of contact between two surfaces.

Azul de Prusia Un pigmento azul; en forma líquida, ayuda en determinar la area de contacto entre dos superficies.

psi Abbreviation for *pounds per square inch*, a measurement of pressure.

Lb/pulg2 Una abreviación de libras por pulgada cuadrada, una medida de la presión.

Puller Generally, a shop tool used to separate two closely fitted parts without damage. Often contains one screw, or several screws that can be turned to apply a gradual force.

Extractor Generalmente, una herramienta del taller que sirve para separar a dos partes apretadas sin incurrir daños. Suele tener una tuerca o varias tuercas, que se pueden girar para aplicar la fuerza gradualmente.

Pulsation To move or beat with rhythmic impulses.

Pulsación Moverse o batir con impulsos rítmicos.

Pulse width Pulse width is the length of time a signal is turned on.

Ancho de pulso El ancho de pulso es la cantidad de tiempo que una señal está encendida.

Race A channel in the inner or outer ring of an antifriction bearing in which the balls or rollers roll.

Pista Un canal en el anillo interior o exterior de un cojinete antifricción en el cual ruedan las bolas o los rodillos.

Radial The direction moving straight out from the center of a circle. Perpendicular to the shaft or bearing bore.

Radial La dirección al moverse directamente del centro de un círculo. Perpendicular a la flecha o al taladro del cojinete.

Radial clearance Clearance within the bearing and between balls and races perpendicular to the shaft. Also called radial displacement.

Holgura radial La holgura en un cojinete entre las bolas y las pistas que son perpendiculares a la flecha. También se llama un desplazamiento radial.

Radial load A force perpendicular to the axis of rotation.

Carga radial Una fuerza perpendicular al cen de rotación.

Radio frequency interference (RFI) An unwanted voltage signal that rides on another voltage signal.

Interferencia de la radiofrecuencia Una señal indeseada del voltaje que interfiere con otra señal del voltaje.

Ratio The relation or proportion that one number bears to another.

Relación La correlación o proporción de un número con respeto a otro.

Reactivity The characteristic of a material that enables it to react violently with water or other materials. Materials that release cyanide gas, hydrogen sulfide gas, or similar gases when exposed to low pH acid solutions are also said to possess reactivity, as are materials that generate toxic mists, fumes, vapors, and flammable gases.

Reactividad Característica de un material que le permite reaccionar violentamente con agua u otros materiales. Los

materiales que emiten gas de cianuro, sulfuro de hidrógeno o gases similares cuando se exponen a soluciones ácidas de bajo pH también se dice que poseen reactividad, como los materiales que generan vapores tóxicos o gases inflamables.

Rear-wheel-drive A term associated with a vehicle in which the engine is mounted at the front and the driving axle and driving wheels are at the rear of the vehicle.

Tracción trasera Un término que se asocia con un vehículo en el cual el motor se ubica en la parte delantera y el eje propulsor y las ruedas propulsores se encuentran en la parte trasera del vehículo.

Relief valve A valve used to protect against excessive pressure in the case of a malfunctioning pressure regulator.

Válvula de seguridad Una válvula que se usa para guardar contra una presión excesiva en caso de que malfulciona el regulador de presión.

Repair order A legal document written for every vehicle brought into the shop for service. It gives detailed information about the customer, the vehicle, the customer's concern or request, and an estimate of the cost of the services and when the services should be completed.

Orden de reparación Documento legal que se emite para cada vehículo que se lleva a un taller para reparación. Da información detallada sobre el cliente, el vehículo, la preocupación o el pedido del cliente, y una estimación del costo de los servicios y el plazo en el que se completarán los mismos.

Retaining ring A removable fastener used as a shoulder to retain and position a round bearing in a hole.

Anillo de retén Un seguro removible que sirve de collarín para sujetar y posicionar a un cojinete en un agujero.

RFI Radio frequency interference. This acronym is used to describe a type of electrical interference that may affect voltage signals in a computerized system.

RFI Interferencia de frecuencias de radio. Esta sigla se usa en describir un tipo de interferencia eléctrica que puede afectar los señales de voltaje en un sistema computerizado.

Right-to-know laws Laws requiring employers to provide employees with a safe work place as it relates to hazardous materials, and information about any and all hazards the employees face while performing their job.

Prerrogativa a saber leyes Los leyes que requieren a patrones proveer de empleados un lugar de trabajo seguro, relacionada con los materiales peligrosos y la información sobre cualesquiera y todos todos los peligros que los empleados pudieron encontrar mientras que realizaban su trabajo.

Rivet A headed pin used for uniting two or more piezas by passing the shank through a hole in each piece and securing it by forming a head on the opposite end.

Remache Una clavija con cabeza que sirve para unir a dos piezas o más al pasar el vástago por un hoyo en cada pieza y asegurarlo por formar una cabeza en el extremo opuesto.

Roller bearing An inner and outer race on which hardened steel rollers operate.

Cojinete de rodillos Una pista interior y exterior en la cual operan los rodillos hecho de acero endurecido.

Rollers Round steel bearings that can be used as the locking element in an overrunning clutch or as the rolling element in an antifriction bearing.

Rodillos Articulaciones redondos de acero que pueden servir como un elemento de enclavamiento en un embrague de sobremarcha o como el elemento que rueda en un cojinete antifricción.

Rotary flow A fluid force generated in the torque converter that is related to vortex flow. The vortex flow leaving the impeller is not only flowing out of the impeller at high speed but is also rotating faster than the turbine. The rotating fluid striking the slower turning turbine exerts a force against the turbine that is defined as rotary flow.

Flujo rotativo Una fuerza fluida producida en el convertidor de torsión que se relaciona al flujo torbellino. El flujo torbellino saliendo del rotor no sólo viaja en una alta velocidad sino también gira más rápidamente que el turbino. El fluido rotativo chocando contra el turbino que gira más lentamente, impone una fuerza contra el turbino que se define como flujo rotativo.

rpm Abbreviation for revolutions per minute, a measure of rotational speed.

RPM Abreviación de revoluciones por minuto, una medida de la velocidad rotativa.

RTV sealer Room-temperature vulcanizing gasket material that cures at room temperature; a plastic paste squeezed from a tube to form a gasket of any shape.

Sellador RTV Una materia vulcanizante de empaque que cura en temperaturas del ambiente; una pasta plástica exprimida de un tubo para formar un empaque de cualquiera forma.

Runout Deviation of the specified normal travel of an object. The amount of deviation or wobble a shaft or wheel has as it rotates. Runout is measured with a dial indicator.

Corrimiento Una desviación de la carrera normal e especificada de un objeto. La cantidad de desviación o vacilación de una flecha o una rueda mientras que gira. El corrimiento se mide con un indicador de carátula.

RWD Abbreviation for rear-wheel-drive.

RWD Abreviación de tracción trasera.

SAE Society of Automotive Engineers.

SAE La Sociedad de Ingenieros Automotrices.

Safety Data Sheets (SDS) Information sheets containing chemical composition and precautionary information for all products that can present a health or safety hazard.

Hoja de datos de seguridad de materiales Hoja informativa que contiene la composición química e información de precaución para todos los productos que puedan presentar un riesgo para la salud o la seguridad.

Scan tool A microprocessor designed to communicate with a vehicle's onboard computer in order to perform diagnosis and troubleshooting.

Herramienta de la exploración Una herramienta con un microprocesador diseñado para comunicarse con el ordenador a bordo de un vehículo para realizar diagnosis y la localización de averías.

Score A scratch, ridge, or groove marring a finished surface.

Entalladura Una raya, una arruga o una ranura que desfigure a una superficie acabada.

Scuffing A type of wear in which there is a transfer of material between parts moving against each other; shows up as pits or grooves in the mating surfaces.

Erosión Un tipo de desgaste en el cual hay una tranferencia de una materia entre las partes que estan en contacto mientras que muevan; se manifesta como hoyitos o muescas en las superficies apareadas.

Seal A material shaped around a shaft, used to close off the operating compartment of the shaft, preventing oil leakage.

Sello Una materia, formado alrededor de una flecha, que sella el compartimiento operativo de la flecha, previniendo el goteo de aceite.

Sealer A thick, tacky compound, usually spread with a brush, that may be used as a gasket or sealant to seal small openings or surface irregularities.

Sellador Un compuesto pegajoso y espeso, comoenmente aplicado con una brocha, que puede usarse como un empaque o un obturador para sellar a las aperturas pequeñas o a las irregularidades de la superficie.

Seat A surface, usually machined, on which another part rests or seats; for example, the surface on which a valve face rests.

Asiento Una superficie, comúnmente maquinada, sobre la cual yace o se asienta otra parte; por ejemplo, la superficie sobre la cual yace la cara de la válvula.

Selective servo apply pins A selective servo apply pin are available in different lengths from the manufacturer to insure proper band to drum clearance as measured with a special tool.

Clavijas selectivas de aplicación de servo El fabricante tiene disponibles clavijas selectivas de aplicación de servo en diferentes longitudes para garantizar el espacio adecuado entre la banda y el tambor, según la medición de una herramienta especial.

Selective thrust washer This type of washer is available in different thicknesses to provide the correct endplay or clearance.

Arandela selectiva Este tipo de arandela está disponible en diversos espesores, para proporcionar a la separación correcta o amplitud correcta.

Serial data The communications to and from the computer are commonly referred to as the system's serial data.

Datos seriales del sistema Las comunicaciones al ordenador y del ordenador se llama comúnmente los datos seriales del sistema.

Servo A device that converts hydraulic pressure into mechanical movement, often multiplying it. Used to apply the bands of a transmission.

Servo Un dispositivo que convierte la presión hidráulica al movimiento mecánico, frequentemente multiplicándola. Se usa en la aplicación de las bandas de una transmisión.

Shank The portion of a shoe that protects the ball of your foot. It is the narrow part of a shoe between the heel and sole.

Cambrillón Porción del zapato que protege la bola del pie. Es la parte angosta del zapato entre la suela y el tacón.

Shift lever The lever used to change gears in a transmission. Also, the lever on the starting motor that moves the drive pinion into or out of mesh with the flywheel teeth.

Palanca del cambiador La palanca que sirve para cambiar a las velocidades de una transmisión. También es la palanca del motor de arranque que mueva al piñon de ataque para engranarse o desegranarse con los dientes del volante.

Shift valve A valve that controls the shifting of the gears in an automatic transmission.

Válvula de cambios Una válvula que controla a los cambios de las velocidades en una transmisión automática.

Shim Thin sheets used as spacers between two parts, such as the two halves of a journal bearing.

Laminilla de relleno Hojas delgadas que sirven de espaciadores entre dos partes, tal como las dos partes de un muñon.

Short An unwanted path for current.

Cortocircuito Un cortocircuito se describe lo más mejor posible como un camino indeseado para la corriente eléctrica.

Side clearance The clearance between the sides of moving parts when the sides do not serve as load-carrying surfaces.

Holgura lateral La holgura entre los lados de las partes en movimiento mientras que los lados no funcionan como las superficies de carga.

Sine wave A smooth repetitive waveform often displayed by A/C signals.

Onda sinusoidal Onda repetitiva y suave que se suelen mostrar por señales A/C.

Sinusoidal The wave shaped like a sine wave.

Sinusoidal El término sinusoidal quiere decir que la onda tiena una formaparecida a una onda senoidal.

Sliding-fit Where sufficient clearance has been allowed between the shaft and journal to allow freerunning without overheating.

Ajuste corredera Donde se ha dejado una holgura suficiente entre la flecha y el muñón para permitir una marcha libre sin sobrecalentamiento.

Snap ring Split spring-type ring located in an internal or external groove to retain a part.

Anillo de seguridad Un anillo partido tipo resorte que se coloca en una muesca interior o exterior para retener a una parte.

Solenoid/pressure switch The solenoid/ pressure switch is a single unit that consists of various solenoids and pressure switches in a Chrysler 41TE transmission.

Interruptor de presión/solenoide El interruptor de presión/solenoide es una única unidad que consiste en diversos solenoides e interruptores de presión en una transmisión Chrysler 41TE.

Spalling A condition in which the material of a bearing surface breaks away from the base metal.

Escamación Una condición en la cual una materia de la superficie de un rodamiento se separa del metal base.

Spindle The shaft on which the wheels and wheel bearings mount.

Husillo La flecha en la cual se montan las ruedas y el conjunto del cojinete de las ruedas.

Spline Slot or groove cut in a shaft or bore; a splined shaft onto which a hub, wheel, gear, and so on, with matching splines in its bore is assembled so that the two must turn together.

Acanaladura (espárrago) Una muesca o ranura cortada en una flecha o en un taladro; una flecha acanalada en la cual se asamblea un cubo, una rueda, un engranaje, y todo lo demás que tiene un acanaladura pareja en el taladro de manera de que las dos deben girar juntos.

Split lip seal Typically, a rope seal sometimes used to denote any two-part oil seal.

Sello hendido Típicamente, un sello de cuerda que se usa a veces para demarcar cualquier sello de aceite de dos partes.

Split pin A round split spring steel tubular pin used for locking purposes; for example, locking a gear to a shaft.

Chaveta hendida Una chaveta partida redonda y tubular hecho de acero para resorte que sirve para el enclavamiento; por ejemplo, para enclavar un engranaje a una flecha.

Spool valve A cylindrically shaped valve with two or more valleys between the lands. Spool valves are used to direct fluid flow.

Válvula de carrete Una válvula de forma cilíndrica que tiene dos acanaladuras de cañon o más entre las partes planas. Las válvulas de carrete sirven para dirigir el flujo del fluido.

Spontaneous combustion Spontaneous combustion is the self-ignition of a substance.

Combustión espontánea La combustión espontánea es la autoinflamación de una sustancia.

Sprag clutch A member of the overrunning clutch family using a sprag to jam between the inner and outer races used for holding or driving action.

Embrague de puntal Un miembro de la familia de embragues de sobremarcha que usa a una palanca de detención trabada entre las pistas interiores e exteriores para realizar una acción de asir o marchar.

Spring A device that changes shape when it is stretched or compressed, but returns to its original shape when the force is removed; the component of the automotive suspension system that absorbs road shocks by flexing and twisting.

Resorte Un dispositivo que cambia de forma al ser estirado o comprimido, pero que recupera su forma original al levantarse la fuerza; es un componente del sistema de suspensión automotívo que absorba los choques del camino al doblarse y torcerse.

Spring retainer A steel plate designed to hold a coil or several coil springs in place.

Retén de resorte Una chapa de acero diseñado a sostener en su posición a un resorte helicoidal o más.

Squeak A high-pitched noise of short duration.

Chillido Un ruido agudo de poca duración.

Squeal A continuous high-pitched noise.

Alarido Un ruido agudo continuo.

Stall A condition in which the engine is operating and the transmission is in gear, but the drive wheels are not turning because the turbine of the torque converter is not moving.

Paro Una condición en la cual opera el motor y la transmisión esta embragada pero las ruedas de impulso no giran porque no mueva el turbino del convertidor de la torsión.

Stall test A test of the one-way clutch in a torque converter.

Prueba de paro Una prueba del embrague de una vía en un convertidor de la torsión.

Static A form of electricity caused by friction.

Estático Una forma de la electridad causada por la fricción.

Stress The force to which a material, mechanism, or component is subjected.

Esfuerzo La fuerza a la cual se somete a una materia, un mecanísmo o un componente.

Sun gear The central gear in a planetary gear system around which the rest of the gears rotate. The innermost gear of the planetary gearset.

Engranaje principal (sol) El engranaje central en un sistema de engranajes planetarios alrededor del cual giran los otros engranajes. El engranaje más interno del conjunto de los engranajes planetarios.

Sweep rate The rate at which the trace moves across the display of an oscilloscope.

Velocidad de barrido Velocidad a la cual la señal se mueve por la pantalla de un osciloscopio.

Tap To cut threads in a hole with a tapered, fluted, threaded tool.

Roscar con macho Cortar las roscas en un agujero con una herramienta cónica, acanalada y fileteada.

Technical Service Bulletin (TSB) A special report detailing the repair of a particular customer concern or updated changes in specifications and repair procedures for a particular model vehicle.

Boletín del servicio técnico (TSB) Informe especial que detalla la reparación hecha a un cliente en particular o la actualización de cambios en las especificaciones y los procedimientos de reparación para un modelo de vehículo en particular.

Temper To change the physical characteristics of a metal by applying heat.

Templar Cambiar las características físicas de un metal mediante una aplicación del calor.

Tension Effort that elongates or "stretches" a material.

Tensión Un esfuerzo que alarga o "estira" a una materia.

Thickness gauge Strips of metal made to an exact thickness, used to measure clearances between parts.

Calibre de espesores Las tiras del metal que se han fabricado a un espesor exacto, sirven para medir las holguras entre las partes.

Thread chaser A device similar to a die that is used to clean threads.

Peine de roscar Un dispositivo parecido a una terraja que sirve para limpiar a las roscas.

Thread pitch The number of threads in 1 inch of threaded bolt length. In the metric system, thread pitch is the distance in millimeters between two adjacent threads.

Paso del filete La cantidad de filetes en una pulgada de la rosca de un tornillo. En el sistema métrico decimal, el paso del filete es la distancia en milímetros entre dos filetes adyacentes.

Threaded insert A threaded coil that is used to restore the original thread size to a hole with damaged threads.

Pieza inserta roscada Una bobina roscada que sirve para restaurar a su tamaño original una rosca dañada.

Throttle position (TP) sensor A potentiometer used to monitor changes in throttle plate opening. The position of the throttle plate determines the voltage drop at the sensor's resistor and the resultant voltage signal is sent to a computer system.

Sensor de posición de la válvula reguladora Un potenciómetro usado para vigilar cambios en la apertura de la placa de la válvula reguladora. La posición de la placa de la válvula reguladora determina la caída de voltaje en el resistor del sensor y la señal resultante del voltaje se envía a un sistema informático.

Thrust bearing A bearing designed to resist or contain side or end motion as well as reduce friction.

Cojinete de empuje Un cojinete diseñado a detener o reprimir a los movimientos laterales o de las extremidades y también reducir la fricción.

Thrust load A load that pushes or reacts through the bearing in a direction parallel to the shaft.

Carga de empuje Una carga que empuja o reacciona por el cojinete en una dirección paralelo a la flecha.

Thrust washer A washer designed to take up end thrust and prevent excessive endplay.

Arandela de empuje Una arandela diseñada para rellenar a la holgura de la extremidad y prevenir demasiado juego en la extremidad.

Tolerance A permissible variation between the two extremes of a specification or dimension.

Tolerancia Una variación permisible entre dos extremos de una especificación o de un dimensión.

Torque A twisting motion, usually measured in ft.-lb. (Nm).

Torsión Un movimiento giratorio, suele medirse en pies/libra (Nm).

Torque capacity The ability of a converter clutch to hold torque.

Capacidad de la torsión La abilidad de un convertidor de embraque a sostener a la torsión.

Torque converter A turbine device utilizing a rotary pump, one or more reactors (stators), and a driven circular turbine or vane, whereby power is transmitted from a driving to a driven member by hydraulic action. It provides varying drive ratios; with a speed reduction, it increases torque.

Convertidor de la torsión Un dispositivo de turbino que utilisa a una bomba rotativa, a un reactor o más, y un molinete o turbino circular impulsado, por cual se transmite la energía de un miembro de impulso a otro arrastrado mediante la acción hidráulica. Provee varias relaciones de impulso; al descender la velocidad, aumenta la torsión.

Torque curve A line plotted on a chart to illustrate the torque personality of an engine. When the engine operates on its torque curve, it is producing the most torque for the quantity of fuel being burned.

Curva de la torsión Una linea delineada en una carta para ilustrar las características de la torsión del motor. Al operar un motor en su curva de la torsión, produce la torsión óptima para la cantidad del combustible que se consuma.

Torque multiplication The result of meshing a small driving gear and a large driven gear to reduce speed and increase output torque.

Multiplicación de la torsión El resultado de engranar a un engranaje pequeño de ataque con un engranaje más grande arrastrado para reducir la velocidad y incrementar la torsión de salida.

Torque steer An action felt in the steering wheel as the result of increased torque.

Dirección la torsión Una acción que se nota en el volante de dirección como resultado de un aumento de la torsión.

Torque wrench A tool used to measure how tight a nut or bolt is.

Llave dinamométrica Herramienta que se usa para medir cuán ajustado está un tornillo o una tuerca.

Traction The gripping action between the tire tread and the road's surface.

Tracción La acción de agarrar entre la cara de la rueda y la superficie del camino.

Transaxle Type of construction in which the transmission and differential are combined in one unit.

Flecha de transmisión Un tipo de construcción en el cual la transmisión y el diferencial se combinan en una unidad.

Transaxle assembly A compact housing most often used in front-wheel-drive vehicles that houses the manual transmission, final drive gears, and differential assembly.

Asamblea de la flecha de transmisión Un cárter compacto que se usa normalmente en los vehículos de tracción delantera que contiene la transmisión manual, los engranajes de propulsión, y la asamblea del diferencial.

Transfer case An auxiliary transmission mounted behind the main transmission. Used to divide engine power and transfer it to both front and rear differentials, either full time or part time.

Cárter de la transferencia Una transmisión auxiliar montada detrás de la transmisión principal. Sirve para dividir la potencia del motor y transferirla a ambos diferenciales delanteras y traseras todo el tiempo o la mitad del tiempo.

Transmission The device in the powertrain that provides different gear ratios between the engine and drive wheels as well as reverse.

Transmisión El dispositivo en el trén de potencia que provee las relaciones diferentes de engranaje entre el motor y las ruedas de impulso y también la marcha de reversa.

Transmission fluid pressure (TFP) switch A switch that indicates to the control module that a shift has taken place.

Interruptor de presión del aceite de transmisión (TFP) Interruptor que indica al módulo de control que se ha realizado un cambio.

Transverse Powertrain layout in a front-wheel-drive automobile extending from side to side.

Transversal Una esquema del tren de potencia en un automóvil de tracción delantera que se extiende de un lado a otro.

U-joint A four-point cross connected to two U-shaped yokes that serves as a flexible coupling between shafts.

Junta de U Una cruceta de cuatro puntos que se conecta a dos yugos en forma de U que sirven de acoplamientos flexibles entre las flechas.

Underdrive The condition that exists when there is speed reduction through the gears.

Bajamarcha Bajamarcha es la condición que existe cuando hay reducción de velocidad a través de los engranajes.

United States Customary System (USCS) A system of weights and measures used in places that do not use the metric system, such as the United States. The USCS system is also known as the English or inch system.

Sistema acostumbrado de Estados Unidos (USCS) Un sistema de los pesos y de las medidas usados en los países que no utilizan la sistema métrico, tal como los Estados Unidos. El sistema de USCS también se conoce como el sistema del inglés o de la pulgada.

Universal joint A mechanical device that transmits rotary motion from one shaft to another shaft at varying angles.

Junta universal Un dispositivo mecánico que transmite el movimiento giratorio desde una flecha a otra flecha en varios ángulos.

Upshift To shift a transmission into a higher gear.

Cambio ascendente Cambiar a la velocidad de una transmisión a una más alta.

Valve body Main hydraulic control assembly of a transmission containing the components necessary to control the distribution of pressurized transmission fluid throughout the transmission.

Cuerpo de la válvula Asamblea principal del control hidráulico de una transmisión que contiene los componentes necessarios para controlar a la distribución del fluido de la transmisión bajo presión por toda la transmisión.

Vehicle identification number (VIN) The number assigned to each vehicle by its manufacturer, primarily for registration and identification purposes.

Número de identificacíon del vehículo El número asignado a cada vehículo por su fabricante, primariamente con el propósito de la registración y la identificación.

Vehicle speed sensor (VSS) A sensor used to track the current speed and the total miles traveled by a vehicle. This input is used by many computer systems in the vehicle.

Sensor de velocidad del vehículo Un sensor utilizado para medir la velocidad actual y las millas totales viajaron por un vehículo. Muchos de los sistemas informáticos del vehículo utilizan estos datos como entrada de información.

Vernier caliper A vernier caliper is a tool used to make inside, outside, and depth measurements by using a stationary and a movable scale.

Calibrador vernier El calibrador vernier es una herramienta para tomar medidas internas, externas y de profundidad al usar una escala fija y otra móvil.

Vibration A quivering, trembling motion; felt in a vehicle at different speed ranges.

Vibración Un movimiento de estremecer o temblar que se siente en el vehículo en varios intervalos de velocidad.

Viscosity The resistance to flow exhibited by a liquid. A thick oil has greater viscosity than a thin oil.

Viscosidad La resistencia al flujo que manifiesta un líquido. Un aceite espeso tiene una viscosidad mayor que un aceite ligero.

Volatility The tendency for a fluid to evaporate rapidly or pass off in the form of vapor. For example, gasoline is more volatile than kerosene because it evaporates at a lower temperature.

Volatilidad Tendencia de un fluido a evaporarse rápidamente o a transformarse en vapor. Por ejemplo, la gasolina es más volátil que el queroseno porque se evapora a menor temperatura.

Voltmeter A test instrument used to measure voltage.

Voltímetro Un aparato utilizado para medir voltaje.

Vortex Path of fluid flow in a torque converter. The vortex may be high, low, or zero depending on the relative speed between the pump and turbine.

Vórtice La vía del flujo de los fluidos en un convertido de torsión. El vórtice puede ser alto, bajo, o cero depende de la velocidad relativa entre la bomba y la turbina.

Vortex flow Recirculating flow between the converter impeller and turbine that causes torque multiplication.

Flujo del vórtice El fluyo recirculante entre el impulsor del convertidor y la turbina que causa la multiplicación de la torsión.

Waveform Pattern of a voltage or current as displayed on an oscilloscope or lab scope.

Ondas Patrón de un voltaje o corriente, como el que se muestra en un osciloscopio.

Wet-disc clutch A clutch in which the friction disc (or discs) is operated in a bath of oil.

Embrague de disco flotante Un embrague en el cual el disco (o los discos) de fricción opera en un baño de aceite.

Wheel A disc or spokes with a hub at the center that revolves around an axle, and a rim around the outside on which the tire is mounted.

Rueda Un disco o rayo que tiene en su centro un cubo que gira alrededor de un eje, y tiene un rim alrededor de su exterior en la cual se monta el neumático.

Yoke In a universal joint, the drivable torque-and-motion input and output member attached to a shaft or tube.

Yugo En una junta universal, el miembro de la entrada y la salida que transfere a la torsión y al movimiento, que se conecta a una flecha o a un tubo.

Note: Figures and tables are denoted by f and t, respectively.